Latin Squares and their Applications

Latin Squares and their Applications

Second Edition

A. Donald Keedwell
University of Surrey
Guildford, Surrey
United Kingdom

József Dénes
Budapest, Hungary

Amsterdam • Boston • Heidelberg • London • New York • Oxford
Paris • San Diego • San Francisco • Singapore • Sydney • Tokyo
North-Holland is an imprint of Elsevier

North-Holland is an imprint of Elsevier
Radarweg 29, PO Box 211, 1000 AE Amsterdam, Netherlands
The Boulevard, Langford Lane, Kidlington, Oxford OX5 1GB, UK

First edition 1974

Second edition 2015

Notices
Knowledge and best practice in this field are constantly changing. As new research and experience broaden our understanding, changes in research methods, professional practices, or medical treatment may become necessary.

Practitioners and researchers must always rely on their own experience and knowledge in evaluating and using any information, methods, compounds, or experiments described herein. In using such information or methods they should be mindful of their own safety and the safety of others, including parties for whom they have a professional responsibility.

To the fullest extent of the law, neither the Publisher nor the authors, contributors, or editors, assume any liability for any injury and/or damage to persons or property as a matter of products liability, negligence or otherwise, or from any use or operation of any methods, products, instructions, or ideas contained in the material herein.

British Library Cataloguing in Publication Data
A catalogue record for this book is available from the British Library

Library of Congress Cataloging-in-Publication Data
A catalog record for this book is available from the Library of Congress

ISBN: 978-0-444-63555-6

For Information on all North-Holland publications
visit our website at www.store.elsevier.com/

Working together
to grow libraries in
developing countries

www.elsevier.com • www.bookaid.org

Foreword to the First Edition

The subject of latin squares is an old one and it abounds with unsolved problems, many of them up to 200 years old. In the recent past one of the classical problems, the famous conjecture of Euler, has been disproved by Bose, Parker, and Shrikhande. It has hitherto been very difficult to collect all the literature on any given problem since, of course, the papers are widely scattered. This book is the first attempt at an exhaustive study of the subject. It contains some new material due to the authors (in particular, in chapters 3 and 7) and a very large number of the results appear in book form for the first time. Both the combinatorial and the algebraic features of the subject are stressed and also the applications to Statistics and Information Theory are emphasized. Thus, I hope that the book will have an appeal to a very wide audience. Many unsolved problems are stated, some classical, some due to the authors, and even some proposed by the writer of this foreword. I hope that, as a result of the publication of this book, some of the problems will become theorems of Mr. So and So.

PAUL ERDÖS

Contents

Preface to the First Edition

The concept of the latin square probably originated with problems concerning the movement and disposition of pieces on a chess board. However, the earliest written reference to the use of such squares known to the authors concerned the problem of placing the sixteen court cards of a pack of ordinary playing cards in the form of a square so that no row, column, or diagonal should contain more than one card of each suit and one card of each rank. An enumeration by type of the solutions to this problem was published in 1723. The famous problem of similar type concerning the arrangement of thirty-six officers of six different ranks and regiments in a square phalanx was proposed by Euler in 1779, but not until the beginning of the present century was it shown that no solution is possible.

It is only comparatively recently that the subject of latin squares has attracted the serious attention of mathematicians. The cause of the awakening of this more serious interest was the realization of the relevance of the subject to the algebra of generalized binary systems, and to the study of combinatorics, in particular to that of the finite geometries. An additional stimulus has come from practical applications relating to the formation of statistical designs and the construction of error correcting codes. Over the past thirty years a great number of papers concerned with the latin square have appeared in the mathematical journals and the authors felt that the time was ripe for the publication in book form of an account of the results which have been obtained and the problems yet to be solved.

Let us analyse our subject a little further. We may regard the study of latin squares as having two main emphases. On the one hand is the study of the properties of single latin squares which has very close connections with the theory of quasigroups and loops and, to a lesser extent, with the theory of graphs. On the other is the study of sets of mutually orthogonal latin squares. It is the latter which is most closely connected with the theory of finite projective planes and with the construction of statistical designs. We have organized our book in accordance with this general scheme. However, each of these two branches of the subject has many links with the other, as we hope that the following pages will clearly show.

We have tried to make the book reasonably self-contained. No prior knowledge of finite geometries, loop theory, or experimental designs has been assumed on the part of the reader, but an acquaintance with elementary group theory and with the basic properties of finite fields has been taken for granted where such knowledge is needed. Full proofs of several major results in the subject have been included for the first time outside the original research papers. These include the

Hall-Paige theorems (chapter 1), two major results due to R. H. Bruck (chapter 9) and the proof of the falsity of Euler's conjecture (chapter 11).

We hope that the text will be found intelligible to any reader whose standard of mathematical attainment is equivalent to that of a third year mathematics undergraduate of a British or Hungarian University. Probably the deepest theorem given in the book is theorem 1.4.8 which unavoidably appears in the first chapter. However, the reader's understanding of the remainder of the book will not be impaired if he skips the details of the proof of this theorem.

Part of the manuscript is based on lectures given by one of the authors at the Loránd Eötvös University, Budapest.

The bibliography of publications on latin squares has been made as comprehensive as possible but bibliographical references on related subjects have been confined to those works actually referred to in the text.

Decimal notation has been used for the numbering of sections, theorems, diagrams, and so on. Thus, theorem 10.1.2 is the second theorem of section 10.1 and occurs in chapter 10. The diagram referred to as Fig. 1.2.3 is the third diagram to be found in section 1.2. The use of lemmas has been deliberately avoided as it seemed to at least one of the authors better for the purposes of cross-reference to present a single numbering system for the results (theorems) of each section.

A list of unsolved problems precedes the bibliography and each is followed by a page reference to the relevant part of the text.

J. Dénes and A. D. Keedwell.

Acknowledgements (First Edition)

The authors wish to express grateful thanks for their helpful comments and constructive criticism both to their official referee, Prof. N. S. Mendelsohn, and also to Profs V.D. Belousov, D.E. Knuth, C. C. Lindner, H. B. Mann, A. Sade and J. Schönheim all of whom read part or all of the draft manuscript and sent useful comments. They owe a particular debt of gratitude to Prof. A. Sade for his very detailed commentary on a substantial part of the manuscript and for a number of valuable suggestions[1].

The Hungarian author wishes also to express thanks to his former Secretary, Mrs. E. Szentes, to Prof. S. Csibi of the Research Institute of Telecommunications[2], to Mr. E. Gergely and to several of his former students who took part in the work of his seminars given at the Loránd Eötvös University since 1964 (some of whom read parts of the manuscript and made suggestions for improvement) and to the staff of the Libraries of the Hungarian Academy of Sciences and its Mathematical Institute.

The British author wishes to express his very sincere thanks for their helpfulness at all times to the Librarian of the University of Surrey and to the many members of his library staff whose advice and assistance were called upon over a period of nearly five years and who sometimes spent many hours locating copies of difficult-to-obtain books and journals. He wishes to thank the Librarian of the London Mathematical Society for granting him permission to keep on loan at one time more than the normal number of Mathematical Journals. He also thanks the Secretarial Staff of the University of Surrey Mathematics Department for many kindnesses and for being always willing to assist with the retyping of short passages of text, etc., often at short notice.

Finally, both authors wish to thank Mr. G. Bernát, General Manager of Akadémiai Kiadó, publishing House of the Hungarian Academy of Sciences, and Messrs B. Stevens and A. Scott of the Editorial Staff of English Universities Press for their advice and encouragement throughout the period from the inception of the book to its final publication. They are also grateful to Mrs. E. Róth, Chief Editor, and to Mrs. E. K. Kállay and Miss P. Bodoky of the Editorial Staff of Akadémiai Kiadó for their care and attention during the period in which the manuscript was being prepared for printing.

---

[1] It is with deep regret that the authors have to announce the sudden death of Prof. Albert Sade on 10th February 1973 after a short illness.

[2] Now appointed to a Chair at the Technical University of Budapest.

Preface to the Second Edition

The original version of this book, although frequently cited as a source book in current literature, has been out of print since 1976. In the intervening years, a huge number of papers have been written and new results obtained so a second edition was overdue.

In this revised edition, much of the original material has been retained but all the chapters have been revised or re-written. However, in order that as many as possible of the large number of citations of the original edition should be valid for the new one, the overall layout of main topics and the order of the first seven chapters has been retained.

Because of the extensive cross-referencing, it is not necessary to read the chapters in the order that they are presented. Indeed, the reader new to the subject may find it useful after he/she has read or glanced through the first two chapters to look at the early sections of chapter 5 in which the concept of orthogonality is introduced and its connection with the existence of transversals explained.

As was the case in the original edition, the needs of the reader new to the subject have been foremost in the author's mind. He hopes that the second edition, like the first, will take such a reader from the beginnings of the subject to the frontiers of research. He also hopes that it will act as a reference book for more knowledgeable readers to the present state of knowledge in the particular topics of interest to them.

In order to keep the book of reasonable length, a few of the topics less central to the subject as now developing which were included in the original book have been omitted from this revised edition: in particular, the chapter on the resolution of the Euler conjecture and part of the section on generalized direct products. Discussion of several other topics has been condensed.

In this connection, we draw our readers' attention to the more recent book "Latin Squares: New Developments in the Theory and Applications" which was edited and part-written by the late J. Dénes and the present author and was published by North Holland, Amsterdam, in 1991. In the present work, we cite this book as [DK2] and it has been our policy not to duplicate relevant results which it contains; for example, on latin squares and codes, latin squares and geometry, on row-complete latin squares and sequencings of groups, and on subsquares in latin squares.

We have retained the narrative style which was much commended in comments on the first edition.

Fortunately, few significant errors have been detected in the original book except for various inconsistencies and mis-statements in the section on enumeration of latin squares in Chapter 4. These have been addressed in the revised version.

The author is very grateful to Ian Wanless for assisting the author in four very significant ways. Firstly, he helped very considerably with the re-writing of several sections of the above chapter and of Chapter 1 in particular. The new version of these sections is substantially his work though the present author takes full responsibility for its final form. Secondly, he has carried out most of the transcribing into $L^A T_E X$ of early versions of the first six chapters. (Without this assistance, the author would not have had the courage to attempt a re-write.) Thirdly, he has made a number of suggestions for improvement in these chapters, most of which the author has adopted and, fourthly, he has promptly answered many queries during the several years which the author has taken to complete the work.

Further remarks.

In a few places, we have wanted to refer to the original edition of the book. In such cases, it is cited as [DK1].

We have omitted the initials of authors cited in the body of the book except in cases when two or more authors have the same surname, in which cases we have included them.

Topics in the Subject Index which begin with a mathematical symbol (such as "N_∞-square") are listed alphabetically in front of those beginning with the letter "A".

Because of the number of pages that it would require, it is no longer possible to provide a comprehensive bibliography of papers concerning latin squares. (Mathematical Reviews has reviewed well over 2000 such papers.) In the present work, only papers which are explicitly referred to in the text are included in the Bibliography. Inevitably, the results of many excellent papers are not mentioned and do not appear in the Bibliography, for which omissions the author apologizes.

When [DK1] was written, the idea of listing Unsolved Problems was a relatively novel one. In this new edition, we have listed these 73 problems again and for each of them given the present state of knowledge so far as we know it. We have also listed a few new Unsolved Problems which we hope will spur progress in the field.

A. D. Keedwell.

Acknowledgements.

The author would like to express special thanks to his departmental colleague Gianne Derks and to Gavin Power of the Faculty Computer Department for help in resolving the many technical computer problems which arose during the book re-write.

CHAPTER 1

Elementary properties

In this preliminary chapter, we introduce a number of important concepts which will be used repeatedly throughout the book. In the first section, we briefly describe the history of the latin square concept and its equivalence to that of a quasigroup. Next, we explain how those latin squares which represent group multiplication tables may be characterized. We mention briefly the work of Ginzburg, Tamari and others on the reduced multiplication tables of finite groups. In the third, fourth and fifth sections respectively, we introduce the important concepts of isotopy, parastrophy[1] and complete mapping, and develop their basic properties in some detail. In the final section of the chapter we discuss the interrelated notions of subquasigroup and latin subsquare.

1.1 The multiplication table of a quasigroup

As we remarked in the preface to the first edition, the concept of the latin square is of very long standing and indeed arose very much earlier than the date of 1723 mentioned there. For details, see Wilson and Watkins(2013) and especially Chapter 6 thereof (written by L.D. Andersen). However, so far as the present author is aware, the topic was first systematically developed by Euler. A *latin square* was regarded by Euler as a square matrix with n^2 entries using n different elements, none of them occurring twice within any row or column of the matrix. The integer n is called the *order* of the latin square. (We shall, when convenient, assume the elements of the latin square to be the integers $0, 1, \ldots, n-1$ or, alternatively, $1, 2, \ldots, n$, and this will entail no loss of generality.)

Much later, it was shown by Cayley, who investigated the multiplication tables of groups, that a multiplication table of a group is in fact an appropriately bordered special latin square. [See Cayley(1877/8) and (1878a).] A multiplication table of a group is called its *Cayley table*.

Later still, in the 1930s, latin squares arose once again in the guise of multiplication tables when the theory of quasigroups and loops began to be developed as a generalization of the group concept. A set S is called a *quasigroup* if there is a binary operation $(\cdot)$ defined in S and if, when any two elements a, b of S are given, the equations $ax = b$ and $ya = b$ each have exactly one solution.[2] A *loop*

---

[1] Also called *conjugacy* but not with the same meaning as in group theory.

[2] Throughout this book, we shall, when convenient, write ax instead of the more formal $a \cdot x$ when the binary operation is $(\cdot)$. Similarly, we may write $a(bc)$ or $a \cdot bc$ instead of $a \cdot (b \cdot c)$. Also, when the quasigroup operation is not stated, it is assumed to be $(\cdot)$.

Latin Squares and their Applications. http://dx.doi.org/10.1016/B978-0-444-63555-6.50001-5

L is a quasigroup with an identity element: that is, a quasigroup in which there exists an element e of L with the property that $ex = xe = x$ for every x of L.

However, the concept of quasigroup had actually been considered in some detail much earlier than the 1930s by Schroeder who, between 1873 and 1890, wrote a number of papers on "formal arithmetics": that is, on algebraic systems with a binary operation such that both the left and right inverse operations could be uniquely defined. Such a system is evidently a quasigroup. A list of Schroeder's papers and a discussion of their significance[3] can be found in Ibragimov(1967).

In 1935, Ruth Moufang published a paper [Moufang(1935)] in which she pointed out the close connection between non-desarguesian projective planes and non-associative quasigroups.

The results of Euler, Cayley and Moufang made it possible to characterize latin squares both from the algebraic and the combinatorial points of view. A number of other authors have studied the close relationship that exists between the algebraic and combinatorial results when dealing with latin squares. Discussion of such relationships may be found in Barra and Guérin(1963a), Dénes(1962), Dénes and Pásztor(1963), Fog(1934), Schönhardt(1930) and Wielandt(1962).

Particularly in practical applications it is important to be able to exhibit results in the theory of quasigroups and groups as properties of the Cayley tables of these systems and of the corresponding latin squares. This becomes clear when we prove:

THEOREM 1.1.1 *Every multiplication table of a quasigroup is a latin square and conversely, any bordered latin square is the multiplication table of a quasigroup.*

PROOF. Let $a_1, a_2, \ldots, a_n$ be the elements of the quasigroup and let its multiplication table be as shown in Figure 1.1.1, where the entry a_{rs} which occurs in the r-th row of the s-th column is the product $a_r a_s$ of the elements a_r and a_s. If the same entry occurred twice in the r-th row, say in the s-th and t-th columns so that $a_{rs} = a_{rt} = b$ say, we would have two solutions to the equation $a_r x = b$ in contradiction to the quasigroup axioms. Similarly, if the same entry occurred twice in the s-th column, we would have two solutions to the equation $y a_s = c$ for some c. We conclude that each element of the quasigroup occurs exactly once in each row and once in each column, and so the unbordered multiplication table (which is a square array of n rows and n columns) is a latin square. $\square$

In fact, a quasigroup has more than one multiplication table because it is always possible to permute the rows and/or columns, together with their bordering elements (an example is given in Figure 1.3.2). So, a given quasigroup defines a number of different (although closely related[4]) latin squares. Conversely, a

---

[3] It is interesting to note that this author was also the first to consider generalized identities. (These are defined and discussed in Section 2.2.)

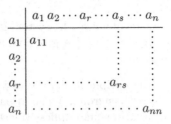

FIG. 1.1.1.

given latin square defines a multiplication table for more than one quasigroup[4] depending upon the order in which its elements are attached to form the borders.

As a simple example of a finite quasigroup, consider the set of integers modulo 3 with respect to the operation defined by $a * b = 2a + b + 1$. A multiplication table for this quasigroup is shown in Figure 1.1.2 and we see at once that it is a latin square.

$(*)$	0	1	2
0	1	2	0
1	0	1	2
2	2	0	1

FIG. 1.1.2.

More generally, the operation $a * b = ha + kb + l$, where addition is modulo n and h, k and l are fixed integers with h and k relatively prime to n, defines a quasigroup on the set $Q = \{0, 1, \ldots, n-1\}$.

As a special case of this, the operation $a * b = 2a - b$ defines a quasigroup for which $a * a = a$. Quasigroups for which $a * a = a$ for all elements a are called *idempotent* (see Section 2.1).

Let us draw attention here to another useful concept.

DEFINITION. A latin square is said to be *reduced* or to be in *standard form* if, in its first row and column, the symbols occur in natural order.

For example, the latin square of Figure 1.1.2 takes reduced form if its first two rows are interchanged.

We end this preliminary section by drawing the reader's attention to the fact that quasigroups, loops and groups are all examples of the primitive mathematical structure called a groupoid.

DEFINITION. A set S forms a *groupoid* $(S, \cdot)$ with respect to a binary operation $(\cdot)$ if, with each ordered pair of elements a, b of S is associated a uniquely determined element $a \cdot b$ of S called their *product*. If a product is defined for only

---

[4]In each case the relationship is that of *isotopy*, which will be discussed in Section 1.3.

a subset of the pairs a, b of elements of S, the system is sometimes called a *half-groupoid*. [See, for example, Bruck(1958).] A groupoid whose binary operation is associative is called a *semigroup*.

Theorem 1.1.1 shows that a multiplication table of a groupoid is a latin square if and only if the groupoid is a quasigroup. Thus, in particular, a multiplication table for a semigroup is not a latin square unless the semigroup is a group.

1.2   The Cayley table of a group

Next, we take a closer look at the internal structure of the multiplication table of a group.

THEOREM 1.2.1 *Any Cayley table of a finite group G (with its bordering elements deleted) has the following properties:*

(1) *It is a latin square, in other words a square matrix $\|a_{ik}\|$ in which each row and each column is a permutation of the elements of G.*

(2) *The quadrangle criterion holds. This means that, for any indices i, j, k, l and i', j', k', l', it follows from the equations $a_{ik} = a_{i'k'}$, $a_{il} = a_{i'l'}$ and $a_{jk} = a_{j'k'}$, that $a_{jl} = a_{j'l'}$.*

Conversely, any matrix satisfying properties (1) and (2) can be bordered in such a way that it becomes the Cayley table of a group.

PROOF.   Property (1) is an immediate consequence of Theorem 1.1.1. Property (2) is implied by the group axioms, since by definition $a_{ik} = a_i a_k$ and hence, using the conditions given, we have

$$a_{jl} = a_j a_l = a_j(a_k a_k^{-1})(a_i^{-1} a_i)a_l = (a_j a_k)(a_i a_k)^{-1}(a_i a_l) = a_{jk} a_{ik}^{-1} a_{il}$$
$$= a_{j'k'} a_{i'k'}^{-1} a_{i'l'} = (a_{j'} a_{k'})(a_{i'} a_{k'})^{-1}(a_{i'} a_{l'}) = a_{j'} a_{l'} = a_{j'l'}$$

To prove the converse, a bordering procedure has to be found which will show that the Cayley table thus obtained is, in fact, a multiplication table for a group. If we use as borders the first row and the first column of the latin square, the invertibility of the multiplication defined by the Cayley table thus obtained is easy to show and is indeed a consequence merely of property (1). For, in the first place, when the border is so chosen, the leading element of the matrix acts as an identity element, e. In the second place, since this element occurs exactly once in each row and column of the matrix, the equations $a_r x = e$ and $y a_s = e$ are soluble for every choice of a_r and a_s.

Now, only the associativity has to be proved. Let us consider arbitrary elements a, b and c. If one of them is identical with e, it follows directly that $(ab)c = a(bc)$. If, on the other hand, each of the elements a, b and c differs from

e, then the submatrix determined by the rows e and a and by the columns b and bc of the multiplication table is

$$\begin{array}{|cc} b & bc \\ ab & a(bc) \end{array}$$

while the submatrix determined by the rows b and ab and by the columns e and c is

$$\begin{array}{|cc} b & bc \\ ab & (ab)c \end{array}$$

Hence, $a(bc) = (ab)c$ because of property (2), and we have associativity.   □

COROLLARY. *If $a_1, a_2, \ldots, a_n$ are distinct elements of a group of order n, and if b is any fixed element of the group, then the sets of products $\{ba_1, ba_2, \ldots, ba_n\}$ and $\{a_1 b, a_2 b, \ldots, a_n b\}$ each comprise all of the n group elements in some order.*

Property (2) was first observed by Frolov(1890a) who remarked that it is valid for any regular latin square (as defined below). Later Brandt(1927) showed that it was sufficient to postulate the quadrangle criterion to hold only for quadruples in which one of the four elements is the identity element. Textbooks on the theory of finite groups [see for example Speiser(1927)] adopted the criterion established by Brandt. Aczél(1969) and Bondesen(1969) have both published papers in which they have rediscovered the quadrangle criterion. Also, Hammel(1968) has suggested some ways in which testing the validity of the quadrangle criterion may in practice be simplified when it is required to test the multiplication tables of finite quasigroups of small orders for associativity.

DEFINITION. We say that a latin square is *group-based* if the quadrangle criterion holds for it. That is, a latin square is group-based if, when appropriately bordered, it becomes a Cayley table for a finite group.

A condition quite different to the quadrangle criterion, for testing whether a latin square is group-based, was given by Suschkewitsch(1929) [see also Siu(1991)]. It is very closely related to Cayley's classic proof that every group of order n is isomorphic to a subgroup of the symmetric group S_n and can be stated as follows:

THEOREM 1.2.2 *Let γ be any fixed column of a latin square L with symbol set Q of cardinality n. For $i = 1, 2, \ldots, n$ let $\sigma_i : Q \to Q$ be the permutation which maps γ to the i-th column of L. Then L is group-based if and only if the set $\Sigma = \{\sigma_i : i = 1, 2, \ldots n\}$ is closed under the usual composition operation for permutations. If the latter is the case then Σ forms a group isomorphic to the group on which L is based.*

PROOF.   Without loss of generality, we can assume that the columns of L have been permuted so as to make γ the first column and that the symbols of L have been replaced by the symbols of the set $\{1, 2, ..., n\} = Q^*$, say, in such a way that γ has these entries in natural order. If we then border L by its own first row and first column (which is γ), we get the Cayley table of a loop $(Q^*, .)$ with identity element 1. The bth column of L is the permutation $\sigma_b : x \to xb(= xR_b)$ of the first column γ. If Σ is closed under composition of permutations then and only then, for each pair $b, c \in Q^*$, we have $R_b R_c = R_d$ for some $d \in Q^*$. So $x R_b R_c = x R_d$ for all $x \in Q^*$. That is, $(xb)c = xd$. In particular, this is true when $x = 1$. So $bc = d$ and we have $(xb)c = x(bc)$ for all $x, b, c \subset Q^*$. Thus, $(Q^*, .)$ is a group and L is group-based. Moreover, in this case, $R_b R_c = R_{bc}$ for all $b, c \in Q^*$ and so the group formed by σ under composition of permutations is isomorphic to $(Q^*, .)$. □

To use the above theorem to test whether a latin square L is group-based it is often convenient to permute either the rows or symbols of L so that the entries in γ are in natural order (assuming the symbols of L are $1, 2, \dots, n$). Then the elements of Σ can be read directly from the columns of L. Of course, the same test will work if rows instead of columns are used throughout.

A third condition for a latin square to be group-based arises from a concept also due to Frolov (1890a,b), who called a reduced latin square "regular" if it has the following property: The squares obtained by raising each row in turn to the top and then re-arranging first the columns and then the remaining rows so that the square is again reduced are all the same.

We shall show (in Theorem 1.2.3 and as a corollary to Theorem 2.4.1 of the next chapter) that a latin square is regular in this sense if and only if it is group-based, though it seems that Frolov did not realize this.[5]

THEOREM 1.2.3 *A reduced latin square is group-based if and only if it is regular.*

PROOF. Let us border the square with its own first row and column so as to form the Cayley table of a loop with identity element 1. We show that, if and only if the square is regular, the quadrangle criterion must hold for all quadrangles which include 1 as one member. (This is sufficient, as we remarked earlier.) Let us choose arbitrarily a quadrangle which contains the element 1 in row h and column k say and suppose that the remaining cells of this quadrangle which are in row h and column k are b (in column v) and a (in row u) respectively. Then the fourth member of the quadrangle is in the cell (u, v). We move row h to row 1 and re-arrange the columns (to make the new first row coincide with the border) so that the k-th column becomes column 1 and so that the element b is in row 1 and column b. Also, the element a is now in column 1. After re-arranging the rows to make the new square reduced, a will be in row a of column 1. So the

---

[5] Frolov commented, without giving an explicit proof, that every regular latin square satisfies the quadrangle criterion but he did not relate either property to that of being group-based. He gave the cyclic latin square as an example of a regular latin square and stated erroneously that every regular latin square is symmetric.

fourth member of the quadrangle will be the entry in the cell (a, b) of the reduced square. But, if and only if the square is regular, this is always the same whatever the initial choices of cells containing 1 and the selected elements a and b. $\quad\square$

Note. If we wish to test whether a latin square is group-based using the Suschke-witsch method, we require n^2 tests since there are n^2 pairs of permutations in the set Σ. If we use the method which Frolov used to test whether a latin square is regular, we need at most n tests. In fact, we shall show in the next chapter that at most n/p tests are needed, where p is the smallest prime which divides the order n of the latin square.

Parker(1959a) proposed an algorithm for deciding whether a loop is a group but that author later found an error in his paper and his method turned out to give only a necessary condition, not a sufficient one.

Wagner(1962) proved that to test whether a finite quasigroup Q of order n is a group it is sufficient to test only about $3n^3/8$ appropriately chosen ordered triples of elements for associativity. However, if a minimal set of generators of Q is known, then it is sufficient to test the validity of at most $n^2 \log_2(2n)$ associative statements provided that these are appropriately selected.

Wagner also showed in the same paper that every triassociative quasigroup Q (that is, every quasigroup whose elements satisfy $xy \cdot z = x \cdot yz$ whenever x, y, z are distinct) is a group, and the same result has been proved independently by D.A.Norton(1960).

These results lead us to ask the question "What is the maximum number of associative triples which a quasigroup may have and yet not be a group?"

Farago(1953) proved that the validity of any of the following identities in a loop guarantees both its associativity and commutativity:

(i) $(ab)c = a(cb),$ (ii) $(ab)c = b(ac),$ (iii) $(ab)c = b(ca),$

(iv) $a(bc) = b(ca),$ (v) $a(bc) = c(ab),$ (vi) $a(bc) = c(ba),$

(vii) $(ab)c = (ac)b,$ (viii) $(ab)c = (bc)a,$ (ix) $(ab)c = (ca)b.$

In fact, as Sade(1962) has pointed out, the identities (iv) and (v) are equivalent and so also are (viii) and (ix). For example, if we permute the elements a, b, c in (v) it becomes $b(ca) = a(bc)$, which is (iv).

More recently, it has been shown with computer aid that there are just four identities of length at most six (if we exclude mirror images and re-labellings) which force a quasigroup to be a group: namely, (A) $a \cdot bc = ab \cdot c$, (B) $a \cdot bc = ac \cdot b$, (C) $a \cdot bc = ca \cdot b$ and (D) $a \cdot bc = b \cdot ca$. Moreover, all but the first of these forces the group to be abelian. See Fiala(2007) and Keedwell(2009a,b).

In fact, (i) is equivalent to (B) and (ii) to the mirror image of (B), (iii) to (C) and to its mirror image, (iv) and (v) to (D) and (viii) and (ix) to the mirror image of (D). (vi) and (vii) do not force a quasigroup to have an identity element.

THEOREM 1.2.4 *A finite quasigroup is commutative if and only if its multiplication table (with row and column borders taken in the same order) has the property that products located symmetrically with respect to the main diagonal represent the same element (i.e. the table is symmetric in the usual matrix sense).*

PROOF. By the commutative law, $ab = ba = c$ for any arbitrary pair of elements a, b and so the cells in the a-th row and b-th column and in the b-th row and a-th column are both occupied by c. If this were not the case for some choice of a and b, we would have $ab \neq ba$ and the commutativity would be contradicted. □

A Cayley table of a group is called *normal* if every element of its main diagonal (from the top left-hand corner to the bottom right-hand corner) is the identity element of the group [see page 4 of Zassenhaus(1958)].

If the notation of Theorem 1.1.1 is used, it follows as a consequence of the definition that a normal multiplication table $\|a_{ij}\|$ of a group has to be bordered in such a way that $a_{ij} = a_i a_j^{-1}$ holds. Thus, if the element bordering the i-th row is a_i, the element bordering the j-th column must be a_j^{-1}.

Obviously, the following further conditions are satisfied: (i) $a_{ij}a_{jk} = a_{ik}$ (since $a_i a_j^{-1} a_j a_k^{-1} = a_i a_k^{-1}$); and (ii) $a_{ji}^{-1} = a_{ij}$ (since $(a_j a_i^{-1})^{-1} = a_i a_j^{-1}$). For example, the normal multiplication table of the cyclic group of order 6, written in additive notation, is shown in Figure 1.2.1.

(+)	0	5	4	3	2	1
0	0	5	4	3	2	1
1	1	0	5	4	3	2
2	2	1	0	5	4	3
3	3	2	1	0	5	4
4	4	3	2	1	0	5
5	5	4	3	2	1	0

FIG. 1.2.1.

(+)	0	5	4	2
0	0	5	4	2
1	1	0	5	3
2	2	1	0	4
4	4	3	2	0

FIG. 1.2.2.

As was first suggested by an example which appeared in Zassenhaus' book on Group Theory [Zassenhaus(1958), page 168, Example 1], the normal multiplication table of a finite group has a certain amount of redundancy since every product $a_i a_j^{-1}$ can be found n times in the table, where n is the order of the group. In fact, $a_i a_j^{-1} = a_{ij} = a_{ik}a_{kj}$ for $k = 0, 1, \ldots, n-1$. Consequently, it is relevant to seek smaller tables that give the same information. A multiplication table having this property is called a *generalized normal multiplication table* if it has been obtained from a normal multiplication table by the deletion of a number of columns and corresponding rows. The idea of such generalized normal multiplication tables was first mentioned by Tamari(1949), who subsequently gave some illustrative examples in Tamari(1951) but without proof. As one of his examples, he stated that the table given in Figure 1.2.2 is a generalized normal multiplication table of the cyclic group of order 6, obtained from the complete table displayed in Figure 1.2.1 by deleting the rows bordered by 3 and 5 and the columns bordered by $3^{-1} = 3$ and $5^{-1} = 1$.

The same idea was mentioned again by Ginzburg(1964), who gave a reduced multiplication table for the quaternion group of order 8. Later, in Ginzburg(1967), he developed the concept in much more detail and gave full proofs of his results. This paper contains, among other things, a complete list of the minimal generalized normal multiplication tables for all groups of orders up to 15 inclusive.

It will be clear to the reader that of special importance to the theory is the determination of the minimal number of rows and columns of a generalized normal multiplication table. If r denotes the minimal number of rows (or columns), then Erdős and Ginzburg(1963) proved that $r < C(n^2 \log n)^{1/3}$ (where C is a sufficiently large absolute constant) while Ginzburg(1967) showed that, in general, $r > n^{2/3}$ and that, for the cyclic group C_n of order n, $r < (6n^2)^{1/3}$.

For further generalizations of the concept of a generalized normal multiplication table and for discussion of some of the mathematical ideas relevant to it, the reader should consult Ginzburg(1960), Ginzburg and Tamari(1969a,b), Tamari(1960) and Specnicciati(1966).

The perceptive reader will realize that these ideas may have application in coding and cryptography.

1.3 Isotopy

Let $(G, \cdot)$ and $(H, *)$ be two quasigroups. An ordered triple (θ, ϕ, ψ) of one-to-one mappings θ, ϕ, ψ of the set G onto the set H is called an *isotopy* or *isotopism* of $(G, \cdot)$ upon $(H, *)$ if $(x\theta) * (y\phi) = (x \cdot y)\psi$ for all x, y in G. The quasigroups $(G, \cdot)$ and $(H, *)$ are then said to be *isotopic*. (It is worth remarking that the same definition holds for any two groupoids.) There is an equivalent notion for latin squares. An *isotopism* of a latin square L permutes the rows of L, permutes the columns of L and permutes the symbols of L. The result is another latin square which is said to be *isotopic* to L.

The concept of isotopy seems to be very old. In the study of latin squares the concept is so natural as to creep in unnoticed and latin squares are simply multiplication tables for finite quasigroups. For example, the concept has already arisen in connection with our comments on Theorem 1.1.1. Also, each latin square is isotopic to a reduced latin square (see page 3) obtained by suitably permuting its rows and columns. The concept was consciously applied by Schönhardt(1930), Baer(1939,1940) and independently by Albert(1943,1944). Albert had earlier borrowed the concept from topology for application to linear algebras; and it had subsequently been virtually forgotten except for applications to the theory of projective planes.

A latin square becomes a multiplication table as soon as it has been suitably bordered. For example the latin square on the left in Figure 1.3.1 becomes a Cayley table of the cyclic group of order 4 if its first row and column are taken as bordering elements as shown on the right in the same figure.

$$
\begin{array}{cccc}
1 & 2 & 3 & 4 \\
2 & 3 & 4 & 1 \\
3 & 4 & 1 & 2 \\
4 & 1 & 2 & 3
\end{array}
\qquad
\begin{array}{c|cccc}
  & 1 & 2 & 3 & 4 \\
\hline
1 & 1 & 2 & 3 & 4 \\
2 & 2 & 3 & 4 & 1 \\
3 & 3 & 4 & 1 & 2 \\
4 & 4 & 1 & 2 & 3
\end{array}
$$

FIG. 1.3.1.

Of the permutations θ, ϕ, ψ introduced in the definition of isotopy, ψ operates on the elements of the latin square which forms the Cayley table of the quasigroup, while θ and ϕ operate on the borders.

Let us suppose, for example, that the elements, row border and column border respectively of the Cayley table exhibited in Figure 1.3.1 are transformed in the manner prescribed by the following permutations

$$
\psi = \begin{pmatrix} 1 & 2 & 3 & 4 \\ 2 & 1 & 4 & 3 \end{pmatrix}, \quad
\theta = \begin{pmatrix} 1 & 2 & 3 & 4 \\ 3 & 2 & 4 & 1 \end{pmatrix}, \quad
\phi = \begin{pmatrix} 1 & 2 & 3 & 4 \\ 2 & 4 & 3 & 1 \end{pmatrix}.
$$

Then the Cayley table in Figure 1.3.1 is transformed into that of an isotopic quasigroup given on the left in Figure 1.3.2. We may re-write the table so that the borders are in natural order as shown on the right in the same Figure. The latin squares in these two Cayley tables are isotopic.

$$
\begin{array}{c|cccc}
  & 2 & 4 & 3 & 1 \\
\hline
3 & 2 & 1 & 4 & 3 \\
2 & 1 & 4 & 3 & 2 \\
4 & 4 & 3 & 2 & 1 \\
1 & 3 & 2 & 1 & 4
\end{array}
\qquad
\begin{array}{c|cccc}
  & 1 & 2 & 3 & 4 \\
\hline
1 & 4 & 3 & 1 & 2 \\
2 & 2 & 1 & 3 & 4 \\
3 & 3 & 2 & 4 & 1 \\
4 & 1 & 4 & 2 & 3
\end{array}
$$

FIG. 1.3.2.

If an isotopism is such that $\theta = \phi = \psi$ then it is an *isomorphism*. For latin squares in which the rows and columns are indexed by the symbols we say that an *isomorphism* is an isotopism which applies the same permutation to the rows, columns and symbols. For example, the latin square exhibited in Figure 1.3.3 is isomorphic to that shown in Figure 1.3.1 with $\theta = \phi = \psi = \begin{pmatrix} 1 & 2 & 3 & 4 \\ 4 & 2 & 1 & 3 \end{pmatrix}$ so that row 1 becomes row 4, row 3 become row 1, row 4 becomes row 3; then column 1 becomes column 4, column 3 becomes column 1, column 4 becomes column 3; and finally symbol 1 becomes symbol 4, etc.

It is easy to see that isotopism and isomorphism are both equivalence relations between quasigroups (or between groupoids) and between latin squares.

DEFINITION. An *isotopy class* of latin squares is an equivalence class for the isotopy relation. That is, it is a maximal set of latin squares every pair of which

$$
\begin{array}{|cccc}
4 & 3 & 2 & 1 \\
3 & 1 & 4 & 2 \\
2 & 4 & 1 & 3 \\
1 & 2 & 3 & 4
\end{array}
$$

FIG. 1.3.3.

is isotopic. Similarly, an *isomorphism class* is a maximal set of latin squares every pair of which is isomorphic.

Geometrically speaking, as we shall show, isotopic quasigroups are quasigroups which co-ordinatize the same 3-net. We shall discuss this aspect of isotopy in detail in Section 8.1. Let us notice at this point that there may also exist isotopisms of a quasigroup onto itself. Such isotopisms are called *autotopisms* and will be discussed in Section 4.1.

The foregoing remarks should make it clear that the concept of isotopy is fundamental to our subject and so we shall need to develop some of its basic properties for future application.

THEOREM 1.3.1 *Every groupoid that is isotopic to a quasigroup is itself a quasigroup.*

PROOF.   Let $(G, \cdot)$ be a quasigroup and $(H, *)$ a groupoid isotopic to $(G, \cdot)$ with $(x\theta) * (y\phi) = (x \cdot y)\psi$ for all $x, y \in G$.

Let a, b be arbitrary elements in H. We require to show that there exists a unique x in H such that $a * x = b$. Since $a\theta^{-1}$ and $b\psi^{-1}$ belong to G and $(G, \cdot)$ is a quasigroup, the equation $a\theta^{-1} \cdot y = b\psi^{-1}$ has a unique solution c in G. Write $x = c\phi$. Then, $a * x = a * c\phi = (a\theta^{-1})\theta * (c\phi) = (a\theta^{-1} \cdot c)\psi = (b\psi^{-1})\psi = b$, so the equation $a * x = b$ is soluble. Further, if $a * x' = b$ we have $(a\theta^{-1} \cdot x'\phi^{-1})\psi = b$ or equivalently, $a\theta^{-1} \cdot x'\phi^{-1} = b\psi^{-1}$. Since the equation $a\theta^{-1} \cdot y = b\psi^{-1}$ has a unique solution, $x'\phi^{-1} = c$ whence $x' = c\phi = x$. Thus, the equation $a * x = b$ is uniquely soluble in H. By a similar argument, we may show that the equation $z * a = b$ is uniquely soluble for z. This proves the theorem.   □

As an alternative to the formal proof above one can see that Theorem 1.3.1 is simply saying that isotopy re-arranges the rows and columns and permutes the elements of a latin square and that the result of applying such operations to a latin square is a latin square again.

DEFINITION. If $(G, \cdot)$ is a given quasigroup (or groupoid) and σ, τ are one-to-one mappings of G onto G, then the isotope $(G, \otimes)$ defined by $x \otimes y = (x\sigma) \cdot (y\tau)$ is called a *principal isotope* of $(G, \cdot)$.

The mappings θ, ϕ, ψ of the general definition are here replaced by σ^{-1}, τ^{-1} and the identity mapping respectively.

THEOREM 1.3.2 *Every isotope $(H, *)$ of a quasigroup $(G, \cdot)$ is isomorphic to a principal isotope of the quasigroup.*

PROOF.   Let θ, ϕ, ψ be one-to-one mappings of G onto H which define the isotopism between $(G, \cdot)$ and $(H, *)$ so that $(x\theta) * (y\phi) = (x \cdot y)\psi$ for all x, y in Q.
Then $\psi\theta^{-1}$ and $\psi\phi^{-1}$ are one-to-one mappings of G onto G, so the operation $\otimes$
given by $x \otimes y = (x\psi\theta^{-1}) \cdot (y\psi\phi^{-1})$ defines a principal isotope $(G, \otimes)$ of G.

Also $(x\psi) * (y\psi) = (x\psi\theta^{-1})\theta * (y\psi\phi^{-1})\phi = [(x\psi\theta^{-1}) \cdot (y\psi\phi^{-1})]\psi = (x \otimes y)\psi$
so $(H, *)$ and $(G, \otimes)$ are isomorphic under the mapping $\psi : G \to H$.   □

THEOREM 1.3.3 *Among the principal isotopes of a quasigroup $(G, \cdot)$ there always
exist loops. [Such loops are called LP-isotopes (loop-principal isotopes) of $(G, \cdot)$.]*

PROOF.   Define mappings σ, τ of G onto G by $x\sigma^{-1} = x \cdot v$, $x\tau^{-1} = u \cdot x$,
where u and v are fixed elements of G, and write $e = u \cdot v$. Then $(G, \otimes)$, where
$x \otimes y = (x\sigma) \cdot (y\tau)$, is a loop with e as identity element. For, let a be in G. Since
$(G, \cdot)$ is a quasigroup, $a = u \cdot a'$ and $a = a'' \cdot v$ for some elements a', a'' in G.
Then
$$e \otimes a = (u \cdot v) \otimes (u \cdot a') = u\sigma^{-1} \otimes a'\tau^{-1} = u \cdot a' = a,$$
and
$$a \otimes e = (a'' \cdot v) \otimes (u \cdot v) = a''\sigma^{-1} \otimes v\tau^{-1} = a'' \cdot v = a.$$

Conversely, every LP-isotope of $(G, \cdot)$ is obtained by mappings σ, τ of the
type defined above. For, if $(G, \otimes)$ is an LP-isotope of $(G, \cdot)$ and has identity
element e then $x \otimes y = x\sigma \cdot y\tau$, so $x\sigma^{-1} = x\sigma^{-1} \otimes e = x \cdot e\tau = x \cdot v$ say, where
$v = e\tau$. Also $x\tau^{-1} = e \otimes x\tau^{-1} = e\sigma \cdot x = u \cdot x$ say, where $u = e\sigma$.   □

The proof of the above theorem can be formulated in terms of latin squares.
It is equivalent to the statement that any latin square can be bordered in such
a way that the borders are identical to one of the rows and one of the columns
of the latin square.

An unsolved problem is that of finding necessary and sufficient conditions
on a loop G in order that every loop isotopic to G be isomorphic to G. [See
Bruck(1958), page 57.] Associativity is sufficient, as our next theorem will show,
but is not necessary.

LP-isotopes of a quasigroup have been further investigated by Bryant and
Schneider(1966).

The preceding three theorems will be found in Albert's paper(1943) on "Quasigroups". The following theorem is due to Bruck(1946) and, independently, N.J.S.
Hughes(1957).

THEOREM 1.3.4 *If a groupoid $(S, \cdot)$ with an identity element e is isotopic to
a semigroup, then the groupoid and semigroup are isomorphic and so both are
associative and both have an identity element.*

PROOF.   Let $(H, *)$ be the semigroup and let the isotopism be defined by map-
pings θ, ϕ, ψ from G onto H such that $(x\theta) * (y\phi) = (x \cdot y)\psi$. Since $(H, *)$ is a
semigroup, we have $(a' * b') * c' = a' * (b' * c')$ for all $a', b', c' \in H$, which implies
$$[(a'\theta^{-1} \cdot b'\phi^{-1})\psi\theta^{-1} \cdot c'\phi^{-1}]\psi = [a'\theta^{-1} \cdot (b'\theta^{-1} \cdot c'\phi^{-1})\psi\phi^{-1}]\psi.$$

Thus,

$$(a'\theta^{-1} \cdot b'\phi^{-1})\psi\theta^{-1} \cdot c'\phi^{-1} = a'\theta^{-1} \cdot (b'\theta^{-1} \cdot c'\phi^{-1})\psi\phi^{-1} \qquad (1.1)$$

for all $a', b', c' \in H$. In particular, when $a'\theta^{-1} = e$ and $c'\phi^{-1} = e$ we get

$$b'\phi^{-1}\psi\theta^{-1} = b'\theta^{-1}\psi\phi^{-1} \qquad (1.2)$$

and this must hold for all $b' \in H$.

Now put $a'\theta^{-1} = e$ in (1.1). We get

$$b'\phi^{-1}\psi\theta^{-1} \cdot c'\phi^{-1} = (b'\theta^{-1} \cdot c'\phi^{-1})\psi\phi^{-1}.$$

Using (1.2),

$$b'\theta^{-1}\psi\phi^{-1} \cdot c'\phi^{-1} = (b'\theta^{-1} \cdot c'\phi^{-1})\psi\phi^{-1}.$$

Therefore,

$$b\psi\phi^{-1} \cdot c = (b \cdot c)\psi\phi^{-1} \qquad (1.3)$$

for all $b, c \in G$.

Next put $c'\phi^{-1} = e$ in (1.1). We get

$$(a'\theta^{-1} \cdot b'\phi^{-1})\psi\theta^{-1} = a'\theta^{-1} \cdot b'\theta^{-1}\psi\phi^{-1}.$$

Using (1.2),

$$(a'\theta^{-1} \cdot b'\phi^{-1})\psi\theta^{-1} = a'\theta^{-1} \cdot b'\phi^{-1}\psi\theta^{-1}.$$

Therefore,

$$(a \cdot b)\psi\theta^{-1} = a \cdot b\psi\theta^{-1} \qquad (1.4)$$

for all $a, b \in G$. Thence,

$$
\begin{aligned}
(a' * b')\phi^{-1}\psi\theta^{-1} &= (a'\theta^{-1} \cdot b'\phi^{-1})\psi\phi^{-1}\psi\theta^{-1} \\
&= (a'\theta^{-1}\psi\phi^{-1} \cdot b'\phi^{-1})\psi\theta^{-1} && \text{using (1.3),} \\
&= (a'\theta^{-1}\psi\phi^{-1} \cdot b'\phi^{-1}\psi\theta^{-1}) && \text{using (1.4),} \\
&= (a'\phi^{-1}\psi\theta^{-1} \cdot b'\phi^{-1}\psi\theta^{-1}) && \text{using (1.2).}
\end{aligned}
$$

Thus, $(a' * b')\sigma^{-1} = (a'\sigma^{-1} \cdot b'\sigma^{-1})$ where $\sigma^{-1} = \phi^{-1}\psi\theta^{-1}$, so σ is a one-to-one mapping of G onto H such that $(a \cdot b) = (a\sigma * b\sigma)\sigma^{-1}$ for all a, b in G. That is, σ maps G isomorphically onto H. In particular,

$$(a \cdot b) \cdot c = [(a\sigma * b\sigma)\sigma^{-1}\sigma * c\sigma]\sigma^{-1} = [(a\sigma * b\sigma) * c\sigma]\sigma^{-1}.$$

Similarly, $a \cdot (b \cdot c) = [a\sigma * (b\sigma * c\sigma)]\sigma^{-1}$, whence $(ab)c = a(bc)$.   □

COROLLARY 1. *If a loop is isotopic to a group then the loop is a group isomorphic to the given group.*

COROLLARY 2. *If groups are isotopic, they are isomorphic as well.*

The first corollary is a consequence of the facts that a quasigroup with identity is a loop and that any isotope of a quasigroup is also a quasigroup as shown in Theorem 1.3.1. It was first proved by Albert(1943). The second corollary follows immediately from the first.

Certain non-invariants of principal isotopy may be illustrated in terms of the two loops $(G, \cdot)$ and $(G, *)$ shown in Figure 1.3.4, as was pointed out by Bruck(1958). [See page 58 of that book. The two loops are related by $x * y = xR_3^{-1} \cdot yL_6^{-1}$, where $R_a : x \to xa$, $L_a : x \to ax$ as before.]

$(\cdot)$	1	2	3	4	5	6
1	1	2	3	4	5	6
2	2	1	6	3	4	5
3	3	4	5	2	6	1
4	4	5	1	6	2	3
5	5	6	4	1	3	2
6	6	3	2	5	1	4

$(*)$	2	3	4	5	6	1
2	2	3	4	5	6	1
3	3	2	6	4	1	5
4	4	6	2	1	5	3
5	5	4	1	2	3	6
6	6	1	5	3	2	4
1	1	5	3	6	4	2

FIG. 1.3.4.

(1) COMMUTATIVITY: $(G, *)$ is commutative but $(G, \cdot)$ is not.

(2) NUMBER OF GENERATING ELEMENTS: $(G, \cdot)$ can be generated by any one of the elements 3, 4, 5, or 6. On the other hand, no single element generates $(G, *)$ but any two of 3, 4, 5, 6, 1 will generate it.

(3) AUTOMORPHISM GROUP: The automorphism group of $(G, \cdot)$ has order 4 and is generated by the permutation (3 4 5 6). In contrast, that of $(G, *)$ has order 20 and is generated by the two permutations (3 4 5 6 1) and (3 4 6 5).

1.4 Conjugacy and parastrophy

The usual representation of a latin square as a matrix has an unfortunate side-effect in that it disguises the symmetry between the rows, columns and symbols. One way to avoid this difficulty is to think instead of a latin square as a set of (row, column, symbol) triples. If $A = \|a_{ij}\|$ is a latin square of order n the corresponding n^2 triples are

$$T_A = \big\{(i, j, a_{ij}) : i, j = 1, 2, \ldots, n\big\}.$$

We refer to the entry occurring in a specific position in a triple as a *co-ordinate*. The latin property of A translates into the observation that no two distinct triples in T_A agree in more than one co-ordinate. The closely related idea of an orthogonal array will be discussed in Section 5.6.

In the previous section we met the concept of isotopy, which applies permutations to each of the three co-ordinates. There is another operation called *parastrophy* or *conjugacy*, which permutes the co-ordinates themselves. Since there are three co-ordinates in a triple there are $6 = 3!$ parastrophes of each square. Each parastrophe can be designated by the permutation which is applied to produce it. For example, the $(2,1,3)$-parastrophe of a latin square is the transpose of that square because it is produced by switching the roles of the first two co-ordinates, namely the rows and columns. Similarly, the $(1,3,2)$-parastrophe is obtained by switching columns and symbols while the $(3,2,1)$-parastrophe is found by switching rows and symbols. The $(1,2,3)$-parastrophe of a square is the square itself, which is therefore included whenever we refer to the parastrophes of a square. An example of a latin square L and its six parastrophes is given in Figure 1.4.1. In fact, in the context of latin squares, the word "conjugate" has, until recently, been used much more frequently than "parastrophe, see below.

$$L = \begin{pmatrix} 3 & 1 & 4 & 2 \\ 2 & 3 & 1 & 4 \\ 1 & 4 & 2 & 3 \\ 4 & 2 & 3 & 1 \end{pmatrix} \qquad \begin{pmatrix} 3 & 2 & 1 & 4 \\ 1 & 3 & 4 & 2 \\ 4 & 1 & 2 & 3 \\ 2 & 4 & 3 & 1 \end{pmatrix} \qquad \begin{pmatrix} 2 & 4 & 1 & 3 \\ 3 & 1 & 2 & 4 \\ 1 & 3 & 4 & 2 \\ 4 & 2 & 3 & 1 \end{pmatrix}$$

$(1,2,3)$-parastrophe of L $(2,1,3)$-parastrophe of L $(1,3,2)$-parastrophe of L

$$\begin{pmatrix} 2 & 3 & 1 & 4 \\ 4 & 1 & 3 & 2 \\ 1 & 2 & 4 & 3 \\ 3 & 4 & 2 & 1 \end{pmatrix} \qquad \begin{pmatrix} 3 & 2 & 1 & 4 \\ 1 & 4 & 2 & 3 \\ 2 & 3 & 4 & 1 \\ 4 & 1 & 3 & 2 \end{pmatrix} \qquad \begin{pmatrix} 3 & 1 & 2 & 4 \\ 2 & 4 & 3 & 1 \\ 1 & 2 & 4 & 3 \\ 4 & 3 & 1 & 2 \end{pmatrix}$$

$(3,1,2)$-parastrophe of L $(2,3,1)$-parastrophe of L $(3,2,1)$-parastrophe of L

FIG. 1.4.1.

An easy way to find the $(1,3,2)$-parastrophe is to consider each row as a permutation from natural order and to replace it by its inverse permutation. For this reason the $(1,3,2)$-parastrophe is sometimes called the *row inverse*; see for example Kolesova *et al* (1990) and Wanless(1999). Similarly the $(3,2,1)$-parastrophe can be found by replacing each column by its inverse permutation and hence can be called the *column inverse*.

Parastrophy (conjugacy) of latin squares extends naturally to quasigroups. Every quasigroup $(Q, \cdot)$ has associated with it five other parastrophic quasigroups on the same set Q, obtained by taking parastrophes of the Cayley table for the operation $(\cdot)$. Stein(1956,1957), Sade(1959a) and Belousov(1965) were among the first to study parastrophic quasigroups.

As mentioned above, most writers on quasigroups use the terminology "parastrophe" (following Sade) and "parastrophy" rather than "conjugate" and "conjugacy". Indeed, the author of the present book considers it essential to do so in order to avoid confusion with the concept of conjugacy between subquasigroups of a quasigroup $(Q, \cdot)$ [which are of course represented by latin subsquares in the Cayley table of $(Q, \cdot)$]. However, as already remarked, the term "conjugate" has been used extensively when discussing latin squares, especially in North America.

The adjective "parastrophic" seems to have been used first by Shaw(1915) while "conjugate" was probably first used by Stein in his papers of the 1950's. In his very well-known book, page 18, Belousov(1967b) used the term "обратных операций" for parastrophic operations (meaning "reverse", "inverse", "reciprocal"). However, in later papers, he too adopted the name parastrophic.

DEFINITION. A *parastrophy class (conjugacy class)* is a maximal set of latin squares each pair of which are parastrophes.

We have stated above that each square has six parastrophes including itself, but it need not be the case that these six parastrophes are distinct. A square may have some symmetry which makes two or more of the parastrophes coincide. For example, if it is symmetric in the usual matrix sense then the $(1, 2, 3)$ and $(2, 1, 3)$-parastrophes coincide. In this case it will also follow that the $(1, 3, 2)$ and $(2, 3, 1)$-parastrophes coincide and, separately, that the $(3, 1, 2)$ and $(3, 2, 1)$-parastrophes coincide. More generally we have:

THEOREM 1.4.1 *The number of latin squares in a parastrophy class is always 1, 2, 3 or 6.*

PROOF. According to their definition, the parastrophes of a square L are produced by the action of the group S_3 on the triples of L. It follows that the distinct parastrophes must be in the orbit of some subgroup of S_3. The only subgroups of S_3 have orders 6, 3, 2 or 1 and indices 1, 2, 3 or 6 respectively, from which the theorem follows.   □

Note that all four of the feasible values given in Theorem 1.4.1 are achieved. We have seen an example with six different parastrophes in Figure 1.4.1 and we shall meet examples of the other types in Chapter 2.

DEFINITION. The set of all parastrophes of the squares in an isotopy class is called a *main class*.[6] A map which combines an isotopy with the taking of a parastrophe is called a *main class isomorphism* or *paratopy*.[7]

Other names for the ideas which we have introduced in this and the previous section were used in the early literature by Fisher, Yates, Norton and Finney. The name *transformation set* was used instead of isotopy class and *species* was used in the place of main class. Also, *adjugacy set* was used instead of parastrophy class.

---

[6]The relationships between parastrophy, isotopy classes and main classes will be studied further in Chapter 4.

[7]The name "paratopy" was introduced by Sade(1959a) and has the virtue of brevity. Belousov(1967a), on the other hand, used "isostrophy" for this concept. More recently, the term "autostrophy" has been used for an isotopism from a quasigroup to one of its parastrophes.

It should also be noted that main class isomorphisms are not the same thing as the isomorphisms defined in Section 1.3. The problem is that the latter notion is the established one for groups and hence the natural one when talking about quasigroups. However, in many problems which deal with latin squares the idea of a main class isomorphism is more natural. This has led authors such as Brown(1968) to call these simply "isomorphisms", which latter then becomes a further source of confusion.

A property which, for each class C, either holds for all members of C or for no member of C is said to be a class invariant. Many important properties of latin squares turn out to be *main class invariants*. A main class is in some deep sense a set of latin squares with the same structure (and is a maximal set with this property). Main class invariants will be found throughout this book. See, for example, Theorems 1.5.5 and 1.6.2 and the Corollary to Theorem 4.2.3.

1.5   Transversals and complete mappings

A *transversal* of a latin square of order n is a set of n cells, one in each row, one in each column, and such that no two of the cells contain the same symbol. This concept has very close connections with the theory and construction of orthogonal latin squares and will be referred to in that connection in Chapter 5.

A *complete mapping* of a group, loop, or quasigroup $(G, \otimes)$ is a bijective mapping $x \to \theta(x)$ of G upon G such that the mapping $x \to \eta(x)$ defined by $\eta(x) = x \otimes \theta(x)$ is again a bijective mapping of G upon G.

The associated mapping $\eta(x)$ is called an *orthomorphism*, a name first used by Johnson, Dulmage and Mendelsohn(1961).

THEOREM 1.5.1 *If Q is a quasigroup which possesses a complete mapping, then its multiplication table is a latin square with a transversal. Conversely, if L is a latin square having a transversal, then at least one of the quasigroups which have L as multiplication table has a complete mapping.*

PROOF.   Let us suppose that Q has a complete mapping, say

$$\theta = \begin{pmatrix} 1 & 2 & \ldots & n \\ a_1 & a_2 & \ldots & a_n \end{pmatrix}, \quad \eta = \begin{pmatrix} 1 & 2 & \ldots & n \\ b_1 & b_2 & \ldots & b_n \end{pmatrix} \tag{1.5}$$

then its multiplication table has at least one transversal since

$$1 \otimes a_1 = b_1$$
$$2 \otimes a_2 = b_2 \tag{1.6}$$
$$\vdots$$
$$n \otimes a_n = b_n$$

implying that the cell of the i-th row and a_i-th column has b_i as entry, for $i = 1, 2, \ldots n$, and these entries are all distinct.

Conversely, if L is a latin square having a transversal comprising the elements $b_1, b_2, \ldots, b_n$ occupying the cells $(1, a_1), (2, a_2), \ldots, (n, a_n)$, then there exists a quasigroup $(Q, \otimes)$ having L as its multiplication table for which (1.6) holds. This quasigroup Q has a complete mapping, characterized by mappings θ and η defined as in (1.5). $\square$

The notion of a transversal was first introduced by Euler(1779) under the title *formule directrix*. The concept was used extensively by H.W.Norton(1939) under the name of *directrix*. It was called a 1-*permutation* by Singer(1960) and it was given the name *diagonal* by Dénes and Pásztor(1963). Modern usage strongly favours the name "transversal", as pioneered by Johnson, Dulmage and Mendelsohn(1961) and Parker(1963), among others.

The concept of complete mapping was introduced by Mann(1942).[8]

In the rest of this section we shall give some of the results concerning these concepts which are contained in these and other papers.

THEOREM 1.5.2 *If L is a latin square of order n which satisfies the quadrangle criterion and possesses at least one transversal, then L has a decomposition into n disjoint transversals.*

PROOF.   By Theorem 1.2.1, L can be written as a multiplication table for some group G.

If L has a transversal formed by taking the symbol c_1 from the first row, c_2 from the second row, $\ldots$, c_n from the n-th row, then it follows easily from the group axioms that another transversal can be obtained by taking $c_1 g$ from the first row, $c_2 g$ from the second row, $\ldots$, $c_n g$ from the n-th row, where g is any fixed element of the group. As g varies through the n elements of the group, we shall thus obtain n disjoint transversals.

To see this, suppose that $c_i = g_i g_{j(i)}$ where the sequences $c_1, c_2, \ldots, c_n$ and $g_1, g_2, \ldots, g_n$ both represent orderings of the elements of G, the latter corresponding to the ordering of the rows and columns of L in the multiplication table of G. In other words, c_i is the element to be found in the cell which occurs in the i-th row and j-th column of L. Also, since the c_i form a transversal, the integer j is a function of i such that $j(i_1) \neq j(i_2)$ if $i_1 \neq i_2$. Then, because G is a group,

$$c_i g = (g_i g_{j(i)}) g = g_i (g_{j(i)} g) = g_i g_{k(i)}$$

where, as g_j varies through the elements of G, so does g_k. Consequently, $c_{i_1} g$ and $c_{i_2} g$ are always in distinct columns and so the $c_i g$ form a transversal. Moreover, $g_{j(i)} \neq g_{k(i)}$ for any value of i and so the transversal formed by the c_i's is disjoint from that formed by the $c_i g$'s. Similarly, the transversals corresponding to two different choices of the element g are disjoint. $\square$

Note that the validity of the associative law is an essential requirement for the proof.

---

[8]Much more recently, Evans(1990) defined a complete mapping to be *strong* if the corresponding orthomorphism is itself a complete mapping.

The converse of Theorem 1.5.2 is not true. We shall exhibit a latin square of order 10 as a counter-example. For the labelled elements in Figure 1.5.1 the quadrangle criterion does not hold, but the square has a decomposition into n disjoint transversals.

$$
\begin{array}{cccccccccc}
0 & 4 & 1 & 7 & 2 & 9 & 8 & 3 & 6 & 5 \\
8 & 1 & 5 & 2 & 7 & 3 & \boxed{9} & 4 & \boxed{0} & 6 \\
9 & 8 & 2 & 6 & 3 & 7 & 4 & 5 & 1 & 0 \\
5 & 9 & 8 & 3 & 0 & 4 & 7 & 6 & 2 & 1 \\
7 & 6 & 9 & 8 & 4 & 1 & \boxed{5} & 0 & \boxed{3} & 2 \\
6 & 7 & 0 & 9 & 8 & 5 & 2 & 1 & 4 & 3 \\
3 & \boxed{0} & 7 & 1 & \boxed{9} & 8 & 6 & 2 & 5 & 4 \\
1 & 2 & 3 & 4 & 5 & 6 & 0 & 7 & 8 & 9 \\
2 & \boxed{3} & 4 & 5 & \boxed{6} & 0 & 7 & 8 & 9 & 7 \\
4 & 5 & 6 & 0 & 1 & 2 & 3 & 9 & 7 & 8 \\
\end{array}
$$

FIG. 1.5.1.

It is also worthwhile to point out that there are many examples of latin squares which have no transversals and which do not satisfy the quadrangle criterion. One of order 6 is shown later in Figure 1.5.2.

THEOREM 1.5.3 *If G is a group of odd order $2n - 1$, then G has a complete mapping.*

PROOF.   If G is a group of odd order, it is well known that every element of G has a unique square root in G. To prove this, let $g \in G$ be an element of (necessarily odd) order $2r - 1$. Then $h = g^r$ satisfies $h^2 = g^{2r} = g$ and so h is a square root of g. Further, if $k \in G$ satisfies $k^2 = g$, we have $h^2 = k^2$ and so $h^{2n} = k^{2n}$. That is, $h = k$ since $h^{2n-1} = e = k^{2n-1}$ (where e is the identity element of G) so h is the unique square root of g.

It follows that, in a group G of odd order, $g_i^2 = g_j^2$ only if $i = j$. Consequently, the mapping $\eta(g_i) = g_i^2$ for $i = 1, 2, \ldots, n$ is a bijective mapping of G upon G. Thus, the identity mapping $\theta(g_i) = g_i$ satisfies the definition of a complete mapping of G. □

In the notation of Theorem 1.5.2, the entries $g_i^2 = g_i g_i$ for $i = 1, 2, \ldots, n$ of the leading diagonal of a multiplication table L of G form a transversal of L.

COROLLARY. *Every finite group of odd order is isotopic to an idempotent quasi-group.*

PROOF.   The proof is due to Bruck(1944), whose argument is as follows. Define a new operation $(*)$ on G by the equation $g_i * g_j = \sigma(g_i g_j)$ where σ is the permutation of G which maps g^2 onto g for every $g \in G$. Then $(G, *)$ is an idempotent quasigroup isotopic to the group $(G, \cdot)$. □

We observe that the proof of Theorem 1.5.3 is solely dependent on the fact that every element of a group of odd order has a unique square root. A quasigroup with the property that every element has an exact square root was called *diagonal* by Sade(1960a). He showed in Sade(1963) that *the necessary and sufficient condition for a commutative quasigroup to be diagonal is that it be of odd order*, by a variant of the following simple argument.

THEOREM 1.5.4 *The entries on the main diagonal of a symmetric latin square are all distinct if and only if the square has odd order.*

PROOF. Suppose L is a symmetric latin square of order n. The entries not on the main diagonal of L can be partitioned into symmetrically placed pairs containing the same symbol. Since each symbol must occur exactly n times in L it follows that the number of occurrences of each symbol on the main diagonal is $n - 2k$ for some $k > 0$. If n is even, this means that any symbol on the main diagonal must occur at least twice on that diagonal. On the other hand, if n is odd it means that every symbol must occur on the main diagonal at least once. Since there are only just as many places as there are symbols, no symbol can be duplicated in this case. □

Sade's result is a corollary of Theorem 1.5.4, since a commutative quasigroup has a symmetric Cayley table (when the same order is chosen for the row and column borders) in which the entries on the main diagonal represent the squares of the elements. Thus, Theorem 1.5.3 holds for all commutative quasigroups of odd order. Certain types of non-commutative quasigroups are also known to be diagonal.

A loop is called a *Bruck loop* if it satisfies the identities $[(xy)z]y = x[(yz)y]$ and $(xy)^{-1} = x^{-1}y^{-1}$. It is called *Moufang* if it satisfies any one of the identities $[(xy)z]y = x[y(zy)]$, $(xy)(zx) = [x(yz)]x$ or $x[y(xz)] = [(xy)x]z$. Such a loop satisfies the identity $(xy)^{-1} = y^{-1}x^{-1}$. Every commutative Moufang loop is a Bruck loop but a non-commutative Moufang loop is not a Bruck loop.

For the justification of the latter statements, see Bruck(1958, 1963b).

It follows from the results of Robinson(1966) and Glaubermann(1964, 1968) that every element of a Moufang loop of odd order or of a Bruck loop of odd order has a unique square root. Consequently, Theorem 1.5.3 remains true for such loops.

In fact, Robinson's results imply that, even in Bol loops (that is, loops satisfying the identity $[(xy)z]y = x[(yz)y]$ alone) of odd order, every element has a unique square root and so Theorem 1.5.3 holds. This follows from the facts that such loops are power associative (see definition below) and satisfy a weak form of Lagrange's theorem (namely, that the order of every element divides the order of the loop).[9] The class of Bol loops includes all Moufang loops and all Bruck

---

[9]It has been proved recently by Grishkov and Zavarnitsine(2005) and independently by Gagola and Hall(2005) that, for Moufang loops, the order of any subloop divides the order of the loop. In fact, finite Moufang loops (and therefore also finite extra loops, defined on page 43)

loops.

Also, conjugacy-closed loops (see definition below) satisfy the strong form of Lagrange's theorem [see Kinyon, Kunen and Philips(2004)] so those which are of odd order and power associative[10] satisfy Theorem 1.5.3. For more information concerning various types of loop and quasigroup, see Section 2.1.

DEFINITION. A loop (or quasigroup) $(Q, \cdot)$ is *power associative* if $a(aa) = (aa)a$ for all $a \in Q$. It is *conjugacy closed* if both its right mappings R_a and its left mappings L_a (where $bR_a = ba$ and $bL_a = ab$) are closed under conjugacy. See Kunen(2000).

Ryser(1967) posed the question of whether there exist any quasigroups of odd order which do not possess a complete mapping. Certainly there exist quasigroups of even order which have no complete mappings. In particular, Mann has proved that if a quasigroup Q of order $4k + 2$ has a subquasigroup of order $2k + 1$, then each multiplication table of Q is without transversals. (For the proof, see Theorem 5.1.5.) An example of such a quasigroup of order 6 is given in Figure 1.5.2.

	1	2	3	4	5	6
1	1	2	3	4	5	6
2	2	3	1	5	6	4
3	3	1	2	6	4	5
4	4	5	6	1	2	3
5	6	4	5	2	3	1
6	5	6	4	3	1	2

FIG. 1.5.2.

In this connection it is relevant to point out that if Q is a quasigroup which has a complete mapping, then any isotopic quasigroup has one also. In fact:

THEOREM 1.5.5 *The number of transversals of a latin square is a main class invariant.*

PROOF. The definition of a transversal is symmetric between rows, columns and symbols and does not depend on the order or labelling chosen for any of these objects. Hence, it is clear that a paratopy maps each transversal to another transversal. Since such a mapping is invertible, it cannot send two distinct transversals to the same transversal and therefore must preserve the number of transversals.   □

have even stronger properties: namely, for such loops, the Sylow properties hold. Kinyon and Kunen(2004) proved this result for extra loops and Gagola(2011) proved it for Moufang loops.

[10]It was shown by Goodaire and Robinson(1990) that there exist CC loops of odd order which are not power associative.

Belousov(1967b) proved the weaker result that if a quasigroup Q has a complete mapping then so do all the parastrophes of Q.

The deep question of which quasigroups possess complete mappings seems a long way from being solved. Even for groups the complete answer has only recently been obtained. For the details, see Section 2.5.

PROLONGATION

Let $(Q, \cdot)$ be a given quasigroup of order n which possesses a complete mapping θ. By a process which Belousov(1967b) has called *prolongation* we shall show how, starting from $(Q, \cdot)$, a quasigroup $(Q', *)$ of order $n+1$ can be constructed, where the set Q' is obtained from Q by the adjunction of one additional element.

Before presenting a formal algebraic definition of a prolongation, we shall explain how the construction may be carried out in practice. We suppose that the elements of Q are $1, 2, \ldots, n$ as usual and let L be the latin square formed by a multiplication table of the quasigroup $(Q, \cdot)$. Since $(Q, \cdot)$ has a complete mapping, L possesses at least one transversal (Theorem 1.5.1). We replace the elements in all the cells of this transversal by the additional element $n+1$ and then, without changing their order, adjoin the elements of the transversal to the resulting square as its $(n+1)$-th row and $(n+1)$-th column. Finally, to complete the enlarged square L', we adjoin the element $n+1$ as the entry of the cell which lies at the intersection of the $(n+1)$-th row and $(n+1)$-th column. The square L' is then latin (see Figure 1.5.3 for an example) and defines a multiplication table for a quasigroup $(Q', *)$ of order one greater than that of $(Q, \cdot)$.

$(\cdot)$	1	2	3
1	1	2	3
2	2	3	1
3	3	1	2

$(*)$	1	2	3	4
1	1	2	4	3
2	4	3	1	2
3	3	4	2	1
4	2	1	3	4

FIG. 1.5.3.

In the example illustrated in Figure 1.5.3, $(Q, \cdot)$ is the cyclic group of order 3 and its prolongation $(Q', *)$ is a quasigroup of order 4.

If L has a second transversal, disjoint from the first, then the process can be repeated; since then the cells of this second transversal of L, together with the cell of the $(n+1)$-th row and column of L', form a transversal of L'.

Algebraically, we may specify a prolongation by defining the products $x * y$ of all the pairs of elements x, y of Q'. If $x \cdot \theta(x) = \eta(x)$ we get the following:

$$x * y = \begin{cases} x \cdot y & \text{if } x, y \in Q, \ y \neq \theta(x) \\ n+1 & \text{if } x, y \in Q, \ y = \theta(x) \\ \eta(x) & \text{if } x \in Q, \ y = n+1 \\ \eta\big(\theta^{-1}(y)\big) & \text{if } y \in Q, \ x = n+1 \\ n+1 & \text{if } x = y = n+1. \end{cases}$$

In our example (Figure 1.5.3) we have

$$\begin{aligned} \theta(1) &= 3, & \eta(1) &= 3, \\ \theta(2) &= 1, & \eta(2) &= 2, \\ \theta(3) &= 2, & \eta(3) &= 1. \end{aligned}$$

The construction of prolongation was first studied by Bruck(1944) who discussed only the case in which $(Q, \cdot)$ is an idempotent quasigroup. The construction for arbitrary quasigroups has been defined by Osborn(1961) and by Dénes and Pásztor (1963). We shall make use of the construction in Section 1.6 and again in Section 6.1. Yamamoto(1961) has used the same concept, under the name 1-*extension*, in connection with the construction of pairs of mutually orthogonal latin squares. He has also defined the inverse construction and called it a 1-*contraction*. We shall mention his work again in Section 9.3.

We mention two further properties of prolongation without proof.

(1) The necessary and sufficient condition that a group G be a prolongation of some quasigroup is that G is an elementary abelian 2-group; that is, a direct sum of cyclic groups each of order two.

(2) Let G be a group which has at least one element of order greater than two. If G possesses two complete mappings θ_1 and θ_2 then the prolongations of G constructed by means of θ_1 and θ_2 are isotopic if and only if θ_1 and θ_2 are themselves isotopic.[11]

The reader will find the proofs lacking here and further results of similar type in Belousov(1967b,c), Belousov and Belyavskaya(1968) and Belyavskaya(1969).

Further constructions related to the concept of prolongation will be found in Belyavskaya(1969,1970a,b,c,d) and Elspas, Minnick and Short(1963).

In particular, in the fourth and fifth of Belyavskaya's papers on this topic, the inverse of a prolongation (which had earlier been defined by Yamamoto under the name 1-contraction, as we have already mentioned above) has been re-introduced under the title of *compression*. Thus, by means of a compression one can obtain a latin square of order $n - 1$ from a given latin square of order n.

More recent papers on this topic are Deriyenko and Dudek(2008,2013) and Deriyenko and Deriyenko(2009). In the first of these papers, the authors show, among other things, that prolongation is still possible when the quasigroup/ latin square (of order n) has only a partial transversal of length $n - 1$. We give their construction below. In the second paper, they give an in-depth investigation of the situations in which a contraction/compression of a latin square is possible.

---

[11]In [DK1], the result was stated as "if and only if $\theta_1 = \theta_2$". Deriyenko(2011) has shown by counter-example that this is false.

The third paper discusses the circumstances in which two prolongations of a quasigroup yield isotopic quasigroups.

A long-standing conjecture of Brualdi and, later independently, of Stein(1975) is that every latin square of order n has a partial transversal of length at least $n-1$. If this proves to be correct (though some recent evidence is to the contrary, see Section 3.5, but no actual counter-example is known), then it would follow that every latin square can be prolonged.

Finally, Elspas et al have introduced a generalization of the notion of prolongation for latin hypercubes (the latter concept will be defined later in Section 5.6).

THE CONSTRUCTION OF DERIYENKO AND DUDEK

Suppose that the latin square L of order n has a partial transversal P of length $n-1$. Then the elements of P occupy $n-1$ rows and $n-1$ columns of L and are all distinct. Let u be the element which is missing from P and let the element in the cell (R, C), where R and C are respectively the row and column of L which do not contain any element of P, be v.

We leave the element v in cell (R, C) unaltered. We replace the entries in the cells of P by a new element w and adjoin a new $(n+1)$th row and column to L. In row r_i of this new column, we place the element which was in row r_i of P for each r_i except $r_i = R$. In row R of the new column we put w and in row $n+1$ of the new column, we put u. In column c_i of the new $(n+1)$th row, we place the element which was in column c_i of P for each c_i except $c_i = C$. In column C of the new row, we put w. Finally, we observe that we have already put u in column $n+1$ of the new row.

We illustrate the construction in Figure 1.5.4. In that figure, the elements of P are in boxes. Also, $u = 2, v = 5$ and $w = 7$.

$(\cdot)$	1	2	3	4	5	6
1	1	[3]	4	2	5	6
2	2	4	[5]	3	6	1
3	4	6	1	5	2	3
4	5	1	2	6	3	[4]
5	[6]	2	3	1	4	5
6	3	5	6	4	[1]	2

$(*)$	1	2	3	4	5	6	7
1	1	7	4	2	5	6	3
2	2	4	7	3	6	1	5
3	4	6	1	5	2	3	7
4	5	1	2	6	3	7	4
5	7	2	3	1	4	5	6
6	3	5	6	4	7	2	1
7	6	3	5	7	1	4	2

FIG. 1.5.4.

To close this section we use prolongation to prove a result which we will need for later constructions.

THEOREM 1.5.6 *There exists an idempotent quasigroup of order m if and only if $m \neq 2$.*

PROOF.   It follows from Theorems 1.5.3 and 1.5.2 that for every odd n, there exists an idempotent group of order n which has a multiplication table composed of disjoint transversals. If $n \neq 1$ then by prolongation using a transversal other than the main diagonal, one can obtain from this an idempotent quasigroup of order $n + 1$. It is easy to check that no quasigroup of order two is idempotent.☐

1.6   Latin subsquares and subquasigroups

The concepts of subquasigroup and latin subsquare are closely connected.

DEFINITION. Let the square matrix A shown in Figure 1.6.1 be a latin square. Then, if the square submatrix B shown in the same figure (where $1 \leq i, j, \ldots, l \leq n$ and $1 \leq p, q, \ldots, s \leq n$) is again a latin square, B is called a *latin subsquare* of A. If B has order 1 or the same order as A then it is said to be *trivial*, otherwise it is *proper*.

$$A = \begin{pmatrix} a_{11} & a_{12} & \cdots & a_{1n} \\ a_{21} & a_{22} & \cdots & a_{2n} \\ \vdots & \vdots & \ddots & \vdots \\ a_{n1} & a_{n2} & \cdots & a_{nn} \end{pmatrix} \qquad B = \begin{pmatrix} a_{ip} & a_{iq} & \cdots & a_{is} \\ a_{jp} & a_{jq} & \cdots & a_{js} \\ \vdots & \vdots & \ddots & \vdots \\ a_{lp} & a_{lq} & \cdots & a_{ls} \end{pmatrix}$$

FIG. 1.6.1.

Thus, the latin square corresponding to a Cayley table of a subquasigroup Q' of any quasigroup Q is a latin subsquare of the latin square defined by a Cayley table of Q. Conversely, any latin subsquare of a latin square derived from a Cayley table of a quasigroup Q becomes, when bordered appropriately, a Cayley table for a subquasigroup of a quasigroup isotopic to Q. (The reason for it being not the same as Q but only isotopic to it is that the bordering elements contained in the rows and columns defining the latin subsquare may be different from those of the latin subsquare itself.)

In Figure 1.6.2, the Cayley table of a quasigroup of order 10 is shown which has a subquasigroup of order 4 (consisting of the elements 1, 2, 3, 4) and also one of order 5 (with elements 3, 4, 5, 6, 7) the intersection of which is a subquasigroup of order 2 (with elements 3, 4). There are also four circled entries which form a latin subsquare on the elements 2 and 4. This is an example of a latin subsquare which is not a subquasigroup because the bordering elements (6 and 0 on the rows and 8 and 6 on the columns) do not coincide with the elements in the latin subsquare. Note that this example also shows that the entries forming a latin subsquare need not be contiguous.

We come now to our first theorem of this section:

THEOREM 1.6.1 *Let S_1 and S_2 be latin subsquares of a Latin square L and let I be the intersection of S_1 and S_2. If I is not empty then it is itself a (possibly trivial) latin subsquare of S_1, S_2 and L.*

	0	8	9	1	2	3	4	5	6	7
5	1	9	2	8	0	6	7	4	5	3
6	8	②	1	0	9	7	5	3	④	6
7	2	1	0	9	8	5	6	7	3	4
3	0	8	9	1	2	3	4	6	7	5
4	9	0	8	2	1	4	3	5	6	7
1	5	6	7	3	4	1	2	0	8	9
2	6	7	5	4	3	2	1	8	9	0
0	7	④	3	5	6	0	9	1	②	8
8	3	5	4	6	7	8	0	9	1	2
9	4	3	6	7	5	9	8	2	0	1

FIG. 1.6.2.

PROOF.   Suppose that the entries of I are in a subset R of the rows of L and let S be the set of symbols that occur in I. Take any $r \in R$ and $s \in S$. Since s is among the symbols used in S_1 we know that in row r it occurs within S_1. Similarly, it occurs within S_2 in row r. But s occurs only once in any given row of L so in row r it must occur in $S_1 \cap S_2 = I$. As r and s were general elements it follows that every symbol in I occurs in every row of I. By a similar argument, every symbol in I occurs in every column of I. Consequently, I is a latin square.   □

This result shows that it is no coincidence that the intersection of the subquasigroups in Figure 1.6.2 is a latin subsquare. However, it does not guarantee that the intersection is a subquasigroup (although it is in this case). But note that if, say, the column labels 1 and 3 were interchanged then we would have an intersection of subquasigroups which is not a subquasigroup.

H.W.Norton(1939) introduced the name *intercalate* for a latin subsquare of order 2, such as that formed by the circled entries in Figure 1.6.2. He made use of intercalates in connection with the enumeration of latin squares of order 7 (see Section 4.3). The reason why subsquares are useful in enumeration problems is explained by the following result.

THEOREM 1.6.2 *For each k, the number of latin subsquares of order k is a main class invariant.*

PROOF.   If we define a latin subsquare by means of ordered triples as in Section 1.4 then it is apparent from the definitions that a paratopy maps each latin subsquare to another latin subsquare of the same order. Since the paratopy is invertible this process must preserve the number of subsquares of each order. □

Our next theorem restricts the possible sizes of latin subsquares and, correspondingly, of subquasigroups. The first proof of the theorem was published by T.Evans(1960). Later, an alternative proof using prolongation was given by Dénes and Pásztor (1963). See also Dénes (1967b). In fact, the theorem is a special case of Theorem 3.1.2 which we prove later so we shall appeal to the latter for our proof. (The proof using prolongation was given in [DK1].)

THEOREM 1.6.3 *Let L_k be an arbitrary latin square of order k. Then, if $n \geq 2k$ there exists at least one latin square L_n of order n such that L_k is a proper latin subsquare of L_n. On the other hand, if $n < 2k$ there is no such L_n.*

COROLLARY. *Let Q_k be an arbitrary quasigroup of order k. Then, if $n \geq 2k$ there exists at least one quasigroup Q_n of order n such that Q_k is a proper subquasigroup of Q_n. On the other hand, if $n < 2k$ there is no such Q_n.*

PROOF. For L_k to be a proper latin subsquare of L_n there must be at least one symbol of L_n which does not occur in L_k. By Theorem 3.1.2 it is necessary and sufficient for L_k to be extendible to L_n that each symbol of L_n occurs at least $2k - n$ times in L_k. Since some symbols of L_n do not occur at all in L_k, this happens if and only if $2k - n \leq 0$. Moreover, the same theorem says that, provided this condition is satisfied, we can extend L_k to a Latin square of order n.

As regards the corollary, if L_k is a latin subsquare of a latin square L_n we can always border L_n using a row and a column of L_n each of which intersects L_k. This will produce a multiplication table for a quasigroup in which L_k corresponds to a subquasigroup. $\square$

In contrast to the above theorem, Hilton(1977) conjectured that, for all sufficiently large n, there exist latin squares of order n which contain no proper latin subsquares. His conjecture is now known to be true when n is not of the form $2^a 3^b$. In the first edition of this book, Hilton's conjecture was mis-reported. For more details of the present situation, see Problem 1.7 in The Present State of the Problems and also Chapter 4 of [DK2].

We can say a great deal about the latin subsquares in group-based latin squares. Firstly, we have:

THEOREM 1.6.4 *Let L be a Cayley table for a finite group G. Let a, b be two fixed elements of G and let H be a subgroup of G. The intersection of the rows bordered by elements of the left coset aH and columns bordered by elements of the right coset Hb is a latin subsquare of L of order $|H|$. Moreover, every subsquare of L can be expressed in this form for some suitable choice of a, b and H.*

PROOF. Let I be the set of entries in the intersection specified in the theorem. Clearly I is a square submatrix of order $|H|$. Also, by construction every element in I belongs to the two sided coset $aHb = aHHb$. But aHb contains exactly $|H|$ entries which is the number of entries in each row and column of I. Since L is latin, I contains no duplicated symbols within any row or column so it must be a subsquare on the elements of aHb.

To prove the last assertion of the theorem, we suppose that S is any latin subsquare of L of order, say, s. Let R and C be the bordering elements of the rows and columns (respectively) of L which intersect to form S, so that each $s_{ij} \in S$ is a product $r_i c_j$, where $r_i \in R$ and $c_j \in C$. Choose any $r \in R$ and $c \in C$ and consider the intersection I of the rows bordered by elements of $r^{-1}R$ and columns bordered by elements of Cc^{-1}. Since $|R| = |C| = |RC| = s$ it follows that $|r^{-1}R| = |Cc^{-1}| = |r^{-1}RCc^{-1}| = s$. Also $r^{-1}R$ contains the identity of G so $Cc^{-1} \subseteq r^{-1}RCc^{-1}$. As these two sets have the same cardinality we must have $Cc^{-1} = r^{-1}RCc^{-1} = H$ say. By a similar argument $r^{-1}R = r^{-1}RCc^{-1} = H$. Thus, $HH = H$ and so H is a subgroup of G. Also, $rHHc = rr^{-1}RCc^{-1}c = RC$ so S is the intersection of the rows bordered by rH and the columns bordered by Hc.   □

From Lagrange's theorem, which states that the order of any subgroup divides the order of the group, we get an immediate corollary:

COROLLARY. *If S is a latin subsquare of order s in a group-based latin square of order n then s divides n.*

Theorem 1.6.4 gives us a neat characterization of the latin subsquares which occur in group-based latin squares. In particular, we note that *a latin square of odd order which satisfies the quadrangle criterion cannot contain any intercalates.*

At a conference held in Prague in 2003 Aleš Drápal asked whether this result could be extended to Moufang loops. That is, how can the latin subsquares which occur in the Cayley tables of Moufang loops be characterized? [It is now known that Moufang loops satisfy Lagrange's theorem. See Gagola(2011).] We may ask the same question with regard to other classes of loops: for example Bruck loops or Bol loops.

Contrasting with the above is a question raised by Fuchs as to whether, given an arbitrary positive integer n, there always exists a latin square of order n (not necessarily group-based) which contains latin subsquares of every order m such that $m \leq n/2$. This question was answered by Heinrich(1977) in the negative except for small values of m. (See Problem 1.8 in The Present State of the Problems).

For the case of group-based latin squares, the question has been answered completely by Hobby, Rumsey and Weichsel(1960) as follows: A finite group G of order n contains elements of every order m which is a proper divisor of n if and only if one of the following conditions is satisfied:

(i) G is cyclic;

(ii) G is a p-group and contains a cyclic subgroup of index p;

(iii) G has order $n = p^\alpha q$ for distinct primes p and q and it contains precisely one Sylow q-subgroup which is the commutator subgroup G' of G. Further, if S is any Sylow p-subgroup of G, then S is cyclic, say $S = \langle g \rangle$, and g^p is in the centre of G.

Let H be a subgroup of index k in a finite group G. Let $a_1, a_2, \ldots, a_k$ be a set of left coset representatives for H and $b_1, b_2, \ldots, b_k$ be a set of right coset representatives for H. Suppose that we order the bordering elements for the rows of a Cayley table for G by writing down first the elements of $a_1 H$, then the elements of $a_2 H$, then the elements of $a_3 H$ and so on, finishing with the elements of $a_k H$. We choose a similar order for the bordering elements of the columns, writing down first the elements of $H b_1$, then the elements of $H b_2$, then the elements of $H b_3$ and so on, finishing with the elements of $H b_k$. This gives us a Cayley table for G which has quite a striking structure in terms of its latin subsquares.

THEOREM 1.6.5 *Let L be the Cayley table just described. If we partition the rows of L into k sets, each consisting of H consecutive rows, and partition the columns similarly, then we get k^2 blocks each of which is a latin subsquare of order $|H|$. If H is a normal subgroup then the k^2 subsquares consist of k latin subsquares whose elements are those of $a_1 H$, k latin subsquares whose elements are those of $a_2 H, \ldots,$ and k latin subsquares whose elements are those of $a_k H$.*

Proof The fact that each block is a latin subsquare follows immediately from Theorem 1.6.4. A general block will contain the elements of some two sided coset $a_i H b_j$. If H is normal then $a_i H b_j = a_m H$ for some m so, in that case, each of our subsquares contains the elements of a left coset. There are exactly k distinct left cosets. Also, there are k blocks which intersect each row and these blocks must contain disjoint sets of elements, so one of them must correspond to each left coset. The result now follows.   □

Latin subsquares can be used to characterize simple groups. We remind the reader that a group G is simple if and only if it has no non-trivial normal subgroup. By Theorem 1.6.5, if G has a subgroup of order h and index k then it decomposes into k^2 disjoint blocks, each of which is a latin subsquare of order h. The subgroup is normal if and only if, among the sets of symbols which occur in these k^2 blocks, there are only k different sets of symbols.

As an example, consider the dihedral group $D_3 = \langle a, b : a^3 = b^2 = e, ab = ba^{-1} \rangle$ of order 6 and its subgroups $H = \{e, b\}$ and $N = \{e, a, a^2\}$, the second of which is normal. The result of Theorem 1.6.5 for this case is exhibited in Figure 1.6.3.

Suppose that N is a normal subgroup of G and consider the latin subsquares in the Cayley table of G identified in Theorem 1.6.5 as corresponding to left cosets of N. If for $i = 1, 2, \ldots, k$ each of the subsquares corresponding to $a_i N$ is replaced by the coset representative a_i then the result is a Cayley table for a group which is isomorphic to the factor group G/N. For full details see Baumgartner(1921).

We note that the same construction was used by Šik(1951) in connection with the generalization of the Jordan-Hölder theorem and by Dénes and Pásztor (1963) for the purpose of disproving Wall's conjecture. (We shall discuss the latter below.)

	e	a	a^2	b	ab	a^2b
e	e	a	a^2	b	ba^2	ba
a	a	a^2	e	ba^2	ba	b
a^2	a^2	e	a	ba	b	ba^2
b	b	ba	ba^2	e	a^2	a
ba	ba	ba^2	b	a^2	a	e
ba^2	ba^2	b	ba	a	e	a^2

	e	b	a	ba	a^2	ba^2
e	e	b	a	ba	a^2	ba^2
b	b	e	ba	a	ba^2	a^2
a	a	ba^2	a^2	b	e	ba
ab	ba^2	a	b	a^2	ba	e
a^2	a^2	ba	e	ba^2	a	b
a^2b	ba	a^2	ba^2	e	b	a

FIG. 1.6.3.

The construction mentioned above was also utilized by Zassenhaus(1958) for the purpose of characterizing the normal Cayley tables of nonsimple groups and by Bruck(1946, page 335) in the construction of an example of a loop L which has a characteristic subloop which is not normal in L.

In Wall(1957), that author suggested a full investigation of the following problem: If a quasigroup Q of order n contains m subquasigroups each of order s and defined on disjoint subsets of Q, what conditions must Q satisfy if the inequality $n \geq (m+1)s$ is to hold? Wall showed that if $m = 2$, the inequality holds for any quasigroup Q, but also pointed out some cases in which it is false with $m > 2$. Later, Dénes and Pásztor (1963) proved the following result:

THEOREM 1.6.6 *If $m > 2$ is a divisor of a given integer n then there exists a quasigroup $(Q, \cdot)$ of order n that contains m subquasigroups which are all of the same order $s = n/m$ and are defined on disjoint subsets of Q.*

PROOF.   Let Q be an idempotent quasigroup of order m. (We have shown existence of such a quasigroup in Theorem 1.5.6.) Let each element a_i of a multiplication table of Q be replaced by a latin square L_i of order s, where if $i \neq j$, the latin squares L_i and L_j are defined over disjoint sets of elements. One thus obtains a latin square which, when bordered by a row consisting of the first rows of the latin squares $L_1, L_2, \ldots, L_m$ which appear along its main left-to-right diagonal taken in order and by a column consisting of the entries in the first columns of these squares taken in order, becomes a multiplication table of a quasigroup of order ms which is a union of m disjoint subquasigroups.   □

Note that the restriction $m > 2$ in Theorem 1.6.6 is crucial. Not only is there no idempotent quasigroup of order 2, but *there is no quasigroup which is the union of two disjoint quasigroups.* An independent proof of this latter result was published by Wall(1957). For the special case of group-based latin squares the result was also proved by Haber and Rosenfeld(1959).

Next, we mention a theorem concerning transversals in quasigroups which have subquasigroups defined on disjoint sets of elements.

THEOREM 1.6.7 *Let Q be a quasigroup of order n and let $Q_1, Q_2, \ldots, Q_r$ be subquasigroups of Q of orders $n_1, n_2, \ldots, n_r$ respectively such that the sets Q_k*

are disjoint and Q is their set theoretical union. If Q_k has a complete mapping for each $k = 1, 2, \ldots, r$ then Q has one too.

PROOF.   Without loss of generality[12] one can suppose that a multiplication table of Q has the following form

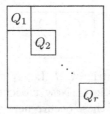

Since $Q_1, Q_2, \ldots, Q_r$ have transversals, Q has at least one too.   □

For further results similar to that of Theorem 1.6.7 see Dénes and Pásztor (1963), Haber and Rosenfeld(1959).

Haber and Rosenfeld(1959) have obtained several results concerning groups which are unions of proper subgroups.

Let us observe first that it is evident that any group G which is not cyclic is expressible as a set-theoretical union of (not necessarily disjoint) proper subgroups because, for example, such a group G is the union of its cyclic subgroups which, by hypothesis, are all proper. In Haber and Rosenfeld(1959) the problem of determining the minimum number of proper subgroups into which a group can be decomposed is discussed. Two of the main results obtained are as follows:

(1) Let G be a finite group of order n, and let p be the smallest prime which divides n. Then G is not the union of p or fewer of its proper subgroups.

(2) A group G is the union of three proper subgroups if and only if Klein's four-group is a homomorphic image of G.

The latter result is not new. It was first obtained by Scorza(1926) for the case of finite groups and was later discovered again by Bruckenheimer, Bryan and Muir(1970).

In Greco(1951,1953), necessary and sufficient conditions that a finite group G be expressible as the union of precisely four and precisely five proper subgroups are given and, in Greco(1956), some extensions of these results to the case of infinite groups are discussed.

The following interesting result is due to McWorter(1964), although Mann(1952) had earlier proved the same result for the case when Q is a group.

THEOREM 1.6.8 *Let Q be a quasigroup and let A and B be two subsets of Q such that not every element of Q has the form ab, with $a \in A$ and $b \in B$. Then $|Q| \geq |A| + |B|$, where $|X|$ denotes the cardinality of a set X.*

---

[12]To obtain this form, it is sufficient to arrange the row and column borders of Q so that the elements of Q_1 come first, then the elements of Q_2, and so on.

PROOF.   If Q is infinite, the result is obvious. Suppose therefore that Q is finite and has multiplication table M. We may re-arrange the rows of M in such a way that the bordering elements for the first $|A|$ rows consist of the elements of A and re-arrange the columns of M so that the bordering elements for the first $|B|$ columns are the elements of B. Then M has four submatrices C, X, Y and Z as follows:

C	Y
X	Z

where C is an $|A| \times |B|$ submatrix of M. Let $N(x)$ denote the number of times that the element $x \in Q$ occurs in C. Now x occurs $|A|$ times in the $|A|$ rows of $C \cup Y$ and at most $|Q| - |B|$ of these occurrences are in the $|Q| - |B|$ columns of Y so $N(x) \geq |A| - (|Q| - |B|)$ for every $x \in Q$. But, by construction, the entries of C consist precisely of the elements of the form ab, with $a \in A$ and $b \in B$. By assumption there is at least one element x of Q which is not of this form and for which $N(x) = 0$. Consequently, $0 \geq |A| + |B| - |Q|$. This completes the proof. $\square$

We use this same argument again in Theorem 3.1.2. It is the condition that C is a proper subrectangle of M. We may also use it to prove the following result:

THEOREM 1.6.9 *A $k \times k$ submatrix of a Cayley table of a quasigroup of order n contains every quasigroup element at least once if $k \geq \frac{1}{2}(n+1)$.*

PROOF.   Theorem 1.6.8 implies that, when $|Q| < |A| + |B|$ holds, then for every $c \in Q$ there exists an element $a \in A$ and an element $b \in B$ such that $ab = c$.

Since $n < \frac{1}{2}(n+1) + \frac{1}{2}(n+1)$, the validity of Theorem 1.6.9 is immediate. $\square$

We end this preliminary chapter with brief mentions of subsquare complete and subsquare avoiding latin squares. The former concept is due to Killgrove(1964,1974) and Hiner and Killgrove(1970) and we have the following definitions.

DEFINITION. A latin square L is *subsquare complete* if for each two distinct cells of L which contain equal elements, there is a proper latin subsquare of L which includes these two cells. We say that L is $\alpha, \beta, \gamma, \ldots$ *subsquare complete* if it is subsquare complete relative to subsquares whose orders lie in the set $\{\alpha, \beta, \gamma, \ldots\}$.

We stress that this definition implies that an α-subsquare complete square is also $\alpha, \beta, \gamma, \ldots$ subsquare complete for arbitrary choice of the integers $\beta, \gamma, \ldots$.

It is easy to check, for example, that the latin square given in Figure 1.6.4 is a 3-subsquare complete latin square. This square arose in an investigation concerning finite projective planes made by the present author [see Keedwell(1965)].

The following theorem was proved by Hiner and Killgrove(19**).

THEOREM 1.6.10 *A multiplication table of any finite non-cyclic group defines a subsquare complete latin square. If the elements of the group have orders $\alpha, \beta, \gamma, \ldots$ then the latin square is $\alpha, \beta, \gamma, \ldots$ subsquare complete.*

PROOF. Let L be a Cayley table for a finite non-cyclic group G with identity ε. Suppose that the two cells (a,b) and (c,d) of L both contain the same element $g \in G$ (that is, $ab = g = cd$). Let H be the subgroup of G generated by the element db^{-1}. Let S be the subsquare of L formed by the rows bordered by elements of aH and columns bordered by elements of Hb, as in Theorem 1.6.4. As G is assumed not to be cyclic H must be a proper subgroup, so S is a proper latin subsquare. Also, by definition, S contains the cells $(a\varepsilon, \varepsilon b) = (a,b)$ and $(a\varepsilon, db^{-1}b) = (a,d)$. But then S contains the symbol g and meets the column d, so it must also include the cell (c,d) by the properties of latin subsquares. Thus the two arbitrary cells (a,b) and (c,d) containing the same symbol are included in the proper subsquare S. As S has order equal to the order of a group element, namely db^{-1}, this completes the proof.   □

```
/ 1   2   3   4   5   6   7   8   9      1'  2'  3'  4'  5'  6'  7'  8'  9'      1"  2"  3"  4"  5"  6"  7"  8"  9"  \
| 1'  2'  3'  6'  4'  5'  8"  9"  7"     1"  2"  3"  6"  4"  5"  8   9   7       1   2   3   6   4   5   8'  9'  7'  |
| 1"  2"  3"  5"  6"  4"  9'  7'  8'     1   2   3   5   6   4   9"  7"  8"      1'  2'  3'  5'  6'  4'  9   7   8   |
|                                                                                                                  |
| 2   3   1   5   6   4   8   9   7      2'  3'  1'  5'  6'  4'  8'  9'  7'      2"  3"  1"  5"  6"  4"  8"  9"  7"  |
| 2'  3'  1'  4'  5'  6'  9"  7"  8"     2"  3"  1"  4"  5"  6"  9   7   8       2   3   1   4   5   6   9'  7'  8'  |
| 2"  3"  1"  6"  4"  5"  7'  8'  9'     2   3   1   6   4   5   7"  8"  9"      2'  3'  1'  6'  4'  5'  7   8   9   |
|                                                                                                                  |
| 3   1   2   6   4   5   9   7   8      3'  1'  2'  6'  4'  5'  9'  7'  8'      3"  1"  2"  6"  4"  5"  9"  7"  8"  |
| 3'  1'  2'  5'  6'  4'  7"  8"  9"     3"  1"  2"  5"  6"  4"  7   8   9       3   1   2   5   6   4   7'  8'  9'  |
| 3"  1"  2"  4"  5"  6"  8'  9'  7'     3   1   2   4   5   6   8"  9"  7"      3'  1'  2'  4'  5'  6'  8   9   7   |
|                                                                                                                  |
| 4   5'  6"  7   8'  9"  1   2'  3"     4'  5"  6   7'  8"  9   1'  2"  3       4"  5   6'  7"  8   9'  1"  2   3'  |
| 5   6'  4"  9   7'  8"  1"  2   3'     5'  6"  4   9'  7"  8   1   2'  3"      5"  6   4'  9"  7   8'  1'  2"  3   |
| 6   4'  5"  8   9'  7"  1'  2"  3      6'  4"  5   8'  9"  7   1"  2   3'      6"  4   5'  8"  9   7'  1   2'  3"  |
|                                                                                                                  |
| 7   8"  9'  1   2"  3'  4   5"  6'     7'  8   9"  1'  2   3"  4'  5   6"      7"  8'  9   1"  2'  3   4"  5'  6   |
| 8   9"  7'  3   1"  2'  4"  5'  6      8'  9   7"  3'  1   2"  4   5"  6'      8"  9'  7   3"  1'  2   4'  5   6"  |
| 9   7"  8'  2   3"  1'  4'  5   6"     9'  7   8"  2'  3   1"  4"  5'  6       9"  7'  8   2"  3'  1   4   5"  6'  |
|                                                                                                                  |
| 4'  5"  6   8'  9"  7   3   1'  2"     4"  5   6'  8"  9   7'  3'  1"  2       4   5'  6"  8   9'  7"  3"  1   2'  |
| 5'  6"  4   7'  8"  9   3"  1   2'     5"  6   4'  7"  8   9'  3   1'  2"      5   6'  4"  7   8'  9"  3'  1"  2   |
| 6'  4"  5   9'  7"  8   3'  1"  2      6"  4   5'  9"  7   8'  3   1   2'      6   4'  5"  9   7'  8"  3   1'  2"  |
|                                                                                                                  |
| 7'  8   9"  3'  1   2"  5"  6'  4      7"  8'  9   3"  1'  2   5   6"  4'      7   8"  9'  3   1"  2'  5'  6   4"  |
| 8'  9   7"  2'  3   1"  5'  6   4"     8"  9'  7   2"  3'  1   5"  6'  4    ·  8   9"  7'  2   3"  1'  5   6"  4'· |
| 9'  7   8"  1'  2   3"  5   6"  4'     9"  7'  8   1"  2'  3   5'  6   4"      9   7"  8'  1   2"  3'  5"  6'  4   |
|                                                                                                                  |
| 4"  5   6'  9"  7   8'  2   3'  1"     4   5'  6"  9   7'  8"  2'  3"  1       4'  5"  6   9'  7"  8   2"  3   1'  |
| 5"  6   4'  8"  9   7'  2"  3   1'     5   6'  4"  8   9'  7"  2   3'  1"      5'  6"  4   8'  9"  7   2'  3"  1   |
| 6"  4   5'  7"  8   9'  2'  3"  1      6   4'  5"  7   8'  9"  2"  3   1'      6'  4"  5   7'  8"  9   2   3'  1"  |
|                                                                                                                  |
| 7"  8'  9   2"  3'  1   6'  4   5"     7   8"  9'  2   3"  1'  6"  4'  5       7'  8   9"  2'  3   1"  6   4"  5'  |
| 8"  9'  7   1"  2'  3   6   4"  5'     8   9"  7'  1   2"  3'  6   4   5"      8'  9   7"  1'  2   3"  6"  4'  5   |
\ 9"  7'  8   3"  1'  2   6"  4'  5      9   7"  8'  3   1"  2'  6   4"  5'      9'  7   8"  3'  1   2"  6'  4   5"  /
```

FIG. 1.6.4.

Thus, for example, the square exhibited in Figure 1.6.5, which represents a multiplication table of the dihedral group D_3 of six elements, is 2,3-subsquare complete. In a group-based subsquare complete latin square every subsquare will have order dividing the order of the square (see the Corollary to Theorem 1.6.4). This need not be the case in non group-based examples, as can be seen from the 2,3-subsquare complete latin square of order 7 shown in Figure 1.6.6.

$$\begin{array}{|cccccc}
1 & 2 & 3 & 4 & 5 & 6 \\
2 & 3 & 1 & 6 & 4 & 5 \\
3 & 1 & 2 & 5 & 6 & 4 \\
4 & 5 & 6 & 1 & 2 & 3 \\
5 & 6 & 4 & 3 & 1 & 2 \\
6 & 4 & 5 & 2 & 3 & 1
\end{array}$$

FIG. 1.6.5.

Hiner and Killgrove(19**) also proved that every 2-subsquare complete latin square is a multiplication table of an elementary abelian 2-group. The same result was obtained independently by Heinrich and Wallis(1981). However, the corresponding result for 3-subsquare complete squares is false, as already shown by the non group-based example in Figure 1.6.4.

Hiner has made use of the principle of 1-extension and 1-contraction introduced by Yamamoto [see Section 1.5 and Section 9.3, and also Yamamoto(1960/61, 1961)] to give algorithms for obtaining 2,3-subsquare complete latin squares of orders $2^\alpha - 1$ and $3^\beta + 1$ for $\alpha > 2$ and $\beta > 1$. These are as follows:

Algorithm A. Take the latin square representing a multiplication table of the elementary abelian 2-group of order 2^α ($\alpha > 2$) and replace the element a_{ii} by the element a_{in} for $i = 1, 2, \ldots, n - 1$, where $n = 2^\alpha$. Then delete the last row and column of the amended square.

Algorithm B. Take the latin square representing a multiplication table of the elementary abelian 3-group of order 3^β ($\beta > 1$) and adjoin to it a new row and column such that $a_{m+1,i} = a_{i,m+1} = a_{ii}$ for $i = 1, 2, \ldots, m$ where $m = 3^\beta$. In the enlarged square replace each element a_{ii} (for $i = 1, 2, \ldots, m + 1$) of the leading diagonal by a new element $m + 1$.

Algorithm A amounts to effecting a 1-contraction of the latin square which represents a multiplication table of the elementary abelian group of order 2^α, while algorithm B amounts to effecting a 1-extension (prolongation) of the latin square which represents a multiplication table of the elementary abelian group of order 3^β. For proofs that these algorithms both yield 2,3-subsquare complete squares, the reader is referred to Hiner and Killgrove(19**). Examples of latin squares of orders 7 and 10 constructed by means of these algorithms are given in Figure 1.6.6 and Figure 1.6.7. (The left-hand squares are those used for the constructions.)

1	2	3	4	5	6	7	8
2	1	4	3	6	5	8	7
3	4	1	2	7	8	5	6
4	3	2	1	8	7	6	5
5	6	7	8	1	2	3	4
6	5	8	7	2	1	4	3
7	8	5	6	3	4	1	2
8	7	6	5	4	3	2	1

8	2	3	4	5	6	7
2	7	4	3	6	5	8
3	4	6	2	7	8	5
4	3	2	5	8	7	6
5	6	7	8	4	2	3
6	5	8	7	2	3	4
7	8	5	6	3	4	2

FIG. 1.6.6.

1	2	3	4	5	6	7	8	9
2	3	1	5	6	4	8	9	7
3	1	2	6	4	5	9	7	8
4	5	6	7	8	9	1	2	3
5	6	4	8	9	7	2	3	1
6	4	5	9	7	8	3	1	2
7	8	9	1	2	3	4	5	6
8	9	7	2	3	1	5	6	4
9	7	8	3	1	2	6	4	5

0	2	3	4	5	6	7	8	9	1
2	0	1	5	6	4	8	9	7	3
3	1	0	6	4	5	9	7	8	2
4	5	6	0	8	9	1	2	3	7
5	6	4	8	0	7	2	3	1	9
6	4	5	9	7	0	3	1	2	8
7	8	9	1	2	3	0	5	6	4
8	9	7	2	3	1	5	0	4	6
9	7	8	3	1	2	6	4	0	5
1	3	2	7	9	8	4	6	5	0

FIG. 1.6.7.

The algorithms can also be applied in the cases $\alpha = 2$ and $\beta = 1$ but in these cases the 2,3-subsquare complete latin squares which they yield are also 3-subsquare complete or 2-subsquare complete respectively.

In the same joint preprint, Hiner and Killgrove have shown that the 2,3-subsquare complete latin squares constructed by Hiner's algorithm A (for $\alpha > 2$) are characterized by the fact that these and only these among the universe of 2,3-subsquare complete squares have the property that, in each two rows, all but three pairs of cells which contain equal elements belong to proper subsquares of order 2. [A proof of this result also appears in Heinrich and Wallis(1981).] In Figure 1.6.6 for example, taking the second and third rows, we find that the only pairs of cells which contain equal elements but do not belong to proper subsquares of order 2 are the pairs (2,2) and (3,5), (2,3) and (3,2), (2,5) and (3,3).

The particular interest in 2,3-subsquare complete latin squares first arose as a consequence of their significance in connection with a hypothesis concerning non-Desarguesian and singly-generated projective planes. For the details, see Killgrove(1964,1974). The results given in Hiner and Killgrove's preprint referred to above show that, up to isomorphism, there is a unique 2,3-subsquare complete latin square for each of the orders 4, 5, 7, 8, 9, and 10 (those of orders 4 and 8 being also 2-subsquare complete and that of order 9 being also 3-subsquare complete) but that there are (for example) at least two 2,3-subsquare complete latin squares of order 12 and at least three of order 27 (each of the latter being also 3-subsquare complete). Later, in Killgrove(1974), it was shown that there are exactly three subsquare complete latin squares of order 12.

DEFINITION. Let S and L be latin squares of orders s and n ($s < n$) respectively defined on the same set Σ. Suppose that S can be superimposed on L in such a way that no cell of S contains the same member of Σ as does the cell of L onto which it is superimposed. Then we say that L avoids the subsquare S.

As an example, the latin square L_1 given in Figure 1.6.8 avoids the cyclic subsquare of order 3 as shown.

$$L_1 = \begin{vmatrix} 2_1 & 1_2 & 4_3 & 3 \\ 1_2 & 2_3 & 3_1 & 4 \\ 4_3 & 3_1 & 1_2 & 2 \\ 3 & 4 & 2 & 1 \end{vmatrix} \qquad L_2 = \begin{vmatrix} 1 & 2 & 3 & 4 & 5 & 6 & 7 & 8 & 9 & 0 \\ 2 & 1 & 4 & 3 & 8 & 9 & 0 & 5 & 6 & 7 \\ 3 & 4 & 1 & 2 & 9 & 0 & 8 & 7 & 5 & 6 \\ 4 & 3 & 2 & 1 & 0 & 8 & 9 & 6 & 7 & 5 \\ 5 & 8 & 9 & 0 & 1 & 7 & 6 & 2 & 3 & 4 \\ 6 & 9 & 0 & 8 & 7 & 1 & 5 & 4 & 2 & 3 \\ 7 & 0 & 8 & 9 & 6 & 5 & 1 & 3 & 4 & 2 \\ 8 & 5 & 7 & 6 & 2 & 4 & 3 & 1 & 0 & 9 \\ 9 & 6 & 5 & 7 & 3 & 2 & 4 & 0 & 1 & 8 \\ 0 & 7 & 6 & 5 & 4 & 3 & 2 & 9 & 8 & 1 \end{vmatrix}$$

FIG. 1.6.8.

We might ask: "What is the largest number of $s \times s$ subsquares for fixed s ($1 < s < n$) which can be avoided by a given $n \times n$ latin square L? Does this number depend on the structure of L: for example, whether L is group-based? Also, we could ask the same question when s is allowed to take all values ($1 < s < n$).

A related, but different question, is to ask which partial latin squares can be avoided by some completed latin square. The latter question has been addressed by Chetwynd and Rhodes(1997), by Cavenagh(2010) and by Kuhl and Denley(2012).

Let us complete our discussion of latin subsquares and subquasigroups by drawing the reader's attention to a particularly interesting latin square L_2 shown in Figure 1.6.8 discovered by D.A. Norton(1952b). It has the following properties:

(1) it is 2,3-subsquare complete;

(2) when bordered by its own first row and column it represents a multiplication table of a commutative loop in which (i) every two independent elements generate a subgroup of order 4, but (ii) no three independent elements associate in any order.

CHAPTER 2

Special types of latin square

In view of the intimate connection between latin squares and quasigroups which the previous chapter has already made clear, a discussion of special types of latin square automatically entails a discussion of special types of quasigroup.

We begin by observing how the type of identity satisfied by a quasigroup is reflected in the structure of the corresponding latin square. We go on to discuss the concept of parastrophy (conjugacy) which we introduced in Section 1.4. In particular, in the second section of the chapter we consider in some detail the parastrophy invariant properties of mediality, idempotency, two-sided self-distributivity and total symmetry. We prove that a medial quasigroup is always isotopic to an abelian group and introduce the concept of generalized identities. The third section is concerned with Steiner and other types of triple system. We give a brief history of Steiner triple systems, pointing out that they predate Steiner himself. We show that a finite idempotent totally symmetric quasigroup is equivalent to a Steiner triple system. We mention Mendelsohn triple systems and Mendelsohn quasigroups and we discuss when and when not a directed triple system is equivalent to a quasigroup.

The next two sections are concerned with a more-detailed account of how to determine whether a latin square is regular and hence group-based and with a more in-depth account of complete mappings in group-based squares.

The last section of the chapter is devoted to the subject of complete latin squares. Squares of this type first became of interest to statisticians in the late 1940s and a number of papers on the subject were subsequently published in journals of chemistry and psychology as well as in more mathematical journals. Consequently, as we point out, several of the constructions have been rediscovered two or three times. We give a detailed account of the work of Gordon and summarize later developments, many of which are dealt with in detail in [DK2].

2.1 Quasigroup identities and latin squares

Two papers are known to the authors which are devoted to a detailed account of quasigroup identities; namely Belousov(1965) and Sade(1957). The aim of the first author was to obtain a description, as complete as possible, of systems of quasigroups which satisfy certain fundamental identities: that is, to determine the structure of quasigroups satisfying a given system of identities by reducing the system to a simpler form. Sade's intention on the other hand, was to give as complete a list as possible of those identities which had been investigated up to the time of publication of his paper. In compiling our own list (given below) we have made extensive use of that of Sade. However we have renamed some of the identities in accordance with current usage and have added a few additional ones.

Two identities are said to be *dual* or to be *mirror images* if one is obtained from the other by a reversal of the order both of the symbols which occur in it and also of the bracketing. Thus, for example, the identities (36) and (37) in our list below are dual to each other. Certain identities such as (8) and (43) below are *self-dual*. For completeness we have usually included both members of each pair of dual identities. Identities for which the dual is not given are marked with a dagger (†).

(A) Identities which involve one element

$\qquad$ (1) $aa = a$ for all $a \in Q$ the idempotent law

(B) Identities which involve two elements (each identity must be valid for all $a, b \in Q$)

(2)	$aa = bb$	the unipotent law
(3)	$ab = ba$	the commutative law
(4)	$(ab)b = a$	Sade's right "keys" law
(5)	$b(ba) = a$	Sade's left "keys" law
(6)	$(ab)b = a(bb)$	the right alternative law[0]
(7)	$b(ba) = (bb)a$	the left alternative law[0]
(8)	$a(ba) = (ab)a$	the medial alternative law[1]
(9)	$a(ba) = b$	the law of right semi-symmetry[0]
(10)	$(ab)a = b$	the law of left semi-symmetry[0]
(11)	$a(ab) = ba$	Stein's first law (or the Stein identity)[†]
(12)	$a(ba) = (ba)b$	Stein's second law[†]
(13)	$a(ab) = (ab)b$	Schröder's first law
(14)	$(ab)(ba) = a$	Schröder's second law
(15)	$(ab)(ba) = b$	Stein's third law

The following two identities, although not quasigroup identities (because they contradict the axioms for a quasigroup) are listed here because they will play a role in the next chapter. (See page 89.)

(16)	$ab = a$	Sade's right translation law
(17)	$ab = b$	Sade's left translation law

(C) Identities which involve three elements (each identity must be valid for all $a, b, c \in Q$)

---

[0]By some authors, the adjectives "left" and "right" have been applied to these identities in reverse order.

[1]Also called the *law of elasticity* or *flexible law*.

(18)	$(ab)c = a(bc)$	the associative law
(19)	$a(bc) = c(ab)$	the law of cyclic associativity[†]
(20)	$(ab)c = (ac)b$	the law of right permutability
(21)	$a(bc) = b(ac)$	the law of left permutability
(22)	$a(bc) = c(ba)$	Abel-Grassmann's law
(23)	$(ab)c = a(cb)$	the commuting product[2]
(24)	$c(ba) = (bc)a$	dual of (23)
(25)	$(ab)(bc) = ac$	Stein's fourth law
(26)	$(ba)(ca) = bc$	the law of right transitivity
(27)	$(ab)(ac) = bc$	the law of left transitivity[3]
(28)	$(ab)(ac) = cb$	Schweitzer's law
(29)	$(ba)(ca) = cb$	dual of (28)
(30)	$(ab)c = (ac)(bc)$	the law of right self-distributivity
(31)	$c(ba) = (cb)(ca)$	the law of left self-distributivity
(32)	$(ab)c = (ca)(bc)$	the law of right abelian distributivity
(33)	$c(ba) = (ca)(bc)$	the law of left abelian distributivity
(34)	$(ab)(ca) = [a(bc)]a$	the Bruck-Moufang identity
(35)	$(ab)(ca) = a[(bc)a]$	dual of (34)
(36)	$[(ab)c]b = a[b(cb)]$	} the Moufang identities
(37)	$[(bc)b]a = b[c(ba)]$	
(38)	$[(ab)c]b = a[(bc)b]$	the Bol identity
(39)	$[b(cb)]a = b[c(ba)]$	dual of (38)
(40)	$[(ab)c]a = a[b(ca)]$	the extra loop law
(41)	$a[b(ca)] = cb$	Tarski's law[†]
(42)	$a[(bc)(ba)] = c$	Neumann's law[†]
(43)	$(ab)(ca) = (ac)(ba)$	the specialized medial law

(D) Identities which involve four elements (each identity must be valid for all $a, b, c, d \in Q$)

(44)	$(ab)(cd) = (ad)(cb)$	the first rectangle rule[†]
(45)	$(ab)(ac) = (db)(dc)$	the second rectangle rule[†]
(46)	$(ab)(cd) = (ac)(bd)$	the medial law[4]

By means of a succession of remarks, we shall point out the structural implications of a number of the above identities in the study of the latin squares which represent multiplication tables of the appropriate types of quasigroups. We shall assume that, in the multiplication tables we are discussing here, the row and column borders are ordered in the same way.

Any latin square which is a multiplication table of a quasigroup satisfying the identity (1) necessarily has a transversal since the cells which contain the products $aa, bb, cc, \ldots$, form a transversal. See also the corollary to Theorem 1.5.3 and the discussion which follows it.

[2]Called the "Eingewandtes Produkt" by Sade(1957).
[3]Called Stein's fifth law by Sade.
[4]Many other names have been used for this identity, see page 48.

If the row and column borders of the Cayley table of a quasigroup which satisfies the identity (2) are both ordered in the same way, then all the elements of the leading diagonal of the resulting latin square are the same. Since, by a re-ordering of its rows, any latin square can be transformed to one in which all the elements of the leading diagonal are the same, it follows that every quasigroup is isotopic to a unipotent quasigroup.

The Cayley table of any quasigroup satisfying identity (3) is symmetric in the usual matrix sense (see Theorem 1.2.4).

A quasigroup which satisfies the identities (1) and (3) (that is, one which is idempotent and symmetric) is necessarily of odd order as we proved in Theorem 1.5.4. On the other hand, one which satisfies the identities (2) and (3) (that is, one which is unipotent and symmetric) is necessarily of even order. We state this as a theorem.

THEOREM 2.1.1 *The entries on the main diagonal of a symmetric latin square are all the same only if the square has even order.*

PROOF.   Suppose that L is a symmetric latin square of order n defined on the symbols $0, 1, 2, \ldots, n-1$ and that the symbol which occurs n times on the main diagonal is 0. Since the square is symmetric, each other symbol occurs an even number of times. Consequently, n must be even.   □

The above theorem was first stated by Elspas, Minnick and Short(1963). Note that, unlike Theorem 1.5.4, the converse of this theorm is false. There exist commutative quasigroups of even order which are not unipotent. For an example, see Figure 2.1.1.

	0	1	2	3	4	5
0	0	1	2	3	4	5
1	1	2	0	4	5	3
2	2	0	4	5	3	1
3	3	4	5	2	1	0
4	4	5	3	1	0	2
5	5	3	1	0	2	4

FIG. 2.1.1.

If $xy = z$ in a quasigroup Q which satisfies identity (4) then the configuration shown in Figure 2.1.2 exists in the Cayley table of Q.

If the identities (4) and (5) both hold in a quasigroup Q then Figure 2.1.2 can be completed to the configuration shown in Figure 2.1.3(a), where $xy = z$. But then, because $xz = y$, we get Figure 2.1.3(b) by interchanging symbols y and z. Combining this with Figure 2.1.3(a), we have that $yz = x = zy$. Consequently we can deduce a result due to Sade(1953b) that the identities (4) and (5) together imply (3). An alternative algebraic proof is given on page 54.

$$
\begin{array}{c|cc}
 & & y \\
 & & \vdots \\
x & \dots\dots\dots & z \\
 & & \vdots \\
z & \dots\dots\dots & x \\
 & & \vdots \\
\end{array}
$$

FIG. 2.1.2.

$$
\begin{array}{c|cc}
 & y & z \\
 & \vdots & \vdots \\
x & \dots\dots z \dots\dots & y \\
 & \vdots & \vdots \\
z & \dots\dots x \dots\dots & \\
 & \vdots & \vdots \\
\end{array}
\qquad
\begin{array}{c|cc}
 & z & y \\
 & \vdots & \vdots \\
x & \dots\dots y \dots\dots & z \\
 & \vdots & \vdots \\
y & \dots\dots x \dots\dots & \\
 & \vdots & \vdots \\
\end{array}
$$

FIG. 2.1.3. (a) and (b)

Among quasigroups which satisfy the identities (6), (7) and (8) are the important class of loops known as *Moufang loops* which we mention again below.

A groupoid or quasigroup which satisfies the identity (9) has been called *demisymétrique* by Sade. We have translated this as "semi-symmetric" although Sade himself preferred the translation "halfsymmetric". In fact, every semi-symmetric groupoid is necessarily a semi-symmetric quasigroup, as was shown by Etherington(1962/63) and Sade(1965a).

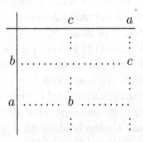

FIG. 2.1.4.

In Figure 2.1.4, the product ba is denoted by c whence, by virtue of the identity (9), $ac = b$. It then follows that $(ac)a = c$ so the validity of (9) implies also identity (10) and justifies the name semi-symmetric.

If $(Q, \cdot)$ is a given quasigroup and a new operation $(*)$ is defined on the set

Q by the relation $z * x = y \Leftrightarrow x \cdot y = z$, then $(Q, *)$ is again a quasigroup which is said to be a parastrophe (or conjugate) of $(Q, \cdot)$, see Section 1.4, and which we shall call the *first translate* of $(Q, \cdot)$. (Confusingly, it has also been called the *transpose* of $(Q, \cdot)$ by Etherington and Sade, but this terminology should be avoided since it does not agree with the usual matrix notion of transposition.) The *second translate* of $(Q, \cdot)$ is defined as being the first translate of $(Q, *)$: that is, the quasigroup $(Q, \otimes)$ such that $y \otimes z = x \Leftrightarrow z * x = y \Leftrightarrow x \cdot y = z$. The first translate of this quasigroup $(Q, \otimes)$ is $(Q, \cdot)$ again. We can, if we wish, give an alternative definition of a semi-symmetric quasigroup as being one which coincides with both its translates [cf. Sade(1965a)].

There is a connection between semi-symmetric quasigroups and balanced incomplete block designs (to be defined in Section 11.4), as has been pointed out by Sade(1965a). The same author has made a very extensive study of other properties and constructions for semi-symmetric quasigroups in his papers Sade(1964/65; 1965a,b; 1967a,b,c; 1968a). Among the many interesting results which he obtained we single out the following two from Sade(1964/65) as being of particular interest.

Sade proved that *if a quasigroup $(Q, \cdot)$ can be mapped isomorphically onto its first translate by a permutation α of its elements (meaning that $ab = c$ implies $(c\alpha)(a\alpha) = b\alpha$) and if α has order $3k$ with k not divisible by 3, then Q is isotopic to a semi-symmetric quasigroup.* He also gave an example of a quasigroup that he claimed was of the lowest possible order having the properties of being isotopic to its first translate but not isotopic to any semi-symmetric quasigroup. This example, which has order 10, was reproduced on page 63 of[DK1]. However, Wanless(2003) has pointed out that Sade's original claim is false since there are 11 main classes of latin squares of order 9 which, when bordered appropriately, become the Cayley tables of quasigroups with the desired property. One example, $(Q, \cdot)$, of such a quasigroup is given in Figure 2.1.5. It is isotopic to its first translate $(Q, *)$ since $x * y = (x \cdot y)\tau$ for all $x, y \in Q$, where τ denotes the permutation $(456)(789)$. It also has an automorphism (τ, τ, τ). These two symmetries generate an autoparatopy group[5] of order 9.

The significance of the identity (11) is that a quasigroup which satisfies it has an orthogonal complement or orthogonal mate. For the definition of this concept and proof of the last statement, see Section 5.4.

If a quasigroup Q which satisfies identity (11) has an identity element we find, on putting b equal to this identity element, that $a^2 = a$ for every $a \in Q$. Thus Q cannot be a group. We may similarly show that a quasigroup which satisfies any one of the identities (12), (13), (14) or (15) cannot be a group.

A quasigroup which satisfies the identity (14) is called a *Schroeder quasigroup*, see Lindner, Mendelsohn and Sun(1980).

An element x of a quasigroup $(Q, \cdot)$ such that, for all $a, c \in Q$, a, x and c satisfy both the identity (19) and its dual [that is, $a(xc) = c(ax)$ and $(cx)a = (xa)c$]

---

[5] Also called a group of *autostrophies*, see page 16

(·)	1	2	3	4	5	6	7	8	9
1	2	1	3	5	6	4	9	7	8
2	1	3	2	6	4	5	8	9	7
3	3	2	1	7	8	9	5	6	4
4	5	4	8	2	7	1	6	3	9
5	6	5	9	1	2	8	7	4	3
6	4	6	7	9	1	2	3	8	5
7	7	8	6	4	9	3	1	5	2
8	8	9	4	3	5	7	2	1	6
9	9	7	5	8	3	6	4	2	1

FIG. 2.1.5.

is called a *centre-associative element*. Quasigroups containing such elements were studied by Guha and Hoo(1965). These authors proved that a quasigroup containing centre-associative elements must be a loop, but not a group unless all its elements are centre-associative. In the latter case it is an abelian group. In any event, the number of centre-associative elements always divides the order of Q.

If a quasigroup has a two-sided identity element (that is, is a loop) and also satisfies any one of the identities (34), (35), (36) or (37) then it satisfies all four identities and is called a *Moufang loop*. It has recently been shown that a Moufang loop not only satisfies Lagrange's theorem and Cauchy's theorem but also has the Sylow properties. See page 20 for the details. Consequently, the orders of all its elements and subloops divide the order of the loop and this has obvious implications in regard to the structure of the latin square which represents the Cayley table of such a loop. A loop which satisfies identity (38) is called a *Bol loop* and such a loop satisfies the weak form of Lagrange's theorem [see Robinson(1966) and also page 20] so again the orders of its elements divide the order of the loop. The same is true for a loop which satisfies the identity (39).[6] There exist Bol loops which are not Moufang, as we remarked in Section 1.5. However, every Moufang loop is also a Bol loop. To see this, notice that if a is put equal to the identity element in the identity (37) we get the identity (8) and this, together with the identity (36) implies the Bol identity (38).

A loop which satisfies the identity (40) is called an *extra loop*. It is clear that identity (18) implies identity (40) and hence that every group is an extra loop. The concept of an extra loop was introduced by Fenyves(1968), who showed that every extra loop is a Moufang loop, while every commutative extra loop is an abelian group. Since there exist extra loops which are not groups and also commutative Moufang loops which are not groups, we infer that the class of extra loops lies properly between the class of groups and the class of Moufang loops. Fenyves also showed that isotopic extra loops are isomorphic. (Compare

---

[6]A loop which satisfies the identity (38) is sometimes called a *right Bol loop*, one which satisfies the identity (39) a *left Bol loop*.

Corollary 2 of Theorem 1.3.4.)

The reader will have noted that each of the identities (34) to (40) has the following form: both sides of the identity contain the same three symbols taken in the same order but one of them occurs twice on each side. Such an identity is said to be of *Bol-Moufang type*. Fenyves(1969) listed all possible identities of Bol-Moufang type and studied their interconnections.

More recently, Kunen(2006a,b) has determined which laws (identities) of Bol-Moufang type force a quasigroup to be a loop or group (cf. page 7 and the work of Farago, Fiala and the present author).

DEFINITION. A loop is called a *C-loop* (central loop) if its elements satisfy the identity $(yx \cdot x)z = y(x \cdot xz)$. It is called an *LC-loop* if its elements satisfy any one of the three equivalent identities $xx \cdot yz = (x \cdot xy)z$, $(x \cdot xy)z = x(x \cdot yz)$, $(xx \cdot y)z = x(x \cdot yz)$. It is called an *RC-loop* if its elements satisfy any one of the duals of these three identities.

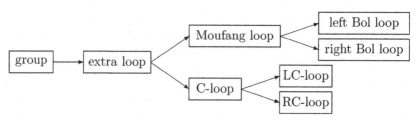

FIG. 2.1.6.

Fenyves showed that *LC-loops and RC-loops are both power associative* and that *a loop is a C-loop if and only if it is both an LC-loop and an RC-loop*. He also showed the validity of the implications shown in Figure 2.1.6, each of which is irreversible. For example, we display in Figure 2.1.7 a multiplication table of an *LC*-loop of six elements which is neither a *C*-loop nor a left Bol loop. In Figure 2.1.8 we give a multiplication table of a *C*-loop of ten elements which is not a Moufang loop. Both of these examples are taken from Fenyves(1969).

The significance of many of the identities given in Sade's list (and the present work) is best seen in the context of parastrophy of quasigroups, as defined in Section 1.4. We remind the reader that every quasigroup $(Q, \otimes)$ has six parastrophes associated with it. If $a \otimes b = c$ is a typical product in the quasigroup, then the corresponding products in the other parastrophes, and the notation we prefer to use for the parastrophic operations[7] can be summarized as follows:

---

[7]A table of notations used by various authors before 1958 can be found on page 76 of Sade(1959a).

	0	1	2	3	4	5
0	0	1	2	3	4	5
1	1	0	5	4	3	2
2	2	4	0	5	1	3
3	3	2	1	0	5	4
4	4	5	3	2	0	1
5	5	3	4	1	2	0

FIG. 2.1.7.

	0	1	2	3	4	5	6	7	8	9
0	0	1	2	3	4	5	6	7	8	9
1	1	0	4	9	2	7	8	5	6	3
2	2	4	0	6	1	9	3	8	7	5
3	3	9	6	0	7	8	2	4	5	1
4	4	2	1	7	0	6	5	3	9	8
5	5	7	9	8	6	0	4	1	3	2
6	6	8	3	2	5	4	0	9	1	7
7	7	5	8	4	3	1	9	0	2	6
8	8	6	7	5	9	3	1	2	0	4
9	9	3	5	1	8	2	7	6	4	0

FIG. 2.1.8.

Parastrophe	Name	Notation	Typical product
$(1,2,3)$	itself	$(Q,\otimes)$	$a \otimes b = c$
$(2,1,3)$	transpose	$(Q,\otimes^{(12)})$	$b \otimes^{(12)} a = c$
$(1,3,2)$	row inverse	$(Q,\otimes^{(23)})$	$a \otimes^{(23)} c = b$
$(3,2,1)$	column inverse	$(Q,\otimes^{(13)})$	$c \otimes^{(13)} b = a$
$(3,1,2)$	first translate	$(Q,\otimes^{(123)})$	$c \otimes^{(123)} a = b$
$(2,3,1)$	second translate	$(Q,\otimes^{(132)})$	$b \otimes^{(132)} c = a$

The first parastrophe is the quasigroup itself, with product $\otimes$. We use $\otimes^{(12)}$ to denote the operation obtained by switching the roles of rows and columns in the Cayley table (transposition). Similarly, we use $\otimes^{(23)}$ to denote the operation obtained by switching the roles of columns and elements (called row inverse, see page 15) and $\otimes^{(13)}$ to denote the operation obtained by switching the roles of rows and elements (column inverse). The remaining two parastrophes are the translates which we encountered on page 42.

In Belousov's classic text [Belousov(1967b)] and in the first edition of this book, the above operations were denoted by $\otimes$, $\otimes^*$, $^{-1}\otimes$, $\otimes^{-1}$, $^{-1}(\otimes^{-1})$ and $(^{-1}\otimes)^{-1}$ respectively and consequently this notation is widely used, especially in Eastern Europe. However, like the many notations used earlier, it has the big drawback that it is not easy to memorize nor does it make obvious that, for example, the operation which we have called the first translate can be denoted by $^{-1}(\otimes^{-1})$ or $^{-1}(\otimes^*)$ or $(\otimes^{-1})^*$. But, in the notation which we have chosen to adopt, the equalities

$$\otimes^{(123)} = (\otimes^{(12)})^{(13)} = (\otimes^{(23)})^{(12)} = (\otimes^{(13)})^{(23)}$$

and

$$\otimes^{(132)} = (\otimes^{(13)})^{(12)} = (\otimes^{(12)})^{(23)} = (\otimes^{(23)})^{(13)}$$

become obvious when the superscripts are multiplied as permutations.

The notation we have chosen to use is in fairly common use, particularly in the variant which writes $A^{(13)}(z, y) = x$, etc., for the parastrophes of the quasigroup (Q, A) in which a typical product is written $A(x, y) = z$.

When the quasigroup operation is $(\cdot)$, a simpler notation for the row and column inverses is commonly used: namely, $a \cdot b = c \Rightarrow a \backslash c = b$ and $c/b = a$.

If the quasigroup $(Q, \otimes)$ satisfies a given identity then in general each of its parastrophes will satisfy a different *parastrophic identity*. Thus, for example, validity of the identity (18) in $(Q, \otimes)$ may be expressed by the statement that $a \otimes b = x$, $x \otimes c = y$ and $b \otimes c = z$ together imply that $a \otimes z = y$. From this statement we deduce that the relations $a \otimes^{(23)} x = b$, $x \otimes^{(23)} y = c$ and $b \otimes^{(23)} z = c$ together imply $a \otimes^{(23)} y = z$. Hence, substituting for b, z and c in the equation $b \otimes^{(23)} z = c$ we get $(a \otimes^{(23)} x) \otimes^{(23)} (a \otimes^{(23)} y) = x \otimes^{(23)} y$. In other words, the parastrophe $(Q, \otimes^{(23)})$ satisfies the identity (27) whenever $(Q, \otimes)$ satisfies the identity (18).

Since a quasigroup which satisfies the identity (18) is a group, it is in some sense true that the theory of groups is equivalent to the theory of quasigroups which satisfy the identity (27), as was remarked by Stein(1957).

The concept of parastrophic identities seems to go back well into the nineteenth century and many other names were used by early writers. For further details, see page 16 and, as we remarked in an earlier footnote, Sade(1959a). In that paper, Sade gave some general rules for determining the identities satisfied by the parastrophes of a quasigroup which satisfies a given identity. Among other results, he proved the following:

(i) If a quasigroup $(Q, \otimes)$ satisfies an identity I then the quasigroup $(Q, \otimes^{(12)})$ satisfies the identity obtained from I by interchanging the pairs of operations $(\otimes, \otimes^{(12)})$, $(\otimes^{(13)}, \otimes^{(123)})$ and $(\otimes^{(23)}, \otimes^{(132)})$ throughout.

(ii) If a quasigroup $(Q, \otimes)$ satisfies an identity I then the quasigroup $(Q, \otimes^{(13)})$ satisfies the identity obtained from I by interchanging the pairs of operations $(\otimes, \otimes^{(13)})$, $(\otimes^{(12)}, \otimes^{(132)})$ and $(\otimes^{(23)}, \otimes^{(123)})$ throughout.

Sade also gave a number of rules for simplifying an identity which involves more than one parastrophic operation.

In Stein(1957), that author listed the parastrophic identities for a number of well-known identities. A similar but slightly more extensive list was given by Belousov(1965), and we reproduce it here in Figure 2.1.9. The occurrence of the symbol $\otimes$ in the body of the table indicates that the identity in question is the same with respect to the parastrophic operation.

It will be noted that the identities (46), (1) and (30), (31) taken jointly, appear to have a special significance in that they are parastrophy invariant. In this connection it is worthwhile to point out, following Stein(1957), that if the identities (46) and (1) both hold so also do the identities (30) and (31); that is, an idempotent medial quasigroup is both left and right self-distributive. We have $cb \cdot ca = cc \cdot ba = c \cdot ba$ by (46) and (1), so (31) holds. Similarly, we may show that (30) holds.

	$\otimes$	$\otimes^{(12)}$	$\otimes^{(13)}$	$\otimes^{(23)}$	$\otimes^{(132)}$	$\otimes^{(123)}$
Associativity	$xy \cdot z = x \cdot yz$	$xy \cdot z = x \cdot yz$	$yx \cdot zx = yz$	$xy \cdot xz = yz$	$xy \cdot xz = yz$	$yx \cdot zx = yz$
Mediality	$xy \cdot uv = xu \cdot yv$					
Idempotence	$xx = x$	$\otimes$	$\otimes$	$\otimes$	$\otimes$	$\otimes$
Commutativity	$xy = yx$	$xy = yx$	$y \cdot yx = x$	$yx \cdot x = y$	$xy \cdot y = x$	$x \cdot xy = y$
Left self-distributivity	$x \cdot yz = xy \cdot xz$	$yz \cdot x = yx \cdot zx$		$x \cdot yz = xy \cdot xz$		$xy \cdot x = zx \cdot yx$
Stein identity	$x \cdot xy = yx$	$yx \cdot x = xy$	$x(y \cdot yx) = yx$	$(x \cdot xy)y = xy$	$(xy \cdot y)x = xy$	$y(yx \cdot x) = x$
Two-sided self-distributivity	$x \cdot yz = xy \cdot xz$ and $yz \cdot x = yx \cdot zx$	$\otimes$	$\otimes$	$\otimes$	$\otimes$	$\otimes$

Fig. 2.1.9.

Medial[8] and idempotent quasigroups play a special role in the theory of latin squares. In the first place, every loop-principal isotope[9] of a medial quasigroup is an abelian group, as was first shown by Murdoch(1941). Consequently, every medial quasigroup is isotopic to some abelian group. That is to say, medial quasigroups arise quite naturally as a consequence of relabelling the elements and re-arranging the rows and columns of the latin square defined by the Cayley table of any abelian group. Moreover, Bruck(1944) has given a general method by which any medial quasigroup may be constructed from the abelian group to which it is isotopic. See Theorem 2.2.2.

Idempotent quasigroups give rise to latin squares which always possess at least one transversal (see page 39). Also, the class of finite totally symmetric (defined on page 53) idempotent quasigroups (which is a parastrophy invariant class) is coextensive with the class of designs known as Steiner triple systems. This will be shown in Section 2.3.

We end this discussion of identities by drawing attention to a remarkable theorem due to Belousov. We say that an identity $W_1 = W_2$ defined on a quasigroup Q is *balanced* if the same variables occur in W_1 as occur in W_2 and if no variable occurs more than once in W_1 or in W_2. This definition is due to Sade(1959c).

The reader can easily check that in our list given at the beginning of this chapter the following identities are balanced: (3), (18), (19), (20), (21), (22), (23), (24), (44), (46).

An identity $W_1 = W_2$ is called *reducible* [see Belousov(1966)] if either (i) each of W_1 and W_2 contains a "free element" x so that W_1 is of the form $U_1 \cdot x$ or $x \cdot V_1$ and W_2 likewise is of the form $U_2 \cdot x$ or $x \cdot V_2$ (where the U_i and V_i are subwords of the W_i) or (ii) W_1 has the product xy of two free elements x and y as a subword and W_2 has one of the products xy or yx as a subword, or the dual of this statement.

An identity which is not reducible is called *irreducible*.

For example, the identity $w(x \cdot yz) = (xy \cdot z)w$ is reducible because each of the two words composing it has w as a free element, the identity $xy \cdot uv = (u \cdot yx)v$ is reducible because the left hand side has xy as a subword and the right hand side has yx as a subword. Of the balanced identities listed above, only the identity (3) is reducible. The remaining identities are irreducible.

In Belousov(1966), that author proved the following very significant theorem, using ideas outside the scope of the present book:

THEOREM 2.1.2 *A quasigroup Q which satisfies any irreducible balanced identity is isotopic to a group*[10]

---

[8]By other authors these quasigroups have been called *abelian quasigroups* [Murdoch(1939,1941)], *alternation quasigroups* [Sholander(1949)] and *entropic quasigroups* [Etherington (1964/65)]. The medial law has been called the *symmetric law* [Frink(1955)] or the *bisymmetric law* [Aczél(1964,1965)].

[9]For the definition of this concept, see Section 1.3

[10]For some other results concerning the conditions under which a quasigroup is isotopic to a group, see T. Evans(1950) and Falconer(1971).

Later, M.A.Taylor(1978) gave a considerably more elementary proof and one which is applicable to a wider class of quasigroups. Then Belousov(1983) himself published an alternative elementary proof by means of reduction to the equation of generalized associativity. (The latter is defined in the next Section.)

We shall explain Taylor's theorem.

DEFINITION. Let W be a word in which the variables x and y occur. We say that x and y are *separated* in W if neither xy nor yx occurs in W: that is, at least one bracket lies between x and y.

THEOREM 2.1.3 *Let $W_1 = W_2$ be a balanced identity such that W_1 contains xy as a subword and x and y are separated in W_2. Then every quasigroup $(Q, \cdot)$ which satisfies this identity is isotopic to a group.*

PROOF.   We shall show that $(Q, \cdot)$ satisfies the quadrangle criterion

$$x_1y_2 = x_2y_1, \; x_1y_4 = x_2y_3, \; x_4y_1 = x_3y_2 \implies x_3y_4 = x_4y_3$$

(see Theorem 1.2.1).

Since x and y are separated in W_2, there must be a subword S of W_2 of length at least two in which only one of x or y occurs. We suppose without loss of generality[11] that S contains x and that y occurs to the right of S so that

$$W_2 = \sim S \sim y \sim,$$

where $\sim$ denotes a (possibly empty) string of subwords and brackets. We shall write $W_1(x, y, u) = W_2(x, y, u)$, where x, y are as just described and u is some third variable which occurs in S so that $W_2 = \sim S(x, u) \sim y \sim$.

By the hypothesis of the quadrangle criterion, $x_1y_2 = x_2y_1$, $x_1y_4 = x_2y_3$ and $x_4y_1 = x_3y_2$. So $W_1(x_1, y_2, u) = W_1(x_2, y_1, u)$, $W_1(x_1, y_4, u) = W_1(x_2, y_3, u)$ and $W_1(x_4, y_1, u) = W_1(x_3, y_2, u)$. Since, $W_1(x, y, u) = W_2(x, y, u)$, these same equalities hold for W_2. That is,

$$\sim S(x_1, u) \sim y_2 \sim \; = \; \sim S(x_2, u) \sim y_1 \sim \tag{2.1}$$

$$\sim S(x_1, u) \sim y_4 \sim \; = \; \sim S(x_2, u) \sim y_3 \sim \tag{2.2}$$

$$\sim S(x_4, u) \sim y_1 \sim \; = \; \sim S(x_3, u) \sim y_2 \sim \tag{2.3}$$

By use of the unique solubility of equations in a quasigroup, we can choose $u_1, u_2 \in Q$ such that $S(x_1, u_1) = S(x_3, u_2)$. Then,

$$\sim S(x_1, u_1) \sim y_2 \sim \; = \; \sim S(x_3, u_2) \sim y_2 \sim$$

and so, from (2.1) and (2.3),

---

[11] By "without loss of generality" here and later in the proof, we mean that we can construct an exactly similar proof if we make the contrary supposition.

$$\sim S(x_2, u_1) \sim y_1 \sim \; = \; \sim S(x_4, u_2) \sim y_1 \sim .$$

Therefore, $S(x_2, u_1) = S(x_4, u_2)$. Hence, putting $u = u_1$ in (2.2), we get

$$\sim S(x_1, u_1) \sim y_4 \sim \; = \; \sim S(x_2, u_1) \sim y_3 \sim .$$

Since $S(x_1, u_1) = S(x_3, u_2)$ and $S(x_2, u_1) = S(x_4, u_2)$, this implies that

$$\sim S(x_3, u_2) \sim y_4 \sim \; = \; \sim S(x_4, u_2) \sim y_3 \sim .$$

We have $W_2(x, y, u) = W_1(x, y, u)$, where W_1 takes the form $\sim xy \sim u \sim$ or $\sim u \sim xy \sim$. Assuming the former again without loss of generality, we get $\sim x_3 y_4 \sim u_2 \sim \; = \; \sim x_4 y_3 \sim u_2 \sim$. Since the $\sim$s on each side do not involve x_3, x_4, y_3, y_4 or u_2 because the identity $W_1 = W_2$ is balanced, it follows that $x_3 y_4 = x_4 y_3$ and that the quadrangle criterion holds in $(Q, \cdot)$. Therefore $(Q, \cdot)$ is isotopic to a group.   □

EXAMPLE. *A quasigroup which satisfies either of the reducible identities $w(x.yz) = (xy.z)w$ or $(xy)(uv) = (u.yx)v$ is isotopic to a group.*

For a further discussion of the topic of balanced identities, see Krapež and M.A.Taylor(1991).

2.2   Quasigroups of some special types and the concept of generalized associativity

THEOREM 2.2.1 *Every LP-isotope of a medial quasigroup $(Q, \cdot)$ is an abelian group.*

PROOF.   We begin by noting that if a medial quasigroup $(Q, \cdot)$ possesses a two-sided identity element e, then it is an abelian group. For, on putting $a = d = e$ in the medial law $ab \cdot cd = ac \cdot bd$ we get $bc = cb$ and so $(Q, \cdot)$ is commutative. Then putting $c = e$, we get $ab \cdot d = a \cdot bd$ and so $(Q, \cdot)$ is associative. The result follows.

Next, let $(Q, \cdot)$ be a given medial quasigroup, and let a new operation $\otimes$ be defined on the elements of Q by $a \otimes b = a\sigma \cdot b$ where $a\sigma^{-1} = av$. Then we shall show that $(Q, \otimes)$ is a medial quasigroup with v as unique right identity element. It is convenient to denote by R_v the unique one-to-one mapping of the set Q onto itself which is defined by $aR_v = av$ for all a in Q. Then $a \otimes b = aR_v^{-1} \cdot b$ and it is evident that either of the relations $a \otimes b = a \otimes c$ or $b \otimes a = c \otimes a$ implies $b = c$, so that $(Q, \otimes)$ is a quasigroup. Also

$$a \otimes v = aR_v^{-1} \cdot v = aR_v^{-1} R_v = a$$

so v is the unique right identity element of $(Q, \otimes)$. To show that $(Q, \otimes)$ is medial, let s be the unique solution of the equation $sv = v$. By the medial law in $(Q, \cdot)$,

$ab\cdot v = ab\cdot sv = as\cdot bv$ for all a, b in Q. That is, $(ab)R_v = aR_s\cdot bR_v$. Replacing a by aR_s^{-1} and b by bR_v^{-1}, we get $(aR_s^{-1}\cdot bR_v^{-1})R_v = ab$ and so $(ab)R_v^{-1} = aR_s^{-1}\cdot bR_v^{-1}$ for all a, b in Q. Using this relation, we have

$$(a \otimes b) \otimes (c \otimes d) = (aR_v^{-1} \cdot b) \otimes (cR_v^{-1} \cdot d) = (aR_v^{-1} \cdot b)R_v^{-1} \cdot (cR_v^{-1} \cdot d)$$
$$= (aR_v^{-1}R_s^{-1} \cdot bR_v^{-1}) \cdot (cR_v^{-1} \cdot d)$$

and similarly

$$(a \otimes c) \otimes (b \otimes d) = (aR_v^{-1}R_s^{-1} \cdot cR_v^{-1}) \cdot (bR_v^{-1} \cdot d).$$

The medial law of $(Q,\cdot)$ shows that these are equal, and so the medial law holds in $(Q,\otimes)$. Thus, $(Q,\otimes)$ is medial and has a unique right identity element, as stated.

By a similar argument, we may show that if $(Q,\cdot)$ is a given medial quasigroup and a new operation $\oplus$ is defined on the elements of Q by $a \oplus b = a\cdot b\tau$, where $b\tau^{-1} = ub$ then $Q,\oplus$ is a medial quasigroup with u as unique left identity.

Now, suppose that $(Q,\cdot)$ is a given medial quasigroup. Any LP-isotope $(Q,*)$ of $(Q,\cdot)$ is obtainable from it by a relation of the form $a * b = a\sigma \cdot b\tau$, where σ, τ are one-to-one mappings of Q onto itself such that $a\sigma^{-1} = av$ and $b\tau^{-1} = ub$ for suitable fixed elements v, u of Q (see Theorem 1.3.3). Also, again by Theorem 1.3.3, the loop $(Q,*)$ then has $e = uv$ as identity element. Now let $a \otimes b = a\sigma \cdot b$ as above. Then $a * b = a\sigma \cdot b\tau = a \otimes b\tau$. Since $(Q,\cdot)$ is a medial quasigroup, $(Q,\otimes)$ is medial, as already proved. Also because $(Q,\otimes)$ is medial, so is the quasigroup $(Q,*)$ defined by $a * b = a \otimes b\tau$ and it has uv as a two-sided identity element, as already stated. [By way of confirmation of this, let us note that $b\tau^{-1} = ub = (uv) \otimes b$ whence, regarding $(Q,*)$ as derived from $(Q,\otimes)$ by the definition $a * b = a \otimes b\tau$, it follows from our earlier analysis that uv is the unique left identity of $(Q,*)$. Symmetry considerations show that it is also the unique right identity.] But a medial quasigroup with a two-sided identity is an abelian group, so every LP-isotope of a medial quasigroup $(Q,\cdot)$ is an abelian group, as required. □

We next give Bruck's method (mentioned in the previous section) for constructing a medial quasigroup from the abelian group to which it is isotopic. This is embodied in the following theorem:

THEOREM 2.2.2 *Every medial quasigroup which is isotopic to a given abelian group $(G,\cdot)$ is isomorphic to some quasigroup $(G,*)$ obtained by a relation of the form $a * b = w \cdot a\sigma \cdot b\tau$, where σ, τ are commuting automorphisms of $(G,\cdot)$ and w is a fixed element of G.*

PROOF. We observe first that if the relation $(*)$ is defined as in the statement of the theorem, then

$$(a * b) * (c * d) = (w \cdot a\sigma \cdot b\tau) * (w \cdot c\sigma \cdot d\tau) = w \cdot (w \cdot a\sigma \cdot b\tau)\sigma \cdot (w \cdot c\sigma \cdot d\tau)\tau$$

$$= w \cdot w\sigma \cdot w\tau \cdot a\sigma^2 \cdot d\tau^2 \cdot b\tau\sigma \cdot c\sigma\tau$$

and similarly

$$(a * c) * (b * d) = w \cdot w\sigma \cdot w\tau \cdot a\sigma^2 \cdot d\tau^2 \cdot c\tau\sigma \cdot b\sigma\tau.$$

Since $\tau\sigma = \sigma\tau$, these two expressions are equal and so every relation of the form given defines a medial quasigroup.

Conversely, let $(G, *)$ be any principal isotope of $(G, \cdot)$ defined by $a*b = a\mu \cdot b\nu$ where μ, ν are one-to-one mappings of G onto itself. Since every isotope of $(G, \cdot)$ is isomorphic to a principal isotope (Theorem 1.3.2), it is sufficient to confine our attention to the latter. If $(G, *)$ is a medial quasigroup, then

$$(a\mu^{-1} * b) * (c * d\nu^{-1}) = (a\mu^{-1} * c) * (b * a\nu^{-1}).$$

This is equivalent to

$$(a \cdot b\nu)\mu \cdot (c\mu \cdot d)\nu = (a \cdot c\nu)\mu \cdot (b\mu \cdot d)\nu$$

for all a, b, c, d in G. Let e be the identity element of $(G, \cdot)$ and define the (fixed) elements u, v, w by $u = e\mu$, $v = e\nu$, $w = uv$. Also let λ be the one-to-one mapping of G onto itself which maps each element a of G onto its inverse $a^{-1} = a\lambda$ in $(G, \cdot)$. Putting $c = e$ in the relation just obtained, we have

$$(a \cdot b\nu)\mu \cdot (ud)\nu = (av)\mu \cdot (b\mu \cdot d)\nu.$$

On multiplying both sides of this by $(ud)\nu\lambda \cdot (av)\mu\lambda$, we get

$$(a \cdot b\nu)\mu \cdot (av)\mu\lambda = (b\mu \cdot d)\nu \cdot (ud)\nu\lambda.$$

Since this must hold for all a, b, d and, since the right-hand side is independent of a, the left-hand side must be also. Similarly, the right-hand side must be independent of d. Therefore, the left-hand side is equal to its expression with $a = e$ and the right-hand side is equal to its expression with $d = e$, whence

$$(a \cdot b\nu)\mu \cdot (av)\mu\lambda = b\nu\mu \cdot v\mu\lambda \text{ and } (b\mu \cdot d)\nu \cdot (ud)\nu\lambda = b\mu\nu \cdot u\nu\lambda.$$

These expresions can be re-written as

$$(ab)\mu = v\mu\lambda \cdot (av)\mu \cdot b\mu \text{ and } (ba)\nu = u\nu\lambda \cdot (ua)\nu \cdot b\nu,$$

where we have postmultiplied the expressions by $(av)\mu$ and $(ud)\nu$ respectively and then replaced $b\nu$ by b in the first expression, $b\mu$ and d by b and a respectively in the second expression.

Since $(G, \cdot)$ is commutative, $(ab)\mu = (ba)\mu$ or $v\mu\lambda \cdot (av)\mu \cdot b\mu = u\mu\lambda \cdot (bv)\mu \cdot a\mu$, whence $(av)\mu \cdot b\mu = (bv)\mu \cdot a\mu$. Also, $(ab)\nu = (ba)\nu$ whence $(ua)\nu bv = (ub)\nu \cdot av$. We can re-write these expressions in the forms $(av)\mu \cdot a\mu\lambda = (bv)\mu \cdot b\mu\lambda$

and $(ua)\nu \cdot a\nu\lambda = (ub)\nu \cdot b\nu\lambda$. Hence, observing that their left-hand sides are independent of a and can be equated to their expressions with $a = e$, we get $(av)\mu \cdot a\mu\lambda = v\mu \cdot u\lambda$ and $(ua)\nu \cdot a\nu\lambda = u\nu \cdot v\lambda$. Therefore, $(av)\mu = a\mu \cdot v\mu \cdot u\lambda$ and $(ua)\nu = u\nu \cdot a\nu \cdot v\lambda$. On substituting these formulae for $(av)\mu$ and $(ua)\nu$ into our earlier expressions for $(ab)\mu$ and $(ba)\nu$ we get

$$(ab)\mu = v\mu\lambda \cdot a\mu \cdot v\mu \cdot u\lambda \cdot b\mu = a\mu \cdot b\mu \cdot u\lambda$$

and similarly $(ab)\nu = (ba)\nu = a\nu \cdot b\nu \cdot v\lambda$. In other words,

$$(ab)\mu \cdot u\lambda = a\mu \cdot u\lambda \cdot b\mu \cdot u\lambda \text{ and } (ab)\nu \cdot v\lambda = a\nu \cdot v\lambda \cdot b\nu \cdot v\lambda.$$

So if we define $a\sigma = a\mu \cdot u\lambda$ and $a\tau = a\nu \cdot v\lambda$ we have that

$$(ab)\sigma = a\sigma \cdot b\sigma \text{ and } (ab)\tau = a\tau \cdot b\tau$$

showing that σ and τ are both automorphisms of $(G, \cdot)$. Thus, we have shown that there exist automorphisms σ and τ of $(G, \cdot)$ such that

$$a * b = a\mu \cdot b\nu = a\sigma \cdot u \cdot b\tau \cdot v = w \cdot a\sigma \cdot b\tau.$$

Because $(G, *)$ is required to satisfy the medial law, we must have

$$w \cdot w\sigma \cdot w\tau \cdot a\sigma^2 \cdot d\tau^2 \cdot b\tau\sigma \cdot c\sigma\tau = w \cdot w\sigma \cdot w\tau \cdot a\sigma^2 \cdot d\tau^2 \cdot c\sigma \cdot b\sigma\tau$$

for all $a, b, c, d \in G$ as in the first part of the theorem, and so

$$b\tau\sigma \cdot b\sigma\tau\lambda = c\tau\sigma \cdot c\sigma\tau\lambda.$$

Thus, the left-hand side is independent of b and is equal to its expression with $b = e$. That is,

$$b\tau\sigma \cdot b\sigma\tau\lambda = (v \cdot v\lambda)\sigma \cdot (u \cdot u\lambda)\tau\lambda = e\sigma \cdot e\tau\lambda = e \cdot e\lambda = e,$$

or $b\tau\sigma = b\sigma\tau$, so $\tau\sigma = \sigma\tau$ is necessary. This completes the proof of the theorem.

□

DEFINITION. A quasigroup $(Q, \cdot)$ is said to be *totally symmetric* if it is commutative and also semi-symmetric.

This implies that all its parastrophes coincide and so each of the equalities $bc = a$, $ca = b$, $ab = c$, $cb = a$, $ac = b$, $ba = c$ implies the other five. Any multiplication table of such a quasigroup defines a latin square which is unaffected when the roles of row, column, and element number are permuted in any way. An example of such a multiplication table is given in Figure 2.2.1

Medial quasigroups which are also totally symmetric play a role in the geometry of plane cubic curves. For the details of this application, see Etherington(1964/65). It is shown in the same paper that the construction of a medial

	1	2	3	4	5	6	7
1	1	3	2	5	4	7	6
2	3	2	1	6	7	4	5
3	2	1	3	7	6	5	4
4	5	6	7	4	1	2	3
5	4	7	6	1	5	3	2
6	7	4	5	2	3	6	1
7	6	5	4	3	2	1	7

FIG. 2.2.1.

quasigroup from the abelian group $(G, +)$ to which it is isotopic which was given in Theorem 2.2.2 can be simplified considerably in the case that the quasigroup is totally symmetric as well as medial.

In any totally symmetric quasigroup, the identities (4) and (5) given at the beginning of this chapter are valid. Their validity follows from the facts that $ab = c$ implies $cb = a$ and $bc = a$ as well as $ba = c$, whence $(ab)b = a$ and $b(ba) = a$ as required. Moreover, it is easy to show that a quasigroup in which the two relations $ab \cdot b = a$ and $b \cdot ba = a$ hold for all a and b is commutative and hence totally symmetric (cf. page 40). We have

$$ab = [(ba) \cdot (ba)a]b = [(ba \cdot b]b = ba,$$

so if $ab = c$, we have $ba = c$ and hence $cb = a$, $bc = a$. Also, from $ba \cdot a = b$ and $a \cdot ab = b$, we get $ca = b$ and $ac = b$. In fact, a totally symmetric quasigroup may be defined as a groupoid obeying any two of the identities $ab = ba$, $ab \cdot b = a$, and $b \cdot ba = a$.

The finite idempotent totally symmetric quasigroups are co-extensive with the class of design known as *Steiner triple systems*, as already mentioned and as we shall see in Section 2.3, where we shall give a brief account of these designs before demonstrating the equivalence of the two concepts.

The identities so far referred to have all been characterized by the fact that only one operation is involved in each. We say that such identities are of *rank* 1. More generally, by the *rank* of an identity $W_1 = W_2$, we understand the number of different binary operations that occur in the expressions W_1 and W_2. By the *length* of an expression W_i we understand the number of elements which occur in W_i. Thus, for example, the expression $a[b(cb)]$ has length 4. Evidently, the rank of the identity $W_1 = W_2$ cannot exceed the number $l = l_1 + l_2 - 2$, where l_1 and l_2 are the lengths of W_1 and W_2 respectively. For given values of l_1 and l_2 we call an identity of highest possible rank a *general identity*. Thus, in a general identity, the number of operations is fixed. In place of the term "general identity", Sade often used the term *identité démosienne* (generalized identity) as for example, in Sade(1957,1960b).

The following are examples of general identities:

(1) $(a \odot b) \oslash c = a \oplus (b \oplus c)$ the general associative law
(2) $(a \odot b) \oplus (c \oslash d) = (a \oplus c) \oplus (b \oplus d)$ the general medial law
(3) $(b \odot a) \oplus (c \oslash a) = b \oplus c$ the general law of right transitivity
(4) $a \odot (b \oslash c) = (a \oplus b) \oplus (a \oplus c)$ the general left distributive law
(5) $a \odot (a \oslash b) = b$ the general (left) keys law
(6) $a \odot b = b \oslash a$ the general commutative law

The idea of a general identity is due to Schauffler who introduced the concept in connection with problems of coding theory. For the details, see Schauffler(1956,1957). Some very comprehensive survey papers on the subject of generalized identities have been written. We refer the reader particularly to Belousov(1958,1965) and Sade(1960b).

As an example of the kind of result that has been obtained, we mention that if the general medial law (identity (2) above) holds with respect to six quasigroups $(G, \odot)$, $(G, \oslash)$, $(G, \oplus)$, $(G, \oplus)$, $(G, \oplus)$ and $(G, \oplus)$ all defined on the same set G, then all six of them are isotopic to one and the same abelian group. [A more general form of this result applicable to the case of multigroupoids has been given by Sade(1959c).] We shall not give the proof here. Instead, we should like to end the present section by giving two theorems which have relevance to the topic of coding theory.

THEOREM 2.2.3   *If four quasigroups* $(Q, \odot)$, $(Q, \oslash)$, $(Q, \oplus)$, $(Q, \oplus)$, *defined on the same set* Q, *are connected by the general associative law, then they are all isotopic to one and the same group.*

PROOF.   Let k be a fixed element of Q and let $\odot$ stand for any one of the four quasigroup operations $\odot$, $\oslash$, $\oplus$, $\oplus$. Let us define mappings L_i and R_i of the set Q onto itself by the statements $xL_i = k \odot x$ and $xR_i = x \odot k$. [These mappings are *translations by* k on the four quasigroups; xL_i is a *left translation* and xR_i a *right translation* of $(Q, \odot)$.)] In the equality $(a \odot b) \oslash c = a \oplus (b \oplus c)$ we put, in turn $a = c = k$, $a = k$, $b = k$ and $c = k$, to get

$$(k \odot b) \oslash k = k \oplus (b \oplus k),$$
$$(k \odot b) \oslash c = k \oplus (b \oplus c),$$
$$(a \odot k) \oslash c = a \oplus (k \oplus c),$$
$$(a \odot b) \oslash k = a \oplus (b \oplus k).$$

The above equations are equivalent to

$$bL_1 R_2 = bR_4 L_3 \tag{2.4}$$
$$bL_1 \oslash c = (b \oplus c)L_3 \tag{2.5}$$
$$aR_1 \oslash c = a \oplus cL_4 \tag{2.6}$$
$$(a \odot b)R_2 = a \oplus bR_4 \tag{2.7}$$

and the last three of these relations show that the quasigroups $(Q, \odot)$ are isotopic to each other; in particular, they are all isotopic to $(Q, \oslash)$.

We define a further isotope $(Q, \cdot)$ of $(Q, ②)$ by $a \cdot b = aR_2^{-1} ② bL_3^{-1}L_4^{-1}$. It then follows that all four of the quasigroups $(Q, ⊙)$ are isotopic to $(Q, \cdot)$. Indeed, we have

$$a ① b = (aR_1R_2 \cdot bR_4L_3)R_2^{-1} \tag{2.8}$$

$$a ② b = aR_2 \cdot bL_4L_3 \tag{2.9}$$

$$a ③ b = aR_1R_2 \cdot bL_3 \tag{2.10}$$

$$a ④ b = (aL_1R_2 \cdot bL_4L_3)L_3^{-1} \tag{2.11}$$

Equation (2.9) follows from the definition of the operation $(\cdot)$. Then, from equation (2.6) we get $a ③ b = aR_1 ② bL_4^{-1}$ and, using (2.9), this gives equation (2.10). From equation (2.5) we get $a ④ b = (aL_1 ② b)L_3^{-1}$ and, again using (2.9), this gives equation (2.11). Finally, from equation (2.7) we get $a ① b = (a ③ bR_4)R_2^{-1}$ and, using (2.10), this gives equation (2.8).

Substituting the expressions for $a ⊙ b$ so obtained in the general identity of associativity, we get

$$[(aR_1R_2 \cdot bR_4L_3)R_2^{-1}]R_2 \cdot cL_4L_3 = aR_1R_2 \cdot (bL_1R_2 \cdot cL_4L_3)L_3^{-1}L_3.$$

That is

$$(aR_1R_2 \cdot bR_4L_3) \cdot cL_4L_3 = aR_1R_2 \cdot (bL_1R_2 \cdot cL_4L_3).$$

But, $bR_4L_3 = bL_1R_2$ by equation (2.4), so $(uv)w = u(vw)$, where $u = aR_1R_2$, $v = bR_4L_3 = bL_1R_2$ and $w = cL_4L_3$.

Since u, v, w may be arbitrary elements in Q, we conclude that $(Q, \cdot)$ is a group, and the equations (2.8), (2.9), (2.10), (2.11) show that all the $(Q, ⊙)$ are isotopic to this group. This proves the theorem.   □

The above theorem was first formulated by Belousov in a lecture given at an algebra conference held at Moscow University in February, 1958. It was subsequently published by Belousov(1958) but without proof. Then in Hosszú(1959), that author re-proved the theorem and pointed out some of its applications. Two years later, Belousov's paper [Belousov(1961)] appeared and this also contained a proof of the theorem. A generalized form of the theorem has been proved by Sade [see Sade(1961), page 334] and he has also extended the theorem to cover the case of multigroupoids in Sade(1959b). See also Milić(1971).

Many authors have considered special cases of the generalized associative law, see, for example, Devidé(1955), T.Evans(1950), Ford(1970), and Suschkewitsch(1929).

DEFINITION. A set Ω_n of quasigroups of order n, defined on the same set Q of elements, is called an *associative system* [see Schauffler(1957)] if, corresponding to every two arbitrarily chosen quasigroups $(Q, ①)$, $(Q, ②)$ of Ω_n, there exist further quasigroups $(Q, ③)$, $(Q, ④)$ in Ω_n of such a kind that, for any three elements $a, b, c \in Q$, the general associative law is satisfied.

It follows from Theorem 2.2.3 that all the quasigroups contained in such a set Ω_n are isotopic to the same group.

In Schauffler(1957), that author raised the following question with regard to such a system. "Under what circumstances does an associative system Ω_n comprise the set of all quasigroups defined on a given set of cardinality n?" The answer he found was this:

THEOREM 2.2.4 *The set Ω_n of all quasigroups of order n is an associative system if and only if $n \leq 3$.*

PROOF. It is easy to show by direct enumeration that all quasigroups of order 2 and of order 3 are isotopic to the cyclic groups of order 2 and 3 respectively. In other words Ω_2 and Ω_3 are associative systems.

Also, by Theorem 2.2.3 we know that all quasigroups belonging to Ω_n are isotopic to one and the same group.

On the other hand Theorem 1.6.3 taken in conjunction with Lagrange's theorem for groups implies that for any $n > 4$, $n \neq 6$, there exists at least one quasigroup which is not isotopic to any group. For $n = 4$ all quasigroups are isotopic either to the cyclic group C_4 or else to $C_2 \times C_2$ but these groups are not themselves isotopic (see Section 4.2), while for $n = 6$, the fact that there exist quasigroups which are not isotopic to any group is shown also in Section 4.2. $\square$

For further information on the subject matter of this section, the reader should consult Belousov(1965) and Sade(1957) and the bibliographies contained therein.

For the general theory of quasigroups, he should consult Aczél(1964,1965), Albert(1943,1944), Belousov(1967a,b), Bruck(1958), Kertész(1964) and Pflugfelder (1990a), each of which contains an extensive bibliography of further papers, and also the papers of Sade.

2.3   Triple systems and quasigroups

We first consider Steiner triple systems and totally symmetric quasigroups.

A *Steiner triple system* of order n is a family of 3-element subsets (called *triples* or *blocks*) of a set V of cardinality n, such that for every pair of distinct elements of V, there is exactly one triple which contains that pair.

If we consider just the triples which contain a particular $v \in V$, we see that the other $n - 1$ elements of V must occur exactly once among these triples and that two other elements occur together with v in each triple. It follows that v must occur in exactly $r = \frac{1}{2}(n - 1)$ triples which implies that n must be odd and also that the number r of triples which contain v is independent of v. So, if t is the total number of triples, we have $nr = 3t$, since each side of this equality represents the total number of elements occurring in all the triples. We deduce that 3 divides r or n. If 3 divides r then $n - 1$ is a multiple of 6, say $n = 6m + 1$. If 3 divides n then n is three times an odd number, say $n = 3(2m + 1) = 6m + 3$. Thus, if n is the order of a Steiner triple system, we necessarily have $n \equiv 1$ or 3 modulo 6.

Before explaining the connections between such systems and quasigroups, we give an outline of their history.

It was in the year 1853 that Steiner(1852/53) posed the problem as to whether the necessary condition $n \equiv 1$ or 3 mod 6, was sufficient for the existence of a triple system having n different elements and of the type described above which now bears his name. The question was answered affirmatively by Reiss(1858/59) in 1859. However, neither of these writers seems to have been aware that the problem had been both posed and solved some twelve years earlier by Kirkman(1847) in an article in the Cambridge and Dublin Mathematical Journal. Indeed, three years after that, in "The Lady's and Gentleman's Diary" of 1850, Kirkman had gone on to pose a more difficult but related problem which is known to this day as Kirkman's schoolgirl problem. This problem requires the construction of Steiner triple systems on $n = 6m + 3$ elements which can be resolved into $r = (n-1)/2$ subsystems each containing every element exactly once. The general case of the latter problem was not solved until 1971. A complete solution will be found in Ray-Chaudhuri and Wilson(1970,1971,1973).

Two systems are *isomorphic* if one can be transformed into the other by a permutation of the n symbols. Netto(1893) showed that for $n = 3$, 7 and 9 there is, up to isomorphism, just one Steiner triple system. In the same year, Moore(1893) proved that for all $n > 13$ ($n \equiv 1$ or 3 mod 6), there are at least two non-isomorphic such systems. See also Hilton(1972). This result was strengthened by Netto(1901) to include the case $n = 13$. It turns out that, for $n = 13$, there are just two Steiner triple systems, up to isomorphism. Cole, White and Cummings(1925) found that there are 80 isomorphically distinct Steiner triple systems on 15 elements. This number was later verified by Hall and Swift(1955) using a computer. Computers were indispensable in settling the next case, that of 19 elements. Kaski, Östergård, Pottonen and Kiviluoto(2009) showed that there are an astonishing 11 084 874 829 systems of that order. Some upper and lower bounds on the number of Steiner triple systems of a given order were obtained by Doyen(1970a,b). Some further results on the same subject will be found in Doyen and Valette(1971), Rokovska(1971,1972), R.M.Wilson(1973/74), Alekseev(1974) and Rosa(1975).

For more information on Steiner triple systems and Kirkman's schoolgirl problem, the interested reader is referred to the book of M.Hall(1967) and to Doyen and Rosa(1973,1978,1980). Also, much additional interesting information on Kirkman's schoolgirl problem and its history will be found in chapter 10 of Ball(1939) and in chapter 10 of R.J.Wilson and Watkins(2013). For a detailed survey of the current state of knowledge on Steiner triple systems and their generalizations, the reader should consult the more recent book of Colbourn and Rosa(1999).

The next two theorems explain how Steiner triple systems are connected with quasigroups and latin squares.

THEOREM 2.3.1 *There exists a one-to-one correspondence between finite idempotent totally symmetric quasigroups*[12] *and Steiner triple systems.*

PROOF.   Let $Q = \{a, b, c, \ldots\}$ be the set of elements of a Steiner triple system. For a and b distinct, define $c = a \cdot b$ to be the third element of the unique triple of the system which contains a and b. Also, define $a \cdot a = a$. Then $(Q, \cdot)$ is an idempotent totally symmetric quasigroup. (The multiplication table of the idempotent totally symmetric quasigroup given in Figure 2.2.1 was obtained in this way.)

Conversely, let $(Q, \cdot)$ be a finite idempotent totally symmetric quasigroup. Since $a \cdot b = c$ implies $b \cdot c = a$, $c \cdot a = b$, $b \cdot a = c$, $c \cdot b = a$, $a \cdot c = b$, the operation $(\cdot)$ separates Q into triples, and these form a Steiner triple system because each pair a, b of distinct elements of Q is associated with a unique third element $c = a \cdot b$, which is distinct from both a and b. $\qquad\square$

Notice that the above correspondence implies that the number of elements in a finite idempotent totally symmetric quasigroup is necessarily an integer congruent to one or three, modulo six. The relation between these quasigroups and the Steiner triple systems was first pointed out by Sade [in Sade(1950, page 4, and Sade (1957), page 159]. It was demonstrated again by Bruck(1963b).

The latter author [in Bruck(1958), page 58, and in Bruck(1963b)] also pointed out that a similar relationship exists between totally symmetric loops and Steiner triple systems as follows:

THEOREM 2.3.2 *There exists a one-to-one correspondence between finite totally symmetric loops and Steiner triple systems.*

PROOF.   If $(G, \cdot)$ is a totally symmetric loop with identity element e then $ae = a = ea$ for each element $a \in G$. Now, because of the totally symmetric property, this implies that $a^2 = e$: that is, each element of $(G, \cdot)$ has order two. If G contains $n + 1$ elements, the set $G \setminus \{e\}$ forms a Steiner triple system of n elements whose triples are given by the statement that the unique triple which contains the pair of elements a, b has $c = ab$ as its third element.

Conversely, let Q be a set of n elements forming a Steiner triple system. We can make Q into a totally symmetric loop by adjoining an additional element e and defining the loop operation $(\cdot)$ by the three rules (i) if $a \neq e$, $b \neq e$, then ab is the third element of the unique triple which contains a and b, (ii) $ae = a = ea$ for all $a \in Q$, and (iii) $a^2 = e^2 = e$ for all $a \in Q$. $\qquad\square$

The above two theorems have recently been put into the more general setting of generalized cubic curves by Buekenhout(2001). [cf. Etherington(1964/65).]

We showed on page 54 that a totally symmetric quasigroup is a groupoid which satisfies any two of the identities (laws) $ab = ba$, $(ab)b = a$, $b(ba) = a$. The laws $(ab)b = a$ and $b(ba) = a$ are Sade's right and left Keys laws. In the

---

[12]Such quasigroups are sometimes called *Steiner quasigroups*. See, for example, Lindner(1971c) or Colbourn and Rosa(1999).

presence of the commutative law, they imply the laws of left and right semi-symmetry: namely, $(ba)b = a$ and $b(ab) = a$. The latter two laws taken together imply that the law of elasticity or *flexible law* $(ba)b = b(ab)$ also holds. Thus, a Steiner quasigroup is idempotent, commutative, semi-symmetric and flexible but is completely defined by $aa = a$ and any two of $ab = ba$, $(ab)b = a$, $b(ba) = a$.

Two other kinds of triple system exist which have connections with quasigroups and latin squares: namely, Mendlesohn triple systems and directed triple systems.

The first of these has its origins in a paper of N.S.Mendelsohn(1971a) who discussed systems of triples $(x\ y\ z)$ which are regarded as being ordered cyclically so that $(b\ c\ a)$ and $(c\ a\ b)$ are regarded as the same triple as $(a\ b\ c)$ and such that each of the adjacent (directed) pairs $(a\ b)$, $(b\ c)$ and $(c\ a)$ occurs just once in the triples of the system. If we use notation similar to that used above for Steiner triple systems and consider the triples which contain a particular element $v \in V$, we see that the other $n - 1$ elements of V must occur twice among these triples; once to the left of v and once to its right. Since each triple in which v occurs contains two ordered pairs which include v, it follows that v must occur in $r = n - 1$ of the triples. If t is the total number of triples, $nr = 3t$ (as for Steiner triple systems) so 3 divides n or $r(= n - 1)$ and so $n \equiv 0$ or 1 mod 3.

An idempotent quasigroup can be constructed from a Mendelsohn triple system by defining $a \cdot b = c$ if c is the third member of the triple which contains the ordered pair $(a\ b)$ when $a \neq b$. In this case, $a \cdot b = c \Rightarrow b \cdot a \neq c$ unless both of the triples $(a\ b\ c)$ and $(c\ b\ a)$ occur in the system so the resulting quasigroup is not in general commutative. [Note: a system T such that, for all triples in T, $(a\ b\ c) \in T \Rightarrow (c\ b\ a) \notin T$ is called *pure*.] Since $a \cdot b = c \Rightarrow b \cdot c = a$ and $c \cdot a = b$, the quasigroup satisfies the laws $b(ab) = a$ and $(ab)a = b$ so it is semi-symmetric. It is also flexible because $b(ab) = a$ and $(ba)b = a$ together imply $(ba)b = b(ab)$. We may call it a *Mendelsohn quasigroup*. Conversely, a Mendelsohn quasigroup defines a Mendelsohn triple system. Such systems exist for $n \equiv 0$ or 1 mod 3 except when $n = 6$. We can also define *Mendelsohn loops*. For more details, see N.S.Mendelsohn(1971a) and Grannel, Griggs and Quinn(1999,2009).

The third type of triple system we might consider is one in which each triple $(a\ b\ c)$ is regarded as consisting of the three ordered pairs $(a\ b)$, $(b\ c)$ and $(a\ c)$ each of which is required to occur exactly once in the triples of the system. Systems of this kind were introduced by Hung and Mendelsohn(1973) and are called *directed triple systems*. Again, a necessary condition for existence of such a system T is $n \equiv 0$ or 1 mod 3. See, for example, Colbourn and Rosa(1992).

If we attempt to define a quasigroup in the same way as before, we find that we are successful if and only if $(x\ y\ z) \in T \Rightarrow (w\ y\ x) \in T$ for some $w \in V$ as we shall show. When this condition holds, we say that the system is a *latin directed triple system*. When such a system exists, the corresponding quasigroup may or may not satisfy the flexible law. Investigation of such systems is quite recent. For more details than we have space to give here, see Drápal, Kozlik and Griggs(2012); also (later) Drápal, Griggs and Kozlik(2014,2015).

THEOREM 2.3.3 *Let $D = (V, T)$ be a directed triple system on a set V of n elements. Denote by $S_{a,b}$ the set of ordered pairs (x, y) in positions a and b respectively of the triples T of D. Then D is a latin directed triple system if and only if $S_{1,2} = S_{3,2}$, $S_{2,3} = S_{2,1}$ and $S_{1,3} = S_{3,1}$.*

PROOF. Suppose that $(x \ y \ z) \in T$. Then $y \cdot z = x$, $x \cdot z = y$ and $x \cdot y = z$. Since the ordered pair y, x occurs in some triple, there exists $w \in T$ such that $y \cdot x = w$ and then one of $(y \ x \ w)$, $(y \ w \ x)$, $(w \ y \ x)$ occurs. If either of the first two of these occurs, we have $y \cdot w = x$. But, since $(x \ y \ z) \in T$, $y \cdot z = x$ implying that $w = z$ and that $(y \ x \ z)$ or $(y \ z \ x)$ is a triple of D. However, the ordered pair y, z already occurs in the triple $(x \ y \ z)$, so this is impossible. Therefore, $(x \ y \ z) \in T \Rightarrow (w \ y \ x) \in T$ (where $w = z$ is possible). From this implication, it follows that $S_{1,2} \subseteq S_{3,2}$.

Again, since the ordered pair z, y occurs in some triple, there exists $v \in T$ such that $z \cdot y = v$ and then one of $(z \ v \ y)$. $(v \ z \ y)$, $(z \ y \ v)$ occurs. If either of the first two of these occurs, we have $v \cdot y = z$. But, since $(x \ y \ z) \in T$, $x \cdot y = z$ implying that $v = x$ and that $(z \ x \ y)$ or $(x \ z \ y)$ is a triple of D. However, the ordered pair x, y already occurs in the triple $(x \ y \ z)$, so this is impossible. Therefore, $(x \ y \ z) \in T \Rightarrow (z \ y \ v) \in T$ (where $v = x$ is possible). From this implication, it follows that $S_{3,2} \subseteq S_{1,2}$.

Since $S_{1,2} \subseteq S_{3,2}$ and $S_{3,2} \subseteq S_{1,2}$, we have $S_{1,2} = S_{3,2}$. Then also $S_{2,1} = S_{2,3}$ since the sets of all pairs in positions 2,1 and 2,3 are the same as those of all pairs in positions 1,2 and 3,2 respectively but oppositely ordered.

However, $S_{1,2} \cup S_{2,3} \cup S_{1,3} = S_{2,1} \cup S_{3,2} \cup S_{3,1}$ because each of these is equal to the set of all ordered pairs taken from V. So, $S_{1,3} = S_{3,1}$. This completes the proof of necessity.

Conversely, we wish to show that, when $S_{1,2} = S_{3,2}$, $S_{2,3} = S_{2,1}$ and $S_{1,3} = S_{3,1}$, $(V, \cdot)$ is a quasigroup.

Suppose that $(x \ y \ z) \in T$. Then $x \cdot y = z$. We require that each of the equations $\alpha \cdot y = z$, $x \cdot \beta = z$ and $x \cdot y = \gamma$ has a unique solution.

Firstly, we have $\gamma = z$ since $x \cdot y = \gamma \Rightarrow$ one of $(\gamma \ x \ y)$, $(x \ \gamma \ y)$, $(x \ y \ \gamma)$ is in T. But $(\gamma \ x \ y)$, $(x \ \gamma \ y)$ are not in T because the ordered pair x, y is in only one triple of T. Therefore, $(x \ y \ \gamma) \in T$ which implies that $\gamma = z$.

Secondly, $x \cdot \beta = z \Rightarrow$ one of $(x \ z \ \beta)$, $(z \ x \ \beta)$, $(x \ \beta \ z)$ is in T. Since the ordered pair x, z occurs in only one triple of T and $(x \ y \ z) \in T$, $(x \ z \ \beta)$ cannot occur. If $(z \ x \ \beta)$ is in T then, because $S_{2,1} = S_{2,3}$, some triple $(u \ x \ z)$ also is in T. But the ordered pair x, z already occurs in $(x \ y \ z)$. Hence, $x \cdot \beta = z \Rightarrow (x \ \beta \ z) \in T$ and so $\beta = y$.

Finally, $\alpha \cdot y = z \Rightarrow$ one of $(z \ \alpha \ y)$, $(\alpha \ z \ y)$, $(\alpha \ y \ z)$ is in T. If $(z \ \alpha \ y)$ is in T, then because $S_{1,3} = S_{3,1}$, some triple $(y \ s \ z)$ also is in T. Similarly, if $(\alpha \ z \ y)$ is in T then because $S_{3,2} = S_{1,2}$, some triple $(y \ z \ t)$ also is in T. But the ordered pair y, z already occurs in $(x \ y \ z)$ so $\alpha \cdot y = z \Rightarrow (\alpha \ y \ z) \in T$ and hence $\alpha = x$.

We may apply similar arguments to each of the equations $x \cdot z = y$ and $y \cdot z = x$ (which also hold when $(x \ y \ z) \in T$) to show that, when any two of the variables are given, the third is uniquely determined. Hence $(V, \cdot)$ is a quasigroup. □

THEOREM 2.3.4 *Let* $D = (V, T)$ *be a directed triple system on a set* V *of* n *elements. Then* D *is a latin directed triple system if and only if* $(x\ y\ z) \in T \Rightarrow$ $(w\ y\ x) \in T$ *for some* $w \in V$.

PROOF.   The stated requirement is that $S_{1,2} \subseteq S_{3,2}$. The previous theorem has shown that this is necessary. We have to show that it is also sufficient.

Since the number of ordered pairs $x, y \in V \times V$, $x \neq y$, is the same as the number of ordered pairs y, x, there is an exact match between the triples $(x\ y\ z)$ and $(w\ y\ x)$ as x and y vary. Thus, $S_{1,2} = S_{3,2}$. Then, as shown in the previous theorem, the equalities $S_{2,1} = S_{2,3}$ and $S_{1,3} = S_{3,1}$ hold also and these equalities together force $(V, \cdot)$ to be a quasigroup. $\square$

For more information about recent work on the three kinds of triple system, the reader is recommended to consult Griggs(2011).

2.4   Group-based latin squares and nuclei of loops

In Section 1.2, we gave three ways of checking whether a given latin square L (which can be assumed to be in reduced form) is group-based: namely, the quadrangle criterion, Suschkewitsch's test and Frolov's regularity test. To be certain that L is group-based using the quadrangle criterion, it is necessary either to check every pair of quadrangles which agree in three corresponding places or else to border the square with its own first row and column so as to form the Cayley table of a loop and then check the subset of all such pairs of quadrangles which have the identity of the loop as top left member. [See Brandt(1927)]. Thus, the number of tests required is of order at least n^3. If Suschkewitsch's test is used, it is necessary to check every pair of row (or column) permutations. (See Theorem 1.2.2.) This requires n^2 tests. If Frolov's test for regularity is used, it follows from Theorem 1.2.3 that at most n tests are required. However, by treating L as the Cayley table of a loop and making use of the concept of loop nuclei, we show in this section that in fact at most $n/2$ tests are necessary and this only when the square is idempotent and of even order.

We shall require two definitions and a theorem:

DEFINITION. The *left nucleus* N_l of a loop $(Q, .)$ comprises all elements $x \in Q$ such that $x(bc) = (xb)c$ for all $b, c \in Q$. The *middle nucleus* N_m comprises all elements $y \in Q$ such that $a(yc) = (ay)c$ for all $a, c \in Q$ and the *right nucleus* N_r comprises all elements $z \in Q$ such that $a(bz) = (ab)z$ for all $b, c \in Q$. The set $N_l \cap N_m \cap N_r$ is called the *nucleus* N.

DEFINITION. The hth row(column) of the Cayley table of a loop $(Q, \cdot)$ is said to have the *Frolov property* if, when the columns(rows) of the latin square formed by the body of the table are re-ordered in such a way that the elements of the hth row(column) are in the same order as that of the row(column) border, each row(column) of the re-ordered square coincides with some row(column) of the body of the original Cayley table.

We note that, if every column (or row) of a latin square has the Frolov property, this is equivalent to saying that the latin square is regular. See page 5.

THEOREM 2.4.1 *Suppose that the cell (i,j) of the Cayley table T of the loop $(Q, \cdot)$ contains the entry $a_i b_j$ for $i, j = 0, 1, \ldots, n-1$, $a_0 = b_0 = e$, where e is the identity element. Then a necessary and sufficient condition that the element b_k belong to the middle nucleus of the loop is that the column of T indexed by b_k has the Frolov property. A necessary and sufficient condition that the element a_h belong to the middle nucleus of the loop is that the row of T indexed by a_h has the Frolov property.*

PROOF. As in Theorem 1.2.2, we may suppose without loss of generality that the symbols used for the loop $(Q, .)$ are $1, 2, \ldots, n$, that 1 is the identity element and that the Cayley table is in reduced form: that is, with the elements of the first row and column in natural order.

Let us suppose that the mapping θ permutes the rows so that the elements $1b, 2b, \ldots, xb, \ldots, nb$ of the column headed by the element b are in the order of the first column: that is, in natural order. Then, $(x\theta)b = x$ for $x = 1, 2, \ldots, n$. (The columns may be re-arranged if we wish so that the new square becomes reduced.)

The elements of the column headed by the element u are xu for $x = 1, 2, \ldots, n$. These become $(x\theta)u$ for $x = 1, 2, \ldots, n$. (See Figure 2.4.1.) Suppose that this is another column of T for each $u \in Q$. Then, for each $u \in Q$, there exists an element $w \in Q$ such that $(x\theta)u = xw$ for all $x \in Q$, where $(x\theta)b = x$. Thus, $x\theta = xR_b^{-1}$ and so $(xR_b^{-1})u = xw$. Putting $x = b$, we get $u = bw$ or $w = uL_b^{-1}$ so $(xR_b^{-1})u = x(uL_b^{-1})$ for all $x, u \in Q$, or, equivalently, $(xR_b^{-1})(bw) = xw$ for all $x, w \in Q$. If we put $x = yb$, this becomes $y(bw) = (yb)w$ for all $y, w \in Q$. Thus, b lies in the middle nucleus of the loop.

To prove the second statement, we need only remark that the loop whose rows are the columns of $(Q, .)$ has middle nucleus the same as that of $(Q, .)$. □

$(\cdot)$	1	.	.	b	.	.	u	.
1θ	1θ	.	.	$(1\theta)b = 1$	.	.	$(1\theta)u$	.
2θ	2θ	.	.	$(2\theta)b = 2$	.	.	$(2\theta)u$	.
.	.			.			.	
.	.			.			.	
$x\theta$	$x\theta$	.	.	$(x\theta)b = x$	.	.	$(x\theta)u$	.
.	.			.			.	
.	.			.			.	
$n\theta$	$n\theta$	.	.	$(n\theta)b = n$	.	.	$(n\theta)u$	.

FIG. 2.4.1. Loop table T after re-arranging rows so that the symbols in column b are in natural order.

Theorem 1.2.3 is an immmediate corollary to Theorem 2.4.1 since, if and only if all rows(columns) have the Frolov property, the middle nucleus is the whole of Q and so $(Q,.)$ is a group.

However, it is well-known that the order of each of the nuclei of a loop divides the order of the loop. [See, for example, Pflugfelder(1990a) for a proof.] So,if p is the smallest prime which divides the order $|Q|$ of a loop $(Q,.)$ (or reduced latin square L formed by its Cayley table) and if $|Q|/p$ rows, excluding the first, of L have the Frolov property, this is sufficient to ensure that L is group-based since it ensures that at least $1 + (|Q|/p)$ elements of Q are in the middle nucleus of $(Q,.)$ whereas the maximum size of the middle nucleus is $|Q|/p$ if it is not the whole loop.

(Note that, in particular, for a latin square of odd order, it is sufficient that at most a third of the rows have the Frolov property to be sure that the square is group-based. For a square of prime order, just one row is sufficient. Thus, for use by hand, our third test for a latin square to be group-based is considerably more economic than its two predecessors.)

But it is also well-known [again see Pflugfelder(1990a)] that each of the nuclei is a group and so all the powers of an element of the middle nucleus and likewise the product of each pair of its elements are themselves members. By exploiting these facts, we may reduce further the number of columns (or rows) which we need to test for the Frolov property to ensure that a given latin square is group-based. (When the square is idempotent, no further reduction is possible.)

In Keedwell(2005), the reader intersted in loop theory will find analogues of Theorem 2.4.1 for testing for elements of the left and right nuclei and will also find examples of loops of small order which have such (proper) nuclei.

2.5   Transversals in group-based latin squares

As we remarked earlier, a latin square L is called *group-based* if it satisfies the quadrangle criterion (or the other criteria mentiond in the previous section) and so can be bordered in such a way that it becomes the multiplication table of a finite group G. For such squares, quite a lot is known about the existence of transversals. As we explained in Section 1.5, L has a transversal if and only if G has a complete mapping.

THEOREM 2.5.1 *A necessary condition for a finite group $(G,\cdot)$ to have a complete mapping is that the product of all its elements in some order is equal to the identity element e of G.*

PROOF.   Let $x \to \theta(x)$ be the complete mapping and $x \to \phi(x) = x\theta(x)$ be the corresponding orthomorphism. Let us write $\phi(x)$ as a product of cycles

$$\phi = (g_{11}\, g_{12}\, \cdots\, g_{1s_1})(g_{21}\, g_{22}\, \cdots\, g_{2s_2})\cdots(g_{h1}\, g_{h2}\, \cdots\, g_{hs_h})\cdots(g_{r1}\, g_{r2}\, \cdots\, g_{rs_r}).$$

Then $g_{h\,k+1} = \phi(g_{hk}) = g_{hk}\cdot\theta(g_{hk})$ for all choices of $h = 1,2,\ldots,r$ and $k \neq s_h$. Also $g_{h1} = \phi(g_{h\,s_h}) = g_{h\,s_h}\cdot\theta(g_{h\,s_h})$. It follows that $\theta(g_{hk}) = g_{hk}^{-1}g_{h\,k+1}$ for all choices of h and $k \neq s_h$ and that $\theta(g_{h\,s_h}) = g_{h\,s_h}^{-1}g_{h1}$. Therefore, the product

$$\theta(g_{h1})\theta(g_{h2})\ldots\theta(g_{h\,s_h}) = g_{h1}^{-1}g_{h2}\cdot g_{h2}^{-1}g_{h3}\cdots g_{h\,s_h}^{-1}g_{h1} = e$$

for each $h = 1, 2, \ldots, r$. Since the complete mapping θ effects a permutation of G, the elements $\theta(g_{hk})$ for $k = 1, 2, \ldots, s_h$ and $h = 1, 2, \ldots, r$ comprise all the elements of G each counted once. The result of the theorem follows. $\square$

Theorem 2.5.1 was first proved by Paige(1951). In the case of abelian groups, the above necessary condition is also sufficient. This also was proved by Paige [see Paige(1947)] and follows as a corollary from Theorem 2.5.3 below. The same author, in Paige(1951), gave a sufficient condition for a finite non-abelian group to have a complete mapping as follows:

THEOREM 2.5.2 *A sufficient condition for a finite group* $(G, \cdot)$ *of order* n *to have a complete mapping is that there exist an ordering* $a_1, a_2, \ldots, a_{n-1}$ *of the non-identity elements of* G *such that the partial products* $b_1 = a_1$, $b_2 = a_1 a_2$, $b_3 = a_1 a_2 a_3$, $\ldots$, $b_{n-2} = a_1 a_2 \ldots a_{n-2}$ *are all distinct and such that the complete product* $b_{n-1} = a_1 a_2 \ldots a_{n-1}$ *is equal to the identity element* e *of* G.

PROOF.   Let c be the element which does not occur among the above partial products. We may define a permutation ϕ of G in cycle form as follows: $\phi = (c)(b_1 \, b_2 \, b_3 \ldots b_{n-1})$. We easily see that ϕ is an orthomorphism of G with corresponding complete mapping θ, where $\theta(b_i) = a_{i+1}$ for $i = 1, 2, \ldots, n - 2$, $\theta(b_{n-1}) = a_1$ and $\theta(c) = e$. We have $c\theta(c) = c = \phi(c)$, $b_i\theta(b_i) = b_i a_{i+1} = b_{i+1} = \phi(b_i)$ for $i = 1, 2, \ldots, n - 2$ and $b_{n-1}\theta(b_{n-1}) = ea_1 = b_1 = \phi(b_{n-1})$ by definition of b_{i+1} and $\phi(b_i)$. $\square$

It is clear that this sufficient condition is not necessary because it requires the existence of an orthomorphism which has a single cycle of length $n - 1$. There exist many groups which have orthomorphisms which consist of several (shorter) cycles. For example, the dihedral group $D_4 = \{a, b : a^4 = b^2 = e, ab = ba^{-1}\}$ of order 8 has an orthomorphism with cycles of lengths 5 and 2.as follows:

$$\phi = (e_{\,ba^2} \, ba^2 \, {}_{a^2} \, b_{\,ba^3} \, a^3 \, {}_b \, ba_{\,ba})(a_{\,a} \, a^2 \, {}_{a^3})(ba^3 \, {}_e).$$

DEFINITION. A group G which satisfies the above sufficient condition is said to be *R-sequenceable*.

THEOREM 2.5.3 *Let* θ *be an arbitrary permutation of the elements* $x_1, x_2, \ldots, x_n$ *of a finite abelian group* $(G, \cdot)$ *of order* n. *Then it is always possible to construct from* θ *another permutation* θ' *such that at least* $n - 1$ *of the elements* $x_i\theta'(x_i) = \phi'(x_i)$ *are distinct.*

PROOF.   If $n - 1$ of the elements $\phi(x_i) = x_i\theta(x_i)$ are distinct, there is nothing to prove, so suppose that only the r elements $S = \{\phi(x_1), \phi(x_2), \ldots, \phi(x_r)\}$, $r < n - 1$, are distinct.

Case 1. If $\exists h, k > r$ such that $x_h\theta(x_k) \notin S$, put $\theta'(x_h) = \theta(x_k), \theta'(x_k) = \theta(x_h)$ and $\theta'(x_i) = \theta(x_i)$ for $i \neq h, k$. Define $\phi'(x_h) = x_h\theta'(x_h) = x_h\theta(x_k) \notin S$ and $\phi'(x_i) = \phi(x_i)$ for $i = 1, 2, \ldots, r$. Then ϕ' has at least $r + 1$ members.

If Case 1 does not hold for any $h, k > r$, we argue as follows:

By hypothesis, $\phi(x_{r+1}) = \phi(x_u)$ for some $u \leq r$.

Case 2a. If $x_u\theta(x_{r+2}) \notin S$, put $\theta'(x_u) = \theta(x_{r+2})$, $\theta'(x_{r+2}) = \theta(x_u)$ and $\theta'(x_i) = \theta(x_i)$ for $i \neq u, r + 2$. Then $\phi'(x_u) = x_u\theta'(x_u) = x_u\theta(x_{r+2}) \notin S$ and $\phi'(x_{r+1}) = x_{r+1}\theta'(x_{r+1}) = x_{r+1}\theta(x_{r+1}) = \phi(x_{r+1}) = \phi(x_u)$ so ϕ' has at least $r + 1$ members.

Case 2b. If Case 2a does not hold, we have $x_u\theta(x_{r+2}) = \phi(x_v) \in S$ for some $v \leq r$. Then $x_v\theta(x_u) \neq \phi(x_u), \phi(x_v)$ since $v \neq u$. (Note that $v = u$ would imply that $\theta(x_{r+2}) = \theta(x_u)$ but θ is a permutation.)

If $x_v\theta(x_u) \notin S$, we can put $\theta'(x_u) = \theta(x_{r+2})$, $\theta'(x_v) = \theta(x_u)$, $\theta'(x_{r+2}) = \theta(x_r)$ and $\theta'(x_i) = \theta(x_i)$ for $i \neq u, v, r + 2$.

Then $\phi'(x_v) = x_v\theta'(x_v) = x_v\theta(x_u) \notin S$, $\phi'(x_u) = x_u\theta'(x_u) = x_u\theta(x_{r+2}) = \phi(x_v)$ and $\phi'(x_{r+1}) = x_{r+1}\theta'(x_{r+1}) = x_{r+1}\theta(x_{r+1}) = \phi(x_{r+1}) = \phi(x_u)$ so again ϕ' has at least $r + 1$ distinct maembers.

If none of the previous cases applies, we shall have

Case 2c. $x_u\theta(x_{r+2}) = \phi(x_v)$ and $x_v\theta(x_u) = \phi(w)$, $u, v, w \leq r$.

We may suppose the notation chosen so that $u = 1, v = 2, w = 3$. Then $x_1\theta(x_{r+2}) = \phi(x_2)$ and $x_2\theta(x_1) = \phi(x_3)$.

By repetition of this construction procedure, we can enlarge S unless or until we reach a point at which $x_1\theta(x_{r+2}) = \phi(x_2)$ and $x_{i+1}\theta(x_i) = \phi(x_{i+2})$ for $i = 1, 2, \ldots, k$, where $k \leq r - 2$.

We omit the rest of the proof (which uses a not-very-obvious induction argument) because of its length and complexity but the above part of the proof is sufficient to show its flavour.

COROLLARY. *If $(G, \cdot)$ is a finite abelian group of order n and if the product of all the elements of G is the identity element e, then G has a complete mapping.*

PROOF. By Theorem 2.5.3 we can assume that we can construct a permutation θ of G such that the elements $\phi(x_i) = x_i\theta(x_i)$ are distinct for $i = 1, 2, \ldots, n - 1$. Let c be the element of G which does not occur in the set $S = \{\phi(x_1), \phi(x_2), \ldots, \phi(x_{n-1})\}$. Then $c\prod_{i=1}^{n-1}\phi(x_i) = e$. Since also $\prod_{i=1}^{n}x_i = e$ and $\prod_{i=1}^{n}\theta(x_i) = e$, we have
$\left(\prod_{i=1}^{n-1}x_i\theta(x_i)\right)\left(x_n\theta(x_n)\right) = e$. That is, $\left(\prod_{i=1}^{n-1}\phi(x_i)\right)\left(x_n\theta(x_n)\right) = e$
so $x_n\theta(x_n) = c \notin S$. This proves that $\phi(x_n) \notin S$ and so ϕ is a permutation of S and θ is a complete mapping. □

THEOREM 2.5.4 *The product of all the elements of a finite abelian group $(G, \cdot)$ is the identity element e unless G contains exactly one element of order two. In this latter case, the product is equal to the unique element of order two.*

PROOF. Consider first the product P of the elements of order greater than two. Since the group is abelian, this product can be re-arranged so that each element is followed by its inverse. So P is equal to e. The remaining non-identity elements form a subgroup $(H, \cdot)$ of G. (H consists of e and all the elements of order two.) Let $a_1, a_2, \ldots, a_r$ be a set of generating elements of H. Then each element $h \in H$ can be written in the form $h = a_1^{\epsilon_1} a_2^{\epsilon_2} \ldots a_r^{\epsilon_r}$, where ϵ_i is 0 or 1. Consider the elements which contain a_i. There are 2^{r-1} such elements since there are two choices of each ϵ_j, $j \neq i$, according as a_j does or does not occur in h. Consequently, when we multiply all these elements together, the 2^{r-1} occurrences of a_i cancel out except when $r = 1$ and there is only one element of order 2. Hence, the product of all the elements of H is equal to e or to h if H has only one element h of order two.   □

COROLLARY. *The Cayley table of an abelian group has a transversal except when that group has exactly one element of order 2.*

Theorem 2.5.4 has been proved by many authors including Miller(1903), Paige(1947), Ramanathan(1947) and M.Hall(1952). See also Remark 2 on page 8 of [DK2]. The corollary has also been proved by Carlitz(1953b).

When a finite group G (assumed non-abelian) has a complete mapping, the product of its elements in some particular order is equal to the identity element e by Theorem 2.5.1. It follows that *the product of these elements in any order is equal to an element of the commutator subgroup of G.* (We shall use this fact in the proof of Theorem 2.5.5.)

More generally, let G be a finite group of order n and let G' be its commutator subgroup. Since the elements of G commute mod G', every product of the n distinct elements of G belongs to the same coset mod G'. Fuchs suggested the study of groups G for which it is true, conversely, that every element of this coset can be represented as a product of n distinct elements. Dénes and Török(1970) called groups which have the latter property *P-groups*. Rhemtulla(1969) proved that every finite soluble group is a P-group and later Dénes and Hermann(1982) proved that every finite group is a P-group. Keedwell(1983b,1984a) has introduced the concept of a *super P-group*. For details of the latter, see [DK2], page 75.

The problem of finding conditions under which a finite group has a complete mapping has been further investigated in M.Hall and Paige(1955). In that paper, these authors proved that for soluble groups[13] the necessary and sufficient, and for groups of even order the necessary, condition for the existence of a transversal in the Cayley table of the group is that its Sylow 2-subgroups should be non-cyclic.They conjectured further that the condition is also sufficient for non-soluble groups. Their main theorems are as follows but, because the proofs require reasonably advanced knowledge of group-theory, we only give outline proofs. The beginning student may prefer to defer reading the proofs until later and the more advanced student can consult the original paper.

THEOREM 2.5.5 *A finite group of order n which has a cyclic Sylow 2-subgroup does not possess a complete mapping.*

---

[13]It is known that every group of odd order is soluble, see Feit and Thompson(1963).

PROOF.   Let S be a cyclic Sylow 2-subgroup of G. We shall show first that S is in the centre of its normalizer $N(S)$.

If S is cyclic of order 2^m, its automorphism group is a 2-group of order 2^{m-1}. [See page 146 of Zassenhaus(1958).] Let Z be the centre of $N(S)$. If $Z = N(S)$ then S lies in the centre of its normalizer and so the claim is true in this case. We may suppose that $Z \neq N(S)$ and, if this is the case, we may choose an arbitrary element $u \in N(S)$ and write $u = vw = wv$ where v, w are some powers of u and so belong to $N(S)$ and the order of v is a power of 2 (and so $v \in S$) while the order of w is prime to 2. To prove our claim in this case, it is sufficient to show that if $w \in N(S)$, where the order of w is r say and r is prime to 2, then $waw^{-1} = a$ for every $a \in S$. Since $wSw^{-1} = S$, the mapping $\phi : a \to waw^{-1}$ is an automorphism of S. The order of the automorphism group of S is 2^{m-1} and hence $w^{2^{m-1}} a w^{-2^{m-1}} = a$ for every $a \in S$. Since r is prime to 2, there exists an integer z such that $(w^{2^{m-1}})^z = w$ and this implies that $waw^{-1} = a$ for every $a \in S$ as required.

By a theorem of Burnside [see page 203 of M.Hall(1959) for a proof], if a Sylow p-subgroup P of a finite group G is in the centre of its normalizer, then G has a normal subgroup H which has the elements of P as its coset representatives. Thus, in the present case, our group G has a normal subgroup H which has the elements of S as its coset representatives. If $G = H \cup Hs_2 \cup \ldots \cup Hs_t$, each $g \in G$ can be written in the form $g = hs_i$ for some $h \in H$ and $s_i \in S$ where $t = \mathrm{ord}(G/H) = \mathrm{ord}\ S$ and ord H is odd. Hence, $\prod_{g \in G} g = h^* \prod_{i=1}^{t} (s_i)^{\mathrm{ord} H}$ where h^* is some element of H. Here, we have used the facts that $G/H \cong S$ is abelian and that exactly ord H elements of G occur in each coset. Since S is abelian and has a unique element of order 2, $\prod_{i=1}^{t} s_i$ is this element of order 2 and so $\prod_{i=1}^{t} (s_i)^{\mathrm{ord} H} \notin H$. But, because G/H is abelian, the commutator subgroup G' of G is contained in H. Thus, $\prod_{g \in G} g \notin G'$ and so, by the remark following the corollary to Theorem 2.5.4, G does not possess a complete mapping.   □

COROLLARY. *If G is an arbitrary group of order $n = 4k + 2$, then G has no complete mapping. In particular, the symmetric group on three elements has no complete mapping.*

The result of the corollary can be deduced from an earlier theorem of Mann (see Theorem 5.1.5) as we shall explain in Section 5.1. Also, it can be obtained as a special case of a more general result due to Bruck(1951) which we discussed in detail in Section 9.2 of [DK1] and which is as follows:

Let G be a finite group of order n. If G contains a normal subloop H of odd order such that the quotient loop G/H is a cyclic group of even order, then the latin square L_G representing the multiplication table of G does not possess a transversal (and consequently L_G has no orthogonal mate).

THEOREM 2.5.6 *A finite soluble group G whose Sylow 2-subgroups are non-cyclic has a complete mapping.*

PROOF.   The proof is obtained in a number of steps. In the first place, Hall and Paige have shown that every non-cyclic 2-group has a complete mapping. They have also shown that if a group G can be written in the form $G = AB$ where A and B are subgroups such that $A \cap B = \{e\}$ and if A and B both have complete mappings, so does G. But, by a theorem of P. Hall(1928), if A is a Sylow 2-subgroup of a soluble group, then A has a 2-complement B and we then have $G = AB$ and $A \cap B = \{e\}$ as desired. Also, B has odd order and so it has a complete mapping (as shown in Theorem 1.5.3) and A, being a non-cyclic 2-group, has a complete mapping also. Hence the result of the theorem follows.
□

Hall and Paige have proved further that all finite alternating groups have complete mappings and that, for order $n > 3$, the symmetric group S_n has a complete mapping. As follows from the corollary to Theorem 2.5.5, the symmetric group S_3 has no complete mapping and so we may deduce from the last result that the property of having a complete mapping is not invariant under homomorphisms.

It was conjectured by Hall and Paige that every finite group whose Sylow 2-subgroups are non-cyclic, soluble or not, has a complete mapping and, after some 60 years of proving the conjecture true for various classes of non-soluble groups one-by-one, it is now known to be true for all groups. For details, see A.B.Evans(2009) and Wilcox(2009). A good summary of the history of the steps which have led to the resolution is contained in the reviews of these two papers in Mathematical Reviews.

From Theorem 2.5.5, Theorem 2.5.6 and a result due to Dénes and Keedwell(1989), we can deduce that the condition that the product of the elements in some order of a group G should be equal to the identity element (call this Condition P) is an alternative sufficient condition for any group, abelian or not, to have a complete mapping. Firstly, it follows from the proof of Theorem 2.5.5 that, if G has a cyclic Sylow 2-subgroup, then the product of its elements in some order (and so in all orders) is not in the commutator subgroup and consequently Condition P is not satisfied. Secondly, it is easy to deduce from Theorem 2.5.6 and Theorem 2.5.1 that Condition P is satisfied for soluble groups whose Sylow 2-subgroups are non-cyclic. Thirdly, Dénes and Keedwell(1989) have given a short direct proof that Condition P holds for all non-soluble groups and have pointed out that all such groups have non-cyclic Sylow 2-subgroups. Consequently, see the remarks which follow Theorem 2.5.6 above, all such groups have complete mappings. For a more recent direct proof, see Vaughan-Lee and Wanless(2003).

In Dénes and Keedwell(1991) two new conjectures related to complete mappings were made. These authors observed that the square of any permutation consists of cycles of odd length (possibly none) and of an even number of cycles of even length (again possibly none). Thus, any permutation of the latter kind has at least one square root. They showed further that each of the permutations representing the rows of the Cayley table L of a non-soluble group has a square root. These square roots define the rows of a square root square (not necessarily latin) which we shall denote by $\sqrt{L}$.

Conjecture 1. If L is the multiplication table of a non-soluble group then, not only does L have at least one, and possibly many, square roots $\sqrt{L}$, but at least one of these square roots is a latin square.

Conjecture 2. A necessary and sufficient condition for a latin square A to have an orthogonal mate is that either A^2 is a latin square or that A can be represented as the product $A = BC$ of two not-necessarily-distinct latin squares B and C.[14]

In Wanless(2001), that author has shown that Conjecture 2 is true without the necessity of the option that A^2 is a latin square. He has also commentd on the difficulty of deciding whether Conjecture 1 is true, so that conjecture remains open.

It has been proved by Bateman(1950) that all infinite groups possess complete mappings.

2.6   Complete latin squares

As is explained more fully in Chapter 10 of [DK2], latin squares are used in the branch of statistics known as the design of experiments. In an agricultural experiment, for example, adjacent cells of a latin square may represent adjacent plots of land which are to receive different treatments, the object of the experiment being to determine the relative efficiency of the various treatments. Treatments applied to adjacent plots might interact and so the problem arose as to whether latin squares exist in which each pair of distinct entries (representing distinct treatments) occur in adjacent cells just once (and necessarily only once) in some row. Such a latin square is called *row complete* or *horizontally complete*. A latin square with the corresponding property when columns are used in place of rows, is said to be *column complete* or *vertically complete*.

Precisely, a latin square with elements $1, 2, \ldots, n$ is called row complete if for any ordered pair of distinct elements α, β there exists a row of the latin square in which α and β appear as adjacent elements: that is, if c_{st} denotes the element at the intersection of the s-th row and t-th column of the square, there exist integers s and t such that $c_{st} = \alpha$ and $c_{s,t+1} = \beta$. Similarly, a latin square is column complete if for any distinct α and β, there exist integers u and v such that $c_{uv} = \alpha$ and $c_{u+1,v} = \beta$. A latin square that is both row and column complete is called a *complete latin square*.

An example of a complete latin square of order 4 is given in Figure 2.6.1. Notice that in this square, the digits 2, 3 and 4 each occur just once in an adjacent position to the left of the digit 1, just once to its right, just once above it, and just once below it. A similar property holds with respect to each other digit.

---

[14]The product of two latin squares is defined in Section 10.2.

$$
\begin{array}{|cccc}
1 & 2 & 3 & 4 \\
2 & 4 & 1 & 3 \\
3 & 1 & 4 & 2 \\
4 & 3 & 2 & 1
\end{array}
$$

FIG. 2.6.1.

Row complete latin squares may also be used advantageously in experiments in which a single subject (experimental unit) is to receive a number of treatments successively, since the effect of each treatment on the subject is likely to be affected both by its immediate predecessor and also by the number of treatments which the subject has previously received.

In an experiment on farm animals, for example, it may be desirable to apply a number of different dietary treatments to each given animal in succession. The effect of a given treatment on an animal may be affected both by the number of treatments which that animal has already received and also by the nature of the immediately preceding treatment. (As another example, in a sequence of psychological experiments on a human being, fatigue after several preceding experiments is likely to influence the reaction of the subject to later experiments. Also, the reaction to a given experiment may be affected by the outcome of the preceding experiment.) If several animals are available for treatment, the first possibility can be allowed for statistically if it can be arranged that the number n of animals to be treated is equal to the number of treatments and if the order in which the treatments are to be applied to these animals is allowed to be determined by the order of entries in the n rows of an $n \times n$ latin square (whose n distinct elements denote the n treatments). Then any particular experiment has a different number of predecessors for each of the n different animals, since a given element of the latin square is preceded by a different number of other elements in each of the n rows of the square. The possibility of interaction between one experiment and the immediately preceding one can also be allowed for if the latin square chosen is row complete. The resulting experiment is then said to be statistically "balanced" both with respect to the effect of the immediately preceding experiment and also with respect to the number of preceding experiments.

So far as the authors are aware, the first people to investigate the existence of row complete latin squares were Bugelski(1949) and E.J.Williams(1949). Both authors were interested in designing balanced experiments of the second type described above: that is, experiments in which a number of different treatments are applied successively to the same experimental unit. Bugelski gave a row complete square of order six. In the same year, Williams published a much more comprehensive paper in which he gave a very simple method for constructing row complete latin squares of any even order, as follows:[15]

---

[15]The graph-theoretical equivalent of this theorem was proved by Beineke(1964).

THEOREM 2.6.1 *Let $n = 2m$ be any even positive integer. The $n \times n$ latin square whose first row is* 0, 1, $2m - 1$, 2, $2m - 2$, 3, $2m - 3$, ..., $m + 1$, m *and whose subsequent rows obey the rule that each element is one greater (modulo n) than the corresponding element of the preceding row, is a row complete latin square.*

PROOF.   Since the differences between the $n - 1$ pairs of adjacent elements in the first row are all different modulo n, it follows that the same is true in each row. Every element occurs just once in each column, so no two successors of a given element differ from it by the same amount. Consequently these successors are all different.   □

It is easy to see from the method of construction just described that, if the rows of our latin square are now re-ordered in such a way that the order of elements in the first column becomes the same as their order in the first row, then the result will be a symmetric latin square. Consequently the amended square will be column complete as well as row complete.

In the first edition of this book, the question was raised as to whether any row complete latin square exists which cannot be made column complete as well by a suitable permutation of its rows. For group-based squares, the answer is "No". This was proved by Keedwell [See Keedwell(1975,1976b) or Theorem 1.2 on page 45 of [DK2]] and the result is mentioned as a footnote on page 83 of [DK1].[16] Shortly afterwards, Owens(1976) proved that the overall answer is "Yes". He devised a construction for an infinite class of row complete latin squares which cannot be made column complete by any permutation of their rows. Much later, Cohen and Etzion(1991) gave an alternative construction for squares with this property.

Theorem 2.6.1 was rediscovered by Dénes and Török(1970) and these authors and others have given an alternative ordering of the elements $0, \ldots, 2m - 1$ which may be used as the first row in the construction. The alternative ordering, which was first described by Bradley(1958), is 0, $2m - 1$, 1, $2m - 2$, 2, $2m - 3$, ..., $m - 1, m$. A generalized form of Theorem 2.6.1, which includes both the above orderings as special cases and is due to Houston(1966), will be given in Theorem 3.1.3. Also, we show there a connection with graph theory and with the Children's Round-dance Problem discussed much earlier by Lucas(1883).

Williams was unable to construct any row complete latin square of odd order n but instead gave a balanced solution to his statistical problem with the aid of a pair of $n \times n$ latin squares, so that for odd n he required $2n$ experimental units (corresponding to the $2n$ rows) for an experiment involving only n successive treatments. For the case $n = 5$, for example, he gave the solution exhibited in Figure 2.6.2, where each of the n treatments is succeeded by every other exactly twice. Much later, Houston(1966) proved the validity of another very similar construction using a pair of latin squares.

---

[16] It was claimed therein that the result is also true for an inverse property loop which satisfies the identity $(gh)(h^{-1}k) = (gk)$ but Belousov later pointed out that any such loop is a group.

$$
\begin{array}{ccccc}
0 & 1 & 3 & 4 & 2 \\
1 & 2 & 4 & 0 & 3 \\
2 & 3 & 0 & 1 & 4 \\
3 & 4 & 1 & 2 & 0 \\
4 & 0 & 2 & 3 & 1
\end{array}
\qquad
\begin{array}{ccccc}
0 & 2 & 1 & 4 & 3 \\
1 & 3 & 2 & 0 & 4 \\
2 & 4 & 3 & 1 & 0 \\
3 & 0 & 4 & 2 & 1 \\
4 & 1 & 0 & 3 & 2
\end{array}
$$

FIG. 2.6.2.

Williams(1949,1950) went on to discuss the design of balanced experiments in which not only the interaction between a treatment and its immediate predecessor is to be allowed for, but also the interaction between a treatment and that which was applied two steps earlier. He then distinguished between cases in which any interaction between these two predecessors themselves could be ignored in assessing their effect on the current treatment and others in which this effect too was allowed for. The former type of experiment requires that each treatment be preceded by each other treatment an equal number of times among the set of sequences applied to the experimental units (represented by the rows of the latin squares providing the solution) and also that the same be true for the next to immediate predecessors. The latter type of experiment requires that every ordered triad of treatments occurs exactly once. Thus, the set of three latin squares of order four given in Figure 2.6.3 provides a balanced experiment of the first kind in that, for example, the treatment 0 is preceded by every other exactly three times among the rows and also has every other treatment as a next to immediate predecessor exactly twice. However, of the 24 ordered triads of treatments six occur twice, twelve occur once and six occur not at all among the rows of the squares, so this experiment does not fulfil the conditions necessary for it to be of the second kind.

$$
\begin{array}{cccc}
0 & 1 & 3 & 2 \\
1 & 2 & 0 & 3 \\
2 & 3 & 1 & 0 \\
3 & 0 & 2 & 1
\end{array}
\qquad
\begin{array}{cccc}
0 & 3 & 2 & 1 \\
1 & 2 & 3 & 0 \\
2 & 0 & 1 & 3 \\
3 & 1 & 0 & 2
\end{array}
\qquad
\begin{array}{cccc}
0 & 2 & 1 & 3 \\
1 & 0 & 3 & 2 \\
2 & 3 & 0 & 1 \\
3 & 1 & 2 & 0
\end{array}
$$

FIG. 2.6.3.

As regards balanced experiments of the second kind, Williams first found a solution for all prime n. That is, he gave a construction for a set of $n - 1$ latin squares of order n in which each ordered triad of elements occurs exactly once among the rows of the set. His construction made use of the complete set of mutually orthogonal latin squares (which we define and consider in detail in Chapter 5) of the same prime order n. He also gave a construction which worked for the smallest prime powers: namely, $n = 4$, 8 and 9. He conjectured that the problem was soluble for every prime power order and this was subsequently confirmed by Niederreiter(1993).

Much later than the paper by Williams, Alimena(1962) discussed the more general situation of an experiment in which the interactions of any number of preceding treatments are allowed for. For an experiment involving $2k$ treatments, he gave a construction for a $2k \times 2k$ latin square which would counterbalance not only immediate sequential effects but all remote ones as well, valid whenever $2k + 1$ is prime: in particular, therefore, for experiments involving $2, 4, 6, 10, 12, 16, 18, 22$ etc. treatments. Later the same problem for more general values of k was discussed by Gilbert(1965). Gilbert also gave a construction for complete latin squares of each even order which is effectively the same as that of Williams and of Gordon(1961), in that it is based on the cyclic group of that order. We describe the work of the latter author next.

THEOREM 2.6.2 *A sufficient condition for the existence of a complete latin square L of order n is that there exist a finite group G of order n with the property that its elements can be arranged into a sequence $a_1, a_2, \ldots, a_n$ in such a way that the partial products $a_1, a_1a_2, a_1a_2a_3, \ldots, a_1a_2 \cdots a_n$ are all distinct.*

PROOF.   Let $b_1 = a_1$, $b_2 = a_1a_2$, $b_3 = a_1a_2a_3$, ..., $b_n = a_1a_2 \cdots a_n$. Then we shall show that the $n \times n$ matrix $\|c_{st}\|$ where $c_{st} = b_s^{-1}b_t$ is a complete latin square.

By hypothesis, the b_i's are all distinct and so $c_{st} \neq c_{su}$ if $t \neq u$. Thus, each row of the matrix contains each element of G exactly once. Similarly, the same is true of the columns, so the matrix is a latin square.

To show that it is row complete we must show that for any ordered pair of distinct elements α and β it is possible to determine s and t uniquely so that $b_s^{-1}b_t = \alpha$ and $b_s^{-1}b_{t+1} = \beta$. By definition of b_t and b_{t+1} we have $\alpha a_{t+1} = \beta$. This equation determines a_{t+1}, and hence t, uniquely. (We have $t > 0$ since the identity element of G is necessarily a_1 if the b_i are all distinct and so $t = 0$ would imply $\alpha = \beta$.) When t has been determined, s is uniquely defined by the equation $b_s^{-1}b_t = \alpha$. Thus, L is row complete.

Finally, we require unique solutions for s and t of the equations $b_s^{-1}b_t = \alpha$ and $b_{s+1}^{-1}b_t = \beta$. When these hold, $\alpha^{-1}a_{s+1} = \beta^{-1}$ which determines s and then the equation $b_s^{-1}b_t = \alpha$ determines t. Thus, L is also column complete.   □

The above result is due to Gordon(1961), who called a group *sequenceable* if it has the property stated in the theorem.[17] We shall say that a complete latin square obtained from a sequenceable group G in this way is *based on the group* G. In the same paper, Gordon proved the following important condition for a finite abelian group to be sequenceable.

THEOREM 2.6.3 *A finite abelian group G is sequenceable if and only if it is the direct product of two groups A and B such that A is a cyclic group of order 2^k, where $k \geq 1$, and B is of odd order.*

---

[17]Very much later in the development of the subject, the elements $b_1, b_2, \ldots, b_n$ were said to form a *directed terrace* corresponding to the sequencing $a_1, a_2, \ldots, a_n$. See [DK2] for details.

PROOF. To see the necessity of the condition, suppose that G is sequenceable and let $b_1 = a_1$, $b_2 = a_1 a_2$, $b_3 = a_1 a_2 a_3$, ..., $b_n = a_1 a_2 \cdots a_n$ be an ordering of the elements of G such that the b_i are all distinct. Then, as above, we necessarily have $b_1 = a_1 = e$, the identity element of G. Therefore $b_n \neq e$. But as we showed in Theorem 2.5.4, an abelian group G in which the product of all the elements is not equal to the identity has a unique element of order two and so has the form given in the statement of the theorem.

To prove the sufficiency of the condition, suppose that $G = A \times B$ with A and B as in the statement of the theorem. We show that G is sequenceable by constructing an ordering $a_1, a_2, \ldots, a_n$ of its elements which has distinct partial products. From the general theory of abelian groups it is known that G has a basis of the form $c_0, c_1, \ldots, c_m$, where c_0 is of order 2^k and where the orders $\delta_1, \delta_2, \ldots, \delta_m$ of $c_1, c_2, \ldots, c_m$ are odd positive integers each of which divides the next. If j is any positive integer, we shall show that there exist unique positive integers $j_0, j_1, \ldots, j_m$ such that

$$0 \leq j_i < \delta_i \text{ for each } i, \tag{2.12}$$

$$j \equiv j_0 \bmod \delta_1 \delta_2 \cdots \delta_m, \tag{2.13}$$

$$j_0 = j_1 + j_2 \delta_1 + j_3 \delta_1 \delta_2 + \cdots + j_m \delta_1 \delta_2 \cdots \delta_{m-1}. \tag{2.14}$$

Firstly, let j_0 be the remainder of j on division by $\delta_1 \delta_2 \cdots \delta_m$. Then $0 \leq j_0 < \delta_1 \delta_2 \cdots \delta_m$. Next by division of j_0 by δ_1, we get $j_0 = j_1 + j_1' \delta_1$ where $0 \leq j_1 < \delta_1$. Then, by division of j_1' by δ_2 we get $j_1' = j_2 + j_2' \delta_2$ where $0 \leq j_2 < \delta_2$ and hence $j_0 = j_1 + j_2 \delta_1 + j_2' \delta_1 \delta_2$. By successive divisions in accordance with this pattern we eventually get

$$j_0 = j_1 + j_2 \delta_1 + j_3 \delta_1 \delta_2 + \cdots + j_{m-1}' \delta_1 \delta_2 \cdots \delta_{m-1}.$$

where $j_1, j_2, \ldots j_{m-1}$ satisfy the inequalities given in (2.12). Now, since $0 \leq j_0 < \delta_1 \delta_2 \cdots \delta_m$ we must have $0 \leq j_{m-1}' < \delta_m$, so we can put $j_m = j_{m-1}'$ to satisfy (2.12) and (2.14).

We are now in a position to define the desired sequencing of G. It is convenient to start by defining the b_i. For each j in the range $0 \leq j < \frac{1}{2}n$ we define

$$b_{2j+1} = c_0^{-j} c_1^{-j_1} c_2^{-j_2} \cdots c_m^{-j_m}$$
$$b_{2j+2} = c_0^{j+1} c_1^{j_1+1} c_2^{j_2+1} \cdots c_m^{j_m+1},$$

where $j_1, j_2, \ldots, j_m$ satisfy (2.12)–(2.14). The elements $b_1, b_2, \ldots, b_n$ thus defined are distinct. For, if $b_s = b_t$ for some $s = 2u + 1$ and $t = 2v + 1$, then

$$u \equiv v \bmod 2^k \tag{2.15}$$

$$u_1 \equiv v_1 \bmod \delta_1$$

$$\vdots$$

$$u_m \equiv v_m \bmod \delta_m$$

where the u_i and v_i are generated from u and v respectively by the algorithm above which generated the j_i from j.

It follows from the inequalities in (2.12) that $u_1 = v_1$, $u_2 = v_2, \ldots, u_m = v_m$ and hence $u_0 = v_0$ and $u \equiv v \bmod \delta_1 \delta_2 \cdots \delta_m$. When combined with (2.15) this gives $u \equiv v \bmod n$ and hence $u = v$. A similar argument shows that $b_{2u+2} = b_{2v+2}$ implies $u = v$, so that the "even" b's are distinct. Next suppose that $b_{2u+1} = b_{2v+2}$. Then

$$-u \equiv v + 1 \bmod 2^k, \tag{2.16}$$
$$-u_1 \equiv v_1 + 1 \bmod \delta_1,$$
$$\vdots$$
$$-u_m \equiv v_m + 1 \bmod \delta_m.$$

Since $0 < u_1 + v_1 + 1 \le 2(\delta_1 - 1) + 1 < 2\delta_1$, we must have $u_1 + v_1 + 1 = \delta_1$. Reasoning similarly, we obtain

$$u_1 + v_1 + 1 = \delta_1,$$
$$u_2 + v_2 + 1 = \delta_2,$$
$$\vdots$$
$$u_m + v_m + 1 = \delta_m.$$

Multiplying the $(i+1)$-th equation of this system by $\delta_1 \delta_2 \cdots \delta_i$ for $1 \le i < m$ and adding the results, we get $u_0 + v_0 + 1 = \delta_1 \delta_2 \cdots \delta_m$. Combining this with (2.16), we find that $u + v + 1 \equiv 0 \bmod n$, which is impossible on account of the inequality $0 < u + v + 1 < n$. Hence, $b_1, b_2, \ldots, b_n$ are all distinct.

Next we calculate $a_1, a_2, \ldots, a_n$. If $i = 2j + 2$, where $0 \le j < \frac{1}{2}n$, then

$$a_i = b_{i-1}^{-1} b_i = c_0^{2j+1} c_1^{2j_1+1} \cdots c_m^{2j_m+1}.$$

These are all distinct by the same argument as above. If $i = 2j + 1$ and $j_1 \neq 0$, then

$$a_i = a_{2j+1} = b_{2j}^{-1} b_{2j+1} = (c_0^j c_1^{j_1} c_2^{j_2+1} \cdots c_m^{j_m+1})^{-1} (c_0^{-j} c_1^{-j_1} c_2^{-j_2} \cdots c_m^{-j_m})$$
$$= c_0^{-2j} c_1^{-2j_1} c_2^{-2j_2-1} \cdots c_m^{-2j_m-1}.$$

since, when j is reduced by 1, so also j_0 and j_1 are reduced by 1 provided that $j_1 \neq 0$. If $j_1 = 0$ and $j_2 \neq 0$, then j_2 is reduced by 1. Hence, if $i = 2j + 1$ and $j_1 = 0$ but $j_2 \neq 0$ then

$$a_i = c_0^{-2j} c_2^{-2j_2} c_3^{-2j_3-1} \cdots c_m^{-2j_m-1},$$

while if $i = 2j + 1$ and $j_1 = j_2 = 0$ but $j_3 \neq 0$, then

$$a_i = c_0^{-2j} c_3^{-2j_3} c_4^{-2j_4-1} \cdots c_m^{-2j_m-1},$$

and so on. These a_i's are obviously distinct from each other by the same reasoning as before. Because of the exponent of c_0 they are also distinct from the a_i with i even. $\square$

Let us note here that Gordon's paper [Gordon(1961)] predates that of Gilbert by four years and that Gordon's construction of complete latin squares includes that of Gilbert as a special case.

The concept of sequenceable groups has given rise to a number of interesting questions some of which were raised by Gordon himself and were posed as "Problems" in the first edition of this book. In particular, he asked

(i) Does any complete latin square of odd order exist?

(ii) What is the necessary and sufficient condition that a non-abelian group be sequenceable?

Also, Rényi(1966) raised the question as to whether all complete latin squares satisfy the quadrangle criterion (that is, are group-based).

Before we discuss these questions in detail, we give some general criteria for a group to be sequenceable following the line of investigation of Dénes and Török (1970). The first two of them are almost direct consequences of Theorem 2.6.2.

THEOREM 2.6.4 *A finite group is sequenceable if and only if there exists a permutation* $a_1, a_2, \ldots, a_n$ *of its elements such that* $a_1 = e$ *(the identity of the group) and such that the product* $a_{i+1}a_{i+2} \cdots a_{i+j} \neq e$ *for any choice of i and j satisfying* $1 < i < i + j < n$.

PROOF. Define $b_h = a_1 a_2 \cdots a_h$ for each positive integer $h \leq n$. Then, since $a_{i+1}a_{i+2} \cdots a_{i+j} = b_i^{-1}b_{i+j}$, the given condition says that $b_i \neq b_{i+j}$ for each choice of i and j such that $1 < i < i + j < n$. Thus if and only if the condition holds the products b_h for $h = 1, 2, \ldots, n$ are all distinct and the given ordering of the elements of the group then provides a sequencing. □

THEOREM 2.6.5 *A finite group is sequenceable if and only if there exists a permutation* $b_1, b_2, \ldots, b_n$ *of its elements such that the equation* $b_{i-1}^{-1}b_i = b_{j-1}^{-1}b_j$ *holds only when* $i = j$.

PROOF. Suppose first that the group G is sequenceable and that the ordering $a_1, a_2, \ldots, a_n$ of its elements is a sequencing. If the elements b_h are defined as in the previous proof, then $b_{i-1}^{-1}b_i = a_i$ and the equation $a_i = a_j$ clearly holds only when $i = j$.

Conversely, suppose that the equation $b_{i-1}^{-1}b_i = b_{j-1}^{-1}b_j$ holds only when $i = j$. Define $a_i = b_{i-1}^{-1}b_i$ for $2 \leq i \leq n$ and let $a_1 = e$, the identity element. Then the a_i are certainly all distinct and $a_1 a_2 \cdots a_h = b_1^{-1}b_h$ for $1 \leq h \leq n$. Since $b_1, b_2, \ldots, b_h$ are the elements of G they are all distinct and so also are the elements $b_1^{-1}b_1, b_1^{-1}b_2, \ldots, b_1^{-1}b_n$. Thus the ordering $a_1, a_2, \ldots, a_n$ of the elements of G provides a sequencing as required. □

THEOREM 2.6.6 *Let G be a group of order n with identity element e and let* S_k *denote the set of all ordered sequences of k distinct elements of* $G \setminus \{e\}$. *We define a subset* E_k *of* S_k *as follows. The sequence* $a_1, a_2, \ldots, a_k$ *is a member of* E_k *if the partial product* $a_h a_{h+1} \cdots a_k$ *is equal to e for some h in the range* $1 \leq h \leq k$ *and if* $a_i a_{i+1} \cdots a_j \neq e$ *whenever* $1 \leq i \leq j < k$. *In this way, we* *define sets $E_1, E_2, \ldots, E_{n-1}$ where E_1 is empty, E_2 comprises all products of the form $a_p a_p^{-1}$ where $a_p \neq e$, E_3 comprises all products of the form $a_p a_q a_q^{-1}$ and $a_p a_q a_r$ where $a_q \neq a_p^{-1}$ and $a_r = (a_p a_q)^{-1}$, and so on. The cardinality of the set E_k will be denoted by ϵ_k. Then the group G is sequenceable if and only if*

$$\epsilon_2(n-3)! + \epsilon_3(n-4)! + \cdots + \epsilon_{n-2}1! + \epsilon_{n-1}0! < (n-1)!.$$

PROOF.   Let us consider the set $G \setminus \{e\}$ and call an arrangement $a_1, a_2, \ldots, a_{n-1}$ of the distinct elements of $G \setminus \{e\}$ *wrong* if there exist integers $1 \leq h < k \leq n-1$ such that $a_h a_{h+1} \cdots a_k = e$.

The total number of distinct arrangements of the set $G \setminus \{e\}$ is $(n-1)!$ and by Theorem 2.6.4 the group is sequenceable if and only if the number of wrong arrangements is less than $(n-1)!$.

Let the ordered sequence $a_1, a_2, \ldots, a_k$ belong to E_k, then this sequence can be completed to a permutation of all the elements of $G \setminus \{e\}$ in $(n-1-k)!$ different ways and all of these permutations are different wrong arrangements. Thus there exist $\epsilon_k(n-1-k)!$ permutations of the $n-1$ distinct elements of $G \setminus \{e\}$ for which the subsequence formed by their first k elements is a member of E_k. Moreover, given any wrong arrangement $a_1, a_2, \ldots, a_{n-1}$, there exists exactly one positive integer k such that $a_1, a_2, \ldots, a_k \in E_k$: for, if $a_1, a_2, \ldots, a_{n-1}$ is a wrong arrangement then $a_h a_{h+1} \cdots a_k = e$ for some integers $1 \leq h < k \leq n-1$. Let us choose the pair h, k for which k is minimal. Then $a_1, a_2, \ldots, a_k \in E_k$ but $a_1, a_2, \ldots, a_l \notin E_l$ for $l > k$ since the condition $a_i a_{i+1} \cdots a_j \neq e$ whenever $1 \leq i \leq j < l$ is violated. Also, since k is minimal, $a_1, a_2, \ldots, a_l \notin E_l$ for $l < k$. Hence, the total number of wrong arrangements is $\epsilon_2(n-3)! + \epsilon_3(n-4)! + \cdots + \epsilon_{n-2}1! + \epsilon_{n-1}0! < (n-1)!$ and both the necessity and the sufficiency of our condition follow.   □

This theorem was used by Dénes and Török(1970) to find a number of sequencings for the dihedral groups D_m for $5 \leq m \leq 8$. Recall that D_m is the non-commutative group of order $2m$ generated by two elements a and b with defining relations $a^m = b^2 = e$ and $ab = ba^{-1}$. They also thought that they had showed that there are no other non-abelian sequenceable groups of order less than or equal to 14. However, there was an error in their computations because it was proved later that both non-abelian groups of order twelve are sequenceable. For the details, see Chapter 3 of [DK2]. In addition, they found a large number of sequencings for the non-commutative group of order 21 generated by two elements a and b with the defining relations $a^7 = b^3 = e$ and $ab = ba^2$.

Mendelsohn(1968) was the first to publish a sequencing for this latter group and thereby disprove an earlier conjecture according to which no latin square of odd order is complete. The author of the present book found that it is not difficult to obtain such sequencings by trial and error, once their existence was known. The sequencings of Mendelsohn and Keedwell are given in Figures 2.6.5 and 2.6.6 respectively (at the end of this chapter) and the complete latin square which corresponds to the fourth of the five sequencings obtained by Mendelsohn is exhibited in Figure 2.6.4.

Complete latin squares also have some connections with graph theory as we shall explain in Section 3.1.

An interesting dichotomy follows from Theorem 2.6.3 and Theorem 2.5.2. That is, the finite abelian groups can be classified into two mutually exclusive kinds: namely, sequenceable groups and groups whose Cayley tables possess a transversal (a group being of the former type if it has a unique element of order two and of the latter type otherwise).

01	02	03	04	05	06	07	08	09	10	11	12	13	14	15	16	17	18	19	20	21
04	03	05	21	11	14	02	09	13	07	18	16	19	08	20	01	06	10	17	12	15
07	09	19	03	06	12	14	21	20	17	08	05	16	01	10	18	15	13	04	02	11
18	14	09	07	19	15	17	01	21	13	06	02	20	12	11	05	04	08	16	10	03
19	04	12	08	21	07	20	02	18	01	16	13	03	11	09	14	05	15	10	06	17
05	17	14	18	09	04	13	12	01	08	19	10	21	15	03	02	16	06	20	11	07
12	10	07	16	02	19	18	06	14	11	03	15	08	17	04	20	13	05	09	21	01
17	21	16	09	15	02	12	03	10	04	01	19	05	18	13	08	11	20	07	14	06
08	12	21	17	16	11	04	18	03	20	15	14	10	02	06	19	07	01	05	13	09
16	07	02	01	03	17	10	14	08	18	05	20	09	06	21	12	19	11	13	15	04
02	13	17	05	14	16	08	15	12	06	09	11	01	04	07	10	20	19	21	03	18
21	05	11	15	18	08	03	13	19	02	10	01	17	09	12	04	14	07	06	16	20
11	06	08	10	13	21	19	16	04	09	17	07	15	20	05	03	01	14	12	18	02
10	08	13	02	17	20	06	04	15	19	14	03	12	16	18	11	21	09	01	07	05
14	20	04	19	12	05	01	11	02	15	21	06	18	07	17	13	10	16	03	09	08
03	19	06	11	08	01	09	20	16	14	13	18	04	21	02	07	12	17	15	05	10
20	18	10	12	07	13	11	17	06	05	02	21	14	19	01	15	09	03	08	04	16
06	15	01	13	20	03	16	05	07	21	04	17	11	10	19	09	18	12	02	08	14
13	01	20	14	04	10	15	07	11	16	12	09	02	05	08	06	03	21	18	17	19
09	16	15	06	01	18	21	10	05	12	20	08	07	03	14	17	02	04	11	19	13
15	11	18	20	10	09	05	19	17	03	07	04	06	13	16	21	08	02	14	01	12

Fig. 2.6.4.

However, no similar result holds for non-abelian groups. On the one hand, the symmetric group S_3 is neither sequenceable nor contains a transversal in its Cayley table, as was pointed out by Gordon himself. On the other hand, the non-abelian group of order 21 (as we remarked above) and the non-abelian group of order 27 on two generators are both sequenceable and both have complete mappings (as is shown by Theorem 1.5.3).

Since the first edition of this book was written, the study of sequenceable groups and of row and/or column complete latin squares has made great progress. The situation as it was in 1990 is described in detail in Chapter 3 of [DK2]. That book also discusses the related concept of R-sequenceability and the more general concepts of quasi-sequencing and quasi-row-complete latin square. A more recent survey is that of Ollis(2002).

We may summarize the present situation with regard to sequencing as follows:

(1) The non-abelian groups D_3, D_4, Q_4 of orders 6, 8 and 8 respectively are not sequenceable but it was conjectured by Keedwell(1983c) that all non-abelian groups of orders greater than 8 are sequenceable. The truth of this conjecture has been verified by B.A.Anderson(1987a) for all orders up to 32 inclusive.

(2) In particular, it has been proved that all dihedral groups except D_3 and D_4 are sequenceable [see Isbell(1990) and Li(1997)] and likewise all dicyclic groups except Q_4. (See [DK2] for details of the proof of the latter result.)

(3) Also, except for Q_4, all finite groups with a unique element of order two[18] which are either soluble or have the alternating group A_5 as their only non-abelian composition factor are sequenceable. For details of the first statement, see B.A.Anderson and Ihrig(1993a) and, for the latter, see Ollis(2002).

(4) However, so far as the present author is aware, there are only two infinite classes of non-abelian groups of odd order for which sequenceability has been proved: namely, (i) the groups of order p^n for $n > 2$ which contain an element of order p^{n-1} [C.Wang(2002)]; and (ii) the groups of order pq when $p < q$ and 2 is a primitive root modulo p [Keedwell(1981a)] and in some cases when the latter requirement is not satisfied [C.Wang(2002)]. But see also B.A.Anderson(1991) and Ollis(2014).

(5) In B.A.Anderson and Ihrig(1993a,b), the existence and application of symmetric sequencings has been discussed. (A sequencing of a group G of even order $2m$ and with a unique element z of order two is called *symmetric* if it takes the form $e, a_1, a_2, \ldots, a_{m-1}, z, a_{m-1}^{-1}, a_{m-2}^{-1}, \ldots, a_1^{-1}$.) A detailed account of earlier results concerning such sequencings can be found on pages 84 to 98 of [DK2].

As regards complete latin squares, Owens(1976) has given a construction for complete latin squares which are not group-based, thus answering the question raised earlier by Rényi (see above), and Higham(1997,1998) has constructed row complete latin squares of every odd composite order except 9. Since row complete latin squares of order 9 and of every even order had previously been constructed, this proves that such squares of every composite order exist. For more details, see Problem 2.3 in "The Present State of the Problems" at the end of this book.

As we remarked in Section 2, Chapter 3 of [DK2], Freeman(1979a,b) introduced the concept of a quasi-row-complete latin square as one for which the

---

[18]Such groups are called *binary groups* in Ollis(2002).

$n(n - 1)$ pairs of adjacent elements which occur in the rows include each unordered pair of distinct elements exactly twice. Also, Bailey(1984) showed that a necessary and sufficient condition that the multiplication table of a group can be written in the form of a quasi-row-complete latin square is that the group has a *terrace*, while B.A.Anderson(1987b) called (effectively) the same concept a *2-sequencing*.

DEFINITION. If $b_0, b_1, \ldots, b_{n-1}$ is a terrace: that is, a sequence comprising all the elements of a group G of order n such that the sequence $a_0 = e$, $a_1 = b_0^{-1}b_1$, $a_2 = b_1^{-1}b_2, \ldots$, $a_{n-1} = b_{n-2}^{-1}b_{n-1}$ includes one occurrence of each element of order 2 in G and, for every other element x of G, includes one occurrence of each of x and x^{-1}, or two occurrences of x, or two occurrences of x^{-1}, then the sequence $a_0, a_1, \ldots, a_{n-1}$ is a 2-sequencing of G.

Shortly after [DK2] was published, B.A.Anderson and Ihrig(1992) proved that all groups of odd order have 2-sequencings. Since that date, a number of papers concerned with the construction of terraces (and especially so-called invertible terraces) for groups of even order have been published. See, in particular, Ollis and Spiga(1995), Ollis and Whitaker(2007), Ollis(2005,2012), Ollis and Willmott(2011). Also, a series of papers concerned with constructing special kinds of terrace for cyclic groups of mainly odd orders by Anderson and Preece have appeared. We cite I.Anderson and Preece(2006,2008a,2008b,2010) as examples of the latter. In Preece(2008), the construction of orthogonal terraces is considered.

Finally, we mention a concept which is related to sequenceability and R-sequenceability.

DEFINITION. A group G of order n is *harmonious* if its elements can be arranged in a sequence $a_1, a_2, \ldots, a_n$ such that the elements $a_1a_2, a_2a_3, \ldots, a_{n-1}a_n, a_na_1$ are all distinct.

A necessary condition for a group to be harmonious is that it should have a complete mapping. Beals, Gallian, Headley and Jungreis(1991) showed, among other things, that all groups of odd order are harmonious. Further results can be found in C.D.Wang(1993) and C.D.Wang and P.A.Leonard(1994,1995).

$$
\begin{array}{ccccccccccccccccccccccccc}
(1)\ e & b & a & b^2 & ba^2 & b^2a^4 & b^2a^2 & a^4 & a^3 & b^2a^2 & a^2 & ba^4 & a^6 & b^2a^6 & a^5 & b^2a^5 & b^2a^6 & ba^4 & b^2a^5 & ba^6 & b^2a^5 & ba & ba^3 & a^4 & b^2a^3 & a^6 \\
\quad\ e & b & ba & a^4 & ba^3 & b^2a^6 & a^2 & b^2a^2 & a & b^2a^4 & a^5 & b^2a^3 & b^2a^6 & b^2a^5 & b^2 & ba^6 & b^2a^2 & a^3 & ba^5 & b^2 & ba^6 & a^3 & ba^2 & a^4 & ba^2 &
\end{array}
$$

$$
\begin{array}{ccccccccccccccccccccccccc}
(2)\ e & a & b & ba^2 & b^2 & b^2a^4 & a^2 & ba^4 & a^4 & a^3 & b^2a^5 & a^6 & b^2a^6 & b^2a^2 & a^5 & ba^6 & b^2a^5 & ba^3 & a^4 & a^6 \\
\quad\ e & a & b & ba^2 & b^2a^6 & ba^3 & a^2 & a^4 & b^2a & b^2a^4 & b^2a^3 & a^3 & b^2a^3 & a^5 & ba & b^2 & b^2a^2 & a^5 & b^2a^5 & a^6
\end{array}
$$

$$
\begin{array}{ccccccccccccccccccccccccc}
(3)\ e & a & b & ba^2 & b^2 & b^2a^4 & a^2 & ba^4 & a^4 & a^3 & b^2a^3 & ba & ba^6 & ba^3 & b^2a^2 & b^2a^5 & b & b^2a^2 & b^2a^6 & b & a^6 \\
\quad\ e & a & b & ba^2 & b^2a^6 & ba^3 & a^2 & a^4 & b^2a & a^3 & a^5 & b^2a^5 & ba^4 & ba & a^5 & b^2a^4 & b^2a^3 & b^2a^6 & b^2a^5 & a^5 & a^6
\end{array}
$$

$$
\begin{array}{ccccccccccccccccccccccccc}
(4)\ e & a^3 & b & a & ba^2 & b^2 & b^2a^4 & b^2a & a^2 & b^2a^6 & ba^5 & b^2a^2 & b^2a^6 & ba^6 & b^2a^2 & a^4 & a^5 & b^2a^6 & b^2a^3 & a^2 \\
\quad\ e & a^3 & ba^6 & b & b^2a^2 & ba & a & b^2a^5 & ba^4 & a^5 & b^2 & a^6 & ba^5 & ba & b^2 & ba^4 & a^5 & b^2a^6 & ba^3 & a^2
\end{array}
$$

$$
\begin{array}{ccccccccccccccccccccccccc}
(5)\ e & a^3 & b & a & ba^2 & b^2 & b^2a^4 & b^2a & a^2 & b^2a^3 & ba^5 & ba & b^2a^5 & ba^3 & b^2a^6 & ba^6 & b^2a^6 & ba^5 & ba^3 & a^4 & a^2 \\
\quad\ e & a^3 & ba^6 & b & b^2a^2 & ba & a & b^2a^4 & ba^4 & a^4 & ba^3 & a^4 & ba & b^2a^3 & ba^4 & b^2a^2 & b^2a^6 & ba & b^2a & a^5 & a^2
\end{array}
$$

Fig. 2.6.5.

$$
\begin{array}{ccccccccccccccccccccccc}
e & b^2 & b^2a^3 & a^4 & b^2a & a^2 & a^5 & b^2a^6 & b^2a^5 & ba^3 & a^3 & b^2a^2 & ba^5 & ba^6 & a & b & ba^4 & b^2a^4 & ba \\
e & b^2 & ba^3 & b^2a & b^2a^5 & b & ba & ba^6 & b^2a^6 & a^2 & a^4 & b^2a^4 & a^6 & ba^4 & ba^5 & a^3 & b^2a^3 & b^2a^2 & a^5
\end{array}
$$

Fig. 2.6.6.

CHAPTER 3

Partial latin squares and partial transversals

The kinds of partial latin square which we discuss in this chapter are latin rectangles, row latin squares and various kinds of incomplete latin square.

3.1 Latin rectangles and row latin squares

A *latin rectangle* of r rows and $n \geq r$ columns is an $r \times n$ array of n symbols such that each row contains all the symbols and no column contains any symbol more than once. Any latin rectangle can be extended to a latin square, as we show in Theorem 3.1.1 below. The number of ways in which this can be done is discussed in Chapter 4.

Our first two theorems both make use of Hall's theorem on representatives of subsets, namely: The necessary and sufficient condition that a system of distinct representatives, one for each member of a set $\{T_1, T_2, \ldots, T_m\}$ of subsets of a given set S, can be chosen simultaneously is that, for each $k = 1, 2, \ldots, m$, any selection of k of the subsets shall contain between them at least k distinct elements of S. [See P.Hall(1935) for the proof.]

THEOREM 3.1.1 *Every $r \times n$ latin rectangle can be extended to a latin square of order n.*

PROOF. Let R be a $r \times n$ latin rectangle. For each $j = 1, 2, \ldots, n$ let us form a set S_j consisting of those $n - r$ symbols which occur in R but which do not occur in the j-th column of R. Since each symbol occurs once in each of the r rows of R and no symbol occurs twice in any column, each symbol must occur exactly $n - r$ times among the S_j.

Any selection of k of the S_j will contain $k(n - r)$ occurrences of symbols and these must involve at least k distinct symbols since each symbol occurs $n - r$ times among the S_j. Thus, the requirements of Hall's theorem are satisfied and distinct representatives $a_1, a_2, \ldots, a_n$ can be chosen so that $a_j \in S_j$. By definition of the sets S_j, these n distinct representatives may be used to form an $(r+1)$-th row of R by placing a_j in column j for $j = 1, 2, \ldots, n$, and thereby yield an $(r + 1) \times n$ latin rectangle.

By repetition of the process $n - r$ times, we eventually obtain an $n \times n$ latin square. $\qquad \square$

The above theorem will be found in M.Hall(1945). The following extension of it is due to Ryser(1951).

Latin Squares and their Applications. http://dx.doi.org/10.1016/B978-0-444-63555-6.50003-9

THEOREM 3.1.2 *An $r \times c$ rectangular matrix M can be extended to a latin square based on a set Σ of n symbols if and only if M contains no symbols other than those in Σ, no symbol appears twice in any row or column of M and the number $N(i)$ of times that any symbol $i \in \Sigma$ appears in M is at least $r + c - n$.*

PROOF.   The necessity of the conditions is easy to see. If we want to extend M to an $r \times n$ latin rectangle then we shall have, in the latter, r occurrences of every symbol. But in the additional $n - c$ columns we cannot include any symbol more than $n - c$ times, and so it must already appear $r - (n - c)$ times in M.

As regards the sufficiency, each of the proofs known to the present authors requires a preceding lemma. [The proof of sufficiency given in the first edition of this book was incorrect because the induction step was not valid.] The original proof given by Ryser(1951), used a lemma concerning matrices of zeros and ones. In the proof given by Lindner on page 222 of [DK2], the lemma used is an extended form of P. Hall's theorem on distinct representatives of subsets.

Here we give a proof which uses graphical ideas. The lemma which we use concerns edge-colouring of a bipartite graph. It was originally proved with the aid of P. Hall's theorem on distinct representatives of subsets. [See, for example, page 94 of Berge(1962).] However, the elementary proof which we give here is due to Hilton(1977). The proof of the theorem itself which we give can be found on page 95 of Berge's book but may date from earlier than this.

LEMMA. *If G is a bipartite graph whose maximum degree[1] is Δ, then G can be edge-coloured[2] with Δ colours.*

PROOF.   We suppose that the two vertex sets of the bipartite graph G are $U = \{u_1, u_2, \ldots, u_r\}$ and $V = \{v_1, v_2, \ldots, v_s\}$.

We colour the edges one by one until we reach an edge for which no suitable colour is available. Suppose that $[u_i, v_j]$ is such an edge. Since at most $\Delta - 1$ of the Δ (or less) edges which are incident at u_i have been coloured, there exists at least one colour α which has not been used to colour any edge incident with u_i. Likewise, there exists at least one colour β which has not been used to colour any edge incident with v_j. If $\beta = \alpha$ or if β has not been used to colour any edge incident with u_i, there is nothing to prove because this colour is then available to colour the edge $[u_i, v_j]$. In the contrary case, there exists a chain of one or more edges $[u_i, v_h], [v_h, u_k], [u_k, v_l], \ldots,$ which are alternately coloured $\beta, \alpha, \beta, \alpha, \beta, \ldots$. This chain cannot pass through v_j since, if so, an edge ending at v_j would have colour β, in contradiction to our choice of β. Nor can it return to u_i, by our choice of α. Consequently, we can re-colour the edges of this chain using $\alpha, \beta, \alpha, \beta, \alpha, \ldots,$ in place of $\beta, \alpha, \beta, \alpha, \beta, \ldots$. Then edge $[u_i, v_j]$ can be coloured with β.

By repetition of this process a sufficient number of times, we obtain a complete edge-colouring of G. □

---

[1] Sometimes called "valency".

[2] An edge-colouring of a graph requires that no two edges which are incident with the same vertex have the same colour.

We can now prove the sufficiency part of Theorem 3.1.2. Let $\rho_1, \rho_2, \ldots, \rho_r$ denote the rows of M and let $\sigma_1, \sigma_2, \ldots, \sigma_n$ denote the n symbols. We form a bipartite graph G whose vertex sets are $\{\rho_1, \rho_2, \ldots, \rho_r\}$ and $\{\sigma_1, \sigma_2, \ldots, \sigma_n\}$. We join vertex ρ_i to vertex σ_j if and only if the symbol σ_j does not occur in row ρ_i of M.

Since exactly c symbols occur in each row of M, each vertex ρ_i of G has degree $n - c$. Since the symbol σ_j occurs $N(j)$ times in M, it does not occur in $r - N(j)$ of the rows of M. Therefore, the degree of the vertex σ_j in G is $r - N(j) \leq n - c$, because $-N(j) \leq n - r - c$. It follows that each vertex of the graph G has degree at most $n - c$. Consequently, G can be edge-coloured with at most $n - c$ colours $k_1, k_2, \ldots, k_{n-c}$. We may use these colours to represent the $n - c$ columns which must be adjoined to M to form an $r \times n$ latin rectangle. If the edge joining vertex ρ_i to vertex σ_j of G has colour k_s, then we put the symbol σ_j in the ρ_i-th row of column k_s. Hence, we can extend M to an $r \times n$ latin rectangle. This in turn can be extended to a latin square by Theorem 3.1.1.

$\square$

Note that Theorem 1.6.3 is an immediate corollary of Theorem 3.1.2. From Theorem 3.1.2, one can also deduce that an incomplete latin square of order $t \geq 4$ with n different elements (that is, a square of order t such that a subset of its t^2 cells are occupied by elements of $\{1, 2, \ldots, n\}$ and no element occurs twice in the same row or column) can be embedded in a latin square of order n provided that $n \geq 2t$. This result can be found in T.Evans(1960) and again in Dénes and Pásztor(1963) where the proof is formulated in terms of the transversals of a latin square (cf. Theorem 1.6.3).

A result in a similar spirit to that of Evans was proved by Treash(1971), who showed that any finite incomplete loop which is also totally symmetric can be embedded in a finite complete totally symmetric loop. The definition of an incomplete loop is given in Section 3.4, where a number of further results concerning the embedding of incomplete latin squares and quasigroups will be found.

A modification of Theorem 3.1.2 obtained by Hilton and Johnson(1988) replaces the requirement that the number $N(i)$ of times that any symbol $i \in \Sigma$ appears in M is at least $r + c - n$ by the requirement that $\sum_{i=1}^{n} t_E(i) = n^2 - rs$, where $t_E(i)$ is the largest size of any i-transversal outside M, defined as follows: Let L be an $n \times n$ matrix which contains the given $r \times c$ rectangular matrix M in its top left corner and such that the cells of $E = L - M$ are all empty. For each symbol i, an i-transversal outside M is a set S of cells of E such that no two cells of S are in the same row or column and no cell of S is in the same row or column as any cell of M which contains i. For the proof of this result, see Hilton and Johnson's paper.[3]

---

[3] A similar condition for a partial Sudoku square (see Section 3.2) to be completable has been obtained by Cameron, Hilton and Vaughan(2012).

In Section 2.6 we defined the concepts of row complete and complete latin square. Analogously, an $r \times n$ latin rectangle, $r \leq n/2$, is said to be *row complete* if the *unordered* pairs of adjacent elements appearing in its rows are all distinct. If we replace "row" by "column" in the preceding definition, it becomes that of a *column complete* latin rectangle. A latin rectangle which is both row and column complete is called *complete*. The following theorem was first proved by Houston(1966) and has already been referred to in Section 2.6. We include it here because of its relevance to latin rectangles (see Theorem 3.1.4 below).

THEOREM 3.1.3 *If there exists a permutation of the integers* $0, 1, 2, \ldots, n-1$ *with the property that the differences (taken modulo n) between pairs of adjacent integers are all distinct, then there exists a row complete latin square of order n.*

The proof is the same as that of Theorem 2.6.1 and will be omitted.

The complete latin square L constructed by the method of Theorem 3.1.3 is a Cayley table for the cyclic group of order n. However, by Theorem 2.6.3, complete latin squares based on abelian groups of odd order do not exist. It follows that a permutation of the integers $0, 1, 2, \ldots, n-1$ to satisfy Theorem 3.1.3 does not exist when n is odd.

From Theorem 3.1.3 we deduce that, in particular, a latin square of order $n = 2m$ whose first row is $0, 1, 2m - 1, 2, 2m - 2, \ldots, k, 2m - k, \ldots, m + 1, m$ and whose i-th row is obtained from the first by adding $i - 1$ modulo n is a row complete latin square. When we construct this latin square, we find that its last m rows are the same as the first but in reverse order. (See Figure 3.1.1.) It follows that the first m rows form a row complete $m \times 2m$ latin rectangle. Consequently, we have:

THEOREM 3.1.4 *Row complete* $m \times 2m$ *latin rectangles exist for every positive integer m.*

0	1	$2m-1$	2	$2m-2$	......	$m+2$	$m-1$	$m+1$	m
1	2	0	3	$2m-1$	......	$m+3$	m	$m+2$	$m+1$
.	.	.	.	.	......	.	.	.	.
$m-1$	m	m_2	$m+1$	$m-3$	......	1	$2m-2$	0	$2m-1$
m	$m+1$	$m-1$	$m+2$	$m-2$	......	2	$2m-1$	1	0
.	.	.	.	.	......	.	.	.	.
$2m-1$	0	$2m-2$	1	$2m-3$	......	$m+1$	$m-2$	m	$m-1$

FIG. 3.1.1.

We note that each such latin rectangle R defines a decomposition of the complete undirected graph K_{2m} on $2m$ vertices into m disjoint Hamiltonian paths. If the vertices of K_{2m} are labelled by the digits $0, 1, \ldots, 2m - 1$, then the rows of R define the Hamiltonian paths.

There are further connections between Theorem 3.1.4 and graph decompositions. If we adjoin an additional digit $2m$ to each row of R to obtain an $m \times 2m+1$ rectangle R' and then read the rows of R' cyclically, they define a decomposition of the complete undirected graph on $2m + 1$ vertices into Hamiltonian cycles. If instead of reading these rows cyclically, we read them successively from left to right with the last row followed by the first, they define an Eulerian circuit of the latter graph. [See Figure 3.1.2 and Keedwell(1974).]

$$
\begin{array}{cccccccccc}
0 & 1 & 2m-1 & 2 & 2m-2 & \ldots\ldots & m+2 & m-1 & m+1 & m & 2m \\
1 & 2 & 0 & 3 & 2m-1 & \ldots\ldots & m+3 & m & m+2 & m+1 & 2m \\
\cdot & \cdot & \cdot & \cdot & \cdot & \ldots\ldots & \cdot & \cdot & \cdot & \cdot & \\
\cdot & \cdot & \cdot & \cdot & \cdot & \ldots\ldots & \cdot & \cdot & \cdot & \cdot & \\
m-2 & m-1 & m-3 & m & m-4 & \ldots\ldots & 0 & 2m-3 & 2m-1 & 2m-2 & 2m \\
m-1 & m & m-2 & m+1 & m-3 & \ldots\ldots & 1 & 2m-2 & 0 & 2m-1 & 2m
\end{array}
$$

Fig. 3.1.2.

There is another interesting application of the augmented rectangle R'. In his book on Recreational Mathematics, Lucas(1883) discussed the problem of devising a performance of successive children's dances in each of which every child would hold hands with his/her two immediate neighbours so that the participants would together form a closed circle. All the children were to take part in every dance and it was required to arrange them in successive dances in such a way that each child would be neighbour to every other exactly once during the complete performance. Evidently the number n of children participating would have to be odd because any particular child would be neighbour to two others in each dance and so would dance with, say $2m$, other children all together. The number of dances is then m and the total number of children is $2m + 1$.

It is easy to see that, if the children are named $0, 1, \ldots, 2m$, then the Hamiltonian cycles described above provide a solution to the problem. Lucas attributes this solution to C.Walecki though that gentleman obtained his solution by a different method. The Walecki solution has prompted Preece(1994) to remark that the Williams' sequencing which we described in Theorem 2.6.1 had been obtained much earlier.

For more recent work on this topic and its application to Statistical designs, see Bailey, Ollis and Preece(2003).

Before changing the topic, we should like to pose two problems connected with complete latin rectangles:

(1) if n is a given odd integer, what is the maximum value of m_n such that a complete latin rectangle of size $m_n \times n$ exists?

(2) for which values of k_n is it true that an arbitrary complete latin rectangle of size $k_n \times n$ can be completed to a complete latin square of order n?

Next, we mention the concept of an (n, d)-complete rectangle.

DEFINITION. An array is called an (n, d)-*complete rectangle* if it contains n different symbols each of which appears exactly d times in the array, and such that no symbol is repeated in any row or column.

As an example, we note that the 6×6 rectangle shown in Figure 3.1.3 is (9,4)-complete.

$$
\begin{array}{|cccccc}
1 & 2 & 3 & 4 & 5 & 6 \\
2 & 6 & 7 & 5 & 9 & 1 \\
3 & 7 & 5 & 6 & 1 & 8 \\
4 & 5 & 9 & 7 & 8 & 3 \\
8 & 9 & 1 & 2 & 3 & 4 \\
6 & 4 & 8 & 9 & 7 & 2 \\
\end{array}
$$

FIG. 3.1.3.

Up to the present, interest seems to have centred on the case $d = 1$. In particular, one of the authors of [DK1] has proved the following theorem, in which a rectangle is considered to be *trivial* if it consists of a single row or a single column:

THEOREM 3.1.5 *If L is the latin square representing a multiplication table of a group G of order n, where n is a composite number, then L can be split into a set of n non-trivial rectangles each of which is $(n, 1)$-complete.*

PROOF. If n is a composite number, G has at least one proper subgroup A_0, of order h say. By Lagrange's theorem, G splits into disjoint cosets of A_0, say $A_0, A_1, \ldots, A_{k-1}$, where $n = hk$.

By selecting one element from each of these k cosets, we get a set of coset representatives of G relative to A_0. It is evident that one can select (in many ways) h such sets of coset representatives which are pairwise disjoint and which together cover G. Call these sets of coset representatives $R_1, R_2, \ldots, R_h$.

Now let a multiplication table for G be formed by taking as row border the elements of the cosets $A_0, A_1, \ldots, A_{k-1}$ in order and as column border the elements of the sets of coset representatives $R_1, R_2, \ldots, R_h$ in order. By the properties of coset representatives, the rectangle whose row border comprises the elements of the coset A_i and whose column border comprises the elements of the set R_j contains each element of G exactly once and is an $(n, 1)$-rectangle. Hence the multiplication table of G is the union of n such $(n, 1)$-rectangles. $\square$

As an example, we show in Figure 3.1.4 the result of carrying out the construction for the cyclic group of order 8 when regarded as the additive group of integers modulo 8, with the subgroup $\{0, 2, 4, 6\}$ as A_0. We have $A_1 = \{1, 3, 5, 7\}$ and we may take (for example) $R_1 = \{0, 1\}$, $R_2 = \{2, 5\}$, $R_3 = \{4, 3\}$, $R_4 = \{6, 7\}$.

The converse of Theorem 3.1.5 is false as is shown by the latin square of order 27 given in Figure 1.6.4. This latter latin square represents a multiplication table

(+)	0	1	2	5	4	3	6	7
0	0	1	2	5	4	3	6	7
2	2	3	4	7	6	5	0	1
4	4	5	6	1	0	7	2	3
6	6	7	0	3	2	1	4	5
1	1	2	3	6	5	4	7	0
3	3	4	5	0	7	6	1	2
5	5	6	7	2	1	0	3	4
7	7	0	1	4	3	2	5	6

FIG. 3.1.4.

for a quasigroup which is not a group but, despite this, it can be split into 27 (27,1)-rectangles.

We turn now to another generalization of the latin square concept which D.A.Norton(1952a) called a row latin square.

A *row latin square* is a square matrix of order n say, each of whose rows is a permutation of the same n elements. A *column latin square* is similarly defined. These concepts are related to that of a latin rectangle by the fact that a union of latin rectangles each having n columns (and such that the total number of rows is n) is obviously a row latin square. However, a row latin square cannot necessarily be separated into latin rectangles except in a trivial way.

It is clear that a square matrix which is both a row latin square and a column latin square is a latin square.

Let R be a groupoid satisfying Sade's left translation law [identity (17) of Section 2.1] then clearly the multiplication table of R is a row latin square. Similarly, the multiplication table of a groupoid R' satisfying Sade's right translation law is a column latin square.

Let R be a row latin square bordered by its elements taken in natural order. Then the i-th row of the square determines a permutation P_i of the top border and the square is completely determined by giving the permutations $P_1, P_2, \ldots, P_n$. The product of two row latin squares A and B, which are represented by the permutations $P_1, P_2, \ldots, P_n$ and $Q_1, Q_2, \ldots, Q_n$ may be defined as the square matrix $C = AB = (P_1Q_1,\ P_2Q_2, \ldots, P_nQ_n)$ whose i-th row is given by the product permutation P_iQ_i. Then C is again a row latin square.

Such a representation seems likely to be very useful for quasigroups since, as Suschkewitsch(1929) observed, there are quasigroups whose right (or left) representation forms a group. Such a quasigroup is isomorphic to a quasigroup $(G, \cdot)$ obtained from an appropriate group $(G, \circ)$ by a relation of the form $x \cdot y = (x\alpha^{-1} \circ y)\alpha$, where α is an arbitrary fixed permutation of G. If both the right and left representations of a quasigroup form groups, the quasigroup is a group.

We shall give an example of a quasigroup Q whose right representation is a group but whose left representation is not a group.

Take Q to be the quasigroup of order four whose multiplication table is given by Figure 3.1.5. Then clearly the right representation of Q consists of the permutations $(1)(2)(3)(4)$, $(1\ 4)(2\ 3)$, $(1\ 2)(3\ 4)$, $(1\ 3)(2\ 4)$ and these form a group. However, the left representation of Q is not a group since it contains an element $(1)(2\ 4\ 3)$ which is of order three. We constructed Q from the Klein group of order four using the rule of Suschkewitsch and by taking

$$\alpha = \begin{pmatrix} 1 & 2 & 3 & 4 \\ 2 & 3 & 1 & 4 \end{pmatrix}.$$

	1	2	3	4
1	1	4	2	3
2	2	3	1	4
3	3	2	4	1
4	4	1	3	2

FIG. 3.1.5.

The representation of a quasigroup by means of the permutations defined by the corresponding latin square has been found useful by the present author in work on finite projective planes.

Some of the results published by Norton(1952a) will now be stated without proof.

The set of all row latin squares of order n forms a group of order $(n!)^n$ under the product operation defined above. This group is isomorphic to the direct product of n symmetric groups of degree n.

Let $p\langle n \rangle$ be given inductively by $p\langle 1 \rangle = 1$, $p\langle 2 \rangle = 2$, and $p\langle n \rangle = p\langle n-1 \rangle + (n-1)p\langle n - 2 \rangle$ for $n > 2$. The number of row latin squares L of order n which have the property that every row of L^2 is represented by the identity permutation, that is,

$$L^2 = \begin{vmatrix} 1 & 2 & \cdots & n \\ 1 & 2 & \cdots & n \\ \vdots & \vdots & \ddots & \vdots \\ 1 & 2 & \cdots & n \end{vmatrix},$$

is $[p\langle n \rangle]^n$. The number of normalized row latin squares which have this same property is $[p\langle n - 2 \rangle]^{n-1}$.

In Norton(1952a), it has been shown that the existence of sets of mutually orthogonal latin squares (see Chapter 5 for the definition of this concept) is dependent upon the parallel problem for row latin squares. Consequently, existence problems concerning sets of the former squares may be studied in terms of the latter.

THEOREM 3.1.6 *A row (respectively column) latin square which satisfies the quadrangle criterion and is such that at least one among its elements occurs in one of the cells of every column (resp. row) is necessarily a latin square.*

PROOF.   This result is a consequence of the well-known theorem that an associative groupoid must be a group if it has a left identity element and each of its elements has a left inverse.

In the present case, we take the element which occurs in every column as the left identity element (by bordering the square appropriately) and we easily see that then every element has a left inverse with respect to this identity.   □

3.2   Critical sets and Sudoku puzzles

In Nelder(1977), in a very short note, that author asked a question about partially completed latin squares which appears deceptively simple to resolve and which we shall now discuss. (He has informed the present writer that he had no particular application in mind but thought the problem interesting.) Let us look at the 3×3 partially filled arrays in Figure 3.2.1. We observe that (i) each array is completable to a unique latin square on the three symbols 0, 1, 2 but that (ii) if any one entry is deleted from either array, that array is no longer uniquely completable to such a latin square. Nelder called a set of entries with these two properties a *critical set*. He asked, for a given latin square of order n, what is the *size* (number of entries) of a smallest critical set and what is the size of a largest critical set? He denoted these two numbers by $scs(n)$ and $lcs(n)$ respectively.

$$
\begin{array}{|ccc|}
0 & \bullet & \bullet \\
\bullet & \bullet & \bullet \\
\bullet & \bullet & 1
\end{array}
\qquad
\begin{array}{|ccc|}
0 & 1 & \bullet \\
1 & \bullet & \bullet \\
\bullet & \bullet & \bullet
\end{array}
$$

FIG. 3.2.1.

The first surprise is that critcal sets of different sizes exist for the same latin square; the second is that different latin square of the same order n have smallest critical sets of different sizes. These remarks are illustrated in Figure 3.2.2 and Figure 3.2.3. The critical sets A and B both complete uniquely to the same latin square L_1 but have different sizes. The critical set C completes uniquely to the latin square L_2. A is a smallest critical set for L_1 and C is a smallest critical set for L_2 but C is larger in size (five entries) than A (four entries). We justify below that A and C are smallest.

$$
A = \begin{array}{|cccc|}
0 & 1 & \bullet & \bullet \\
1 & \bullet & \bullet & \bullet \\
\bullet & \bullet & \bullet & \bullet \\
\bullet & \bullet & \bullet & 2
\end{array}
\qquad
B = \begin{array}{|cccc|}
0 & 1 & 2 & \bullet \\
1 & 2 & \bullet & \bullet \\
2 & \bullet & \bullet & \bullet \\
\bullet & \bullet & \bullet & \bullet
\end{array}
\qquad
C = \begin{array}{|cccc|}
\bullet & 1 & \bullet & \bullet \\
1 & \bullet & \bullet & 2 \\
\bullet & \bullet & 0 & \bullet \\
3 & \bullet & \bullet & \bullet
\end{array}
$$

FIG. 3.2.2.

$$L_1 = \begin{vmatrix} 0 & 1 & 2 & 3 \\ 1 & 2 & 3 & 0 \\ 2 & 3 & 0 & 1 \\ 3 & 0 & 1 & 2 \end{vmatrix} \qquad L_2 = \begin{vmatrix} 0 & 1 & 2 & 3 \\ 1 & 0 & 3 & 2 \\ 2 & 3 & 0 & 1 \\ \mathbf{3} & 2 & 1 & 0 \end{vmatrix} \qquad L_3 = \begin{vmatrix} 0 & 1 & 2 & 3 \\ 1 & 0 & 3 & 2 \\ 2 & 3 & 1 & 0 \\ 3 & 2 & 0 & 1 \end{vmatrix}$$

FIG. 3.2.3.

However, a smallest critical set for the square L_3 (which differs from L_2 in just one intercalate) has only four entries. In fact, the latin squares L_1 and L_3 are both isotopes of the cyclic group Z_4 (and so have smallest critical sets of the same size) whereas L_2 is isotopic to the 2-group $Z_2 \times Z_2$.

Let us now explain why A, C are respectively smallest critical sets for L_1, L_2. In any Cayley table of the cyclic group, each element belongs to exactly one intercalate. There are four intercalates all together and no two of them have any cell in common. If the cells of a critical set did not include at least one cell of each intercalate, it would be possible to obtain two completions of that set, one to L_1 and one to a latin square which differs from L_1 in having the two distinct symbols of the "uncovered" intercalate interchanged. Since there are four non-overlapping intercalates, there must be at least four cells in a critical set.

The latin square L_2 has a different arrangement of intercalates. Each cell lies in three different intercalates of L_2. Since there are 16 cells altogether but each intercalate has four cells, the total number of intercalates is $(16 \times 3)/4 = 12$. The cells of a critical set must contain at least one cell of each of these twelve intercalates. This can be achieved by means of a transversal of the square. The cells of L_2 which are shown in bold in Figure 3.2.3 form such a transversal. However, these cells alone (or those of any other transversal) are not sufficient for unique completion of L_2. Figure 3.2.4 shows two different ways in which the remaining twelve cells can be filled, so we need at least one further cell in our critical set. It is easy to check that one appropriately chosen further cell is sufficient to ensure unique completion.

$$\begin{vmatrix} 0_2 & \bullet & 2_3 & 3_0 \\ 1_0 & 0_3 & 3_1 & \bullet \\ 2_1 & 3_2 & \bullet & 1_3 \\ \bullet & 2_0 & 1_2 & 0_1 \end{vmatrix}$$

FIG. 3.2.4.

Figure 3.2.4 is an example of a concept which is now called a *latin bitrade*. (In the early days of this subject, several other terms were used, see later.) This consists of two (juxtaposed) partial latin squares P_1 and P_2 such that (i) the same cells are filled in each; (ii) no cell has the same entry in P_1 as it does in P_2; (iii) the same entries occur in each particular row of P_1 as occur in the corresponding row of P_2; and (iv) the same entries occur in each particular column of P_1 as occur in the corresponding column of P_2.

Each of P_1 and P_2 is called a *latin trade*. (In earlier papers, the term latin trade was often used for the pair P_1, P_2.)

Suppose that a given latin square L contains P_1 as a part. Then, if the entries of P_1 are deleted, the remaining partial square H can be completed in two ways according as the entries of P_1 or the entries of P_2 are placed in the empty cells of H. It follows that a critical set for L must include at least one cell of every latin bitrade that has its P_1 (or P_2) part included in L. Thus, the study of latin bitrades plays a crucial role in the investigation of critical sets. An intercalate is just the P_1 part of a 2×2 latin bitrade.

Before proceeding further, we need a few formal definitions. It is convenient for this purpose to think of a latin square of order n as consisting of n^2 ordered triples (row, column, symbol) as we did in Section 1.4. Usually, we shall use $0, 1, \ldots, n - 1$ as our symbols so that rows and columns will be labelled from 0 to $n - 1$.

A *uniquely completable* set (*UC* set) U of triples is such that it characterizes only one latin square. That is, there is a unique latin square of assigned order n which has U as a subset of its triples. Such a set U of triples is said to be a *critical set* if no subset of U is uniquely completable. Thus, as we showed above, the sets $U_1 = \{(0,0,0),(0,1,1),(1,0,1),(3,3,2)\}$ and $U_2 = \{(0,0,0),(0,1,1),(0,2,2),(1,0,1),(1,1,2),(2,0,2)\}$ are both critical sets for the Cayley table K of the cyclic group Z_4 and U_1 is a *minimal critical set* for K.

In the process of completing a *UC* set U to the latin square L which it characterizes, we say that adjunction of a triple $t = (r, c, s)$ is *forced* in the process of completion of a set T of triples ($|T| < n^2, U \subseteq T \subset L$) to the complete set of triples which represents L (and which we also write as L), if either

(i) $\forall r' \neq r, \exists z \neq c$ such that $(r', z, s) \in T$ or $\exists z \neq s$ such that $(r', c, z) \in T$ (that is, in the partial completion F of L, where F is the partial latin square formed by the triples of T, each cell of column c except that in row r is either in a row of F which already contains the symbol s or else is already filled with an element z distinct from s), or

(ii) $\forall c' \neq c, \exists z \neq r$ such that $(z, c', s) \in T$ or $\exists z \neq s$ such that $(r, c', z) \in T$ such that (r, c, z)eT (that is, in the partial completion F of L, each cell of row r except that in column c is either in a column of F which already contains the symbol s or else is already filled with an element z distinct from s), or

(iii) $\forall s' \neq s, \exists z \neq r$ such that $(z, c, s') \in T$ or $\exists z \neq c$ such that $(r, z, s') \in T$ (that is, in the partial completion F of L, every symbol except s already occurs either in column c or else in row r of F).

A *UC* set U is called *strong* if we can define a sequence of sets of triples $U = F_1 \subset F_2 \subset \ldots, \subset F_r = L$ such that each triple $t \in F_{v+1} - F_v$ is forced in F_v. It is *super strong* if the adjunction of each triple is forced by virtue of (iii) alone.

A *UC* set which is not strong is called *weak*. In particular, a critical set may be weak or strong. A recent refinement of the latter concept is to say that a critical set is *totally weak* if there is no triple whose adjunction is forced initially rather than only at a later stage of the completion to a unique latin square. See Adams and Khodkar(2001).

Clearly, if the subset U of cells of a latin square L form a UC set for L then those cells of any latin square L' isotopic to L onto which the cells of U are mapped form a set U' which is UC for L' and is of the same type relative to L' as U is relative to L: that is, weak or strong, critical or minimal critical or neither. We shall regard the sets U and U' as equivalent.

In order to determine whether a given set S of triples is UC for a given latin square L, we need to check that the triples of S include at least one cell/triple of every latin trade which L contains. Thus, the study of latin trades plays a crucial role in the investigation of UC and/or critical sets as we have already remarked. An intercalate is a latin trade of smallest possible size.

Let us mention here that among earlier names used for a latin trade were *critical partial latin square* and *latin interchange*. The latter was used by Drápal [see Drápal and Kepka(1985), for example] and the former in several papers of the present author.

In our further discussion of this topic, we shall look first at critical sets in group-based latin squares.

Very early in the development of the subject of critical sets, in a private letter to Jennifer Seberry, Nelder(1979) conjectured that, for a latin square based on the cyclic group $(Z_n, +)$, a set consisting of two appropriately sized triangles in the top left and bottom right corners of the square would be a smallest critical set. For n odd, these two triangles would each have $(n^2 - 1)/8$ cells; for n even, one would have $(n^2 + 2n)/8$ cells and the other $(n^2 - 2n)/8$ cells. (See Figure 3.2.5 for an illustration.) Thus, the size of a smallest critical set based on the cyclic group would be $\lfloor n^2/4 \rfloor$.

0	1	2	•	•	•
1	2	•	•	•	•
2	•	•	•	•	•
•	•	•	•	•	•
•	•	•	•	•	3
•	•	•	•	3	4

0	1	2	•	•	•	•
1	2	•	•	•	•	•
2	•	•	•	•	•	•
•	•	•	•	•	•	•
•	•	•	•	•	•	3
•	•	•	•	•	3	4
•	•	•	•	3	4	5

FIG. 3.2.5.

Also, Nelder conjectured that a set of $n(n-1)/2$ cells consisting of the upper left triangular portion of the square but excluding its main right-to-left diagonal (as in the square of Figure 3.2.6) would be a largest critical set. It has been proved that the first conjecture is correct for cyclic latin squares of even order and for strongly completable critical sets in cyclic latin squares of odd order.

This conclusion is arrived at as follows:

The critical sets suggested by Nelder are strongly completable and it was proved by Bate and Van Rees(1999) that the sizes of strongly completable critical sets are bounded below by $\lfloor n^2/4 \rfloor$. Also, Curran and Van Rees(1979) had proved much earlier that, when n is even, a critical set must have size at least $n^2/4$ (in order to cover all the intercalates) and that the set comprising triangles of sizes $(n^2 + 2n)/8$ and $(n^2 - 2n)/8$ is critical. Then, Cooper, Donovan and Seberry(1991) proved that, when n is odd, the set consisting of two triangles each of size $(n^2 - 1)/8$ is likewise critical.

The possibility remains that, for a cyclic latin square of odd order, a weakly completable critical set of size less than $\lfloor n^2/4 \rfloor$ might exist.

$$
\begin{array}{|cccccc}
0 & 1 & 2 & 3 & 4 & \bullet \\
1 & 2 & 3 & 4 & \bullet & \bullet \\
2 & 3 & 4 & \bullet & \bullet & \bullet \\
3 & 4 & \bullet & \bullet & \bullet & \bullet \\
4 & \bullet & \bullet & \bullet & \bullet & \bullet \\
\bullet & \bullet & \bullet & \bullet & \bullet & \bullet \\
\end{array}
$$

FIG. 3.2.6.

In a long paper involving the use of many latin trades, Donovan and Cooper (1996) succeeded in showing that Nelder's candidate for a largest critical set in a cyclic latin square is indeed critical though it is still not known whether it is of largest size for such a square. However it is known and had been shown much earlier by Stinson and van Rees(1982) that this critical set is not in general largest for a given size of latin square. In particular, there exist latin squares of orders 6, 7, 8 with critical sets of sizes at least 18, 24, 37 respectively. Questions regarding the sizes of largest critical sets are mostly open.

To summarize, Nelder's construction for a smallest strongly completable critical set in a cyclic latin square is a construction valid for all orders n.

However, attempts to provide similar constructions of minimal critical sets for other classes of latin square have, so far as the present author is aware, been unsuccessful. Constructions which give critical sets of latin squares based on the dihedral and dicyclic groups have been obtained but these are not minimal for general values of n. For details, see Keedwell(2004).

Next, we turn to the problem of weakly completable critical sets.

A problem which until recently seemed particularly difficult to solve was that of finding, and finding the minimal sizes of, weakly completable critical sets. In Keedwell(1999), it was shown by exhaustive argument that weakly completable critical sets do not exist in the latin squares of orders three and four. Later, by computer [see Bedford and Johnson(2000)], it was also shown that they do not exist in the isotopy class of latin squares based on the cyclic group of order five. On the other hand, they do exist in the only other isotopy class of latin squares of order five. Examples were found by Burgess(2000) who also showed how to construct, for any chosen order, a latin square which contains a weakly completable critical set.

More recently, Bedford and Johnson(2000,2001) showed that all latin squares based on cyclic groups of orders greater than five have weakly completable critical sets, thus disproving a conjecture of the present author [see page 237 of Keedwell(1996)]. Also, in the first paper, they proved the much stronger result that a weak uniquely completable set exists in every group-based latin square of order greater than five. Clearly, by successive deletion of unnecessary entries, such a uniquely completable set can be reduced to a critical set which is weakly completable. However, the initial steps in the completion of such a critical set to the unique latin square which it defines may then be forced. Observation of this fact led Adams and Khodkar(2001) to introduce the notion of a *totally weak critical set* that is, one for which initially no empty cell has its entry forced.

For further details of the work of Adams and Khodkar, see Keedwell(2004) and the relevant papers cited therein.

In or about 1979, the freelance puzzle-maker Howard Garns invented a new puzzle originally called "Number Place" for Dell Magazine of New York. This consisted of a 9×9 square in the form of nine 3×3 subsquares and in which a subset of the 81 cells were already filled with numbers from the set $S = \{1, 2, \ldots, 9\}$. The puzzle-solver was required to fill the remaining cells with numbers from the set S in such a way that the completed square would be a latin square L and so that each of the nine 3×3 subsquares would contain each of the numbers in the set S just once. In our language, the given filled cells formed a UC set for L which was intended to be strongly completable. Two examples of such puzzles are given in Figure 3.2.7. Later, the puzzle became very popular in Japan and later still, it was promoted under the Japanese name of "Sudoku" to "The Times" newspaper in Great Britain by a New Zealander by the name of Wayne Gould (who had come across the puzzle on a visit to Japan). An initial example of this "Sudoku puzzle" was published in "The Times" in November 2004.

Subsequently, many versions of the puzzle have appeared in many publications throughout the World. In particular, the size of the puzzle square has been generalized to $n^2 \times n^2$ and such a square is separated into n $n \times n$ subsquares. (Usually, $n = 3, 4$ or 5.) When all the cells of such a square have been filled, it becomes an example of a latin square of order n^2 separated into n^2 subsquares of order n each of which is $(n^2, 1)$-complete. We shall call such a square a *Sudoku latin square* and such squares will arise again in Section 5.3 and Section 6.3.

Sudoku latin squares which are group-based can be constructed by the method of Theorem 3.1.5. However, squares obtained in this way are too regular to be used in Sudoku puzzles.

A key question for mathematicians is "What is the size of a minimal critical set for a Sudoku latin square L whether group-based or not?" For the original 9×9 puzzle, it is believed that 17 given entries is minimal. Recently, McGuire(2012)

·	·	9	·	·	1	·	5	·
·	4	1	·	7	5	·	·	·
2	·	·	4	·	·	·	·	3
·	·	6	9	·	·	3	·	·
3	·	·	2	·	7	·	·	1
·	7	·	·	·	3	8	·	6
4	·	5	·	·	2	·	·	8
·	·	·	8	·	4	9	6	·
·	9	·	5	·	·	7	·	·

·	·	·	·	·	·	1	·	3
9	·	·	·	5	·	·	·	·
·	·	·	·	·	·	8	·	·
·	6	·	·	2	·	·	7	·
·	·	1	·	·	·	·	·	·
·	·	·	3	·	·	·	·	·
·	·	·	·	·	1	4	6	·
7	2	·	·	·	·	·	5	·
·	·	·	8	·	·	·	·	·

Fig. 3.2.7.

has claimed that he has proved by exhaustive enumeration of cases with the aid of a computer that this is so but, so far as the present author is aware, that claim has not been independently checked. In most published puzzles, the given filled cells form a strongly completable UC set which is neither minimal nor critical. In Figure 3.2.7, the first example is of a typical puzzle square while the second is one of many obtained by computer or otherwise which has just 17 filled cells. We give the following hint for commencing the completion of the latter. The cell (3, 9) must contain 5 because 5 cannot occur in the second row or eighth column but must occur in the subsquare (1, 3). Also, the cell (8, 9) must contain 1 because 1 cannot go elsewhere in the eighth row. [Note that 1 already occurs in the subsquare (3, 2).] Observe that completion would be impossible without the extra requirements for the completed square to be a Sudoku square.

In the 1990's, Sudoku latin squares with the additional property that both main diagonals contain each symbol just once were introduced and studied (under the name of *perfect latin squares*) outside the Sudoku puzzle environment in connection with the electronic storage and retrieval of $n^2 \times n^2$ arrays (e.g. pictures). If such an array is to be stored using only n^2 memory modules, then retrieval of a particular set of items (e.g. pixels) is said to be "conflict free" if the required data can be accessed without the necessity to collect more than one item from any particular memory module. It is usually particularly important that retrieval of the items which form a row, column, diagonal or subsquare of the array should be conflict free. This can be achieved by using the symbols of a perfect latin square to specify for each location in the array into which memory module it should be put.

Full details of this application can be found in Kim and Prasanna Kumar(1993) and in Heinrich, Kim and Prasanna Kumar(1992). Prasanna Kumar *et al* devised a fairly complicated method for constructing perfect latin squares but a somewhat simpler one was given later in Keedwell(2007) although at that time the latter author was unaware of the application just described.

Let us return to the general topic of critical sets and latin bitrades.

A lot of work has been done on trying to determine for which sizes critical sets exist in particular kinds of latin square. In Donovan(1999), that author gave a reference list of known results (up to the date of her paper) on the possible sizes of critical sets for various types of latin square. Later, in a joint paper, Donovan and Howse(2000) gave constructions by means of which they showed that, in suitable latin squares of order n, critical sets exist of all sizes between $\lfloor n^2/4 \rfloor$ and $(n^2 - n)/2$. [See also Donovan and Bean(2000) for a missing case.] A number of questions remain open in regard to this topic. In particular, what is the largest size which a critical set in a latin square may have? For more details, see Keedwell(2004).

A related question is to find the spectrum for the sizes and types (or *shapes*) of latin bitrades. (By the *size* of a latin trade, we mean the number of its filled cells. By its *shape*, we mean the way in which these filled cells are arranged.) In addition, there may be several kinds of latin trade of the same size and type/shape.[4] For example, in Figure 3.2.8 we exhibit four kinds of latin bitrade of the same size and type. (Three of these have the same shape.) An early attempt to resolve this question was made by the present author [see Keedwell(1994)] who set out to enumerate first the possible types and then the possible kinds of each type for sizes m up to $m = 10$. However, there were a few omissions in the latter listing which were later filled by Donovan, Howse and Adams(1997). In fact, the same question had been investigated from a quite different point of view much earlier by Drápal and Kepka(1983,1985). These authors had treated a latin trade and its mate as a pair of partial groupoids. Again we refer the reader to Keedwell(2004) for more details.

$$
\begin{vmatrix}
1_2 & 2_1 & \bullet & \bullet \\
2_1 & 1_2 & \bullet & \bullet \\
\bullet & \bullet & 1_2 & 2_1 \\
\bullet & \bullet & 2_1 & 1_2
\end{vmatrix}
\quad
\begin{vmatrix}
1_2 & 2_1 & \bullet & \bullet \\
2_1 & 1_2 & \bullet & \bullet \\
\bullet & \bullet & 1_3 & 3_1 \\
\bullet & \bullet & 3_1 & 1_3
\end{vmatrix}
\quad
\begin{vmatrix}
1_2 & 2_1 & \bullet & \bullet \\
2_1 & 1_2 & \bullet & \bullet \\
\bullet & \bullet & 3_4 & 4_3 \\
\bullet & \bullet & 4_3 & 3_4
\end{vmatrix}
\quad
\begin{vmatrix}
1_2 & 2_1 & \bullet & \bullet \\
\bullet & 1_2 & 2_1 & \bullet \\
\bullet & \bullet & 1_2 & 2_1 \\
2_1 & \bullet & \bullet & 1_2
\end{vmatrix}
$$

FIG. 3.2.8.

A considerable amount of work has been done on finding upper and lower bounds for the minimal size $gdist(n)$ of a latin trade in a group-based latin square of order n. For an integer $n \geq 2$, let $gdist(n)$ denote the minimum distance between two groupoids $(Q, \circ)$ and $(Q, \cdot)$ as $(Q, \circ)$ varies through all quasigroups of order n and $(Q, \cdot)$ varies through all groups of order n. It is immediate to see that $gdist(n)$ is the quantity whose bounds we are seeking. Drápal and Kepka(1989) obtained the result $gdist(n) \geq 3 + e \log_e p$ for n equal to the prime p (which implies that the latin square is cyclic) and, later, Cavenagh(2003) obtained the almost identical result $gdist(n) \geq \lceil 2 + e \log_e p \rceil$ by a simpler and more direct method. Again, we refer the reader to Keedwell(2004) for details.

---

[4]For the precise definition of type, see Keedwell(1994,2004).

More recent work on latin trades and/or bitrades has focussed on the geometrical and topological representation of the latter. This work also was initiated by Drápal(1991,2001) who conceived the idea of representing a latin bitrade by an orientable surface[5] and hence giving it a genus.

In order to introduce the idea, we shall think of a latin square in its guise as the multiplication table of a quasigroup. Let $(Q, \star)$ and $(Q, \circ)$ be two quasigroups of order n defined on the same set Q and put $M = \{(a, b) \in Q \times Q, a \star b \neq a \circ b\}$. That is, the elements of M are the entries in the cells of the latin bitrade that transforms $(Q, \star)$ to $(Q, \circ)$ and conversely. This is made clear by Figure 3.2.9 wherein, for example, the elements $(2, 0)$ and $(0, 2)$ of M arise from the second row of the latin bitrade. Thus, M is the size of the latin bitrade as defined earlier.

We define a set $\Delta(\star, \circ)$ of triangles such that their vertices are elements of M and each triangle has vertices of the form $(a, b), (c, b), (a, d)$, where $a \star b = c \circ b = a \circ d$. We define a *row permutation* ρ by $(a, b)\rho = (a, d)$, where $a \star b = a \circ d$, and a *column permutation* μ by $(a, b)\mu = (c, b)$, where $a \star b = c \circ b$. Lastly, we define an *element permutation* γ by $\gamma = \rho^{-1}\mu$.

The following facts are easily proved:

(1) If $(a, d) \in M$ and $(a, d)\gamma = (c, b)$, then $c \circ b = a \circ d$.

[$(a, d)\gamma = (c, b) \Rightarrow (a, d)\rho^{-1} = (c, b)\mu^{-1}$ and $(a, d) \in M$ implies that there is an element $b \in Q$ such that $a \circ d = a \star b$ $(b \neq d)$ Then $(a, d)\rho^{-1} = (a, b)$. It follows that $(c, b)\mu^{-1} = (a, b)$ so $c \circ b = a \circ d$.]

(2) For every triangle $\Delta \in \Delta(\star, \circ)$, there is exactly one ordered pair $l \in M$ such that $\Delta = \{l, l\mu, l\rho\}$, where $l = (a, b)$ say.

Hence, there is exactly one triangle of $\Delta(\star, \circ)$ with vertices $\{l, l\mu, l\rho\}$ or $\{m, m\rho^{-1}, m\gamma\}$ where $m = l\rho$, or $\{n, n\gamma^{-1}, n\mu^{-1}\}$ where $n = l\mu$.

It follows that each vertex $m \in M$ occurs in at most three triangles of $\Delta(\star, \circ)$ and that these triangles when they exist and are distinct are $\{m, m\mu, m\rho\}$, $\{m, m\rho^{-1}, m\gamma\}$ and $\{m, m\gamma^{-1}, m\mu^{-1}\}$.

It is easy to show that, for a fixed m, no two of these triangles coincide.

THEOREM 3.2.1 *All connected components of the polyhedron induced by* $\Delta(\star, \circ)$ *are orientable surfaces.*

PROOF. We orient each triangle in the order $(m, m\mu, m\rho)$. Then we define the faces of the polyhedron to be these triangles and also the orbits of the permutations ρ^{-1}, μ and γ^{-1} (where cycles of length 2 are represented as "faces" with two edges). Each edge is then a side of exactly one triangle $\{m, m\mu, m\rho\}$ and exactly one orbit and adjacent faces are oriented in the same way (clockwise or anti-clockwise). □

Since three triangles and three orbits meet at each vertex, the vertices have valency six. There are $|M|$ vertices, so the number of triangles is $|M|$ and the number of edges is $3|M|$. Since each cycle of an orbit of ρ, μ^{-1} and γ defines a

---

[5]Most of the papers published in the present century have concerned themselves with latin bitrades rather than with the original concept of critical sets for latin squares.

face of the polyhedron, the total number of faces (including those formed by the triangles) is $|M| + \omega(\rho) + \omega(\mu) + \omega(\gamma)$, where $\omega(\theta)$ denotes the number of cycles of the permutation θ.

$\left(\genfrac{}{}{0pt}{}{\star}{\circ}\right)$	c_1	c_2	c_3	c_4
b_1	0_0	1_1	2_2	3_3
b_2	1_1	2_0	3_3	0_2
b_3	2_3	3_2	0_1	1_0
b_4	3_2	0_3	1_0	2_1

$\left(\genfrac{}{}{0pt}{}{\star}{\circ}\right)$	c_1	c_2	c_3	c_4
b_1	$\bullet$	$\bullet$	$\bullet$	$\bullet$
b_2	$\bullet$	2_0	$\bullet$	0_2
b_3	2_3	3_2	0_1	1_0
b_4	3_2	0_3	1_0	2_1

FIG. 3.2.9.

EXAMPLE. *Let $Q = \{0,1,2,3\}$ and the quasigroups $(Q, \star)$ and $(Q, \circ)$ be defined as shown in the left-hand diagram of Figure 3.2.9.*

We find that the triangles are

$$
\begin{array}{cccc}
& m & \to \quad m\mu & \to \quad m\rho \\
(1) & b_2 \star c_2 = 2 & = b_3 \circ c_2 & = b_2 \circ c_4 \\
(2) & b_2 \star c_4 = 0 & = b_3 \circ c_4 & = b_2 \circ c_2 \\
(3) & b_3 \star c_1 = 2 & = b_4 \circ c_1 & = b_3 \circ c_2 \\
(4) & b_3 \star c_2 = 3 & = b_4 \circ c_2 & = b_3 \circ c_1 \\
(5) & b_3 \star c_3 = 0 & = b_4 \circ c_3 & = b_3 \circ c_4 \\
(6) & b_3 \star c_4 = 1 & = b_4 \circ c_4 & = b_3 \circ c_3 \\
(7) & b_4 \star c_1 = 3 & = b_3 \circ c_1 & = b_4 \circ c_2 \\
(8) & b_4 \star c_2 = 0 & = b_2 \circ c_2 & = b_4 \circ c_3 \\
(9) & b_4 \star c_3 = 1 & = b_3 \circ c_3 & = b_4 \circ c_4 \\
(10) & b4_{\star}c_4 = 2 & = b_2 \circ c_4 & = b_4 \circ c_1
\end{array}
$$

The permutations are

(a) row orbits (ρ):

$(b_2, c_2) \to (b_2, c_4) \to (b_2, c_2)$
$(b_3, c_1) \to (b_3, c_2) \to (b_3, c_1)$
$(b_3, c_3) \to (b_3, c_4) \to (b_3, c_3)$
$(b_4, c_1) \to (b_4, c_2) \to (b_4, c_3) \to (b_4, c_4) \to (b_4, c_1)$

(b) column orbits (μ):

$(b_2, c_2) \to (b_3, c_2) \to (b_4, c_2) \to (b_2, c_2)$
$(b_2, c_4) \to (b_3, c_4) \to (b_4, c_4) \to (b_2, c_4)$
$(b_3, c_1) \to (b_4, c_1) \to (b_3, c_1)$
$(b_3, c_3) \to (b_4, c_3) \to (b_3, c_3)$

(c) element orbits (γ) in $(Q, \circ)$:

$(b_2, c_2) \to (b_3, c_4) \to (b_4, c_3) \to (b_2, c_2)$
$(b_2, c_4) \to (b_3, c_2) \to (b_4, c_1) \to (b_2, c_4)$
$(b_3, c_1) \to (b_4, c_2) \to (b_3, c_1)$
$(b_3, c_3) \to (b_4, c_4) \to (b_3, c_3)$

Hence we obtain the polyhedron shown in Figure 3.2.10.

In general, if g is the genus of a surface, then $v - e + f = 2 - 2g$. We have
$$v = |M|, \ e = 3|M|, \ f = |M| + \omega(\rho) + \omega(\mu) + \omega(\gamma).$$
Thence, $2g = 2 - v + e - f = 2 + |M| - \omega(\rho) - \omega(\mu) - \omega(\gamma)$.

If $\omega(\rho) \geq 1 + |M|/3$, $\omega(\mu) \geq 1 + |M|/3$ and $\omega(\gamma) \geq 1 + |M|/3$, then $g < 0$ so at least one of one of $\omega(\rho)$, $\omega(\mu)$, $\omega(\gamma)$ is less than $1 + |M|/3$.

If $\omega(\rho)$, $\omega(\mu)$ and $\omega(\gamma)$ are all less than $|M|/3$, then $g \geq 1$.

In the case when $g = 0$, $\omega(\theta) \geq |M|/3$ for at least one θ ($\theta = \rho$, μ or γ) so at least one cycle has length 2 (since θ acts on $|M|$ symbols): that is, $m\theta^2 = m$ for some $m \in M$. This fact is illustrated by Figure 3.2.10.

In Drápal(2003) and in subsequent publications, that author has pointed out that any latin bitrade can be replaced by a canonical latin bitrade (called a *separated latin trade* by Drápal) in which the rows, columns and symbols are in one-to-one correspondence with the cycles. For example, the latin bitrade in Figure 3.2.9 can be replaced by one in which the second row is replaced by two rows, one having the entries 2_3 and 3_2 in the first two columns and the other 0_1 and 1_0 in the last two columns. The column and symbol maps which involve more than one cycle can be separated in the latin bitrade in a similar manner. This is illustrated by Figure 3.2.11 in which, reading from left to right, the diagrams show firstly the original latin bitrade, secondly the same bitrade "element separated"[6], thirdly "element and row separated", and finally "fully separated".

For any separated latin bitrade, we have the interesting result that the sum of the numbers of rows, columns and symbols is less than $|M| + 3$ (since $g \geq 0$). Moreover, the numbers of rows, columns and symbols in a separated latin bitrade are equal to $\omega(\rho)$, $\omega(\mu)$, $\omega(\gamma)$ respectively, so it is not necessary to construct the corresponding polyhedron in order to calculate the genus of such a bitrade.

Let us note also that the concept of genus only makes sense if the latin bitrade is *connected* or *indecomposable* or *primary* (all three terms have been used): that is, only if the latin bitrade cannot be decomposed into two or more latin bitrades of smaller size.

Two interesting questions arise. Firstly, what is the size of the smallest latin bitrade which has genus greater than zero; and, secondly, how can we characterize those latin trades which have a particular genus?

The answer to the first question is "size 9 if the bitrade is itself a latin square or size 12 otherwise". The author is not sure who first observed these facts but Figure 3.2.12 provides examples of such bitrades.

As regards the second question, it had been conjectured for some while that the two component trades of every latin bitrade whose genus is zero could be

---

[6]The cycles corresponding to the element 0 are $(b_1, c_3) \to (b_4, c_1) \to (b_1, c_3)$ and $(b_3, c_4) \to (b_5, c_2) \to (b_3, c_4)$.

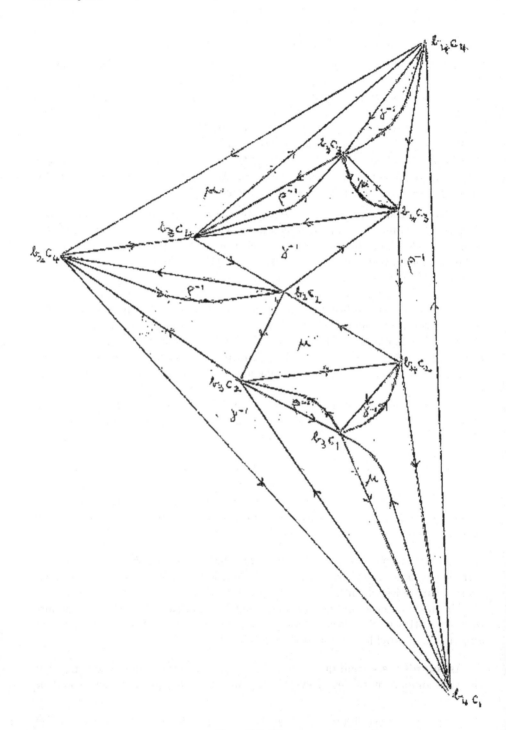

FIG. 3.2.10.

$$
\begin{array}{c|ccccc}
\binom{\star}{\circ} & c_1 & c_2 & c_3 & c_4 & c_5 \\
\hline
b_1 & 0_2 & \bullet & 2_0 & \bullet & \bullet \\
b_2 & \bullet & 3_4 & 4_3 & \bullet & \bullet \\
b_3 & 2_4 & 0_2 & \bullet & 4_0 & \bullet \\
b_4 & 3_0 & 4_3 & 0_2 & 1_4 & 2_1 \\
b_5 & 4_3 & 2_0 & 3_4 & 0_1 & 1_2 \\
\end{array}
\qquad
\begin{array}{|ccccc}
\hline
0_2 & \bullet & 2_0 & \bullet & \bullet \\
\bullet & 3_4 & 4_3 & \bullet & \bullet \\
2_4 & 5_2 & \bullet & 4_5 & \bullet \\
3_0 & 4_3 & 0_2 & 1_4 & 2_1 \\
4_3 & 2_5 & 3_4 & 5_1 & 1_2 \\
\end{array}
$$

$$
\begin{array}{|ccccc}
\hline
0_2 & \bullet & 2_0 & \bullet & \bullet \\
\bullet & 3_4 & 4_3 & \bullet & \bullet \\
2_4 & 5_2 & \bullet & 4_5 & \bullet \\
3_0 & 4_3 & 0_2 & 1_4 & 2_1 \\
4_3 & \bullet & 3_4 & \bullet & \bullet \\
\bullet & 2_5 & \bullet & 5_1 & 1_2 \\
\end{array}
\qquad
\begin{array}{|ccccccc}
\hline
0_2 & \bullet & \bullet & 2_0 & \bullet & \bullet & \bullet \\
\bullet & 3_4 & \bullet & \bullet & 4_3 & \bullet & \bullet \\
2_4 & \bullet & 5_2 & \bullet & \bullet & 4_5 & \bullet \\
3_0 & 4_3 & \bullet & 0_2 & \bullet & 1_4 & 2_1 \\
4_3 & \bullet & \bullet & \bullet & 3_4 & \bullet & \bullet \\
\bullet & \bullet & 2_5 & \bullet & \bullet & 5_1 & 1_2 \\
\end{array}
$$

FIG. 3.2.11.

$$
\begin{array}{|cccccc}
\hline
0_2 & \bullet & 2_4 & \bullet & 4_0 & \bullet \\
\bullet & \bullet & \bullet & \bullet & \bullet & \bullet \\
2_4 & \bullet & 4_0 & \bullet & 0_2 & \bullet \\
\bullet & \bullet & \bullet & \bullet & \bullet & \bullet \\
4_0 & \bullet & 0_2 & \bullet & 2_4 & \bullet \\
\end{array}
\qquad
\begin{array}{|cccc}
\hline
\bullet & 1_2 & 2_3 & 3_1 \\
1_3 & 0_1 & 3_0 & \bullet \\
2_1 & \bullet & 0_2 & 1_0 \\
3_2 & 2_0 & \bullet & 0_3 \\
\end{array}
$$

FIG. 3.2.12.

embedded in abelian groups (as distinct from merely being embeddable in quasi-groups) and the truth of this conjecture was proved almost simultaneously by Cavenagh and Wanless(2009) and by Drápal, Hämäläinen and Vitězslav(2010) but using different methods. Later, it was further shown by Blackburn and Mc-Court(2013) that, for a bitrade $L = (T, T')$ of genus zero, the abelian groups into which T and T' embed are isotopic. So far as the present author is aware, it is not known whether latin bitrades of genus one or any other genus $g > 0$ can be characterized in some similar way.

There are a number of facts which are easy to check:

(i) For T (and T') to be embeddable in the Cayley table of an abelian group, it is sufficient that $L = (T, T')$ has genus zero but is certainly not necessary.

(ii) Some latin trades can be embedded in the Cayley table of more than one group (none or several of which may be abelian).

(iii) There exist latin bitrades $L = (T, T')$ of genus $g > 0$ either only one or neither of whose components T and T' can be embedded in groups.

(iv) If $L = (T, T')$ has genus $g > 0$, T' and T may be embeddable in different groups but, if T and T' are isotopic, they can certainly be embedded in isotopes of the same group.

Cavenagh and Wanless(2009) have obtained the smallest connected and separable latin bitrade $L = (T, T')$ with the property that neither T nor T' embed in any group since both fail the quadrangle criterion (defined on page 4). We exhibit L in Figure 3.2.13. It is easy to see that the quadrangles whose cells are (b_2, c_1), (b_2, c_2), (b_3, c_1), (b_3, c_2) and (b_1, c_2), (b_1, c_3), (b_2, c_2), (b_2, c_3) fail the quadrangle criterion in both T and T'. These authors have also obtained the smallest (which has size 14) connected and separated bitrade in which T embeds in a group but not in any abelian group. They have given examples of other interesting situations. Drápal, Hämäläinen and Vitězslav(2010) have given an example of a latin trade of genus one and size 18 which does not itself fail the quadrangle criterion but nonetheless does not embed in any group.

$(\star)$	c_1	c_2	c_3	c_4
b_1	0	1	2	3
b_2	1	2	3	0
b_3	2	0	1	$\bullet$

$(\circ)$	c_1	c_2	c_3	c_4
b_1	1	2	3	0
b_2	2	0	1	3
b_3	0	1	2	$\bullet$

FIG. 3.2.13.

The above remarks raise two further questions. First, the reader will observe that the latin bitrade exhibited in Figure 3.2.14 can be regarded as having been obtained as the union (in some sense) of the latin bitrade shown in the left-hand diagram of Figure 3.2.12 and the intercalate $\begin{vmatrix} 4_3 & 3_4 \\ 3_4 & 4_3 \end{vmatrix}$ by coalescing the cells which contain 2_4 and 4_3 in the respective bitrades to give an amended cell 2_3.

$$\begin{vmatrix} 0_2 & \bullet & 2_4 & \bullet & 4_0 & \bullet \\ \bullet & \bullet & \bullet & \bullet & \bullet & \bullet \\ 2_4 & \bullet & 4_0 & \bullet & 0_2 & \bullet \\ \bullet & \bullet & \bullet & \bullet & \bullet & \bullet \\ 4_0 & \bullet & 0_2 & \bullet & 2_3 & 3_4 \\ \bullet & \bullet & \bullet & \bullet & 3_4 & 4_3 \end{vmatrix}$$

FIG. 3.2.14.

This idea of combining (or "adding") latin bitrades has been generalized in several different ways. See, in particular, Donovan and Mahmoodian(2002,2003), Drápal(2003) and Cavenagh, Donovan and Drápal(2004).

Secondly, the latin square (group table) of smallest order into which the component T of the latin trade of Figure 3.2.14 can be embedded is cyclic of order six. We may thence ask the general question: "What is the minimal order of a latin square into which a given separated (or non-separated) latin trade T of size h may be embedded?"

A latin bitrade is called *minimal* if it contains no latin bitrade of smaller size. Lefevre, Donovan, Cavenagh and Drápal(2007) have investigated the minimum size h of a latin bitrade of genus g and, in particular, they have shown that a minimal latin bitrade of genus g and size $8g + 8$ exists for each non-negative integer g.

A latin bitrade is called *k-homogeneous* if it has the same number k of rows, columns and symbols. The latin bitrades exhibited in Figure 3.2.12 are both 3-homogeneous. Because a derangement of three objects necessarily consists of a single cycle, every connected 3-homogeneous latin bitrade has genus one (and also is necessarily in separated form).

Cavenagh, Donovan and Drápal(2005a) have enumerated all 3-homogeneous latin bitrades. [See also Cavenagh(2006).] The latter paper contains another interesting result: namely that every 3-homogeneous latin trade can be partitioned into three disjoint transversals. This fact has subsequently been re-proved several times.

For results concerning 4-homogeneous latin bitrades, see Cavenagh, Donovan and Drápal(2005b) and for more general results concerning k-homogeneous latin bitrades, see Bean, Bidkhori, Kosravi and Mahmoodian(2005) and Behrooz Bagheri and Mahmoodian(2011).

More generally, a latin bitrade has been called (r, c, s)-*homogeneous* if each row has r entries, each column has c entries and each symbol occurs s times. In Drápal and Griggs(2010), these authors have proved that (r, c, s)-homogeneous latin bitrades of genus one exist only when $(r, c, s) = (3,3,3)$, $(4,4,2)$ or $(6,3,2)$.

There are many further papers on latin bitrades. A fairly comprehensive list of those published up to 2007 is given in a survey paper written by Cavenagh(2008). Also, there exist alternative ways of representing latin bitrades topologically and so obtaining their genus. The simplest of these is as follows:

For the connected and separated latin bitrade $L = (T, T')$ with rows $b_1, b_2, \ldots,$ b_u, columns $c_1, c_2, \ldots, c_v$ and symbols $s_1, s_2, \ldots, s_w$, construct a triangulated pseudo-surface as follows. The vertices are the "names" of the rows, columns and symbols, so there are $V = u + v + w$ vertices. Corresponding to each occupied cell (b_i, c_j) of T, construct a triangle with vertices b_i, c_j, s_k, where s_k is the symbol in the cell (b_i, c_j), and with edges $[b_i, c_j]$, $[c_j, s_k]$, $[s_k, b_i]$ oriented in that order. We call them "black" triangles. Corresponding to each cell of T', we construct a "white" triangle according to the same rule. Since each symbol which occurs in row b_i of T also occurs in row b_i of T' and since the same statement is true for the columns, each edge is a side of two triangles, one black and one white, so the total number E of edges is one half of three times the number of triangles. The total number F of triangles (faces) is $2|M|$, where $|M|$ is the size of T (or T'). Hence, the genus g satisfies $2g = 2 - (V - E + F) = 2 - (u + v + w - 3|M| + 2|M|) = 2 + |M| - \omega(\rho) - \omega(\mu) - \omega(\gamma)$ which is the same as we obtained for the earlier representation.

The black and white triangle representation of a latin bitrade has been, for example, used by Blackburn and McCourt(2013) in their paper mentioned above.

We might ask: "Are there other combinatorial structures which can be represented topologically and hence assigned a genus?"

In the first part of this section, we discussed uniquely completable and critical sets for assigned latin squares. Cameron(1994) has solved a different but related question: "How few positions (cells) in an $n \times n$ array have the property that *any* latin square of order n is uniquely determined by its entries in these particular cells?"

For example, the six cells $(1,1)$, $(1,2)$, $(1,3)$, $(2,1)$, $(2,2)$, $(3,1)$ of a 4×4

array do not satisfy the Cameron requirement because, although the left-hand latin square of Figure 3.2.15 is completely determined by its entries in these particular cells (they form a critical set for it), neither of the remaining two squares of Figure 3.2.15, is uniquely determined by its entries in these six cells since the same entries occur in both.

$$\begin{vmatrix} 1\,2\,3\,4 \\ 2\,3\,4\,1 \\ 3\,4\,1\,2 \\ 4\,1\,2\,3 \end{vmatrix} \quad \begin{vmatrix} 1\,2\,3\,4 \\ 3\,4\,1\,2 \\ 2\,1\,4\,3 \\ 4\,3\,2\,1 \end{vmatrix} \quad \begin{vmatrix} 1\,2\,3\,4 \\ 3\,4\,1\,2 \\ 2\,3\,4\,1 \\ 4\,1\,2\,3 \end{vmatrix}$$

FIG. 3.2.15.

Cameron has given a structural characterization for such a set of cells valid for all $n \geq 3$ and has shown that the minimum number of cells in such a set is $n(n-2)$ except when $n = 4$. For $n = 2, 3, 4$, the numbers of cells required are 1, 3, 7 respectively.

In Bartlett(2013), yet another problem concerning completion of partial latin squares has been considered. A partial latin square of order n is ϵ-*dense* if each row and column contains at most ϵn filled cells and each symbol occurs at most ϵn times. It is conjectured that every $\frac{1}{4}$-dense partial latin square is completable. (cf. Section 3.4.)

In the next section of this chapter, we consider, for the case of latin squares which are isotopic to group tables, the largest number of *randomly chosen* elements which we may delete without the resulting partial square losing the property of being uniquely completable.

3.3   Fuchs' problems

In his book Fuchs(1958), that author raised the following problem: "Let k elements be deleted at random in the Cayley table of a finite Abelian group G of order n. Determine the greatest $k = k(n)$ for which

(a) the rest of the table always determines G up to isomorphism;

(b) the table can be reconstructed uniquely from its remaining part.

The first author of [DK1] gave a solution to problem (b) without the restriction that the group is abelian [see Dénes(1962)]. His result should have been that, for any given group of order $n \neq 4$ or 6, we have $k(n) = 2n - 1$. (In fact, there was an error in his proof and he omitted to exclude $n = 6$ from the statement, see Theorem 3.3.1 below.)

Thus, unexpectedly, $k(n)$ does not depend on the structure of the group but depends only on its order. For the case $n = 4$, he obtained the value $k(n) = 3$.

For the proof, let us begin by pointing out that an abstract group is completely known when each of its elements has been represented by a symbol and the product of any two symbols in each order has been exhibited.

For a finite group, the products may be exhibited conveniently by means of the Cayley table of the group and we may represent the elements of the group by means of the natural numbers $1, 2, \ldots, n$. Moreover, as proved in Section 1.1, a matrix $\|a_{ik}\|$ whose entries are the natural numbers $1, 2, \ldots, n$ will represent the Cayley table of such a group if and only if (1) it is a latin square and (2) the quadrangle criterion holds, that is, for all subscripts $i, j, \ldots$, the equalities

$$a_{ik} = a_{i_1 k_1}, \quad a_{il} = a_{i_1 l_1}, \quad a_{jk} = a_{j_1 k_1} \quad \text{imply} \quad a_{jl} = a_{j_1 l_1}.$$

Further, we may choose an arbitrary row and column of the unbordered Cayley table, say the j-th row and l-th column, and consider them as being the products of the elements by the group identity from the left and from the right respectively by bordering the table with this row and column. Then a_{jl} will be the identity of the system G_{jl} thus arising, and necessarily $a_{il} a_{jk} = a_{ik}$. Now (1) ensures that G_{jl} is a loop and (2) implies associativity, thus G_{jl} is a group. Clearly, every group with the same Cayley table arises in this way.

All these G_{jl} are isomorphic, for a transition from G_{jl} to G_{rs} means simply that we take three permutations θ, ϕ, ψ such that $ab = c$ in G_{rs} if and only if $a\theta \cdot b\phi = c\psi$ in G_{jl}: that is, G_{jl} and G_{rs} are isotopic and hence, by Theorem 1.3.4, isomorphic groups.

Note that different multiplication tables of a group can be transformed into one another by row and column interchanges.

DEFINITION. If $\|a_{ik}\|$ and $\|b_{ik}\|$ are two matrices of the same dimensions, then we call the i-th rows *corresponding rows*, the k-th columns *corresponding columns*, and a_{ik} and b_{ik} *corresponding elements*. Also, we define the *distance* between $\|a_{ik}\|$ and $\|b_{ik}\|$, denoted by $d(\|a_{ik}\|, \|b_{ik}\|)$, to be equal to the number of cells in which the corresponding elements a_{ik} and b_{ik} are not equal.

The distance so defined is a generalization of the original notion of Hamming distance between vectors having binary components which is used in coding theory. See Section 11.4. The above notion of distance was introduced by Lee(1958).

Any permutation may be written as a product of disjoint cycles. If all these cycles have the same length, the permutation is called *regular*. If, for the permutations σ, τ we have $a\sigma \neq a\tau$ for exactly k symbols a, then we say that they differ in k places.

We shall need the following lemma concerning permutations.

LEMMA. *Two regular permutations which differ in exactly two places must be of the form* $(i_1 \, i_2 \, \cdots i_n)$ *and* $(i_1 \, i_2 \, \cdots \, i_{n/2})(i_{(n/2)+1} \, i_{(n/2)+2} \, \cdots \, i_n)$.

PROOF. Suppose that the permutations ρ and σ differ from each other by a transposition $\tau = (i_a \, i_b)$ so that $\sigma = \rho\tau$ and $\rho = \sigma\tau$. Without loss of generality we may assume that the number of cycles in ρ is no more than the number of cycles in σ. Then ρ necessarily has a cycle containing both i_a and i_b and this cycle gets split into two by the action of τ. It follows from the regularity that ρ must consist of a single cycle, say

$$(i_1 \ i_2 \ \dots \ i_a \ \dots \ i_b \ \dots \ i_n)$$

and hence that $\sigma = \rho\tau$ consists of two cycles

$$(i_1 \ i_2 \ \dots \ i_{a-1} \ i_b \ i_{b+1} \ \dots \ i_n)(i_a \ i_{a+1} \ \dots \ i_{b-1}).$$

Since σ is a regular permutation, $b - a = n - (b - a)$ so $b - a = n/2$ and this, after appropriate re-labelling, proves the lemma. □

THEOREM 3.3.1 *Two different Cayley tables, A and A', for the (not necessarily distinct) groups G and G' of order $n \notin \{4,6\}$, differ from each other in at least $2n$ places.*

[The authors are grateful to Frisch for pointing out the fact that Theorems 3.2.1 and 3.2.2 as stated in Dénes (1962) and in the first edition of this book were not correct and should be replaced by the one above.]

PROOF.   Suppose firstly that A and A' do not have any pair of corresponding rows equal. Then, each pair of corresponding rows differ in at least two places, so in this case A and A' differ in at least $2n$ places altogether. Thus, without loss of generality, we may suppose that the x-th row of A is equal to the x-th row of A' and, by a similar argument, that the y-th column of A is equal to the y-th column of A'. The rows of A (likewise those of A') are permutations of the x-th row of A (or A') and these permutations form regular permutation groups P and P' isomorphic to G and G' respectively. (This is a variant of the classical result of group theory known as Cayley's theorem.) Every element of $P \cap P'$ represents one of the equal rows in A and A'. It follows that the number s of rows that are the same in A and A' is equal to the order of the subgroup $H = P \cap P'$ and so either $s = n/2$ or $s \leq n/3$ (by Lagrange's theorem for groups).

Case 1. No pair of corresponding permutations in A and A' has the form described in the lemma: namely,

$$\rho = (i_1 \ i_2 \ \dots \ i_n) \quad \text{and} \quad \sigma = (i_1 \ i_2 \ \dots \ i_{n/2})(i_{(n/2)+1} \ i_{(n/2)+2} \ \dots \ i_n).$$

(This means that every pair of distinct corresponding permutations differ in at least three places.)

Case 1.1. $s \leq n/3$. (When n is odd, only this case can occur.)

In this case, $d(A, A') \geq 3(n - n/3) = 2n$ since there are at least $n - n/3$ pairs of corresponding rows each pair of which differ in at least three places.

Case 1.2. $s = n/2$. (This case occurs only when n is even.)

We shall show that two regular permutations ϕ and ψ (of degree n) which differ in exactly three places must have one of the two forms below. We use ξ to denote the cycles which ϕ and ψ have in common. Then, either,

(a) $\phi = (i_1\ i_2\ \cdots\ i_h\ i_{h+1}\ \cdots\ i_{h+j}\ i_{h+j+1}\ \cdots\ i_{h+j+k})\xi,$

$\quad \psi = \phi(i_1\ i_{h+1}\ i_{h+j+1})$

$\quad = (i_1\ i_2\ \cdots\ i_h\ i_{h+j+1}\ i_{h+j+2}\ \cdots\ i_{h+j+k}\ i_{h+1}\ i_{h+2}\ \cdots\ i_{h+j})\xi$

where $h, j, k \geq 1$, or

(b) $\phi = (i_1\ i_2\ \cdots\ i_{n/3}\ i_{(n/3)+1}\ \cdots\ i_{2n/3}\ i_{(2n/3)+1}\ \cdots\ i_n),$

$\quad \psi = \phi(i_1\ i_{(2n/3)+1}\ i_{(n/3)+1})$

$\quad = (i_1\ i_2\ \cdots\ i_{n/3})(i_{(n/3)+1}\ i_{(n/3)+2}\ \cdots\ i_{2n/3})(i_{(2n/3)+1}\ i_{(2n/3)+2}\ \cdots\ i_n).$

To see this, assume that $\psi = \phi\theta$ where θ consists of a single 3-cycle. We first observe that, if ϕ is regular and the three elements re-arranged by θ come from two or more different cycles of ϕ then ψ is not regular. If they come from three different cycles in ϕ then they lie in the same cycle of ψ and in this case we interchange the roles of ϕ and ψ. So henceforth we may suppose that they come from the same cycle. We can also suppose without loss of generality that the labelling of the symbols is chosen so that

$$\phi = (i_1\ i_2\ \cdots\ i_h\ i_{h+1}\ \cdots\ i_{h+j}\ i_{h+j+1}\ \cdots\ i_{h+j+k})\xi$$

and the three elements re-arranged by θ are i_1, i_{h+1} and i_{h+j+1}. There are two cases to consider; either $\theta = (i_1\ i_{h+1}\ i_{h+j+1})$ or $\theta = (i_1\ i_{h+j+1}\ i_{h+1})$. The first case leads directly to the form (a) above. In the second case,

$$\psi = (i_1\ i_2\ \cdots\ i_h)(i_{h+1}\ i_{h+2}\ \cdots\ i_{h+j})(i_{h+j+1}\ i_{h+j+2}\ \cdots\ i_{h+j+k})\xi,$$

which can only be regular if $h = j = k$. Then, because both ψ and ϕ are regular, we find that ξ must be trivial and this gives us the form (b).

Next, we shall show that, if $n > 6$, neither of the forms (a) or (b) is possible. Since, in both cases, $[P : H] = [P' : H] = 2$, both ϕ^2 and ψ^2 lie in H (since the square of every element of P or P' must lie in H if the index is 2). Hence also $\phi^2\psi^{-2} \in H$ and so is regular.

Now, if $n > 6$ and case (b) holds, we have $i_1\phi^2\psi^{-2} = i_3\psi^{-2} = i_1$ and, because P and P' consist of regular permutations, we must therefore have $\phi^2\psi^{-2}$ equal to the identity. But $i_{n/3}\phi^2\psi^{-2} = i_{(n/3)+2}\psi^{-2} = i_{2n/3} \neq i_{n/3}$ which is a contradiction. (If $n = 6$, $i_1\phi^2\psi^{-2} = i_3\psi^{-2} = i_3$ so $\phi^2\psi^{-2}$ is not equal to the identity and no contradiction arises.)

If case (a) holds, ϕ and ψ must consist of only one cycle since otherwise $\phi^2\psi^{-2} \in H$ would fix some symbols and, being regular, would then be equal to the identity whereas $i_h\phi^2\psi^{-2} = i_{h+2}\psi^{-2} = i_{h+j+k}$. If $n > 6$, then one of h, j, k is ≥ 3. Let us suppose that $h \geq 3$. Then $i_1\phi^2\psi^{-2} = i_3\psi^{-2} = i_1$ whereas $i_h\phi^2\psi^{-2} = i_{h+2}\psi^{-2} = i_{h+j+k} \neq i_h$ which is contradictory. (Again, if $n = 6$, no contradiction arises.) If $j \geq 3$, $i_{h+1}\phi^2\psi^{-2} = i_{h+1}$ whereas $i_{h+j}\phi^2\psi^{-2} = i_h$. If $k \geq 3$, $i_{h+j+1}\phi^2\psi^{-2} = i_{h+j+1}$ whereas $i_{h+j+k}\phi^2\psi^{-2} = i_{h+j}$.

Thus, when $n > 6$ and Case 1.2 holds, two different corresponding permutations must differ from each other in at least 4 places and so $d(A, A') \geq 4(n - n/2) = 2n$.

When $n = 6$ and Case 1.2 holds, it is shown in Note 1 below that a minimum distance of 9 is possible.

Case 2. There exist two corresponding permutations $\rho = (i_1 \ i_2 \ \ldots \ i_n) \in P$ and $\sigma = (i_1 \ i_2 \ \ldots \ i_{n/2})(i_{(n/2)+1} \ i_{(n/2)+2} \ \ldots \ i_n) \in P'$ as in the lemma. (This case occurs only when n is even.)

Then $P = \langle \rho \rangle$ and $[P' : \langle \sigma \rangle] = 2$. Also, using ϵ to denote the identity element of H, we have $P \cap \langle \sigma \rangle = \langle \epsilon \rangle$ because $i_{(n/2)-1}\rho^k = i_{(n/2)+k-1}$ and so each cycle of a non-trivial power of ρ contains elements with suffices $\leq n/2$ and $> n/2$. Consequently, $H \cap \langle \sigma \rangle = \langle \epsilon \rangle$ and so all products of the form $\sigma^k\gamma$ (where $\gamma \in H$ and $0 \leq k < n/2$) are different. These products all lie in P'; whence $(n/2)|H| \leq |P'| = n$. Therefore, $|H| \leq 2$. With the exception of the pairs (ρ, σ) and (ρ^{-1}, σ^{-1}), there is no other pair of permutations $\rho^k = (j_1 \ j_2 \ \ldots \ j_n) \in \langle \rho \rangle = P$ and $\tau = (j_1 \ j_2 \ \ldots \ j_{n/2})(j_{(n/2)+1} \ j_{(n/2)+2} \ \ldots \ j_n) \in P'$. For suppose there were such a pair (ρ^k, τ). Then each cycle of τ would contain elements i_u, i_v with $u \leq n/2$ and $v > n/2$ and consequently the non-identity elements of $\langle \tau \rangle$ would all be distinct from the non-identity elements of $\langle \sigma \rangle$. Because $\sigma, \tau \in P'$, the order of P' would be at least $(n/2)^2$. Since ord $P' \leq n$, this cannot happen except when $n = 4$.

Also, it is clear that it is only possible for there to be another pair of permutations $\theta = (j_1 \ j_2 \ \ldots \ j_{n/2})(j_{(n/2)+1} \ j_{(n/2)+2} \ \ldots \ j_n) \in P$ and $\eta = (j_1 \ j_2 \ \ldots \ j_n) \in P'$ if both A and A' are cyclic (because $(i_1 \ i_2 \ \ldots \ i_n) \in P$ implies that P, and so A, is cyclic and similarly $(j_1 \ j_2 \ \ldots \ j_n) \in P'$ implies that P', and so A', is cyclic). When such a pair of permutations θ, η, does exist, at most four pairs of corresponding rows of A and A' have distance two: namely, those corresponding to the pairs (ρ, σ), (ρ^{-1}, σ^{-1}), (θ, η), and (θ^{-1}, η^{-1}). In all other cases, at most two pairs of corresponding rows of A and A' have distance two.

Thus, remembering that $|H| \leq 2$, we see that at most two corresponding pairs of rows are the same and at most four pairs of corresponding rows differ in only two places and so $d(A, A') \geq 3(n - 6) + 4.2 = 3n - 10$ if both groups are cyclic and $d(A, A') \geq 3(n - 4) + 2.2 = 3n - 8$ otherwise. We note that $3n - 10 \geq 2n$ when $n \geq 10$ and that $3n - 8 \geq 2n$ when $n \geq 8$. However, one can check that two different tables of the cyclic group $\mathbb{Z}_8$ can have at most two pairs of corresponding rows which are at distance two. (See Note 3 below.) Consequently, $d(A, A') \geq 2n$ for all $n \geq 8$.

When $n = 6$, the minimal distance $d(A, A') = 3n - 10 = 8$ and this distance can be achieved as the following example shows (but only for two tables of $\mathbb{Z}_6$ as is shown in Note 2 below):

1	2	3	4	5	6
2	1	4	3	6	5
3	4	6	5	1	2
4	3	5	6	2	1
5	6	1	2	4	3
6	5	2	1	3	4

1	2	3	4	5	6
2	1	4	3	6	5
3	4	5	6	1	2
4	3	6	5	2	1
5	6	1	2	3	4
6	5	2	1	4	3

When $n = 4$, any pair of group tables with one pair of corresponding rows and one pair of corresponding columns equal (in all other cases, the tables differ in at least 8 places) can be transformed to standard form using the isotopy operation without changing their distance apart.

The four tables in standard form are as follows:

1	2	3	4
2	1	4	3
3	4	1	2
4	3	2	1

1	2	3	4
2	1	4	3
3	4	2	1
4	3	1	2

1	2	3	4
2	3	4	1
3	4	1	2
4	1	2	3

1	2	3	4
2	4	1	3
3	1	4	2
4	3	2	1

Of these, the first is the multiplication table of $\mathbb{Z}_2 \times \mathbb{Z}_2$ and the other three are multiplication tables of $\mathbb{Z}_4$. The tables of the two different groups differ in four places and the different tables of $\mathbb{Z}_4$ differ in seven places, so $d(A, A') < 2n = 8$. This completes the proof of the theorem. $\qquad\square$

REMARK. In Drápal(2004), that author has given an alternative proof of this theorem and has dedicated it to the memory of József Dénes. Drápal's proof makes use of work on the Hamming distance between group tables developed by himself and others: for example, in Drápal(1992,2001) and Donovan, Oates-Williams and Praeger(1997). There is also some connection with Problem 1.1 of [DK1].

In her Ph.D. thesis "Lateinische Quadrate" (Vienna, 1988), Frisch has made the following additional comments [see also Frisch(1997]:

Note 1. Distance 9 between two Cayley tables of the cyclic group $\mathbb{Z}_6$ can be attained as shown:

1	2	3	4	5	6
2	3	1	5	6	4
3	1	2	6	4	5
4	5	6	2	3	1
5	6	4	3	1	2
6	4	5	1	2	3

1	2	3	4	5	6
2	3	1	5	6	4
3	1	2	6	4	5
4	5	6	1	2	3
5	6	4	2	3	1
6	4	5	3	1	2

In this example, both $(163524) \in P$, $(162435) \in P'$ and $(142536) \in P$, $(14)(25)(36) \in P'$ are pairs of corresponding permutations so both of the forms (a) and (b) of Case 1.2 occur.

However, Case 1.2 with $d(A, A') < 12$ can only occur between two different tables of the cyclic group $\mathbb{Z}_6$. If $G \cong D_3$, $G' \cong \mathbb{Z}_6$ and $|H| = 3$ (as required by Case 1.2) so that $P = \langle \epsilon, (1\ 2\ 3)(4\ 5\ 6), (1\ 3\ 2)(4\ 6\ 5), (1\ 4)(2\ 6)(3\ 5), (1\ 5)(2\ 4)(3\ 6),$ $(1\ 6)(2\ 5)(3\ 4) \rangle$, then the only possibilities for P' are $\langle (1\ 4\ 2\ 5\ 3\ 6) \rangle$, $\langle (1\ 5\ 2\ 6\ 3\ 4) \rangle$ and $\langle (1\ 6\ 2\ 4\ 3\ 5) \rangle$. One can check that, in each of these cases, every pair of corresponding permutations outside H differ in at least 4 places.

Note 2. Case 2 for $n = 6$ with $d(A, A') < 12$ can only occur between two different tables of the cyclic group $\mathbb{Z}_6$. For, suppose, if possible, that Case 2 holds with $G \cong \mathbb{Z}_6$, $G' \cong D_3$. Then $|H| \leq 2$ and we may suppose without loss of generality that $\mathbb{Z}_6 = \langle (1\ 2\ 3\ 4\ 5\ 6) \rangle$ and that $(1\ 2\ 3)(4\ 5\ 6) \in D_3$. In that case, D_3 cannot contain $(1\ 4)(2\ 5)(3\ 6)$ because $(1\ 2\ 3)(4\ 5\ 6) \cdot (1\ 4)(2\ 5)(3\ 6) = (1\ 5\ 3\ 4\ 2\ 6) \notin D_3$. Therefore, $|H| = 1$ and so $d(A, A') \leq 3(n - 3) + 2 \cdot 2 = 13$.

Note 3. If, in two different tables for $\mathbb{Z}_8$, $(1\ 2\ 3\ 4\ 5\ 6\ 7\ 8) \in P$ and $(1\ 2\ 3\ 4)(5\ 6\ 7\ 8) \in P'$ form a pair of corresponding permutations which differ in two places, then each of the possible cycles of length 8 which can generate P' contains at most two integers in a sequence of consecutive even integers and at most two integers in a sequence of consecutive odd integers. (An example is the cycle $(1\ 6\ 2\ 7\ 3\ 8\ 4\ 5)$.) However, the elements of order 4 in P are $(1\ 3\ 5\ 7)(2\ 4\ 6\ 8)$ and $(1\ 7\ 5\ 3)(2\ 8\ 6\ 4)$. It follows from the lemma that any 8-cycle which differs from one of these in only two places must contain sequences of more than two consecutive even integers and more than two consecutive odd integers.

To summarize, if A, A' are distinct Cayley tables of groups of order n, we have $d(A, A') \geq 2n$ in all cases except the following:

$d(A, A') \geq 4$ if the tables are those of the distinct groups $\mathbb{Z}_2 \times \mathbb{Z}_2$ and $\mathbb{Z}_4$;

$d(A, A') \geq 7$ if the tables are both tables of the group $\mathbb{Z}_4$;

$d(A, A') \geq 8$ if the tables are both tables of the group $\mathbb{Z}_6$ and Case 2 holds;

$d(A, A') \geq 9$ if the tables are both tables of the group $\mathbb{Z}_6$ and Case 1.2 holds.

THEOREM 3.3.2 *For a group of order n ($n \neq 4$ or 6), we have $k(n) = 2n - 1$ [where $k(n)$ is defined as at the beginning of this Section].*

PROOF. Let us delete $2n - 1$ arbitrary elements in a Cayley table A of the group G of order n ($n \neq 4$ or 6). Suppose that there is a Cayley table $A' \neq A$ of G having the property that the rest of A may be completed to A'. Then, clearly, A and A' differ in no more than $2n - 1$ places, which is impossible, by Theorem 3.3.1.

We have to prove further that we can delete $2n$ elements of a Cayley table A of a group G of order n in such a way that the rest of the table may be completed to a Cayley table A' different from A. If we interchange two arbitrary symbols, a and b, throughout A, then we shall obtain a new Cayley table differing from A in exactly $2n$ places and which still satisfies the quadrangle criterion. So the proof of our statement is completed. □

THEOREM 3.3.3 *An arbitrary Cayley table of the cyclic group of order 4 differs in at least four places from an arbitrary Cayley table of Klein's 4-group.*

PROOF. This has already been shown as part of the revised proof of Theorem 3.3.1. □

The result of Theorem 3.3.1 remains the same if we restrict the class of groups to any one of the following classes of finite groups: (i) soluble groups; (ii) nilpotent groups; (iii) abelian groups; (iv) cyclic groups.

It seems to be natural to raise the following problem:

What is the maximum number of squares which a set of latin squares satisfying the quadrangle criterion and all of the same order n can contain if each pair of squares in the set are to differ from each other in at most m places?

At the beginning of this section, we stated the problem of Fuchs as a problem concerning groups. The analogous result for quasigroups, which was first proved by Dénes and Pásztor(1963), may be deduced as a corollary to the theorem which follows. For this, we need to remember the fact that the multiplication table of an arbitrary quasigroup is a latin square (proved in Theorem 1.1.1).

THEOREM 3.3.4 *For $n = 2$ and $n \geq 4$ there exist two latin squares of order n which differ in precisely four entries.*

PROOF.   Given two different symbols a and b there are two possible latin squares that can be formed using those symbols, namely,

$$\begin{array}{cc} a & b \\ b & a \end{array} \quad \text{and} \quad \begin{array}{cc} b & a \\ a & b \end{array}.$$

These two squares are clearly at distance four from each other, settling the case $n = 2$. Moreover, from Theorem 1.6.3 it follows that for all $n \geq 4$ there exists a latin square L_n of order n having a latin subsquare U_n of order 2. Let us form L'_n from L_n by replacing U_n by the other possible subsquare on the same two symbols. Then $d(L_n, L'_n) = 4$ as desired.   □

For an analogous result concerning semigroups, see Dénes(1967a).

As regards problem (a) of Fuchs, Frisch(1997) has proved that the Cayley tables of non-isomorphic groups always differ in strictly more than $2n$ places.

3.4   Incomplete latin squares and partial quasigroups

A latin rectangle of size $r \times s$ is called *incomplete* or *partial* if less than rs of its cells are occupied. An *incomplete* or *partial* latin square is analogously defined. Precisely, we have:

DEFINITION. An $n \times n$ *incomplete* or *partial* latin square defined on a set S of n symbols is an $n \times n$ array such that in some subset of the n^2 cells of the array each of the cells is occupied by an element from the set S and such that no element from the set S occurs more than once in any row or column.

In Section 3.2 and Section 3.3, we investigated incomplete latin squares whose elements were not arbitrarily assigned but had been obtained by deleting elements from a given latin square. The example of an incomplete latin square shown in Figure 3.4.2 will make the distinction clear.

In this section we shall consider incomplete latin squares and rectangles more generally and investigate the question of when and whether they can be completed to a latin square or latin rectangle.

$$
\begin{array}{cccc}
1 & \cdot & \cdot & \cdot \\
\cdot & 2 & 3 & 4
\end{array}
$$

FIG. 3.4.1.

Figure 3.4.1 shows an incomplete latin rectangle of size 2×4 which cannot be completed to a latin rectangle of the same size. By generalizing, it is easy to see that, for any $n \geq 2$, there exists an incomplete latin rectangle with $2n$ cells occupied which cannot be completed to an $n \times 2n$ latin rectangle[7]. On the other hand, Lindner(1970a) proved that an incomplete $n \times 2n$ latin rectangle with $2n - 1$ cells occupied can always be completed to an $n \times 2n$ latin rectangle.

In T.Evans(1960), the following problem was posed:
"What conditions suffice to enable an incomplete $n \times n$ latin square to be embedded in a fully completed[8] latin square of order n? In particular, can an $n \times n$ incomplete latin square which has $n - 1$ or less places occupied be completed to a latin square of order n?"

Exactly the same problem was posed by Klarner on page 1167 of Erdös, Rényi and Sös(1970) and independently by Dénes in a lecture given in the late 1960s at the University of Surrey.

It is easy to see that there do exist incomplete latin squares with n cells occupied which cannot be so completed, as Figure 3.4.1 and Figure 3.4.2 illustrate.

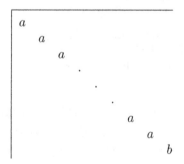

FIG. 3.4.2.

The general case of Evans' problem had not been solved when the first edition ([DK1]) of this book was published but it is now known that his conjecture that any $n \times n$ incomplete latin square which has $n - 1$ or less places occupied can be completed to a latin square of order n is true. Two proofs of this fact are

---

[7]An example is the rectangle whose only filled cells are (1,1), (2,2), ... ,(n,n) and $(n, n+r)$ for $r = 1, 2, \ldots n$ and in which these cells have distinct entries.

[8]The temptation is to use the adjective "complete" to emphasize that a latin square has all of its entries filled in. This is best avoided because the term "complete latin square" is widely used with a different meaning as described in Section 2.6.

discussed in detail in Chapter 8, Section 10, of [DK2]. We merely remark here that the simplest proof is that of Smetaniuk(1981).

In [DK1], the next pages of this section were devoted to the discussion of partial solutions to Evans' conjecture and to related embedding problems. Since a much more up-to-date account of this topic is available in [DK2], we refer the reader to that book for more recent results. However, some of the earlier results are not mentioned in [DK2].

Firstly, Marica and Schönheim(1969) proved that an incomplete latin square containing $n - 1$ arbitrarily chosen elements can be completed to a latin square of order n provided that the chosen elements are in different rows and columns.

Secondly, in Lindner(1970b), that author (using the same technique as Marica and Schönheim) proved that an incomplete latin square containing $n - 1$ *distinct* elements can be completed to a latin square of order n provided that the chosen elements are either in different rows or in different columns.

Thirdly, Lindner(1970a) also proved the following theorem: Let L be an $n \times n$ incomplete latin square with $n - 1$ cells occupied. Let r and s denote respectively the number of rows and the number of columns in which occupied cells occur. Then, if $r \leq \lfloor n/2 \rfloor$ or $c \leq \lfloor n/2 \rfloor$, L can be completed to a latin square of order n.

In the same connection, one can ask for the solution of the following problem: "How many elements of a latin square of order n and which satisfies the quadrangle criterion can be located arbitrarily subject only to the condition that no row or column shall contain any element more than once?" Since a latin square of odd order which satisfies the quadrangle criterion cannot contain a latin subsquare of order two (see the Corollary to Theorem 1.6.4), it is easy to see that, for n odd, this number is at most three. Thus, for example, no completion of the partial latin square shown in Figure 3.4.3 can satisfy the quadrangle criterion, since it is of odd order.

$$
\begin{array}{cccccc}
\cdot & \cdot & \cdot & \cdot & \cdot & \\
\cdot & \cdot & 1 & 2 & \cdot & \\
\cdot & \cdot & \cdot & \cdot & \cdot & \\
\cdot & \cdot & 2 & 1 & \cdot & \\
\cdot & \cdot & \cdot & \cdot & \cdot &
\end{array}
$$

FIG. 3.4.3.

Moreover, not even *two* arbitrary elements can be arbitrarily placed in a symmetric latin square of odd order: for Theorem 1.5.4 implies that no element can appear more than once in the main diagonal of such a square.

Csima(1972) has translated Evans' problem of finding under what conditions a partial latin square can be embedded in a fully completed latin square into a problem about a combinatorial structure which he has called a *pattern*. For details, the reader is referred to Csima's paper.

In his paper T.Evans(1960), already referred to, that author posed the following further question: "Can a pair of $n \times n$ incomplete latin squares which are orthogonal (insofar as the condition for orthogonality applies to the incomplete squares) be respectively embedded in a pair of $t \times t$ orthogonal latin squares; and, if so, what is the smallest value of t for each value of n?" The first part of this question has been answered in the affirmative by Lindner(1976) but the smallest value of t for a given value of n remains unknown so far as the present author is aware.

Somewhat related to this question is an interesting result proved by Lindner(1971b): namely, *any finite collection of mutually orthogonal $n \times n$ partial latin squares can be embedded in a complete set of mutually orthogonal infinite latin squares.*

By an *infinite latin square* we mean a countably infinite array of rows and columns such that each positive integer occurs exactly once in each row and exactly once in each column. If P is a finite (partial) latin square, we shall denote by C_p the set of all the cells which are occupied in P. If P and Q are finite (partial) latin squares of the same size, we say that P and Q are *orthogonal* if the cardinalities of the sets $C_p \cap C_q$ and $\{(p_{ij}, q_{ij}) : (i,j) \in C_p \cap C_q\}$ are equal. If P and Q are infinite latin squares, we say that P and Q are orthogonal provided that the set $\{(p_{ij}, q_{ij}) : (i,j) \in C_p \cap C_q\}$ has the same cardinality as the Cartesian product $\mathbb{Z}^+ \times \mathbb{Z}^+$ (where $\mathbb{Z}^+$ denotes the set of positive integers) and that each pair of cells in different rows and columns is occupied by the same symbol in at most one of P and Q. If $\{P_i\}_{i \in I}$ is a collection of mutually orthogonal latin squares of the same size, we say that this collection is a *complete set* of mutually orthogonal latin squares provided that each pair of cells in different rows and columns is occupied by the same symbol in exactly one member of the collection. If the latin squares in the collection are finite of order n, the set I has cardinality $n - 1$, while, if they are infinite, I has cardinality equal to that of the set $\mathbb{Z}^+$.

The reader new to the subject is advised that he will find the above concepts easier to understand if he first reads Section 5.1 and Section 5.2.

For the proof of his result stated above, Lindner used the geometrical concepts of *projective plane* and *partial projective plane*. (The connection between orthogonal latin squares and projective planes is explained in Section 5.2.)

DEFINITION. An $n \times n$ *partial idempotent latin square* is an $n \times n$ partial latin square with the additional property that, for each $i = 1, 2, \ldots, n$, the cell (i, i) is either empty or else is occupied by the integer i. An $n \times n$ *partial symmetric latin square* is an $n \times n$ partial latin square with the additional property that, if the cell (i, j) is occupied by the symbol k so also is the cell (j, i).

In a series of papers, Lindner solved a number of embedding problems concerning square of these kinds. In particular,

(1) Any finite partial idempotent latin square can be embedded in a finite idempotent latin square [Lindner(1971a)].

(2) Any finite partial symmetric latin square can be embedded in a finite symmetric latin square [Lindner(1972a,1976)].

(3) Any finite partial idempotent and symmetric latin square can be embedded in a finite idempotent and symmetric latin square [Lindner(1972a)].

Subsequently, these results were made more precise and these and other results were proved by more elegant methods by Cruse, Hilton, Lindner and others. A more up-to-date account and more details are given in Sections 4 and 5 of [DK2], chapter 8. Also, in Sections 6 and 7 of that chapter, an account of embedding theorems for partial quasigroups is given.

By a *partial quasigroup* is meant a half-groupoid G such that, if the equations $ax = b$ and/or $ya = b$ have solutions for x and y in G, then these solutions are unique. A *partial loop* is a partial quasigroup with an identity element such that the product of this element with each element a of the loop is defined and is equal to a. [By other authors, partial quasigroups have been called *incomplete quasigroups* or *half-quasigroups*, see Bruck(1958) and T.Evans(1960).]

In Brandt(1927), that author introduced a special kind of partial groupoid B (which has subsequently been called a *Brandt groupoid*) satisfying the following postulates:

(1) If any three elements a, b, c satisfy an equation $ab = c$ then each of the three elements is uniquely determined by the other two.

(2) If ab and bc both exist, then $(ab)c$ and $a(bc)$ also exist; if ab and $(ab)c$ both exist, then bc and $a(bc)$ also exist; if bc and $a(bc)$ both exist, then ab and $(ab)c$ also exist; and, in all of these cases, the equation $(ab)c = a(bc)$ is valid and consequently the expression abc has an unambiguous meaning.

(3) To any given element a there corresponds a unique right identity element e such that $ae = a$, a unique left identity element e' such that $e'a = a$ and a left inverse $\bar{a}$ of a such that $\bar{a}a = e$.

(4) If e and e' are any two members of the set of one-sided identity elements, there exists an element a such that e and e' are respectively right and left identities with respect to a.

Postulate (1) implies that the multiplication table of a Brandt groupoid is an incomplete latin square and we shall devote the remainder of this section to mentioning some results and conjectures concerning such multiplication tables.

THEOREM 3.4.1 *The multiplication table of a Brandt groupoid is an incomplete latin square satisfying the quadrangle criterion.*

PROOF. As we have just remarked, postulate (1) implies that the multiplication table of a Brandt groupoid is an incomplete latin square and it then follows immediately from postulate (2) that the multiplication table of a Brandt groupoid satisfies the quadrangle criterion. □

We should like to point out by means of an example that incomplete latin squares exist which satisfy the quadrangle criterion but which can be completed to latin squares in which the quadrangle criterion does not hold. The incomplete latin square exhibited in the left-hand square of Figure 3.4.4 has these properties, as is indicated by the right-hand square.

$$
\begin{array}{ccccc}
3 & 4 & 2 & 1 & \bullet \\
1 & 2 & \bullet & 3 & 4 \\
\bullet & \bullet & 1 & 2 & \bullet \\
\bullet & \bullet & 3 & 4 & \bullet \\
\bullet & \bullet & \bullet & \bullet & \bullet
\end{array}
\qquad
\begin{array}{ccccc}
3 & 4 & 2 & 1 & 5 \\
1 & 2 & 5 & 3 & 4 \\
4 & 5 & 1 & 2 & 3 \\
5 & 1 & 3 & 4 & 2 \\
2 & 3 & 4 & 5 & 1
\end{array}
$$

FIG. 3.4.4.

On the other hand, the authors conjecture that every Brandt groupoid can be embedded in a quasigroup Q which is group isotopic. The conjecture is certainly valid, for example, for the Brandt groupoid $(B, \circ)$ given as follows.

The elements of $(B, \circ)$ are all binary triplets and we shall denote them as follows:

$$
\begin{array}{llll}
1 = (000) & 2 = (100) & 3 = (010) & 4 = (110) \\
5 = (001) & 6 = (011) & 7 = (101) & 8 = (111)
\end{array}
$$

If $(H, +)$ denotes the cyclic group of order two with elements 0, 1, the product of two triplets $(a\ b\ c)$ and $(d\ e\ f)$ is defined to be the triplet $(a\ b + e\ f)$ if $c = d$ and is not defined otherwise.

Then Figure 3.4.5 shows that the multiplication table of $(B, \circ)$ can be embedded in the multiplication table of a quasigroup Q which is isotopic to the generalized Klein group of order 8. (In the diagram, the products which are defined in $(B, \circ)$ are enclosed in squares.)

$(\circ)$	1	2	3	4	5	6	7	8
1	1	7	3	8	5	6	4	2
2	2	5	4	6	7	8	3	1
3	3	8	1	7	6	5	2	4
4	4	6	2	5	8	7	1	3
5	7	1	8	3	4	2	5	6
6	8	3	7	1	2	4	6	5
7	5	2	6	4	3	1	7	8
8	6	4	5	2	1	3	8	7

FIG. 3.4.5.

Let us remark that it is well-known that an arbitrary Brandt groupoid B can be embedded in a semigroup S with one additional element 0 such that $a0 = 0a = 00 = 0$ for all $a \in B$ and such that $ab = 0$ in S if a, b are in B and ab is not defined in B. [See, for example, page 35 of Bruck(1958).] This fact leads the authors of this book to propose the alternative conjecture that in fact any Brandt groupoid can be embedded in a group.

3.5   Partial transversals and generalized transversals

As has been shown earlier, many latin squares do not have any transversals. For example, see Theorem 2.5.1 and Theorem 2.5.4. This fact led Koksma(1969) to ask "What is the largest number of cells of an arbitrarily chosen latin square (or rectangle) of given order which can be contained in a single partial transversal?"

DEFINITION. By a *partial transversal* of a latin square or rectangle is meant a collection of cells taken from different rows and different columns and containing different entries. The number of cells in the collection is called the *length* of the partial transversal.

Koksma proved that an arbitrary latin square of order n ($n \geq 7$) has at least one partial transversal of length $t \geq (2n+1)/3$. We outlined his proof in [DK1]. However, this lower bound has subsequently been much improved. An account of more recent work on the question up to 1990 has been given in [DK2] so we shall not repeat it here. However, so far as the present author is aware, the best lower bounds for t so far obtained are those in Hatami and Shor(2008) and Fu, Lim and Fu(2002) which both give bounds of the form $t \geq n - O(\log n)^2$.

Of course, neither of these bounds is anywhere near the bound $t \geq n - 1$ conjectured by Brualdi (see Section 1.5) and independently by Stein(1975) nor to the bound $t = n$ for latin squares of odd order conjectured by Ryser. However, in Wanless and Webb(2006), evidence is offered to suggest that the latter conjecture may be false: namely, these authors have proved that, for every order $n > 3$, there exists a latin square which contains a cell that is not included in any transversal. [In fact, stronger results have been obtained. See Egan and Wanless(2012) or page 408 of Wanless(2011).] For even order, this result was shown by the corollary to Theorem 2.5.4. It follows that, for every order $n > 3$, there exist latin squares with no orthogonal mate. [See Section 5.1 for the definition of this concept. Such squares were called *bachelor squares* by Van Rees(1990).] Because the proof (for squares of odd order) is short and elegant, we give it here.

Before doing so, it is useful to have the following definition.

DEFINITION. In a latin square of order n, a collection of n cells, one from each row and one from each column is called a *chain*. The entries in the cells of a chain are its *elements* and the number of these which are distinct is the *rank* of the chain.[9]

The number of chains of rank i in a particular latin square L is denoted by t_i.

---

[9] In Wanless(2011), that author has called this concept a *diagonal* but that conflicts with the normal usage of that term as, for example, in *diagonal latin squares* which we discuss in Section 6.1 so we shall follow Belyavskaya (see [DK2], page 177) and call the concept a chain.

LEMMA. *Let L be a latin square of order n with rows and columns indexed by the set $\mathbf{Z}_n^+ = \{0, 1, \ldots, n-1\}$ and with symbols also from $\mathbf{Z}_n^+$. Let s denote the symbol in row r and column c and let T be a transversal of L. Then $\sum_{(r,c,s)\in T}(r+c-s) \equiv 0$ or $n/2$ mod n according as n is odd or even.*[10]

THEOREM 3.5.1 *For every odd order $n = 2m+1$, there exists a latin square which contains a cell that is not contained in any transversal.*

PROOF. By definition of a transversal, each symbol of $\mathbf{Z}_n^+$ occurs once in each of $\sum r$, $\sum c$ and $\sum s$ and so

$$\sum_{(r,c,s)\in T}(r+c-s) = (n-1)n/2$$

$= m(2m+1) \equiv 0$ if $n = 2m+1$ or $= (2m-1)m \equiv n/2$ if $n = 2m$. □

Let us now consider the latin square $L = (l_{ij})$ which is a slight modification of the cyclic latin square of order $n = 2m+1$ and is defined as follows:

$$l_{ij} = \begin{cases} 1 & \text{if } (i,j) = (0,0) \text{ or } (1, n-1) \\ 0 & \text{if } (i,j) = (1,0) \text{ or } (2, n-1) \\ j+2 & \text{if } i = 0 \text{ and } j = \{1,3,5,\ldots,n-2\} \\ j & \text{if } i = 2 \text{ and } j = \{1,3,5,\ldots,n-2\} \\ i+j & \text{otherwise} \end{cases}$$

Then the entry 0 in the cell $(1,0)$ is not in any transversal because $r+c-s = 1$ for this cell while $r+c-s = -2$ for each of the modified cells in row 0 except the cell $(0,0)$ and $r+c-s = 2$ for each of the modified cells in row 2 while $r+c-s = 0$ for all other cells and so $\sum(r+c-s) = 0$ is impossible if the summation is over the cells of a chain. □

We exhibit this modified cyclic square in the left hand square of Figure 3.5.1 for the case $n = 7$.

1	3	2	5	4	0	6
0	2	3	4	5	6	1
2	1	4	3	6	5	0
3	4	5	6	0	1	2
4	5	6	0	1	2	3
5	6	0	1	2	3	4
6	0	1	2	3	4	5

0	**2**	**3**	**4**	5	6	1
5	**3**	**4**	**2**	6	1	0
1	**4**	**2**	**3**	0	5	6
2	5	6	1	3	0	4
4	0	5	6	1	3	2
3	6	1	0	2	4	5
6	1	0	5	4	2	3

FIG. 3.5.1.

In fact, four different proofs of the above result exist. The above proof is taken from Egan(2011). Alternative proofs are in A.B.Evans(2006), Wanless and Webb(2006), Egan and Wanless(2012).

[10] This has been called the *Delta lemma* by Wanless.

If the entries in the cells of a chain are all different, we have a transversal. If not, the cells whose entries are all different form a partial transversal as defined above.

It has been shown in Cameron and Wanless(2005) that every latin square possesses a chain in which no symbol occurs more than twice.

Another question which we may ask is "What is the shortest length of a non-extendible partial transversal?" (The longest length of such a partial transversal is covered by the conjectures of Brualdi-Stein and Ryser.)

The shortest length is not less than $n/2$ because otherwise there would not be sufficient symbols used in the partial transversal to fill the subsquare formed by the rows and columns not containing cells of the partial transversal. On the other hand, for all $n > 4$, non-extendible partial transversals of length $\lceil n/2 \rceil$ do exist since latin squares of order n which contain a subsquare S of order $\lfloor n/2 \rfloor$ and a partial transversal containing the symbols of S but not involving any of the rows and columns of S can easily be constructed. See the right hand square of Figure 3.5.1 for an example. These observations are in Wanless(2011).

In contrast to this result, Cavenagh, Hämäläinen and Nelson(2009) have proved that, for any prime p, every partial transversal of length 3 in the Cayley table of the cyclic group of order p can be extended to a transversal of length p.

In [DK2], we discussed several generalizations of the concept of a transversal. Of these, the one which has been investigated most fully in recent years is the k-plex.

DEFINITION. A k-plex in a latin square of order n is a set of nk cells, k from each row, k from each column and including k occurrences of each symbol. When convenient, we shall call such a k-plex a plex of order k.

Such objects for the case $k = 2$ were first studied by Finney(1945) and Freeman(1985) who called them *duplexes* (from which epithet came the name k-plex when the concept was generalized). Duplexes have statistical applications because, for example, although the cells of a 6×6 cyclic latin square cannot be partitioned into six disjoint transversals, they can be partitioned into three disjoint duplexes.

A conjecture which has been attributed to Rodney [see Vaughan-Lee and Wanless(2003)] is that every latin square contains a duplex. It follows from the truth of the Hall-Paige conjecture that Rodney's conjecture is true for all group-based latin squares, as is shown in Vaughan-Lee and Wanless(2003). In Wanless(2002), that author has shown that, for latin squares based on soluble groups, a more general result is true: namely, that such squares have a $2c$-plex for each possible integer c. He has obtained several other related results in the same paper.

We note that the entries not in a k-plex of a latin square form an $(n-k)$-plex. More generally, it may be possible to partition the cells of a latin square L of order n into a set of plexes of orders $k_1, k_2, \ldots, k_s$, where $\sum_{i=1}^{s} k_i = n$. When this is possible, we shall say that L has a $(k_1, k_2, \ldots, k_s)$-plex partition. Of particular interest is the case when all the plexes have the same order k (necessarily a divisor of n). For example, the cyclic latin square $\mathbf{Z}_6$ of order 6 is the union of three duplexes as shown in the left-hand square of Figure 3.5.2. Thus, $\mathbf{Z}_6$ has a *duplex partition*. More generally, we shall call a $(k, k, \ldots, k)$-partition a *k-plex partition*. For example, the right-hand square of Figure 3.5.2, given in Wanless(2011), has a 3-plex partition. In particular, if L has a 1-plex partition, then it has an orthogonal mate. (See Section 5.1 for the latter concept.)

0_a	1_a	2_b	3_b	4	5
1	2_a	3_a	4_b	5_b	0
2	3	4_a	5_a	0_b	1_b
3_b	4	5	0_a	1_a	2_b
4_b	5_b	0	1	2_a	3_a
5_a	0_b	1_b	2	3	4_a

0_a	1	2	3_a	4	5_a
1_a	0	3	2_a	5_a	4
2	4_a	0	5	1_a	3_a
3	5	1_a	4	2_a	0_a
4_a	3_a	5_a	1	0	2
5	2_a	4_a	0_a	3	1

FIG. 3.5.2.

The union of an h-plex and an l-plex is an $(h + l)$-plex. However, it is not always possible to split an $(h + l)$-plex into an h-plex and an l-plex. If a k-plex contains no h-plex for $0 < h < k$, we say that it is *indivisible*. For example, since the cyclic latin square of order six has no transversals (that is, no 1-plexes), then clearly the duplexes shown in Figure 3.5.2 are indivisible.

It has been shown that, for every $k \geq 2$ and $m \geq 2$, there exists a latin square of order mk with an indivisible k-plex partition. Also, if $n = 2k + 1 \geq 5$, there is a latin square of order n with an indivisible $(k, k, 1)$-plex partition and an indivisible $(k, k + 1)$-partition. See Bryant, Egan, Maenhaut and Wanless(2009) and Egan and Wanless(2011) for the details.

As regards the question "For which integers k and n is there a latin square of order n which contains an indivisible k-plex?", it has been shown in the first of the two papers just cited that, if $5k \leq n$, such an indivisible k-plex always exists.

The reader will have observed that, in each of the first three chapters of this book, we have devoted whole sections to the topic of transversals and their generalizations. (We discuss their enumeration in Chapter 4 and they also play a prominent role in [DK2].) It would be easy to devote our whole book to this topic. Instead, we refer the reader to Chapters 2, 3 and 6 of [DK2] and to a recent survey paper by Wanless(2011) which we have already cited several times.

CHAPTER 4

Classification and enumeration of latin squares and latin rectangles

This chapter is devoted to the classification and enumeration of latin squares and latin rectangles. We begin by giving an introductory account of the notion of autotopism, which plays an important role in obtaining results on the subject of this chapter. Many of our results first appeared in Schönhardt(1930). However, the terminology of that paper is very antiquated and so we have reformulated the results using present day terms.

The second section of the chapter is on classification. It includes a table which classifies the reduced latin squares of all orders up to and including six into their main, isotopism and isomorphism classes. The third section, on enumeration, contains a historical account of the development of the subject of classification and enumeration over the last two centuries.

The fourth section of the chapter is devoted to the enumeration of latin rectangles, covering both exact enumerations and asymptotic formulae.

The final sections are on enumeration of transversals and subsquares of latin squares.

4.1 The autotopism group of a quasigroup

We begin by reminding the reader that the concepts of isotopism and autotopism for quasigroups were defined in Section 1.3. These concepts may be applied more generally to groupoids.

THEOREM 4.1.1 *The set of isotopisms of a groupoid of order n form a group I_n of order $(n!)^3$.*

PROOF. An isotopism of a groupoid, which latter we defined at the end of Section 1.1, can be characterized by three permutations of degree n exactly as for quasigroups. Since each of the permutations may be chosen arbitrarily from the $n!$ possibilities, the number of isotopisms is $(n!)^3$.

An operation on a pair of the isotopisms, which we call their product, can be defined by stating that the product of the isotopisms (α, β, γ) and $(\alpha', \beta', \gamma')$ is $(\alpha\alpha', \beta\beta', \gamma\gamma')$. It is obvious that the product so defined is again an isotopism. Moreover, it is easy to check that the remaining group axioms are valid. Consequently, the isotopisms of a groupoid of order n form a group I_n under this product operation. □

Latin Squares and their Applications. http://dx.doi.org/10.1016/B978-0-444-63555-6.50004-0

COROLLARY 1. $I_n \cong S_n \times S_n \times S_n$ where S_n denotes the symmetric group of degree n.

COROLLARY 2. The set of all isotopisms of a quasigroup of order n form a group of order $(n!)^3$.

DEFINITION. If (α, β, γ) is an isotopism of a groupoid Q onto a groupoid Q^* and if $Q^* = Q$ then the isotopism (α, β, γ) is called an *autotopism* of Q.

THEOREM 4.1.2 The set A_Q of all autotopisms of a groupoid Q of order n form a group which is a subgroup of I_n.

PROOF.   The identity element of I_n is clearly an autotopism of Q. Hence the autotopisms of Q are a non-empty subset of the finite group I_n. To show that they form a subgroup using the product operation inherited from I_n, it suffices to prove closure. Namely, we must establish that the product of two autotopisms will be an autotopism again, but this is obvious from the definition.   □

COROLLARY. The order of A_Q is a divisor of $(n!)^3$.

By analogy with the concept of principal isotopism (see Section 1.3) one can define a *principal autotopism* as an autotopism (α, β, γ) such that γ is the identity permutation.

It is an immediate consequence of Theorem 4.1.2 that the principal auto-topisms of a groupoid of order n form a group which is a subgroup of the group formed by the principal isotopisms contained in I_n.

DEFINITION. An autotopism (α, β, γ) is called an *automorphism* if $\alpha = \beta = \gamma$.

The following results follow almost immediately from this definition:

  (i) The group of automorphisms and the group of principal autotopisms are contained in the group of autotopisms.
  (ii) The group of principal autotopisms is a normal subgroup of the group of autotopisms.
  (iii) The group of principal autotopisms and the group of automorphisms have no common elements other than the identity autotopism.

THEOREM 4.1.3 If Q_1 and Q_2 are two isotopic groupoids then $A_{Q_1} \cong A_{Q_2}$.

PROOF.   Suppose that (α, β, γ) is an isotopism mapping Q_1 onto Q_2. It follows that if $(\rho, \sigma, \tau) \in A_{Q_1}$, then $(\rho, \sigma, \tau)(\alpha, \beta, \gamma) = (\rho\alpha, \sigma\beta, \tau\gamma)$ also maps Q_1 onto Q_2 and hence $(\alpha^{-1}, \beta^{-1}, \gamma^{-1})(\rho, \sigma, \tau)(\alpha, \beta, \gamma)$ is an autotopism of A_{Q_2}. It follows immediately that

$$(\alpha^{-1}, \beta^{-1}, \gamma^{-1})A_{Q_1}(\alpha, \beta, \gamma) \subseteq A_{Q_2}.$$

Since $(\alpha^{-1}, \beta^{-1}, \gamma^{-1}) = (\alpha, \beta, \gamma)^{-1}$ is an isotopism mapping Q_2 onto Q_1, a similar argument with the roles of Q_1 and Q_2 interchanged gives

$$(\alpha, \beta, \gamma)A_{Q_2}(\alpha^{-1}, \beta^{-1}, \gamma^{-1}) \subseteq A_{Q_1}$$

and so

$$A_{Q_2} \subseteq (\alpha^{-1}, \beta^{-1}, \gamma^{-1})A_{Q_1}(\alpha, \beta, \gamma).$$

The equality $A_{Q_2} = (\alpha^{-1}, \beta^{-1}, \gamma^{-1})A_{Q_1}(\alpha, \beta, \gamma)$ is an immediate consequence of the above statements and implies that $A_{Q_1} \cong A_{Q_2}$. $\square$

THEOREM 4.1.4 *Two components of an autotopism of a quasigroup determine the third one uniquely.*

PROOF. Let L be the latin square which represents the multiplication table of the quasigroup Q. Let $(\alpha_1, \beta_1, \gamma_1)$ and $(\alpha_2, \beta_2, \gamma_2)$ be autotopisms of Q. Then

$$(\alpha_1, \beta_1, \gamma_1)(\alpha_2, \beta_2, \gamma_2)^{-1} = (\alpha_1\alpha_2^{-1}, \beta_1\beta_2^{-1}, \gamma_1\gamma_2^{-1}) \qquad (4.1)$$

is also an autotopism of Q. The components of an autotopism rearrange the row border, the column border, and the elements of L respectively. It is clear that if one of these items is altered but not the other two, then the new multiplication table cannot represent Q. Thus, if two of the three components in (4.1) are the identity, then so is the third. That is, any two of the equalities $\alpha_1 = \alpha_2$, $\beta_1 = \beta_2$ and $\gamma_1 = \gamma_2$ implies the other one. $\square$

COROLLARY 1. *The order of the group of autotopisms of a quasigroup of order n cannot exceed $(n!)^2$.*

COROLLARY 2. *Any one of the non-identity components of a principal autotopism of a quasigroup determines the autotopism uniquely.*

COROLLARY 3. *The order of the group of principal autotopisms of a quasigroup of order n cannot exceed $n!$.*[1]

A detailed study of the autotopism groups of quasigroups has been made by Sade(1968b). He has shown, for example, that the upper bound given in Corollary 1 to Theorem 4.1.4 can be improved to $n(n!)$. This follows from the fact that each autotopism of a quasigroup Q can be represented as the product of a so-called fundamental autotopism [see page 6 of Sade(1968b)], of which there can be at most n^2, and an automorphism of a loop L isotopic to Q. Such a loop evidently has at most $(n-1)!$ distinct automorphisms and so the order of the automorphism group of Q, which is equal to that of L by Theorem 4.1.3, is at most $n^2(n-1)! = n(n!)$.

A number of further results on the autotopism groups of particular kinds of quasigroup will be found in Belousov(1967b). For an investigation of the properties of the autotopism group of a group, see also Sade(1967d).

---

[1]This upper bound has been improved to $n^{O(\log n)}$ in Browning, Stones and Wanless(2013). These authors have shown further that this bound is achieved infinitely often.

4.2   Classification of latin squares

For the purposes of enumeration it is desirable to separate the set Ω_n of all latin squares of order n into smaller classes. The equivalence relations induced by isomorphism and isotopy (defined in Section 1.3) and paratopy (defined in Section 1.4) provide the natural way to do this. As mentioned in Section 1.4, they separate Ω_n into isomorphism, isotopism and main classes respectively. The relation between these is hierarchical as summarized by our first theorem.

THEOREM 4.2.1   *For each n, the set Ω_n is the union of one or more disjoint main classes. Each main class is the union of one or more disjoint isotopy classes. Each isotopy class is the union of one or more disjoint isomorphism classes.*

PROOF.   Every isomorphism is an isotopy and every isotopy is a paratopy so the result follows.   □

The following definition (which we first gave in Section 1.1) can be applied to any latin square for which there is a natural order on the symbols. In particular, it applies whenever the symbols are integers.

DEFINITION.   A latin square is said to be *reduced* or to be in *standard form* if in the first row and column these symbols occur in natural order.

It follows from the definition of isotopism that each main or isotopy class contains one or more reduced latin squares belonging to that class and so it is always possible to choose one such square as a class representative. In contrast, not all isomorphism classes contain reduced latin squares. For example, only one of the five isomorphism classes in Ω_3 discussed on page 134 contains a reduced square.

Further, we remind the reader that in Section 1.4 and Section 2.1 we showed that, with any quasigroup $(Q, \otimes)$ are associated six parastrophes (or conjugates), the first of which is the quasigroup itself. The other five parastrophes we denoted by $(Q, \otimes^{(12)})$, $(Q, \otimes^{(23)})$, $(Q, \otimes^{(13)})$, $(Q, \otimes^{(123)})$ and $(Q, \otimes^{(132)})$. The six parastrophic quasigroups define six parastrophic latin squares and we now consider the relationships between these squares.

DEFINITION.   A loop $(Q, \otimes)$ is said to have the *inverse property* [see Bruck(1958)] if each element $a \in Q$ has a two-sided inverse a^{-1} such that $a^{-1} \otimes (a \otimes b) = b$ and $(b \otimes a) \otimes a^{-1} = b$ for all $b \in Q$. In such a loop, the mapping J defined by $aJ = a^{-1}$ for all $a \in Q$ is a one-to-one mapping of Q onto itself.

THEOREM 4.2.2   *If $(Q, \otimes)$ is a group or a loop with the inverse property, then all six parastrophes of $(Q, \otimes)$ are isotopic to each other.*

PROOF.   If $a \otimes b = c$ in a loop $(Q, \otimes)$ with the inverse property, then $b = a^{-1} \otimes c$ and $a = c \otimes b^{-1}$. That is, $(aJ) \otimes c = b$ and $c \otimes (bJ) = a$. Thus, the isotopisms (J, I, I) and (I, J, I) where I denotes the identity permutation on the set Q, map $(Q, \otimes)$ onto its parastrophes $(Q, \otimes^{(23)})$ and $(Q, \otimes^{(13)})$ respectively. (We have $a \otimes b = c \Leftrightarrow a \otimes^{(23)} c = b$ and $c \otimes^{(13)} b = a$, by definition of the operations $\otimes^{(23)}$ and $\otimes^{(13)}$.)

Further, the transpose $(Q, \otimes^{(12)})$ is isomorphic to $(Q, \otimes)$ since $a \otimes b = c \Rightarrow b = a^{-1} \otimes c \Rightarrow b \otimes c^{-1} = a^{-1} \Rightarrow c^{-1} = b^{-1} \otimes a^{-1} \Rightarrow (bJ) \otimes (aJ) = cJ$ showing that the isotopism (J, J, J) maps $(Q, \otimes)$ onto $(Q, \otimes^{(12)})$.

Similarly, the transposes $(Q, \otimes^{(123)})$ and $(Q, \otimes^{(132)})$ of, respectively, $(Q, \otimes^{(23)})$ and $(Q, \otimes^{(13)})$ are isomorphic to these loops. Since the product of an isotopism and an isomorphism is again an isotopism, this completes the proof. □

From Theorem 4.1.3 it is obvious that two latin squares contained in the same isotopy class have autotopism groups of the same order. In fact, a stronger result than this is true, as we show next.

THEOREM 4.2.3 *The autotopism groups of parastrophic quasigroups are isomorphic and so have equal orders.*

PROOF. Let $(Q, \otimes)$ be a given quasigroup and let us consider for example the parastrophe $(Q, \otimes^{(13)})$. If $\sigma = (\alpha, \beta, \gamma)$ is an autotopism of $(Q, \otimes)$, then $a \otimes b = c \Leftrightarrow (a\alpha) \otimes (b\beta) = c\gamma$. But $a \otimes b = c$ if and only if $c \otimes^{(13)} b = a$ and $(a\alpha) \otimes (b\beta) = c\gamma$ if and only if $(c\gamma) \otimes^{(13)} (b\beta) = a\alpha$. Therefore, $c \otimes^{(13)} b = a \Leftrightarrow (c\gamma) \otimes^{(13)} (b\beta) = a\alpha$ and this implies that $\sigma^{(13)} = (\gamma, \beta, \alpha)$ is an autotopism of $(Q, \otimes^{(13)})$.

In the same way it can be shown that $\sigma^{(23)} = (\alpha, \gamma, \beta)$ is an autotopism of $(Q, \otimes^{(23)})$, $\sigma^{(12)} = (\beta, \alpha, \gamma)$ is an autotopism of $(Q, \otimes^{(12)})$, $\sigma^{(123)} = (\gamma, \alpha, \beta)$ is an autotopism of $(Q, \otimes^{(123)})$ and $\sigma^{(132)} = (\beta, \gamma, \alpha)$ is an autotopism of $(Q, \otimes^{(132)})$.

Suppose that $\sigma = (\alpha, \beta, \gamma)$ and $\tau = (\alpha', \beta', \gamma')$ are autotopisms of $(Q, \otimes)$. As just shown, they correspond one-to-one to autotopisms $\sigma^{(13)} = (\gamma, \beta, \alpha)$ and $\tau^{(13)} = (\gamma', \beta', \alpha')$ of $(Q, \otimes^{(13)})$. Also, in this correspondence, the product $\sigma\tau = (\alpha\alpha', \beta\beta', \gamma\gamma')$ corresponds to the product $\sigma^{(13)}\tau^{(13)} = (\gamma\gamma', \beta\beta', \alpha\alpha')$. That is, $\sigma^{(13)}\tau^{(13)} = (\sigma\tau)^{(13)}$ and so the correspondence defines an isomorphism between the autotopism groups of $(Q, \otimes)$ and $(Q, \otimes^{(13)})$. It is clear that similar correspondences can be established between the autotopism group of $(Q, \otimes)$ and those of $(Q, \otimes^{(23)})$, $(Q, \otimes^{(12)})$, $(Q, \otimes^{(132)})$ and $(Q, \otimes^{(123)})$. □

COROLLARY. *Latin squares contained in the same main class have autotopism groups of the same order.*

We note that it follows from Theorem 4.2.2 that, if a main class of latin squares contains a latin square satisfying the quadrangle criterion, then the main class comprises a single isotopy class. In fact we can prove an even stronger result:

THEOREM 4.2.4 *If a main class contains a latin square which satisfies the quadrangle criterion then the reduced latin squares in it are contained in a single isomorphism class.*

PROOF. Suppose that a main class M contains a latin square G satisfying the quadrangle criterion. By Theorem 4.2.2, M consists of a single isotopy class. Hence every loop in M is isotopic to G, which is enough, by Corollary 1 to Theorem 1.3.4, to show that every loop in M is isomorphic to G. □

We shall next give the classification of reduced latin squares of all orders up to and including six into their main, isotopy and isomorphism classes.

We shall give a reduced form class representative of each class and shall use decimal notation to denote the various classes. A quadruple $j.k.l.m$ of integers will denote a reduced latin square of order j belonging to the k-th main class M_k, l-th isotopy class I_l of M_k, and m-th isomorphism class of I_l.

The list given on the following pages is a modified version of that given by Schönhardt(1930). It is followed by a number of explanatory remarks which the reader may find helpful to look at first. In particular, we emphasize that isomorphism classes that contain no reduced latin squares are excluded from our list. Also, as an illustration of the fact that a main, isotopy or isomorphism class may contain more than one reduced latin square, let us note that there exist two reduced latin squares of order 4 in addition to those given in our classification list below but that these both belong to the class labelled 4.2.1.1 in that list.

$$
\begin{array}{|cccc|}
1 & 2 & 3 & 4 \\
2 & 4 & 1 & 3 \\
3 & 1 & 4 & 2 \\
4 & 3 & 2 & 1 \\
\end{array}
\qquad
\begin{array}{|cccc|}
1 & 2 & 3 & 4 \\
2 & 1 & 4 & 3 \\
3 & 4 & 2 & 1 \\
4 & 3 & 1 & 2 \\
\end{array}
$$

Fig. 4.2.1.

The two squares in question are displayed in Figure 4.2.1 and can be mapped onto the square exhibited as representative of the class 4.2.1.1 by the isomorphisms $\big((34),(34),(34)\big)$ and $\big((23),(23),(23)\big)$ respectively.

Classification list for reduced latin squares of orders 2 to 6.

$$
\begin{array}{|cc|}
1 & 2 \\
2 & 1 \\
\end{array}
\qquad
\begin{array}{|ccc|}
1 & 2 & 3 \\
2 & 3 & 1 \\
3 & 1 & 2 \\
\end{array}
\qquad
\begin{array}{|cccc|}
1 & 2 & 3 & 4 \\
2 & 1 & 4 & 3 \\
3 & 4 & 1 & 2 \\
4 & 3 & 2 & 1 \\
\end{array}
\qquad
\begin{array}{|cccc|}
1 & 2 & 3 & 4 \\
2 & 3 & 4 & 1 \\
3 & 4 & 1 & 2 \\
4 & 1 & 2 & 3 \\
\end{array}
$$

2.1.1.1   3.1.1.1   4.1.1.1   4.2.1.1

```
1 2 3 4 5        1 2 3 4 5        1 2 3 4 5
2 3 4 5 1        2 1 4 5 3        2 1 5 3 4
3 4 5 1 2        3 5 1 2 4        3 4 2 5 1
4 5 1 2 3        4 3 5 1 2        4 5 1 2 3
5 1 2 3 4        5 4 2 3 1        5 3 4 1 2
    5.1.1.1          5.2.1.1          5.2.1.2
```

```
1 2 3 4 5        1 2 3 4 5        1 2 3 4 5
2 1 4 5 3        2 1 4 5 3        2 3 4 5 1
3 4 5 1 2        3 4 5 2 1        3 5 2 1 4
4 5 2 3 1        4 5 1 3 2        4 1 5 3 2
5 3 1 2 4        5 3 2 1 4        5 4 1 2 3
    5.2.1.3          5.2.1.4          5.2.1.5
```

```
1 2 3 4 5 6        1 2 3 4 5 6        1 2 3 4 5 6
2 3 4 5 6 1        2 1 5 6 3 4        2 3 1 5 6 4
3 4 5 6 1 2        3 6 1 5 4 2        3 1 2 6 4 5
4 5 6 1 2 3        4 5 6 1 2 3        4 6 5 2 1 3
5 6 1 2 3 4        5 4 2 3 6 1        5 4 6 3 2 1
6 1 2 3 4 5        6 3 4 2 1 5        6 5 4 1 3 2
      6.1.1.1            6.2.1.1            6.3.1.1
```

```
1 2 3 4 5 6        1 2 3 4 5 6
2 1 4 3 6 5        2 1 4 3 6 5
3 4 5 6 1 2        3 4 5 6 1 2
4 3 6 5 2 1        4 3 6 5 2 1
5 6 1 2 4 3        5 6 2 1 4 3
6 5 2 1 3 4        6 5 1 2 3 4
      6.4.1.1            6.4.1.2
```

```
1 2 3 4 5 6        1 2 3 4 5 6        1 2 3 4 5 6
2 1 5 6 3 4        2 1 4 5 6 3        2 1 4 3 6 5
3 6 2 5 4 1        3 5 1 6 2 4        3 5 1 6 2 4
4 5 6 2 1 3        4 6 5 1 3 2        4 6 2 5 1 3
5 4 1 3 6 2        5 3 6 2 4 1        5 3 6 2 4 1
6 3 4 1 2 5        6 4 2 3 1 5        6 4 5 1 3 2
      6.5.1.1            6.5.1.2            6.5.1.3
```

```
1  2  3  4  5  6        1  2  3  4  5  6
2  1  5  6  3  4        2  1  4  3  6  5
3  4  1  2  6  5        3  5  1  6  4  2
4  3  6  5  1  2        4  6  2  5  3  1
5  6  4  3  2  1        5  3  6  1  2  4
6  5  2  1  4  3        6  4  5  2  1  3
       6.5.1.4                 6.5.1.5
```

```
1  2  3  4  5  6      1  2  3  4  5  6      1  2  3  4  5  6
2  1  4  5  6  3      2  1  4  5  6  3      2  3  4  1  6  5
3  6  2  1  4  5      3  4  2  6  1  5      3  6  2  5  1  4
4  5  6  2  3  1      4  6  5  3  2  1      4  5  6  2  3  1
5  3  1  6  2  4      5  3  6  1  4  2      5  4  1  6  2  3
6  4  5  3  1  2      6  5  1  2  3  4      6  1  5  3  4  2
      6.6.1.1                6.6.1.2                6.6.1.3
```

```
1  2  3  4  5  6      1  2  3  4  5  6      1  2  3  4  5  6
2  1  4  5  6  3      2  1  4  6  3  5      2  5  4  6  3  1
3  4  5  6  2  1      3  4  5  1  6  2      3  1  5  2  6  4
4  6  1  2  3  5      4  5  6  2  1  3      4  6  2  3  1  5
5  3  6  1  4  2      5  6  2  3  4  1      5  4  6  1  2  3
6  5  2  3  1  4      6  3  1  5  2  4      6  3  1  5  4  2
      6.6.1.4                6.6.1.5                6.6.1.6
```

```
1  2  3  4  5  6      1  2  3  4  5  6      1  2  3  4  5  6
2  5  1  6  4  3      2  3  4  5  6  1      2  3  6  1  4  5
3  4  5  2  6  1      3  6  5  2  1  4      3  4  5  2  6  1
4  6  2  3  1  5      4  1  2  6  3  5      4  5  2  6  1  3
5  3  6  1  2  4      5  4  6  1  2  3      5  6  1  3  2  4
6  1  4  5  3  2      6  5  1  3  4  2      6  1  4  5  3  2
      6.6.1.7                6.6.1.8                6.6.1.9
```

```
1  2  3  4  5  6      1  2  3  4  5  6      1  2  3  4  5  6
2  1  4  3  6  5      2  1  4  3  6  5      2  1  5  6  3  4
3  5  1  6  4  2      3  5  2  6  4  1      3  6  4  5  1  2
4  6  5  1  2  3      4  6  5  2  1  3      4  5  6  3  2  1
5  3  6  2  1  4      5  3  6  1  2  4      5  4  1  2  6  3
6  4  2  5  3  1      6  4  1  5  3  2      6  3  2  1  4  5
      6.7.1.1                6.7.1.2                6.7.1.3
```

```
1 2 3 4 5 6          1 2 3 4 5 6          1 2 3 4 5 6
2 1 6 5 4 3          2 1 5 6 4 3          2 1 4 3 6 5
3 5 4 6 1 2          3 4 1 5 6 2          3 5 1 6 4 2
4 6 5 3 2 1          4 3 6 2 1 5          4 6 5 2 3 1
5 3 1 2 6 4          5 6 4 3 2 1          5 4 6 1 2 3
6 4 2 1 3 5          6 5 2 1 3 4          6 3 2 5 1 4
     6.7.1.4              6.7.1.5              6.7.1.6
```

```
1 2 3 4 5 6          1 2 3 4 5 6          1 2 3 4 5 6
2 1 4 5 6 3          2 1 5 6 3 4          2 5 6 1 4 3
3 5 2 6 1 4          3 4 2 5 6 1          3 4 5 2 6 1
4 6 5 3 2 1          4 5 6 3 1 2          4 6 2 3 1 5
5 3 6 1 4 2          5 6 1 2 4 3          5 3 1 6 2 4
6 4 1 2 3 5          6 3 4 1 2 5          6 1 4 5 3 2
     6.7.1.7              6.7.1.8              6.7.1.9
```

```
1 2 3 4 5 6          1 2 3 4 5 6          1 2 3 4 5 6
2 5 4 6 3 1          2 1 6 5 3 4          2 1 5 3 6 4
3 6 5 2 1 4          3 6 1 2 4 5          3 6 2 5 4 1
4 1 2 3 6 5          4 3 5 6 2 1          4 3 6 2 1 5
5 4 6 1 2 3          5 4 2 1 6 3          5 4 1 6 3 2
6 3 1 5 4 2          6 5 4 3 1 2          6 5 4 1 2 3
    6.7.1.10             6.7.1.11             6.7.1.12
```

```
        1 2 3 4 5 6          1 2 3 4 5 6
        2 1 6 5 3 4          2 1 5 6 4 3
        3 6 1 2 4 5          3 6 4 5 1 2
        4 5 2 1 6 3          4 5 6 3 2 1
        5 3 4 6 1 2          5 3 2 1 6 4
        6 4 5 3 2 1          6 4 1 2 3 5
             6.8.1.1              6.8.1.2
```

Isomorphism class 6.8.2.1 is obtained by taking the $(1, 3, 2)$-parastrophe of the square labelled 6.8.1.1 above, then reordering the rows to get a reduced square. Then isomorphism class 6.8.3.1 is obtained by taking the transpose of the representative of 6.8.2.1. Isomorphism classes 6.8.2.2 and 6.8.3.2 are derived from 6.8.1.2 in exactly the same way. (Thus, in main class 6.8 the total number of isomorphism classes which contain reduced squares is six.) Indeed, for each of the 24 remaining isomorphism classes which we list below, there are two parastrophic isomorphism classes which we have not listed but which can easily be found in the way just outlined.

1	2	3	4	5	6
2	3	1	6	4	5
3	1	2	5	6	4
4	6	5	1	2	3
5	4	6	2	3	1
6	5	4	3	1	2

6.9.1.1

1	2	3	4	5	6
2	3	1	5	6	4
3	1	2	6	4	5
4	6	5	1	2	3
5	4	6	3	1	2
6	5	4	2	3	1

6.9.1.2

1	2	3	4	5	6
2	3	1	5	6	4
3	1	2	6	4	5
4	6	5	2	3	1
5	4	6	1	2	3
6	5	4	3	1	2

6.9.1.3

1	2	3	4	5	6
2	1	6	5	4	3
3	5	1	2	6	4
4	6	2	1	3	5
5	3	4	6	2	1
6	4	5	3	1	2

6.10.1.1

1	2	3	4	5	6
2	1	5	6	3	4
3	6	4	5	1	2
4	5	6	3	2	1
5	4	2	1	6	3
6	3	1	2	4	5

6.10.1.2

1	2	3	4	5	6
2	1	4	3	6	5
3	5	1	6	4	2
4	6	5	2	3	1
5	3	6	1	2	4
6	4	2	5	1	3

6.10.1.3

1	2	3	4	5	6
2	1	5	6	3	4
3	6	2	5	4	1
4	5	6	3	1	2
5	4	1	2	6	3
6	3	4	1	2	5

6.10.1.4

1	2	3	4	5	6
2	1	4	5	6	3
3	4	2	6	1	5
4	5	6	2	3	1
5	6	1	3	2	4
6	3	5	1	4	2

6.11.1.1

1	2	3	4	5	6
2	1	4	5	6	3
3	6	5	2	4	1
4	3	1	6	2	5
5	4	6	1	3	2
6	5	2	3	1	4

6.11.1.2

1	2	3	4	5	6
2	1	6	5	3	4
3	6	1	2	4	5
4	5	2	1	6	3
5	3	4	6	2	1
6	4	5	3	1	2

6.11.1.3

1	2	3	4	5	6
2	1	5	6	4	3
3	6	4	5	1	2
4	5	6	3	2	1
5	3	1	2	6	4
6	4	2	1	3	5

6.11.1.4

1	2	3	4	5	6
2	1	5	6	4	3
3	6	2	5	1	4
4	5	6	3	2	1
5	3	4	1	6	2
6	4	1	2	3	5

6.11.1.5

1	2	3	4	5	6
2	5	6	1	4	3
3	1	5	2	6	4
4	6	2	3	1	5
5	3	4	6	2	1
6	4	1	5	3	2

6.11.1.6

1	2	3	4	5	6
2	5	1	6	3	4
3	6	5	2	4	1
4	1	2	3	6	5
5	4	6	1	2	3
6	3	4	5	1	2

6.11.1.7

1	2	3	4	5	6
2	1	5	6	4	3
3	5	4	2	6	1
4	6	2	3	1	5
5	4	6	1	3	2
6	3	1	5	2	4

6.12.1.1

1	2	3	4	5	6
2	1	5	6	4	3
3	6	2	5	1	4
4	5	6	2	3	1
5	3	4	1	6	2
6	4	1	3	2	5

6.12.1.2

1	2	3	4	5	6
2	1	5	3	6	4
3	4	1	6	2	5
4	6	2	5	1	3
5	3	6	2	4	1
6	5	4	1	3	2

6.12.1.3

1	2	3	4	5	6
2	1	5	3	6	4
3	5	2	6	4	1
4	3	6	2	1	5
5	6	4	1	3	2
6	4	1	5	2	3

6.12.1.4

1	2	3	4	5	6
2	1	4	6	3	5
3	4	5	1	6	2
4	6	2	5	1	3
5	3	6	2	4	1
6	5	1	3	2	4

6.12.1.5

1	2	3	4	5	6
2	3	4	5	6	1
3	5	2	6	1	4
4	6	1	2	3	5
5	4	6	1	2	3
6	1	5	3	4	2

6.12.1.6

1	2	3	4	5	6
2	5	6	1	4	3
3	6	5	2	1	4
4	1	2	3	6	5
5	3	4	6	2	1
6	4	1	5	3	2

6.12.1.7

1	2	3	4	5	6
2	5	6	1	3	4
3	6	5	2	4	1
4	1	2	3	6	5
5	4	1	6	2	3
6	3	4	5	1	2

6.12.1.8

1	2	3	4	5	6
2	3	1	5	6	4
3	6	5	2	4	1
4	1	2	6	3	5
5	4	6	1	2	3
6	5	4	3	1	2

6.12.1.9

1	2	3	4	5	6
2	3	6	1	4	5
3	1	5	2	6	4
4	5	2	6	1	3
5	6	4	3	2	1
6	4	1	5	3	2

6.12.1.10

Further to our comments on page 132, we emphasize that, for each isomorphism class $6.x.1.y$ given above, where $x \in \{8, 9, 10, 11, 12\}$ and y is any of the applicable values, there are two other isomorphism classes for which we have not given representatives. A representative of the class $6.x.2.y$ can be obtained from the representative of $6.x.1.y$ by taking the row-inverse and permuting the rows to get a reduced square. A representative of the class $6.x.3.y$ can be obtained by taking the transpose of the representative of $6.x.2.y$.

Let us observe that the latin square which corresponds to the Cayley table of the cyclic group of order 2 serves as class representative of the single main class of latin squares of order 2 and that Ω_3 again consists of a single main class which is represented by the Cayley table of the cyclic group of order 3. In both cases, the single main class is also the sole isotopy class (by Theorem 4.2.2) and consequently, by virtue of Theorem 4.2.4, there is only one isomorphism class containing reduced squares. In fact, Ω_3 splits into five isomorphism classes containing 1, 2, 3, 3, 3 members. The single reduced latin square of order 3 belongs to one of the isomorphism classes which contains three members.

We note that, by definition of isomorphism, any isomorph of an idempotent latin square is again idempotent. Since only one idempotent latin square of order 3 exists, this square forms an isomorphism class of one member. Also, if $a.a \neq a$ for all elements a of a quasigroup, then this statement remains true for each isomorph. There exist just two latin squares (quasigroups) of order 3 with this property and they form an isomorphism class of two members.

Turning next to the order 4 squares, we observe that Ω_4 splits into two main classes both of which are represented by latin squares which satisfy the quadrangle criterion. The main class 4.1.1.1 is represented by the Cayley table of the Klein group and the main class 4.2.1.1 by that of the cyclic group of order 4. Once again, each main class consists of a single isotopy class (by Theorem 4.2.2) and consequently (by Theorem 4.2.4), it contains only one isomorphism class which includes reduced latin squares.

For $n = 5$, there are two main classes but only the first of these is represented by a latin square satisfying the quadrangle criterion, namely the square 5.1.1.1. Each main class again comprises a single isotopy class.

For $n = 6$, two of the 12 main classes, namely those represented by the squares 6.1.1.1 and 6.2.1.1, contain latin squares satisfying the quadrangle criterion. Main class 6.1 contains the Cayley table of the cyclic group of order 6, while main class 6.2 can be represented by the Cayley table of the dihedral group of order 6. We note also that this is the smallest value of n for which some of the main classes consist of more than one isotopy class. The first seven main classes that we have listed contain a single isotopy class, which implies that every square in them is isotopic to all of its parastrophes. However, each of the squares in main classes 6.8 to 6.12 is isotopic to only one of its parastrophes other than itself. Consequently, each of these main classes contains three isotopy classes. The representatives listed for 6.8.1.1, 6.9.1.1, 6.11.1.1 and 6.12.1.1 are all symmetric squares. By contrast, main class 6.10 does not include any square for which two of its parastrophes are isotopic. However, the listed representative of 6.10.1.1 is isomorphic to its transpose by the isomorphism $((56), (56), (56))$.

To assist the reader to determine to which main class a given latin square belongs, we provide various data concerning these main classes in Figure 4.2.2. In particular, we give the numbers of transversals, intercalates and subsquares of order 3 and also the order of the autotopism group. Note that by Theorem 1.5.5 and Theorem 1.6.2 and the corollary to Theorem 4.2.3, all of these quantities are main class invariants.

Sade(1970/71) made use of techniques previously developed by him in his

Main Class	Number of Transversals	Number of Intercalates	Number of Order 3 Subsquares	Size of Autotopism Group
6.1	0	9	4	72
6.2	0	27	4	216
6.3	0	0	4	108
6.4	32	9	0	24
6.5	0	19	0	8
6.6	8	4	0	4
6.7	8	11	0	4
6.8	24	15	0	120
6.9	0	9	4	36
6.10	0	15	0	12
6.11	8	7	0	8
6.12	8	5	0	4

FIG. 4.2.2.

papers Sade(1958a,1962,1968b) to provide a systematic tabulation of all quasigroups of orders 2 to 6 inclusive by means of reduced form representatives for each main class (somewhat similar to that of Schönhardt given above) and then went on to summarize the main properties of each of these loop representatives of the main classes. Thus, for example, he stated for each the orders of its automorphism and autotopism groups, whether or not it is isotopic to some or all of its parastrophes (conjugates), the number of distinct loops in the main class which contains it, the number of isomorphically distinct semi-symmetric quasigroups (see Section 2.1) which are isotopic to it and whether it is (a) commutative, (b) isomorphic to its transpose, or (c) neither. For the orders $n = 2$, 3 and 4 he explored the whole universe of quasigroups and listed a number of interesting properties possessed by these quasigroups. In this way, he has provided a valuable compendium of results which previously could only be found by searching among his many earlier papers on the subject of quasigroups.

Cayley(1889) listed class representatives, for $n \leq 12$, of all the main classes of Ω_n for which the quadrangle criterion holds.

A general account of the historical development of the classification problem will be found in the next section.

4.3   History of the classification and enumeration of latin squares

In the first part of this section we give a short historical account of the many contributions which have been made to the problem of classifying and enumerating latin squares.

The problem of the enumeration of latin squares was first discussed by Euler(1779). He showed that the number of distinct reduced latin squares of order n (that is, latin squares with first row and first column in natural order) is 1 if

$n = 2$ or 3, 4 if $n = 4$, and 56 if $n = 5$. He also discussed latin squares of order 6, but did not succeed in enumerating these.

The next authors to write on the subject were Cayley(1890) and Frolov(1890a), both of whom published papers in 1890. Cayley stated that if the first row of a latin square is in natural order then the number of possible second rows is

$$n!\left(1 - \frac{1}{1!} + \frac{1}{2!} - \cdots + (-1)^n \frac{1}{n!}\right) \tag{4.2}$$

but that to calculate the total number of possibilities for the second and third rows when considered jointly is considerably more difficult, since the number of choices for the third row varies with the particular choice of the second. He then discussed the determination of the number of distinct reduced latin squares for values of n up to 5 and obtained the same values as had been given earlier by Euler. Frolov, writing independently and undaunted by the difficulties mentioned by Cayley, attempted an enumeration of all reduced latin squares of orders 6 and 7. He obtained 9408 reduced squares of order 6 and 221 276 160 reduced squares of order 7. The former value is correct and was confirmed later by a number of authors, but the number 221 276 160 was seriously in error. However, this did not become apparent until much later on.

Frolov [see Frolov(1890b), page 30] also stated two remarkable formulae[2]. The first is a recurrence relation for R_n, the number of reduced latin squares of order n. Frolov claimed that

$$\frac{R_n}{R_{n-1}} = \left(\frac{R_{n-1}}{R_{n-2}}\right)^2 - \frac{R_{n-1}}{2}. \tag{4.3}$$

With the initial conditions of $R_n = 1$ for $n \le 3$ and $R_4 = 4$, this formula gives the correct value for $n = 5$ and 6. Frolov was under the impression that it was also valid for $n = 7$, where it gives $R_7 = 221\,276\,160$. Curiously, (4.3) does not even define an infinite sequence since it gives $R_{12} = 0$ and hence it cannot be used to define R_n for $n \ge 14$.

Frolov's second formula purported to give the total number U_n of latin squares of order n as a function of R_n and the number C_n of reduced "regular" latin squares of order n. The reader will recall that we showed in Theorem 1.2.3 and Theorem 2.4.1 that a latin square is "regular" in this sense if and only if it is group-based though it is unlikely that Frolov was aware of this latter fact. The formula was

$$U_n = (n!)^2(R_n - C_n) + n!(n-1)!C_n = (n!)^2\left(R_n - \frac{n-1}{n}C_n\right).$$

This formula is incorrect for all $n > 4$. For $n \le 4$ every latin square is group-based so $C_n = R_n$ and in this special case Frolov's formula is correct and agrees with that obtained later by MacMahon (see below).

---

[2]In the original paper, Frolov used T_n for the number of reduced squares and R_n for the number of regular squares. However, we prefer to use R_n for the number of reduced squares for the sake of consistency with later parts of this chapter.

About the year 1899, Tarry became interested in Euler's conjecture (see Section 5.1) and, by separating the reduced latin squares of order 6 into 17 basic "families", he was able to show that no latin square of order 6 has an orthogonal mate. In other words, the problem of the thirty-six officers (described in Section 5.1) is insoluble. Of the 17 basic "families" obtained by Tarry, 12 were isotopy classes and the remaining 5 were unions of pairs of isotopy classes, the squares of one class of each pair being mirror images in the main left-to-right diagonal of the other. For further details of this work, see Tarry(1899,1900a,b,c,d).

We should mention here that H.W.Norton(1939) and Sade(1951b) state that reports exist of enumerations of latin squares having been attempted much earlier in the nineteenth century, but none of this earlier work seems to have survived. In particular, both Norton and Sade remark that, from the evidence of a letter mentioned by Gunther(1876a), it seems likely that Clausen, an assistant of the German astronomer Schumacher, had correctly enumerated the 6×6 latin squares as early as 1842 and had also shown the impossibility of any of them having an orthogonal mate.

About the same time (1899) as Tarry was enumerating the latin squares of order 6, MacMahon published a complete algebraic solution to the problem of enumerating latin squares of finite order n. He expressed this algebraic solution in two different forms [see MacMahon(1898,1900)] both of which involve the action of differential operators on an expanded operand. If his algebraic apparatus is actually put into operation it will be found that different terms are written down corresponding to all the different ways in which each row of the square could conceivably be filled up, that those arrangements which conflict with the requirements for the formation of a latin square are ultimately eliminated and that those which conform to these requirements survive the final operation and each contribute unity to the result. The manipulation of the algebraic expression, therefore, is considerably more laborious than the systematic enumeration of the squares by a simple backtracking algorithm. It is probably this fact which has forced most other authors to abandon this line of investigation, although an attempt to simplify MacMahon's procedure was later made by Saxena(1950,1951).

Using his own method, MacMahon again obtained the values 1, 1, 1, 4 for the numbers of reduced latin squares of orders 1, 2, 3, 4 respectively, but for the number of reduced squares of order 5 he obtained the value 52. The falsity of this value was subsequently pointed out to MacMahon by Fisher. As a result, MacMahon detected an error in his calculation and the corrected value of 56 was incorporated into later editions of MacMahon(1915).

In Jacob(1930), that author carried out an enumeration of $3 \times n$ latin rectangles and in the latter part of his article he also attempted an enumeration of the 5×5 and 6×6 reduced latin squares. For this purpose, he first separated the squares into families according to the nature of the permutation which transforms the first row into the second. He obtained the correct value of 56 for the number of 5×5 reduced latin squares but found only 8192 reduced 6×6 latin squares. Later, Sade(1948a) explained the error of Jacob which had led him to obtain the latter incorrect result.

In the same year, Schönhardt(1930) wrote a very comprehensive article on the subject of latin squares and loops. This included a detailed investigation of latin squares of orders 5 and 6. Schönhardt showed correctly that there exist 2 isotopy classes of 5×5 latin squares and 22 isotopy classes of 6×6 latin squares. He showed that there are 6 isomorphism classes of reduced 5×5 squares and 109 isomorphism classes of reduced 6×6 squares: further, that the total numbers of reduced squares of orders 5 and 6 are 56 and 9408 respectively. All these results are correct.

The next two papers on the subject of enumeration were those of Fisher and Yates(1934) and H.W.Norton(1939). These two papers seem to be the ones best known to, and most often quoted by, statisticians. Both papers made use of the same basic idea: namely, to make a preliminary classification of latin squares of given order according to the nature of their main diagonal. [See also, Fisher(1942a) for a later development of the same idea.] They also introduced some new terminology.

An *intramutation* of a reduced latin square L which has the integers $1, 2, \ldots, n$ as its elements is obtained by permuting the symbols $2, 3, \ldots, n$ and then rearranging the rows and columns so as to put the new square back into reduced form. Such a transformation preserves a certain property (called the "type") of the main diagonal.

The concepts of isotopy class and main class were respectively called *transformation set* and *species* by Fisher, Yates and Norton. The six parastrophes of a latin square were said to form an *adjacency set* by the latter author, who also introduced the name *intercalate* for a 2×2 latin subsquare (see Section 1.6). Interchange of the two elements in an intercalate of a latin square transforms it into another latin square, usually (but not always) from a different main class. [For examples where any such interchange produces a square in the same main class, see Wanless(2004a).] Note also that the number of intercalates is a main class invariant by Theorem 1.6.2, which in many cases provides a quick means of distinguishing main classes.

Using the above ideas, Fisher and Yates showed that there are 9408 reduced latin squares of order 6 which can be arranged into 22 isotopy classes or 12 main classes (species). Of the 22 isotopy classes, 10 can be arranged into pairs such that one member of each of these pairs is obtained from the other by interchange of rows and columns (that is, by transposition). If the squares of each such pair of isotopy classes are regarded as forming a single "family", the total number of families is 17, a result previously obtained both by Tarry and by Schönhardt.

Norton classified the main classes of 7×7 squares according to their numbers of intercalates and the numbers of isotopy classes and of adjacency sets contained in each. He found 146 distinct species (main classes) of 7×7 latin squares and a total of $16\,927\,968$ reduced latin squares of that order, although he admitted the possibility that his enumeration might be incomplete.

Somewhat later, in 1948, Sade(1948b) carried out an independent enumeration of reduced 7×7 latin squares which avoided the necessity of separating them into species and he obtained $16\,942\,080$ such squares. Three years later, in Sade(1951a,b), he gave an explanation of the discrepancy between this result and that of Norton, which was due to the fact that one species had been overlooked by the latter author. [See also Wanless(2004a) for an analysis of Norton's method.] Thus, the correct number of species is 147. Sade also pointed out that this confirmed an earlier conjecture of Ghurye(1948).

Sade's method was to calculate successively for $k = 1, 2, \ldots, l$, where l is a definite integer less than or equal to 7, a complete set of reduced $k \times 7$ latin rectangles inequivalent under any combination of permutations of rows, permutations of columns, and permutations of symbols, keeping track in so doing of the number of different rectangles in each equivalence class. The $(k+1)$-rowed rectangles were formed from the k-rowed rectangles by adding a row to each k-rowed rectangle in all possible ways, eliminating equivalent rectangles as they appeared. Sade pointed out that it was not necessary, or efficient to continue the process until $k = 7$. When k reached the value 4 (being the first value which is at least half the order of the square), Sade summed the products of the number of rectangles in an equivalence class and the number of ways a representative of that class could be completed to a square, thus obtaining the total number of reduced 7×7 squares.

Let us illustrate Sade's method by applying it to the enumeration of 4×4 reduced latin squares and taking $l = 2$. There are two equivalence classes of 2×4 latin rectangles. The rectangles L_1 and L_2 given in Figure 4.3.1 belong to one class and L_3 to the other. L_2 may be obtained from L_1 by first interchanging the symbols 3 and 4 and then interchanging the third and fourth columns.

$$
L_1 = \begin{vmatrix} 1 & 2 & 3 & 4 \\ 2 & 3 & 4 & 1 \end{vmatrix} \qquad
L_2 = \begin{vmatrix} 1 & 2 & 3 & 4 \\ 2 & 4 & 1 & 3 \end{vmatrix} \qquad
L_3 = \begin{vmatrix} 1 & 2 & 3 & 4 \\ 2 & 1 & 4 & 3 \end{vmatrix}
$$

$$
L_1' = \begin{vmatrix} 1 & 2 & 3 & 4 \\ 2 & 3 & 4 & 1 \\ 3 & 4 & 1 & 2 \\ 4 & 1 & 2 & 3 \end{vmatrix} \qquad
L_3' = \begin{vmatrix} 1 & 2 & 3 & 4 \\ 2 & 1 & 4 & 3 \\ 3 & 4 & 1 & 2 \\ 4 & 3 & 2 & 1 \end{vmatrix} \qquad
L_3'' = \begin{vmatrix} 1 & 2 & 3 & 4 \\ 2 & 1 & 4 & 3 \\ 3 & 4 & 2 & 1 \\ 4 & 3 & 1 & 2 \end{vmatrix}
$$

Fig. 4.3.1.

L_1 can be completed to a reduced 4×4 latin square in only one way, since the entry in the first column of row three must be 3 and then the entry in the second column of this row must be 4. (Correspondingly, L_2 can be completed in only one way because the entry in the first column of row four must be 4 and then the entry in the second column of this row must be 3). L_3 can be completed to a reduced 4×4 latin square in either of two ways, as shown. Thus, there exist

$2 \times 1 + 1 \times 2 = 4$ reduced latin squares of order 4.

Shortly after the publication of Sade's paper, Yamamoto(1952) showed how, using the Erdős and Kaplansky formula [given in Erdős and Kaplansky(1946)] for the number of ways of extending a latin rectangle by one row and by a more detailed classification of the $k \times 7$ latin rectangles, Sade's method could be made self checking. This check confirmed the accuracy of Sade's calculations but revealed a few minor errors.

Later, Brant and Mullen(1985) used a computer to determine the number of isomorphism classes of reduced latin squares of order 7.

In Wells(1967), that author used an adaptation of Sade's method suitable for computer calculation and, after confirming Sade's result for the number of reduced 7×7 latin squares, used it to calculate the number of reduced 8×8 latin squares. He obtained 535 281 401 856 as the number of such squares, thereby confirming Sade's earlier conjecture [see Sade(1948b)] that the number lay between 45×10^{10} and 6×10^{11}. He also estimated that the number of main classes of order 8 must be more than a quarter of a million.

Using a method which is substantially equivalent to that of Schönhardt, Brown(1968) computed (incorrectly) that the number of isotopy classes of latin squares of order 8 is 1 676 257. He also stated that there are 563 isotopy classes of latin squares of order 7; an error which was incorporated in many subsequent works including [DK1]. This is despite the correct value of 564 having been published by Preece(1966) prior to Brown's paper. Also, Arlazarov et al.(1978) computed (incorrectly) that there are 283 640 main classes of latin squares of order 8. Kolesova, Lam and Thiel(1990) corrected the results for order 8 squares, finding that there are 1 676 267 isotopy classes and 283 657 main classes. These numbers have been independently confirmed by a number of people including, in particular, Wanless. The paper of Kolesova et al. also gives a breakdown of the order 8 squares according to the size of their autotopy and paratopy groups.

The number of reduced latin squares of order 9 was first calculated by Bammel and Rothstein(1975). There are 377 597 570 964 258 816 such squares. Working independently but reporting their results jointly, McKay and Rogoyski(1995) counted the latin squares of order 10. They found that there are 7 580 721 483 160 132 811 489 280 reduced squares of that order. They also confirmed the numbers computed by Wells and Bammel and Rothstein [as had Mullen and Purdy(1993) slightly earlier] and used sampling to obtain estimates for the number of reduced Latin squares for orders in the range $11 \leq n \leq 15$. Their prediction for $n = 11$ was confirmed by McKay and Wanless(2005) who established that there are 5 363 937 773 277 371 298 119 673 540 771 840 reduced latin squares of order 11.

More recently, McKay, Meynert and Myrvold(2007) studied the main class automorphisms of the latin squares up to order 10 in order to obtain counts of various categories of squares. As well as counting main and isotopy classes and reduced squares, they also counted the number of quasigroups (up to isomorphism) and the number of loops (again, up to isomorphism). These results are summarized in Figure 4.3.2. The number of reduced latin squares of each order up to 11 is given later in this chapter, in Figure 4.4.3 and Figure 4.4.4.

In Hulpke, Kaski and Östergård(2011), these authors obtained counts for the numbers of main and isotopy classes of latin squares of order 11. Their results are in agreement with those of McKay and Wanless obtained earlier.

Since a reduced latin square is one which has its first row and column in natural order, any such latin square is the multiplication table of a loop and so the enumeration of loops (without taking account of isomorphisms) of a given order is equivalent to enumerating the reduced latin squares of that order. From this point of view, two further papers should be mentioned as having a place in the above history of the subject: namely Albert(1944) and Bryant and Schneider(1966).

In the first of these papers, and probably unaware of Schönhardt's earlier work, Albert initiated a general discussion on the enumeration of loops. He pointed out first that it is easy to check that the only loops of orders 2, 3 and 4 are the groups of those orders. He went on to give a complete enumeration of the loops of order 5 and showed that there are just six isomorphism classes of loops of order 5. Multiplication tables for representatives of the six classes can be obtained by bordering the latin squares of order 5 given on page 129 by their first row and column (although these representatives differ from those chosen by Albert). Albert proved that every loop of order five is either isomorphic to the cyclic group of order 5 or belongs to the isotopy class we have labelled 5.2.1. Furthermore, he showed that, with the exception of 5.2.1.3 and 5.2.1.4 which are anti-isomorphic (meaning that each is isomorphic[3] to the transpose of the other), no other pair of isomorphism classes is anti-isomorphic.

Albert went on to discuss the enumeration of loops of order 6 but he did not attempt to do this exhaustively. Instead, he first showed the existence of simple loops of order 6 (that is, loops with no proper subloops) and then confined his further investigations to the subclass of loops of order 6 which contain one or more subloops of order 3. He was able to establish, in particular, that every loop of order 6 with a subloop of order 3 has only a single subloop of that order. This work was published in 1944.

Later, in 1966, Bryant and Schneider carried out a complete computer aided enumeration of loops of order 6 and showed that there exist 109 isomorphism classes of loops of order 6. This confirmed Schönhardt's result. Their method was first to develop theorems (of a similar nature to those given in Section 4.1) which described successively the principal classes, the isotopy classes, and the isomorphism classes until they had developed the theory to a point at which a computer could be employed effectively.

To complete this historical account, we should mention the paper of Sade(1970 /71), already referred to in the previous section, in which he summarized his own

---

[3] For the representatives that we have chosen this isomorphism is actually the identity. That is, one square is the transpose of the other. The squares chosen by Albert are exhibited on page 145 of [DK1].

Order	Number of main classes	Number of isotopy classes	Number of loops	Number of quasigroups
1	1	1	1	1
2	1	1	1	1
3	1	1	1	5
4	2	2	2	35
5	2	2	6	1411
6	12	22	109	1 130 531
7	147	564	23 746	12 198 455 835
8	283 657	1 676 267	106 228 849	2 697 818 331 680 661
9	19 270 853 541	115 618 721 533	9 365 022 303 540	15 224 734 061 438 247 321 497
10	34 817 397 894 749 939	208 904 371 354 363 006	20 890 436 195 945 769 617	2 750 892 211 809 150 446 995 735 533 513

Fig. 4.3.2.

enumerative work for quasigroups of all orders up to 6 inclusive and also tabulated many interesting properties of particular quasigroups of these orders.

Most of the remaining part of this section will be devoted to enumeration results concerning groups. However, before pursuing this subject, we make two further general remarks.

If the number of reduced latin squares of order n is denoted by R_n and the total number of latin squares of order n by U_n then $U_n = n!(n-1)! R_n$. This was pointed out by MacMahon [see page 248 of MacMahon(1915)] and by a number of later authors and is a special case of Theorem 4.4.1, which we shall prove in the next section. M.Hall(1948) showed that $U_n \geq n!(n-1)! \cdots 2! 1!$ and so $R_n \geq (n-2)!(n-3)! \cdots 2! 1!$ for all n.

Despite the fact that the aim of the present book is to describe properties of latin squares of finite order, we think it worthwhile to give the following result of Mano(1960) as a curiosity. Mano proved that the number of latin squares of infinite order is equal to the cardinal number of the continuum. We remark in passing that very few authors have studied latin squares of infinite order.

The number of non-isomorphic abelian groups of order $n = p_1^{\alpha_1} p_2^{\alpha_2} \cdots p_r^{\alpha_r}$ is well known and will be found, for example, in Fuchs(1958), page 53. The number of such groups is $\prod_{i=1}^{r} P(\alpha_i)$ where the p_i are distinct primes, the α_i are positive integers and $P(\alpha_i)$ denotes the number of distinct partitions of the integer α_i into positive integers. For asymptotic results on the same subject, the reader is referred to Erdős and Szekeres(1934/35), Kendall and Rankin(1947) and Krätzel(1970).

A remarkable result of Rédei(1947) can be formulated as follows: If and only if $n = p_1 p_2 \cdots p_u q_1^2 q_2^2 \cdots q_v^2$, where $p_1, p_2, \ldots, p_u$ and $q_1, q_2, \ldots, q_v$ are distinct primes each of which is relatively prime to

$$\prod_{i=1}^{u}(p_i - 1) \prod_{j=1}^{v}(q_j^2 - 1),$$

then all groups of order n are abelian.

A general formula for the number of non-abelian groups of order n is not yet known. However, a table giving the number of groups for each order up to 100 was published by Miller(1930). This was extended to 162 (except the order 128) by Senior and Lunn(1934) and later to 215 (except the order 192). [See also Hall and Senior(1964) and Sloane's Online Encyclopaedia of Integer Sequences(2012), sequence number A060689.]

In Figure 4.3.3, we give the number of non-abelian groups of each order $n \leq 32$. The structure of these groups is described by Thomas and Wood(1980). The generating relations for the groups of orders $n < 32$ can be found in Coxeter and Moser(1965).

Szele(1947) showed that a necessary and sufficient condition that the only group of a given order n is the cyclic group of that order is that n is relatively prime to $\phi(n)$, where ϕ denotes Euler's function.

Order	Number of non-abelian groups	Order	Number of non-abelian groups	Order	Number of non-abelian groups	Order	Number of non-abelian groups
1	0	9	0	17	0	25	0
2	0	10	1	18	3	26	1
3	0	11	0	19	0	27	2
4	0	12	3	20	3	28	2
5	0	13	0	21	1	29	0
6	1	14	1	22	1	30	3
7	0	15	0	23	0	31	0
8	2	16	9	24	12	32	44

FIG. 4.3.3.

A detailed description of all groups up to order 64 inclusive has been given by Hall and Senior(1964). The numbers of groups of orders 16, 32 and 64 were found to be 14, 51 and 267, respectively. The number 267 [which was confirmed later by Thomas and Wood(1980), among others] conflicts with the count of 294 obtained earlier by Miller(1930).

It has been shown by Gallagher(1967) that, if $M(n)$ denotes the number of non-isomorphic finite groups (abelian and non-abelian) of order n then $M(n) \leq n^{cn^{2/3}\log_2 n}$, where $c = 2/(1 - 2^{-2/3})$. A result of Higman(1960) shows that, for the special case of soluble groups, the above estimate of Gallagher is "best possible". An exact method of enumeration for soluble groups has been given by Lunn and Senior(1934).

Greenberg and Newman(1970) considered a related question concerning the number and distribution of soluble groups generated by elements of specified odd orders $a_1, a_2, \ldots, a_r$. They proved the following result:

For each positive integer n, define the value of the function $s(n)$ to be 1 if a soluble group of order n exists generated by a set of elements $x_1, x_2, \ldots, x_r$ of orders $a_1, a_2, \ldots, a_r$ respectively, and 0 otherwise. Let $S(x) = \sum_{n \leq x} s(n)$. Then $S(x) = O\big(x(\log x)^{-1/(2h)}\big)$, where $h = \phi(a_1, a_2, \ldots, a_r)$ and ϕ is Euler's function. Consequently, $\lim_{x \to \infty} S(x)/x = 0$. Soon after the work of Greenberg and Newman was pubished, a slight improvement in the power of $\log x$ in the bound on $S(x)$ was obtained by Indlekofer(1973).

For some results concerning the enumeration of p-groups, see Davies(1962) and Sims(1965).

In Gilbert(1965), that author found a formula for the number of latin squares of order n which are multiplication tables of quasigroups isotopic to the cyclic group of order n. Such latin squares were called *addition squares* by Gilbert. The number of addition squares of order n is $n!\big((n-1)!\big)^2/\phi(n)$, where $\phi(n)$ is

Euler's function. In the same paper, Gilbert found a formula for the number of addition squares of order n which are complete latin squares. (For the definition of a complete latin square, see Section 2.6).

His result is as follows: *Let Q_n denote the number of permutations $a_1, a_2, \ldots, a_n$ of the integers $1, 2, \ldots, n$ which are such that the differences $a_2 - a_1, a_3 - a_2, \ldots, a_n - a_{n-1}$ are all distinct modulo n. Then there are exactly $n! \, Q_n^2 / (n^2 \phi(n))$ complete addition squares of order n.*

Finally, we mention some enumerative results concerning special kinds of latin square.

Phelps(1980) obtained a lower bound for the number of symmetric latin squares of a particular order. Gross, Mullen and Wallis(1973) obtained a lower bound for the number $\nu(r)$ of pairwise perpendicular symmetric idempotent latin squares of order r, where r is a prime power: namely, $\nu(p^k) \geq \lfloor (t-1)/2^{n-1} \rfloor + 1$. Graham and Roberts(2006) enumerated the self-orthogonal latin squares up to order 9. See also Burger, Kidd and van Vuuren(2010) for some further enumerative results for such squares.

4.4   Enumeration of latin rectangles

Since most of the methods for enumerating latin squares are dependent on the enumeration of $k \times n$ latin rectangles we devote this section to that topic. We first discuss exact enumeration and then the known asymptotic results.

To determine $L(k,n)$, the number of $k \times n$ latin rectangles, for $k \leq n$ it is most practical to enumerate some canonical subset of the latin rectangles in such a way that the total number of rectangles can be determined from this partial enumeration. The choice of which canonical subset to use has varied from author to author, as we shall see in the following pages. We collect the definitions here at the outset for the sake of clarity and to facilitate comparisons between them.

DEFINITION. A $k \times n$ latin rectangle is said to be *normalized* if the n symbols of its first row are in ascending order.[4] It is said to be *semi-reduced* if its first row and column are in ascending order. It is said to be *reduced* if its first row and column are in ascending order and consist of consecutive symbols (usually integers). We denote the numbers of normalized, semi-reduced and reduced $k \times n$ latin rectangles by $K(k,n)$, $S(k,n)$ and $R(k,n)$ respectively. We shall sometimes refer to the $k \times n$ latin rectangles counted by $L(k,n)$ as *unrestricted* to emphasize that they are not necessarily normalized, semi-reduced or reduced.

Note that semi-reduced rectangles may seem to be the obvious generalization of reduced squares as defined on page 126. However, it is the reduced latin rectangles which can most easily be extended to reduced latin squares by appending (as opposed to inserting) rows. We should also warn the reader that there is no consistency in the literature concerning the terms we have just defined. For

---

[4] We assume that we are using a set of symbols which can be ordered such as $0, 1, 2, \ldots, n-1$ or $1, 2, 3, \ldots, n$.

example, Riordan(1946,1952), Yamamoto(1949) and others call our normalized rectangles "reduced", while McKay and Rogoyski(1995) call our reduced rectangles "normalized". To make matters worse, some authors state our definition of "semi-reduced" when they actually mean to define reduced latin rectangles. The distinction is demonstrated by the semi-reduced latin rectangle L_3 in Figure 4.4.1 below. Its first column is in natural order, however the entries are not consecutive integers, so the rectangle is not reduced.

We next consider the relationship between the numbers $L(k,n)$, $K(k,n)$, $S(k,n)$ and $R(k,n)$. It is easy to see that the columns of an unrestricted latin rectangle may be permuted in a unique way to obtain a normalized latin rectangle. Conversely, the columns of a normalized latin rectangle may be permuted in $n!$ ways to obtain an unrestricted latin rectangle. Hence $L(k,n) = n! \, K(k,n)$. By a similar argument, but permuting the rows (except the first row) rather than permuting the columns, we see that $K(k,n) = (k-1)! \, S(k,n)$.

EXAMPLE. *Let L_1 be defined as in Figure 4.4.1. We may transform L_1 to a normalized latin rectangle L_2 by re-arranging its columns so that its first row becomes 1 2 3 4. We can then convert L_2 into a semi-reduced rectangle L_3 by re-arranging the rows, excluding the first, so that the elements of the first column form an increasing sequence as shown. $L_1 \to L_2 \to L_3$ (Figure 4.4.1).*

$$L_1 = \begin{vmatrix} 2 & 4 & 1 & 3 \\ 3 & 1 & 4 & 2 \\ 4 & 2 & 3 & 1 \end{vmatrix} \quad L_2 = \begin{vmatrix} 1 & 2 & 3 & 4 \\ 4 & 3 & 2 & 1 \\ 3 & 4 & 1 & 2 \end{vmatrix} \quad L_3 = \begin{vmatrix} 1 & 2 & 3 & 4 \\ 3 & 4 & 1 & 2 \\ 4 & 3 & 2 & 1 \end{vmatrix}$$

FIG. 4.4.1.

Note that an alternative way to transform L_1 to reduced form is by first permuting its symbols so that its first row becomes 1 2 3 4 and then re-arranging the rows, excluding the first, as before. $L_1 \to L_2' \to L_3'$ (Figure 4.4.2). The isotopism $(\alpha_1, \beta_1, \gamma_1)$ which effects the first transformation has $\alpha_1 = (2\ 3), \beta_1 = (1\ 2\ 4\ 3), \gamma_1 = \epsilon$. That which effects the second transformation has $\alpha_2 = (2\ 3), \beta_2 = \epsilon, \gamma_2 = (2\ 1\ 3\ 4)$.

Either one of these two transformation methods separates the universe of 3×4 latin rectangles into $L(3,4)/[4!(3-1)!]$ families of rectangles such that the members of a family all reduce to the same semi-reduced form but the families are different in the two cases.

$$L_1 = \begin{vmatrix} 2 & 4 & 1 & 3 \\ 3 & 1 & 4 & 2 \\ 4 & 2 & 3 & 1 \end{vmatrix} \quad L_2' = \begin{vmatrix} 1 & 2 & 3 & 4 \\ 4 & 3 & 2 & 1 \\ 2 & 1 & 4 & 3 \end{vmatrix} \quad L_3' = \begin{vmatrix} 1 & 2 & 3 & 4 \\ 2 & 1 & 4 & 3 \\ 4 & 3 & 2 & 1 \end{vmatrix}$$

FIG. 4.4.2.

As regards reduced rectangles, there is only one general way in which to transform a given $k \times n$ latin rectangle L to reduced form. We have to ensure that the symbols used for the first column are $1\ 2\ \ldots\ k$ so we must first transform the symbols $a_1, a_2, \ldots, a_k$ used in the first column of L to $1, 2, \ldots, k$ respectively. Secondly, we have to re-order the remaining $n - 1$ columns so that the first row is normalized. The required isotopism (α, β, γ) is such that γ maps $a_1, a_2, \ldots, a_k$ to $1, 2, \ldots, k$ respectively, $\alpha = \epsilon$ and β permutes the last $n - 1$ columns.

Conversely, there exist $(n-1)!$ permutations of $1, 2, \ldots, n$ which fix the symbol 1. Let β be one of these. Also, let γ be an injection from the set $1, 2, \ldots, k$ into the set $1, 2, \ldots, n$. There are $n(n - 1)(n - 2) \ldots (n - k + 1) = n!/(n - k)!$ such injections. Let R be a reduced $k \times n$ latin rectangle. We may first permute its columns by means of β and, secondly, we may replace the symbols 1 to k by those specified by γ to obtain an unrestricted $k \times n$ latin rectangle $R\beta\gamma$. Each unrestricted $k \times n$ latin rectangle L can be written in the form $R\beta\gamma$ for a unique pair of permutations $\beta\gamma$ since there is only one β which could produce the first column of L and then only one γ which could produce the first row. Therefore, $L(k, n) = [n!/(n - k)!](n - 1)!R(k, n)$.[5] Thus, we have proved:

THEOREM 4.4.1 *Any one of the numbers $L(k, n)$, $K(k, n)$, $S(k, n)$ and $R(k, n)$ (representing the numbers of unrestricted, normalized, semi-reduced and reduced $k \times n$ latin rectangles respectively) determines the others by the relationship:*

$$L(k, n) = n!\, K(k, n) = n!\, (k - 1)!\, S(k, n) = \frac{n!\, (n - 1)!}{(n - k)!}\, R(k, n).$$

It is clear from the definitions that every reduced rectangle is semi-reduced and every semi-reduced rectangle is normalized. Hence, of the options we have considered above, the most efficient way to determine $L(k, n)$ is to count the reduced rectangles because there are fewest of them. In practice the computer enumerations of latin squares by Wells(1967), Bammel and Rothstein(1975), McKay and Rogoysky(1995) and McKay and Wanless(2005) have made use of a further observation: namely that the number of possible extensions to a given latin rectangle depends only on the symbols which occur in each particular column, not on the order of those symbols within the column. This idea leads to a way of expressing the enumeration problem for latin rectangles in graph theoretic terms. See the above-mentioned papers for further details.

Among the earliest writers on the subject of enumerating latin rectangles after Cayley and MacMahon (see the previous section) was Jacob(1930). This author attempted an enumeration of $3 \times n$ normalized rectangles and tabulated the results for values of $n \leq 15$. The first five of Jacob's results are in agreement with those obtained later by Kerewala(1941) and Riordan(1946) and are as follows:

---

[5]This modification of the author's original proof was suggested by Brendan McKay to answer an objection raised by another expert in this topic.

n	3	4	5	6	7
$K(3,n)$	2	24	552	21280	1073760

These values can be compared, using Theorem 4.4.1, with the values for $R(k,n)$ computed by McKay and Rogoyski(1995), whose results are reproduced in Figure 4.4.3. The values of $R(k,n)$ for $k < n = 8$ were published by Mullen and Purdy(1993). The values of $R(k,11)$ were computed by McKay and Wanless(2005) and are quoted in Figure 4.4.4. These authors noted that as k grows, $R(k,n)$ seems to be divisible by an increasing power of each small prime. The effect, which was subsequently explained in Stones and Wanless(2010), is particularly striking with powers of 2, as can be seen in Figure 4.4.4. Similar observations had been made earlier by Alter(1975) and by Mullen(1978).

More recently, Stones(2010) has computed $R(5,n)$ for $n \le 28$ and $R(6,n)$ for $n \le 13$ using about two months of computer time.

A latin rectangle with only two rows defines a permutation displacing all symbols. Such a permutation is traditionally called a *derangement* and the number of derangements is given by formula (4.2) on page 136. Alternatively, the two rows can be thought of as two separate permutations from natural order. Two such permutations have the property that they are *discordant* [see Riordan(1944)]: that is, they do not agree in any position. The enumeration of permutations discordant with a given permutation is the famous *problème des rencontres*. The enumeration of permutations discordant with each of two permutations, one of which is obtained from the other by a cyclic permutation of the symbols of the form $(2\ 3\ 4\ \cdots\ n\ 1)$ is known as the reduced *problème des ménages*. The latter problem was considered by Kaplansky(1943), Riordan(1944), Kaplansky and Riordan(1946), and Touchard(1934,1953). See also Moser(1982).

The next case in this hierarchy, the enumeration of permutations discordant with three permutations of the form given in Figure 4.4.5, has been examined by Riordan(1954). More generally, we can consider the number $M(k,n)$ of permutations discordant with a $k \times n$ latin rectangle whose first row is $1\ 2\ \ldots\ n$ and each subsequent row is obtained from its predecessor by applying the permutation $(2\ 3\ 4\ \cdots\ n\ 1)$. This number $M(k,n)$ was called the *generalized ménage number*[6] by Godsil and McKay(1990) and will be discussed later in this section.

Generalizing in a different direction, Riordan(1954) found the numbers $N_{n,k}$ of permutations which have exactly $n - k$ places in which they disagree with all three permutations of Figure 4.4.5. His results are reproduced in Figure 4.4.6 which gives the values of $N_{n,k}$ for small values of n and k. When $k = 0$, the numbers

$$N_{n,0} = \sum_{k=1}^{n} (-1)^k (n-k)!\, r_k$$

enumerate the permutations discordant with the permutations in Figure 4.4.5, that is, they count $M(3,n)$. Here r_k is the number of ways of putting k elements in

---

[6]The phrase "generalized ménage number" was used in a different sense by Nechvatal(1981a).

n	k	R(k,n)	n	k	R(k,n)	n	k	R(k,n)
1	1	1	7	1	1	9	1	1
2	1	1		2	309		2	16 687
	2	1		3	35 792		3	103 443 808
3	1	1		4	1 293 216		4	207 624 560 256
	2	1		5	11 270 400		5	112 681 643 083 776
	3	1		6	16 942 080		6	12 952 605 404 381 184
4	1	1		7	16 942 080		7	224 382 967 916 691 456
	2	3	8	1	1		8	377 597 570 964 258 816
	3	4		2	2 119		9	377 597 570 964 258 816
	4	4		3	1 673 792	10	1	1
5	1	1		4	420 909 504		2	148 329
	2	11		5	27 206 658 048		3	8 154 999 232
	3	46		6	335 390 189 568		4	147 174 521 059 584
	4	56		7	535 281 401 856		5	746 988 383 076 286 464
	5	56		8	535 281 401 856		6	870 735 405 591 003 709 440
6	1	1					7	177 144 296 983 054 185 922 560
	2	53					8	4 292 039 421 591 854 273 003 520
	3	1064					9	7 580 721 483 160 132 811 489 280
	4	6552					10	7 580 721 483 160 132 811 489 280
	5	9408							
	6	9408							

FIG. 4.4.3.

forbidden positions, subject to the compatibility conditions that no two elements may be in the same position and no two positions have the same element.

A recurrence relation for the number of three-line normalized latin rectangles was obtained by Riordan(1952). If we define the sequence $\{k_n\}$ by $k_0 = 1$ and $k_n = -nk_{n-1} - (n-1)2^n$. Then Riordan's recurrence is

$$K(3,n) = n^2K(3,n-1) + n(n-1)K(3,n-2) + 2n(n-1)(n-2)K(3,n-3) + k_n$$

Formulae for $L(3,n)$ were given by Dulmage and McMaster(1975), Bogart and Longyear(1976) and Gessel(1985). The later author expressed the number of reduced $3 \times n$ latin rectangles in terms of the coeficients of a power series in an indeterminant x. Athreya, Pranesachar and Singhi(1980) also derived a formula for $L(3,n)$ and went on to give a complicated expression for $L(4,n)$. A

n	k	Factorisation of $R(k,n)$
11	1	1
	2	1 468 457
	3	$2^7 \cdot 13 \cdot 23 \cdot 20\,851\,549$
	4	$2^{10} \cdot 3^2 \cdot 1823 \cdot 8\,569\,184\,461$
	5	$2^{13} \cdot 3^2 \cdot 29 \cdot 168\,293 \cdot 20\,936\,295\,857$
	6	$2^{17} \cdot 3^2 \cdot 5 \cdot 31 \cdot 2\,334\,139 \cdot 225\,638\,611\,943$
	7	$2^{21} \cdot 3^2 \cdot 5 \cdot 9437 \cdot 269\,623\,520\,098\,467\,133$
	8	$2^{28} \cdot 3^2 \cdot 5 \cdot 97 \cdot 73\,488\,673\,152\,815\,765\,447$
	9	$2^{32} \cdot 3^3 \cdot 5 \cdot 61 \cdot 7487 \cdot 260\,951 \cdot 42\,053\,669\,617$
	10	$2^{35} \cdot 3^4 \cdot 5 \cdot 2801 \cdot 2\,206\,499 \cdot 62\,368\,028\,479$
	11	$2^{35} \cdot 3^4 \cdot 5 \cdot 2801 \cdot 2\,206\,499 \cdot 62\,368\,028\,479$

FIG. 4.4.4.

$$
\begin{array}{ccccc}
1 & 2 & \cdots & n-1 & n \\
2 & 3 & \cdots & n & 1 \\
3 & 4 & \cdots & 1 & 2
\end{array}
$$

FIG. 4.4.5.

direct, though fairly complicated, method for enumerating $K(4,n)$ was given by Light(1973).

In Riordan(1952), that author showed that $K(3, n + p) \equiv 2\,K(3, n)$ mod p, where p is any prime greater than two. Carlitz(1953a) extended this result to show that for arbitrary m, $K(3, n + m) \equiv 2^m K(3, n)$ mod m.

$k \setminus n$	3	4	5	6	7	8	9	10
0	0	1	2	20	144	1265	12072	125655
1	0	0	15	72	609	4960	46188	471660
2	0	6	20	180	1106	9292	82980	831545
3	6	8	40	176	1421	10352	93114	912920
4		9	30	180	980	8326	70272	695690
5			13	72	595	4096	39078	379760
6				20	154	1676	14292	155690
7					31	304	4230	43880
8						49	576	9905
9							78	1060
10								125

FIG. 4.4.6.

Recently, a more far-reaching congruence relation expressed in terms of $R(k,n)$ has been obtained by Stones and Wanless(2010): namely,

$$R(k, n+d) \equiv (-1)^{k-1}(k-1)!R(k,n)R(k,d) \bmod d$$

for all $k \le n$ and positive integers d.

Formulae giving the values of both $M(3,n)$ and $M(4,n)$ are known[7]. The first,

$$M(3,n) = \sum_{i=0}^{n}(-1)^i \frac{2n}{2n-i}\binom{2n-i}{i}(n-i)! \quad \text{for } n \ge 3,$$

was obtained by Kaplansky(1943). The second is due to Moser(1967) who showed that

$$M(4,n) = \sum_{i=0}^{n}(-1)^i g(n,3,i)(n-i)! \quad \text{for } n \ge 4,$$

where

$$g(n,3,i) = \sum_{\alpha=0}^{\lfloor i/2 \rfloor}\sum_{\beta=0}^{m} \frac{n}{n-i}\binom{n+\alpha-i-1}{\alpha}\binom{n-i}{\beta}2^\beta\binom{n-\alpha-1}{i-2\alpha-\beta}$$

for $0 \le i < n$, where $m = \min(n-i, i-2\alpha)$, and

$$g(n,3,n) = 3 + \sum_{\alpha=1}^{\lfloor n/2 \rfloor} \frac{n}{\alpha}\binom{n-\alpha-1}{\alpha-1}.$$

A complicated formula for $L(k,n)$, valid for all k, was obtained by Nechvatal(1981b) and another similar one by Gessel(1987).

However, exact enumeration of $L(k,n)$ (or, for that matter, $R(k,n)$, $K(k,n)$, $S(k,n)$ or $M(k,n)$) for larger values of k and n seems very difficult. Besides, judging from the known formulae such as those just mentioned, any results are likely to become increasingly cumbersome as k grows. An alternative to exact enumeration is to study the asymptotic growth of these functions.

Using two different methods Riordan(1944,1946) obtained the asymptotic result $K(3,n) \sim (n!)^2 e^{-3}$. An asymptotic formula for the number of latin rectangles was obtained by Erdős and Kaplansky(1946). Kerewala(1947a,b) also published some results on this subject. For confirmation of the results given in Kerewala(1947a), see Yamamoto(1949,1953). Some further interesting results connected with the enumeration of latin rectangles will be found in Yamamoto(1956) and Riordan(1958).

---

[7]In the derivation of these results both Kaplansky and Moser used the term "very reduced" for a latin rectangle whose rows are the same as those of Figure 4.4.5 but with leading entries $1, n, n-1, \ldots, \ldots$ so that the leading diagonal consists entirely of 1s). This nomenclature is not ideal since any latin rectangle can be "reduced" by an appropriate isotopism to a reduced rectangle (see page 126) but "very reduced rectangles" lack any similar useful property.

The asymptotic relation obtained by Erdős and Kaplansky for the total number $L(k,n)$ of $k \times n$ latin rectangles is

$$L(k,n) \sim (n!)^k \exp\left(\frac{-k(k-1)}{2}\right) \tag{4.4}$$

and they showed that this relation is valid as $n \to \infty$ not only for fixed k but also for any $k < (\log n)^{(3/2)-\epsilon}$, where ϵ is an arbitrarily small positive constant. They conjectured that (4.4) is valid for values of k up to nearly $n^{1/3}$. This conjecture was confirmed by Yamamoto(1951) who showed the validity of (4.4) whenever $k = o(n^{1/3})$ and who later [in Yamamoto(1969)] showed that

$$L(k,n) \sim (n!)^k \exp\left(\frac{-k(k-1)}{2} - \frac{k^3}{6n}\right) \tag{4.5}$$

when $k = O(n^{(5/12)-\epsilon})$ for a fixed $\epsilon > 0$. C.M.Stein(1978) extended Yamamoto's result by proving (4.5) for $k = o(n^{1/2})$. Then Godsil and McKay(1990) proved that

$$L(k,n) \sim (n!)^k \left(\frac{n(n-1)(n-2)\cdots(n-k+1)}{n^k}\right)^n \left(1 - \frac{k}{n}\right)^{-n/2} e^{-k/2} \tag{4.6}$$

for $k = o(n^{6/7})$ as $n \to \infty$. The latter authors conjecture that (4.6) is valid for $k = O(n^{1-\delta})$ for any arbitrary constant $\delta > 0$.

For a fairly recent, well-written and comprehensive survey of methods and results in the enumeration of latin rectangles, see Stones(2010).

4.5   Enumeration of transversals

Firstly, we remind the reader that some latin squares have no transversals at all. For example, it follows from Theorem 2.5.1 and Theorem 2.5.4 that the Cayley tables of abelian groups which have a unique elemnet of order two have no transversals. Also, it follows from Theorem 2.5.5 that the same is true for the Cayley tables of all groups which have cyclic Sylow 2-subgroups (including those just mentioned) and, in particular, for groups of orders $n \equiv 2 \mod 4$.

A more general class of latin squares of even order which have no transversals are those which Maillet(1894b) called latin squares of *q-step type*.

DEFINITION. A latin square L of order $n = mq$ is said to be of q-step type if it can be represented by a matrix C of the form shown in Figure 4.5.1 where, for each fixed choice of k, the $A_{ij}^{(k)}$ are latin subsquares of L all of which contain the same q elements.

For example, the latin square shown in Figure 4.5.2 is of 2-step type while both the squares shown in Figure 4.5.3 are of 3-step type.

$$C = \begin{vmatrix} A_{00}^{(0)} & A_{01}^{(1)} & A_{02}^{(2)} & \cdots & \cdots & A_{0,m-2}^{(m-2)} & A_{0,m-1}^{(m-1)} \\ A_{10}^{(1)} & A_{11}^{(2)} & A_{12}^{(3)} & \cdots & \cdots & A_{1,m-2}^{(m-1)} & A_{1,m-1}^{(0)} \\ A_{20}^{(2)} & A_{21}^{(3)} & A_{22}^{(4)} & \cdots & \cdots & A_{2,m-2}^{(0)} & A_{2,m-1}^{(1)} \\ \cdot & \cdot & \cdot & \cdots & \cdots & \cdot & \cdot \\ \cdot & \cdot & \cdot & \cdots & \cdots & \cdot & \cdot \\ \cdot & \cdot & \cdot & \cdots & \cdots & \cdot & \cdot \\ A_{m-1,0}^{(m-1)} & A_{m-1,1}^{(0)} & A_{m-1,2}^{(1)} & \cdots & \cdots & A_{m-1,m-2}^{(m-3)} & A_{m-1,m-1}^{(m-2)} \end{vmatrix}$$

FIG. 4.5.1.

As early as 1779, Euler was able to prove that a cyclic latin square of even order cannot have any transversals and so has no orthogonal mate.[8] In the same paper [Euler(1779)], he showed that the same is true for latin squares of order 6 or 12 which are of 3-step type and that it is also true for all latin squares of order $2q$ and of q-step type when q is odd. In Maillet(1894b), that author noted that a cyclic latin square is a square of 1-step type and that all the above mentioned results of Euler are special cases of the following general theorem.

0	1	2	3	4	5
1	0	3	2	5	4
2	3	5	4	1	0
3	2	4	5	0	1
4	5	0	1	2	3
5	4	1	0	3	2

FIG. 4.5.2.

0	1	2	3	4	5		0	1	2	3	4	5
1	2	0	4	5	3		1	2	0	5	3	4
2	0	1	5	3	4		2	0	1	4	5	3
3	4	5	0	1	2		3	5	4	0	2	1
4	5	3	1	2	0		4	3	5	2	1	0
5	3	4	2	0	1		5	4	3	1	0	2

FIG. 4.5.3.

THEOREM 4.5.1 *A latin square L of order $n = mq$ and of q-step type has no transversals if m is even and q is odd.*

PROOF. Our proof substantially follows that of Maillet. Since L is of q-step type we may represent it by means of the matrix C given in Figure 4.5.1, where

[8]Much later, Hedayat and Federer(1969) claimed this as a new result.

each $A_{ij}^{(k)}$ represents a $q \times q$ latin subsquare of L. If we take the elements of L to be the integers $0, 1, 2, \ldots, n - 1$, there will be no loss of generality in supposing that (by a change of labelling if necessary) all the subsquares which correspond to the same fixed value of k contain the same q elements $kq + b$, where k has this fixed value and $0 \le b \le q - 1$: for every one of the integers $0, 1, 2, \ldots, n - 1$ has a unique representation in the form $aq + b$ with $0 \le a \le m - 1$ and $0 \le b \le q - 1$. Let us note for use later that, for each latin subsquare A of the matrix C, k is determined by the relation $k \equiv i + j \bmod m$.

Let us suppose that the theorem is false and that τ is a transversal of L. Then τ will contain just q cells belonging to the set of latin subsquares $A_{00}, A_{10}, \ldots, A_{m-1,0}$ which form the first column of the matrix C. This is because the first column of matrix C represents the first q columns of L and each of these columns contains exactly one cell of τ. Let the entries in these q cells be the integers $a_{00}q + b_{00}, a_{10}q + b_{10}, \ldots a_{q-1,0}q + b_{q-1,0}$ and suppose that the cells belong to the latin subsquares $A_{c_{00}0}, A_{c_{10}0}, \ldots, A_{c_{q-1},0}$, respectively, the integers $c_{00}, c_{10}, \ldots, c_{q-1,0}$ being not necessarily all distinct.

The transversal τ will also contain just q cells belonging to the latin subsquares of the jth column of the matrix C for the same reason as before. (This is true for each fixed j in the range $0 \le j \le m-1$.) Let the entries in these q cells be the integers $a_{0j}q + b_{0j}, a_{1j}q + b_{1j}, \ldots, a_{q-1,j}q + b_{q-1,j}$ and suppose that the cells belong to the latin subsquares $A_{c_{0j}j}, A_{c_{1j}j}, \ldots, A_{c_{q-1,j}j}$, respectively, the integers $c_{0j}, c_{1j}, \ldots, c_{q-1,j}$ being again not necessarily all distinct. Since $a_{ij}q + b_{ij}$, is an element of $A_{c_{ij}j}$, we have $a_{ij} \equiv c_{ij} + j \bmod m$ in consequence of the special choice of labelling which we established at the beginning.

The entries in the n cells of the transversal τ are equal in some order to the n integers $0, 1, \ldots, n - 1$. When these n integers are expressed in the form $a_{ij}q + b_{ij}$, there exist just q of them which have the same fixed value of a_{ij} in the range $0 \le a_{ij} \le m - 1$ but have different values of b_{ij} since b_{ij} varies through the integers $0, 1, \ldots, q - 1$. Hence,
$$\sum_{\tau} a_{ij} = q(0 + 1 + \ldots + m - 1) = qm(m - 1)/2$$
where the summation is over the entries in all the cells of τ. Also, $\sum_{\tau} a_{ij} \equiv \sum_{\tau}(c_{ij} + j) \bmod m$. The right-hand side of this congruence is equal to $\sum_{\tau} c_{ij} + q(0 + 1 + \ldots + m - 1)$ since j varies between 0 and $m - 1$ and takes each of these values for exactly q of the cells of τ: namely, it takes the value j for all the cells occurring in the q columns of L which are included in the latin subsquares of the jth column of the matrix C. Further, since τ has exactly one cell in each row of L, it has exactly q cells belonging to the set of latin subsquares $A_{h0}, A_{h1}, \ldots, A_{h,m-1}$, which form the hth row of the matrix C. For each of these cells, $c_{ij} = h$. Hence, as h varies between 0 and $m - 1$, c_{ij} takes each of these values q times, and so $\sum_{\tau} c_{ij} = q(0 + 1 + \ldots + m - 1)$ where the summation is over the entries in all the cells of τ as before. Hence,
$$\sum_{\tau}(c_{ij} + j) = qm(m - 1)/2 + qm(m - 1)/2 = qm(m - 1).$$
It follows that $\sum_{\tau} a_{ij} \equiv 0 \bmod m$. But, we have shown already that $\sum_{\tau} a_{ij} = qm(m - 1)/2$. These conditions are not consistent unless $q(m - 1)$ is divisible by

2, so no transversal exists if q is odd and m is even. $\square$

Maillet's theorem was re-discovered by Hedayat and Federer(1969) and by Parker(1971). Astute eaders will notice that Theorem 5.1.5, described on page 162, is a special case of this theorem although it was obtained much later. A shorter (but more sophisticated) proof of the theorem due to Drake(1977)and using the concept of a k-net is given on pages 26-28 of [DK2].

Let t_n denote the number of transversals that exist in the reduced latin square K_n which represents the multiplication table of the cyclic group of order n. Then Singer(1960) noted that $t_n = 0$ if n is even (see above) and he obtained the following values for small odd values of n: $t_1 = 1$, $t_3 = 3$, $t_5 = 15$, $t_7 = 133$, $t_9 = 2025$, and $t_{11} = 37851$. He also showed that $t_n \equiv 0 \bmod n$ for all values of n. Later, Belyavskaya(1971) described a computer algorithm for finding all the transversals of a latin square which she used to confirm the value of t_7 obtained by Singer. Over the intervening three decades faster computers and better algorithms have revealed more terms of the sequence $\{t_n\}$. At the time of writing it had been calculated up to t_{23} by Shieh, Hsiang and Hsu(2000). The sequence is number A006717 in Sloane's encyclopedia of integer sequences(1973,1994). Not much is known about the asymptotic growth rate of $\{t_n\}$, although McKay, McLeod and Wanless(2006) have shown that $t_n = o(0.62^n n!)$ (and also that t_n is always odd when n is odd). See also Cavenagh and Wanless(2010).

The result $t_n \equiv 0 \bmod n$ is valid for a wider class of latin squares than that discussed by Singer. In particular, it is valid for all latin squares which satisfy the quadrangle criterion. More generally, Belyavskaya and Russu(1975) have proved that the number t_n of transversals of an arbitrary standard form latin square L_n of order n is congruent to zero modulo the number of elements in the left nucleus N_λ (or the right nucleus N_μ) of the loop of which L_n (bordered by its own first row and column) is the multiplication table. For the proof, see Section 3, Chapter 2, of [DK2]. (When G is a group, N_λ coincides with G and then the latin square L_n satisfies the quadrangle criterion.)

Since any latin square is isotopic to one in standard form which represents the multiplication table of a loop G and since the orders of its left and right nuclei are divisors of the order of G, then the number of transversals of such a square L is congruent to zero modulo a divisor of its order. Balasubramanian(1990) has shown further that, if L has even order then the number of its transversals is even. However, the corresponding statement for odd order is false since there exist (non-cyclic) latin squares of odd order with an even number of transversals.

Later, Glynn(2011) gave a very short proof of Balasubramanian's result using a new concept which he called a *latin array*.

More recently, there has been interest in counting the number of transversals and, more generally, chains for group-based latin squares of small order.

We defined the concept of a chain and its rank in Section 3.4. For clarity, we shall denote the number of chains of rank i in a latin square L of order n by $t_n^{(i)}$.

In Belyavskaya(1979), that author computed the number of chains in each of the two non-abelian groups of order 8. For the dihedral group, she obtained $t_8^{(1)} = 8$, $t_8^{(2)} = 296$, $t_8^{(3)} = 1568$, $t_8^{(4)} = 8368$, $t_8^{(5)} = 16640$, $t_8^{(6)} = 12544$, $t_8^{(7)} = 512$ and $t_8^{(8)} = 384$. For the group $\langle a, b : a^4 = e, a^2 = b^2, ab = ba^3 \rangle$, she obtained $t_8^{(1)} = 8$, $t_8^{(2)} = 104$, $t_8^{(3)} = 2208$, $t_8^{(4)} = 7408$, $t_8^{(5)} = 17408$, $t_8^{(6)} = 12288$, $t_8^{(7)} = 512$ and $t_8^{(8)} = 384$.

Surprisingly, both $t_8^{(7)}$ and $t_8^{(8)}$ are the same for both groups so, in particular, both groups have the same number $t_8^{(8)}$ of transversals. Later, Bedford and Whitaker(1999) offered an explanation for the remarkable fact that all four non-cyclic groups of order 8 have 384 transversals.

In Bedford(1993), that author enumerated the left neofields of all orders up to 9 inclusive. Since, for each group G of order n, the number of left neofields which have G as multiplicative group is equal to the number of complete mappings and near-complete mappings (equivalent to chains of rank $n - 1$) of that group, see Section 7.6 or Hsu and Keedwell(1984), it follows that the enumeration of left neofields provides an enumeration of transversals (complete mappings) and chains of rank $n - 1$ of G. Thus, in effect, Bedford gave an enumeration of these items for groups of orders 2 to 9. This list was reproduced in the first edition of the CRC Handbook of Combinatorial Designs [Eds. Colbourn and Dinitz (1996)]. Also, as much as 20 years earlier, Zhang and Dai(1964) and Zhang, Xiang and Dai(1964) had obtained the numbers of orthomorphisms possessed by all groups of orders less than 15.

More recently, the number of transversals in all groups of orders $n \leq 23$ have been computed. See Wanless(2011) for the details.

Also, some work on counting the number of transversals in general latin squares of small order has been done. Let $t(n)$ and $T(n)$ denote respectively the minimum and maximum number of transversals among the latin squares of order n. In McKay, McLeod and Wanless(2006), an exhaustive computation of the number of transversals in all latin squares of orders up to 9 has been carried out. The values of $t(n)$ and $T(n)$ for these orders are given in Wanless(2011).

There has also been work done on counting sets of disjoint transversals. Following Wanless(2011), we describe such a set as *maximal* if it is not contained in a larger set of disjoint transversals. Let $\Lambda(L)$ be the largest cardinality of any set of disjoint transversals in the latin square L and let $\alpha(L)$ be the smallest cardinality of any maximal set of disjoint transversals of L. Then $0 \leq \alpha \leq \Lambda \leq n$. In Egan and Wanless(2011) both Λ and α have been computed for all main classes of latin squares of order 9 and Λ has been counted for all latin squares of orders less than 9. Tables giving the results are in Wanless(2011).

In contrast to this, Clark and Lewis(1997) were interested in counting so-called *double transversals*: that is transversals of the (left) cyclic latin square which are also transversals of an alternative version of this square which they called (for reasons unclear to the present author) the right cyclic latin square.

Two further results relating to enumeration of transversals are (i) if G is a group of order $n \not\equiv 1 \mod 3$, then the number of transversals in the Cayley table of G is divisible by 3 [McKay, McLeod and Wanless(2006)]; and (ii) for all even $n \geq 10$, there exists a latin square of order n which has transversals but in which every transversal coincides on one particular cell [Egan and Wanless(2012)]. (Contrast this with the fact that, for all odd $m \geq 3$, there exists a latin square L of order $3m$ which contains an $(m - 1) \times m$ latin sub-rectangle none of whose entries is in any transversal of L [Egan and Wanless(2012)]. This generalizes a statement made in Section 3.5.)

A result obtained by Hedayat(1972) which is related to property (ii) is that the number of transversals of an arbitrary latin square of order n which have exactly one cell in common cannot exceed $n - 2$ and he has given examples for which this bound is attained.

Questions concerning numbers of transversals remain a fruitful topic for new research.

Let K_n denote the standard form latin square which represents the multiplication table of the cyclic group of order n as before. In the first edition of this book, the authors conjectured that for odd n no two distinct sets of n disjoint transversals of K_n have a transversal in common. This is true for $n \leq 5$ but for larger n it is a long way from the truth. In fact, a more likely conjecture is that for odd $n \geq 7$ every transversal of K_n can be extended in at least two distinct ways to a decomposition of K_n into transversals (cf. Theorem 1.5.2). This has been confirmed for $n = 7$, $n = 9$ and $n = 11$ by Wanless (unpublished).

In Figure 4.5.4 we display three latin squares each of which is orthogonal to K_7. The three squares have the symbols 1, 2 and 3 occurring in identical positions, demonstrating that K_7 has 3 distinct decompositions into transversals which share three common transversals between them.

Also, in Figure 4.5.5 we give two latin squares which are orthogonal to K_9 and which differ only in the six entries shown in **bold**. From these we deduce that K_9 has two decompositions into disjoint transversals which share seven of their nine transversals!

1	2	3	4	5	6	7		1	2	3	4	5	6	7		1	2	3	4	5	6	7
4	1	7	6	3	2	5		6	1	5	7	3	2	4		5	1	7	6	3	2	4
6	5	2	1	4	7	3		4	7	2	1	6	5	3		7	6	2	1	4	5	3
2	7	5	3	6	1	4		2	6	4	3	7	1	5		2	4	5	3	7	1	6
3	4	1	2	7	5	6		3	5	1	2	4	7	6		3	7	1	2	6	4	5
7	6	4	5	2	3	1		7	4	6	5	2	3	1		4	5	6	7	2	3	1
5	3	6	7	1	4	2		5	3	7	6	1	4	2		6	3	4	5	1	7	2

FIG. 4.5.4.

1	2	3	4	5	6	7	8	9
3	1	2	6	7	5	9	4	8
2	5	1	3	6	7	8	9	4
7	9	8	1	2	3	4	5	6
8	4	9	5	1	2	6	7	3
9	8	4	2	3	1	5	6	7
4	3	6	7	**8**	**9**	1	2	5
6	7	5	9	4	**8**	3	1	2
5	6	7	**8**	**9**	4	2	3	1

1	2	3	4	5	6	7	8	9
3	1	2	6	7	5	9	4	8
2	5	1	3	6	7	8	9	4
7	9	8	1	2	3	4	5	6
8	4	9	5	1	2	6	7	3
9	8	4	2	3	1	5	6	7
4	3	6	7	**9**	**8**	1	2	5
6	7	5	**8**	4	**9**	3	1	2
5	6	7	**9**	**8**	4	2	3	1

FIG. 4.5.5.

4.6   Enumeration of subsquares

Until recently, most of the work done on this topic has concerned enumeration of intercalates. In Heinrich and Wallis(1981), these authors showed that the maximum number of intercalates that any latin square of order n can contain is $n^2(n-1)/4$ if n is even and $n(n-1)(n-3)/4$ if n is odd. They showed that these bounds are attained if and only if $n = 2^h$ in the first case or $n = 2^h - 1$ in the second case. They also investigated lower bounds for the number of intercalates which latin squares of specified orders can contain. More recently, in Browning, Cameron and Wanless(2014), these authors have obtained the value $\frac{1}{8}(n-1)(n-3)(n-15)$ for this lower bound.

A number of authors have dealt with the question of constructing latin squares which have no intercalates. We discussed this topic in [DK2] and mention it again in Section 9.2.

In McKay and Wanless(1999), the number of intercalates in a randomly chosen $k \times n$ latin rectangle R is investigated and it is proved that, for most latin squares of order n, $N(R) \geq n^{3/2-\epsilon}$ where $\epsilon > 0$.

Van Rees(1990) proved that the maximum number of 3×3 subsquares which any latin square of order n can contain is $\frac{1}{18}n^2(n-1)$ and conjectured that this bound is attained only if $n = 3^h$. In a more recent paper, Kinyon and Wanless(2015), these authors have proved a number of equivalent conditions. In particular, the conditions (1) L has $\frac{1}{18}n^2(n-1)$ subsquares of order 3; (2) for any two occurrencies of the same symbol in L, there is a subsquare of order 3 containing these two occurrencies; (3) every cell in L is in $(n-1)/2$ subsquares of order 3; are equivalent. The paper contains many other interesting equivalencies though Van Rees's original conjecture remains unresolved.

Let L be an $n \times n$ latin square. In Browning, Stones and Wanless(2013), bounds on the maximum number $I_k(L)$ of $k \times k$ latin subsquares which L can contain are obtained including exact values for $I_2(L)$ and $I_3(L)$ when $n \leq 9$. The authors also investigate the same question for the numbers of subsquares of orders between k and $2k-1$ (inclusive) and between 1 and n (inclusive) which L can contain. In Browning, Cameron and Wanless(2014), asymptotic values for $I_k(L)$ are obtained for $k = 2, 3$ and 5.

CHAPTER 5

The concept of orthogonality

In this chapter, we introduce the concept of orthogonality between latin squares. First, we consider the case of latin squares which represent the multiplication tables of groups and then we go on to give a historical account of the famous Euler conjecture and its eventual resolution. We show that the maximum possible number of latin squares of order n in a mutually orthogonal set is $n - 1$ and that any such complete mutually orthogonal set represents a finite projective plane.

Next, we show how the concept of orthogonality between latin squares leads on to that of orthogonality between quasigroups, groupoids and triple systems. We end the chapter with a short discussion of various extensions of the idea of orthogonality to other related structures: notably to latin rectangles, latin cubes and permutation cubes, and we also introduce orthogonal arrays.

5.1 Existence questions for incomplete sets of orthogonal latin squares

We begin with the definition of orthogonality. Two latin squares $L_1 = ||a_{ij}||$ and $L_2 = ||b_{ij}||$ on n symbols are said to be *orthogonal* if every ordered pair of symbols occurs exactly once among the n^2 pairs (a_{ij}, b_{ij}), $i, j = 1, 2, \ldots, n$.

It is easy to see by trial that the smallest value of n for which two orthogonal squares exist is three. A pair of orthogonal squares of this order is shown in Figure 5.1.1 and the corresponding ordered pairs (a_{ij}, b_{ij}) are exhibited alongside.

$$
L_1 = \begin{array}{|c|c|c|} \hline 2 & 3 & 1 \\ \hline 1 & 2 & 3 \\ \hline 3 & 1 & 2 \\ \hline \end{array}
\qquad
L_2 = \begin{array}{ccc} 2 & 1 & 3 \\ 1 & 3 & 2 \\ 3 & 2 & 1 \end{array}
\qquad
\begin{array}{ccc} 2,2 & 3,1 & 1,3 \\ 1,1 & 2,3 & 3,2 \\ 3,3 & 1,2 & 2,1 \end{array}
$$

FIG. 5.1.1.

As another example, the reader will find a pair of orthogonal latin squares of order ten displayed in Figure 5.1.2.

If we consider the n cells of the latin square L_2 all of which contain the same fixed entry h say, then the entries in the corresponding cells of L_1 must all be different, otherwise the squares would not be orthogonal. Since the symbol h occurs exactly once in each row and once in each column of L_2, we see that the n entries of L_1 corresponding to the entry h in L_2 form a transversal of L_1

(which we defined in Section 1.5). In Figure 5.1.1, the elements of a transversal are shown enclosed in boxes.

It is immediately obvious from these remarks that

THEOREM 5.1.1 *A latin square of order n possesses an orthogonal mate[1] if and only if it has n disjoint transversals.*

As well as searching for pairs of orthogonal latin squares we can consider the problem of constructing sets of more than two latin squares (of the same order) with the property that each pair of the set is an orthogonal pair. Such a set is called a set of *mutually orthogonal latin squares*, often abbreviated in the literature to MOLS. The name *pairwise orthogonal latin squares* (POLS) has also been used. See, for example, Heinrich(1979) and Owens(1992).

It is worthwhile to comment at this point on the effect of applying isotopisms, which we defined in Section 1.3, to the squares of a set of MOLS. Firstly, orthogonality of the set is unaffected by relabelling the symbols in any or all of the squares. This is because the definition of orthogonality does not depend on the symbols used, only on their positions. But, while it is legitimate to permute the symbols, reordering the rows or columns of one square in a set of MOLS will usually destroy the orthogonality property. However, orthogonality will be preserved if the same isotopism is applied to all the squares simultaneously. For, when any one square is superimposed on any other, each ordered pair of symbols occurs exactly once if the squares are orthogonal, and these ordered pairs are preserved by the simultaneous reordering of the rows or columns. Thus, by changing the symbols in each of the squares separately to 1, 2, ... , n in suitable order, we can make the first row of each square take natural order of these integers. Then, by re-arranging the rows (other than the first) in all the squares simultaneously, we can arrange that the first column of one of the squares is also in natural order $1, 2, \ldots, n$.

Hence, we have:

DEFINITION. A set of MOLS is said to be *a standardized set* [see Bose and Nair(1941)] when, in the first rows of all the squares, the symbols are in natural order, and when, in addition, the symbols of the first column of one of the squares are in natural order. A single latin square is said to be in *standard form* or to be *reduced* when the symbols of both its first row and its first column are in natural order, as we have already mentioned in Section 1.1.

As an example, we may standardize the squares of Figure 5.1.1 in the following way: Rename the symbols 2, 3, 1 of L_1 as 1, 2, 3 respectively and also rename the symbols 2, 1, 3 of L_2 as 1, 2, 3 respectively. This gives a standardized set with the square L_2 in standard form. If it is desired to have the square L_1 in standard form, interchange the second and third rows of both the squares simultaneously.

---

[1] The descriptive term *orthogonal mate* for a latin square L_2 which is orthogonal to a given latin square L_1 was first used by Parker(1963).

A natural question to ask is how large a set of MOLS can be. We can answer this as follows:

THEOREM 5.1.2 *Not more than $n-1$ mutually orthogonal latin squares of order n exist.*

PROOF.   Each of the squares may have its symbols renamed without affecting the orthogonality of the set. By such renamings, we may arrange that the symbols which occur in the first rows of all the squares are $1, 2, \ldots, n$ in natural order as above. [See also Mann(1942).] The symbols in the first cells of the second rows of the squares must then all be different: for suppose two of them were the same, both containing the symbol r, say. Then the ordered pair (r, r) would occur in both the $(1, r)$-th position and the $(2, 1)$-th position in the two squares and the squares could not be orthogonal. None of the squares can have the symbol 1 as the entry in the first cell of the second row, otherwise this symbol would occur twice in the first column of that square. Thus, at most $n-1$ mutually orthogonal squares can exist corresponding to the $n-1$ different symbols distinct from 1 which can appear in the first cells of their respective second rows.[2]   □

Since no larger set is possible, a set of MOLS achieving the bound in Theorem 5.1.2 is said to be *complete*, see Bose and Nair(1941). We study complete sets of MOLS in the next section.

If a latin square is the multiplication table of a group then we know more about its structure and hence we can strengthen Theorem 5.1.1. In that case, the existence of a single transversal is sufficient, by Theorem 1.5.2, to guarantee existence of an orthogonal mate.

THEOREM 5.1.3 *Let L be a Latin square based on a finite group G. The following statements are equivalent.*

 (i) *G has a complete mapping (as defined in Section 1.5),*
 (ii) *L has a transversal,*
 (iii) *L can be decomposed into disjoint transversals,*
 (iv) *There exists a Latin square orthogonal to L.*

Moreover, in the Cayley table of a group of odd order, the entries of the leading diagonal always form a transversal if the row and column borders are ordered in the same way by Theorem 1.5.3, so we have:

THEOREM 5.1.4 *The multiplication table of any group of odd order forms a latin square which possesses an orthogonal mate.*

COROLLARY. *There exist pairs of orthogonal latin squares of every odd order.*

A result due to Mann(1944) which applies to all quasigroups, not only to groups, and which is in contrast to this is the following:

---

[2] An interesting alternative proof of this theorem is in Liang(201?).

THEOREM 5.1.5 *If a latin square L of order $4k + 2$ represents the Cayley table of a quasigroup which contains a subquasigroup of order $2k + 1$ then L has no orthogonal mate.*

PROOF.   This is just a special case of Theorem 4.5.1 because such a square is of 2-step type.   □

However, Mann's theorem of 1944, which we give next, was a refinement of Theorem 5.1.5 and contained results applicable to both even and odd orders.

THEOREM 5.1.6 *(a) Let L be a latin square of order $4n + 2$ whose entries are the symbols $1, 2, \ldots, 4n + 2$. Then if L contains a $(2n + 1) \times (2n + 1)$ submatrix A such that less than $n + 1$ of its cells contain elements distinct from the symbols $1, 2, \ldots, 2n + 1$, L has no orthogonal mate.*
    (b) Let L be a latin square of order $4n + 1$ whose entries are the symbols $1, 2, \ldots, 4n + 1$. Then, if L contains a $2n \times 2n$ submatrix A such that less than $n/2$ of its cells contain elements distinct from the symbols $1, 2, \ldots, 2n$, L has no orthogonal mate.

PROOF.   (a) We may suppose that the rows and columns of L have been re-arranged so that the submatrix A occurs in the first $2n + 1$ rows and columns. Let $L = \left(\frac{A \mid B}{C \mid D} \right)$, where each of A, B, C, D is a $(2n + 1) \times (2n + 1)$ submatrix.
    Suppose that the symbol x occurs r times in the submatrix A. In that case, since x must appear exactly once in each of the first $2n + 1$ rows of L, it must appear $(2n + 1) - r$ times in the submatrix B. But then, since x must appear exactly once in each of the last $2n + 1$ columns of L, it must appear r times in the submatrix D. Thus, each symbol of L appears as many times in the submatrix D as it does in the submatrix A and an even number of times among the cells of A and D combined.
    Let k $(< n + 1)$ be the number of cells of A which contain entries different from the symbols $1, 2, \ldots, 2n + 1$. From what we have just said, it follows that there must be just k cells of D also which contain entries distinct from the symbols $1, 2, \ldots, 2n + 1$. (Each "foreign" symbol which occurs in A occurs in D an equal number of times.) Since each of the symbols $1, 2, \ldots, 2n + 1$ occurs in L altogether $4n + 2$ times and since A and D combined contain $(2n + 1) \times (4n + 2)$ cells, there exist exactly $2k$ cells not in A or D whose entries are among the subset of symbols $1, 2, \ldots, 2n + 1$. Now let us suppose that L has an orthogonal mate L^* and consequently has $4n + 2$ disjoint transversals. Then the preceding remarks imply that at least $4n + 2 - 2k$ transversals have the $2n + 1$ of their cells which contain the symbols $1, 2, \ldots, 2n + 1$ included among the cells of A or D. Also, since only $2k$ of the cells of A and D combined contain symbols distinct from $1, 2, \ldots, 2n + 1$, at most $2k$ transversals of L have cells containing symbols other than $1, 2, \ldots, 2n + 1$ in A or D. Therefore, not less than $(4n + 2 - 2k) - 2k$ transversals have the $2n + 1$ of their cells which contain the symbols $1, 2, \ldots, 2n + 1$ included among the cells of A and D *and have no other cells in A or D.* For

$k < n + 1$, this number of transversals is at least two and is even. The cells of L^* which correspond to the cells of such a transversal of L all contain the same symbol w, say. So, in L^*, w occurs $2n + 1$ times (that is, an odd number of times) among the cells of A^* and D^* combined. (We suppose that L^* has been partitioned in the same way as L.) But, as shown above for L, this is impossible. This contradiction shows that no orthogonal mate L^* can exist for L if $k < n+1$.

(b) As before, we may suppose that the rows and columns of L have been re-arranged so that the submatrix A occurs in the first $2n$ rows and columns. Let $L = \left(\begin{array}{c|c} A & B \\ \hline C & D \end{array} \right)$, where A, B, C, D are submatrices of sizes $2n \times 2n$, $2n \times (2n + 1)$, $(2n + 1) \times 2n$, $(2n + 1) \times (2n + 1)$ respectively. Suppose that the symbol x occurs r times in the submatrix A. In that case, by an argument similar to that of (a), it must occur $2n - r$ times in the submatrix B and $(2n + 1) - (2n - r) = r + 1$ times in the submatrix D. That is, any symbol occurs an odd number of times among the cells of A and D combined. In particular, a symbol x that does not occur in the submatrix A at all occurs exactly once in the submatrix D.

Let k ($< n/2$) be the number of cells of A which contain entries distinct from the symbols $1, 2, \ldots, 2n$. If these entries are all the same, equal to the symbol x say, the symbol x occurs $k + 1$ times in D and the $(4n + 1) - 2n - 1$ symbols of L which do not occur at all in A, each occur just once in D, so there exist $(k+1) + 2n$ cells of D which contain symbols distinct from $1, 2, \ldots, 2n$ and $2k + 1 + 2n$ cells of A and D combined which contain such symbols. If, on the other hand, the k cells of A which contain entries distinct from $1, 2, \ldots, 2n$ all contain different symbols, say the symbols $x_1, x_2, \ldots, x_k$, then each of these symbols occurs twice in D and D also has $(4n+1) - 2n - k$ further cells which contain symbols distinct from $1, 2, \ldots, 2n$, equal to the number of symbols of L which do not occur at all in A. Thus, in this case, there exist $k + 2k + [(4n + 1) - 2n - k] = 2n + 2k + 1$ cells of A and D combined which contain entries distinct from $1, 2, \ldots, 2n$. If the symbols in the k cells of A under discussion are some different and some the same, we shall still get the number $2n + 2k + 1$ of cells of A and D combined which contain symbols distinct from $1, 2, \ldots, 2n$.

Since each of the symbols $1, 2, \ldots, 2n$ occurs $4n + 1$ times in L and since A and D have $(2n)^2 + (2n + 1)^2$ cells all together of which at most $2n + 2k + 1$ contain symbols distinct from this subset, there exist at most $(4n + 1)2n - [(2n)^2 + (2n + 1)^2 - (2n + 2k + 1)] = 2k$ cells not in A or D whose entries are among the subset of symbols $1, 2, \ldots, 2n$. If L had an orthogonal mate L^*, it would have $4n + 1$ disjoint transversals. Of these, at least $(4n + 1) - 2k$ would have the $2n$ of their cells which contain the symbols $1, 2, \ldots, 2n$ included among the cells of A and D. Also, at most $2n + 2k + 1$ transversals could have cells containing symbols other than $1, 2, \ldots, 2n$ included among the cells of A or D and so at least $(4n + 1) - 2k - (2n + 2k + 1) = 2n - 4k$ of the transversals would have the $2n$ of their cells which contain the symbols $1, 2, \ldots, 2n$ included among the cells of A and D *and have no other cells in A or D*. For $k < n/2$, this number of transversals is at least two [$= 2n - 4(n/2 - 1/2)$]. The cells of

L^* which correspond to the cells of such a transversal of L all contain the same symbol x. So, in L^*, x occurs $2n$ times (that is, an even number of times) among the cells of A^* and B^*, which (compare part (a)) gives us a contradiction.   □

M.Hall(1967) contains an interesting alternative proof of the above theorem, using orthogonal arrays.

It has been shown by Ostrowski and Van Duren(1962) that Mann's result (a) is "best possible". With the aid of a computer, these authors have constructed a pair of orthogonal latin squares L and L^* of order 10 such that one square L of the pair contains a 5×5 submatrix A with only 3 entries distinct from the symbols 1, 2, 3, 4, 5, thus showing that for a square L with $n = 2$ and $k = n + 1 = 3$, an orthogonal mate can exist. The squares are as exhibited in Figure 5.1.2

A result similar to that just proved but applicable to MOLS of order $4n + 3$ has been given by Drake(1977) and is proved on page 30 of [DK2].

1 2 3 4 5	6 7 8 9 0		1 2 0 3 4	9 5 7 6 8
4 5 1 2 3	8 0 9 7 6		7 8 9 0 6	3 4 2 1 5
5 4 2 3 1	0 8 7 6 9		0 4 8 5 7	6 9 3 2 1
2 3 5 1 8	9 6 4 0 7		4 9 3 6 5	8 0 1 7 2
3 1 4 8 6	7 9 0 5 2		2 5 6 1 8	4 7 0 9 3
6 8 7 0 9	4 5 2 3 1		3 6 7 2 0	5 1 9 8 4
9 0 8 6 7	2 3 1 4 5		5 1 2 4 9	7 3 8 0 6
7 6 0 9 2	5 4 3 1 8		6 7 5 9 1	2 8 4 3 0
0 9 6 7 4	1 2 5 8 3		9 3 1 8 2	0 6 5 4 7
8 7 9 5 0	3 1 6 2 4		8 0 4 7 3	1 2 6 5 9

FIG. 5.1.2.

Also, Parker(1962b) proved the following interesting theorem:

THEOREM 5.1.7 *If a set of t MOLS of order n has a set of t mutually orthogonal latin subsquares of order u ($u \leq n$), then $n \geq (t+1)u$. Moreover, if a latin square orthogonal to all t squares exists, then $u \leq \frac{n-u}{t+1} + \lfloor \frac{u^2}{n} \rfloor$.*

In the next section, we explain a deduction from this theorem which has a close connection with a well-known result of Bruck concerning projective planes.

Let us remark here that a latin square which has no orthogonal mate has been called a *bachelor square*. Such squares are discussed in more detail in Section 9.1.

Notice that the result of Theorem 5.1.5 alone ensures that the Cayley table of a group of order $4k+2$ cannot possess a transversal, because such a group always has a subgroup of order $2k + 1$ as we may easily see by considering the regular permutation representation of the group. For, certainly the group has elements of order two. These are represented by products of $2k+1$ transpositions; that is, by odd permutations. The product of two odd permutations is even, so the regular representation contains both odd and even permutations. But, in a permutation group which contains both odd and even permutations, the even permutations form a subgroup of index two.

The question of which group-based latin squares possess an orthogonal mate has only recently been answered completely. Existence of a complete mapping in the group is sufficient (and necessary) as already remarked and it is known that such mappings exist if and only if the group has non-cyclic Sylow 2-subgroups. See Section 2.5 for details.

It seems that the basic result given in the statement of Theorem 5.1.5 itself was known to Euler. Euler(1779) had successfully constructed pairs of orthogonal latin squares of odd order and also had a construction for pairs of order equal to any multiple of four. However, he had failed to find a construction which would yield pairs of order an odd multiple of two: in particular, of orders six and ten. He therefore posed the following problem.

"Thirty-six officers of six different ranks and taken from six different regiments, one of each rank in each regiment, are to be arranged, if possible, in a solid square formation of six by six, so that each row and each column contains one and only one officer of each rank and one and only one officer from each regiment."[3]

As will readily be seen, the problem is soluble if and only if there exists a pair of orthogonal latin squares of order six. When no solution was forthcoming, Euler made (in 1782) the bold conjecture that no pairs of MOLS exist of any order n which is an odd multiple of two.

In 1900, by systematic enumeration of cases, Tarry(1900a,b,c,d) proved that no pair of MOLS of order six can exist. In more recent years, several much shorter proofs of the same result have been published. See Fisher and Yates(1934), Yamamoto(1954), Rybnikov and Rybnikova(1966), McCarthy(1976), Betten(1983, 1984), Stinson(1984), Dougherty(1994) and Appa, Magos and Mourtos(2004).

It was almost 180 years before the original conjecture made by Euler was finally resolved although fallacious proofs of the truth of his conjecture were published by several authors: notably by Petersen(1902), Wernicke(1910) and MacNeish(1922). [See also Fleischer(1934).] An explanation of the error in Wernicke's proof was given by MacNeish(1921) and also by Witt(1939). [Further details of the early history of the problem will be found in H.W.Norton(1939).]

The reader may also be interested to look at a series of notes concerning the problem which were published in the Intermédiaire des Mathématiciens in the 1890s. These are, in date order, as follows: Loriga(1894), Maillet(1894a,1895), Akar(1895), Laugel(1896), Brocard(1896), Heffter(1896), Barbette(1898), Dellanoy and Barbette(1898), Tarry(1899), and Lemoine(1899,1900).

The first step in the resolution came in 1958 when Bose and Shrikhande(1959)

---

[3]For a generalization of the problem of the 36 officers, see Rao(1961).

managed to construct an orthogonal mate for a certain latin square of order 22.[4] Shortly afterwards, Parker(1959a,b) used a different construction to obtain an orthogonal mate for a square of order 10. Then, in a combined paper, Bose, Shrikhande and Parker(1960) proved that the Euler conjecture is false for all odd multiples of two except the values $n = 2$ and $n = 6$. All these results were obtained with the aid of statistical designs and we gave a full account of them in [DK1].

For interesting comment on the steps in the resolution of Euler's conjecture and short biographies of Bose, Parker and S.S. Shrikhande, see M. Shrikhande(2010).

5.2   Complete sets of orthogonal latin squares and projective planes

The question, for which values of n do sets of $n - 1$ mutually orthogonal latin squares exist, is still an open one. However, three important facts bearing upon this question are known. In the first place, a set of $n-1$ mutually orthogonal latin squares of order n exists if and only if there exists a finite projective plane of order n. More precisely, every finite projective plane of order n defines and is defined by such a set of squares. This result was first proved by Bose(1938). Alternative proofs will be found in Mann(1944), Keedwell(1966) and Martin(1968). Secondly, it is well known that a finite projective plane of order n exists whenever n is a prime power but, up to the present, no finite projective plane of non-prime power order has been discovered. Thirdly, Bruck and Ryser(1949) have shown that, for a certain infinite set of values of n, there cannot exist any projective plane of order n. The proofs of these three results appear in theorems 5.2.2, 5.2.3 and 5.2.6 below.

As regards the non-existence of projective planes, MacInnes(1907) proved by direct combinatorial arguments that no finite projective plane of order 6 could exist though in fact, of course, this result was already implicit in Tarry's proof of seven years earlier that no orthogonal pair of latin squares of order 6 exists. The non-existence of a plane of order 6 was also implicit in a paper of Safford(1907), published in the same year, and an elementary proof of the non-existence has been given by Rybnikova and Rybnikov(1966).

Safford's paper gave a proof that the following problem, proposed by Veblen and arising from an earlier problem of diophantine analysis, had no solutions. The problem was that of arranging, if possible, 43 distinct objects in 43 sets of seven each in such a way that every pair of objects should lie in one and only one set of seven. It would then follow that each two of the sets of seven would have in common one and only one object. In effect Veblen was asking for a projective plane of order 6 since such a plane would necessarily have 43 lines with seven points on each line.

No further results concerning non-existence were obtained until 1949. In that year, Bruck and Ryser(1949) showed that, when n is congruent to 1 or 2 modulo

---

[4]In Bose and Shrikhande(1960), the original counter-example for $n = 22$ was generalized to give pairs of MOLS of all orders n of the form $36m + 22$.

4 and the square free part of n contains at least one prime factor of the form $4k + 3$, there does not exist a finite projective plane of order n. This result excludes the possibility of the existence of a complete set of latin squares for the orders $n = 6, 14, 21, 22, 30$ and so on. The only other non-existence result known at the present time is the non-existence of a projective plane of order 10. The reader is referred to Chapter 11 of [DK2] for a discussion of this result, and other results on existence and non-existence of planes of small orders. But see also page 174 of this section.

We remind the reader that a *projective plane* comprises a set of elements called *points* and a set of elements called *lines* (which may conveniently be thought of as subsets of the points) with a relation called *incidence* connecting them such that each two points are incident with (belong to) exactly one line, and each two lines are incident with (have in common) exactly one point. The plane is *non-degenerate* if there are at least four points no three of which belong to the same line: that is, if the plane contains a proper quadrangle. It will be assumed from now on that all projective planes to be discussed are non-degenerate. If a non-degenerate plane has a finite number of points on one line, it is called a *finite* projective plane.

THEOREM 5.2.1 *A finite projective plane π necessarily has the same number $n + 1$ of points on every line, has $n + 1$ lines through every point, and has $n^2 + n + 1$ points and $n^2 + n + 1$ lines altogether.*

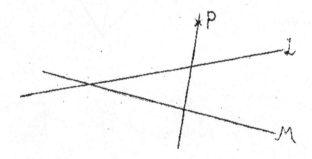

FIG. 5.2.1.

PROOF. Since π is a finite projective plane it has some line $\mathcal{L}$ containing a finite number, say $n + 1$, of points. Let $\mathcal{M}$ be any other line (Figure 5.2.1) and P a point not on $\mathcal{L}$ or $\mathcal{M}$. (Such a point exists because π contains a proper quadrangle.) The lines joining P to the $n + 1$ points of $\mathcal{L}$ intersect $\mathcal{M}$ in $n + 1$ distinct points. $\mathcal{M}$ cannot contain further points otherwise there would be more than $n + 1$ lines through P and they would not all meet $\mathcal{L}$. Since every point of π is on one of the $n + 1$ lines through P (because there is a unique line joining any two points) and since each such line contains n points other than P, there are $n(n + 1) + 1$ points all together. Since there are $n + 1$ lines through the point P, repetition of the argument with the roles of point and line interchanged shows that there are $n+1$ lines through every point and $n(n+1)+1$ lines all together.

□

DEFINITION. A finite projective plane having $n+1$ points on every line is said to be of *order n*.

It may seem perverse not to define the order of the plane to be the number n of points on each line but the definition just given is more natural in consequence of our next result.

THEOREM 5.2.2 *Every finite projective plane of order n defines at least one complete set of MOLS of order n; and conversely a complete set of MOLS of order n defines a finite projective plane.*

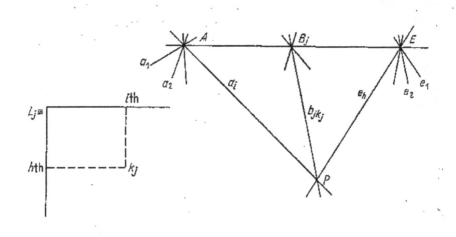

FIG. 5.2.2.

PROOF. Let G be a finite projective plane with $n+1$ points on every line. We pick[5] any line $\mathcal{L}$ of G and call it the line at infinity. Let A, E, $B_1, B_2, \ldots, B_{n-1}$ be the points of $\mathcal{L} = l_\infty$. Through each of these points there pass n lines other than l_∞. We label these lines as follows: $a_1, a_2, \ldots, a_n$ are the lines through A; $e_1, e_2, \ldots, e_n$ are the lines through E; $b_{j1}, b_{j2}, \ldots, b_{jn}$ are the lines through B_j. Every finite point $P(h, i)$ can then be identified with a set of $n+1$ numbers $(h, i, k_1, k_2, \ldots, k_{n-1})$ describing $n+1$ lines $e_h, a_i, b_{1k_1}, b_{2k_2}, \ldots, b_{n-1,k_{n-1}}$ with which it is incident, one through each point of l_∞. A complete set of orthogonal latin squares can now be formed in the following way: in the j-th square, put

---

[5]The ambiguity in this choice is one reason why a projective plane may possibly define more than one set of MOLS up to equivalence. See below for the definition of equivalence.

k_j in the (h,i)-th place. Each square is latin since, as h varies with i fixed, so does k_j and, as i varies with h fixed, so does k_j. Each two squares L_p and L_q are orthogonal: for suppose that k were to appear in the (h_1, i_1)-th place of L_p and also in the (h_2, i_2)-th place and that l were to appear in the corresponding places of L_q. This would imply that the lines b_{pk} and b_{ql} both passed through the two points (h_1, i_1) and (h_2, i_2) in contradiction to the axioms of G. This construction is depicted in Figure 5.2.2.

Conversely, we may show that a complete set of mutually orthogonal latin squares of order n defines a plane of order n. From the given complete set of latin squares, we may define a set of n^2 "finite" points (h, i), $h = 1, 2, \ldots, n$; $i = 1, 2, \ldots, n$; where the point (h, i) is to be identified with the $(n+1)$-tuple $(h, i, k_1, k_2, \ldots, k_{n-1})$, k_j being the entry in the h-th row and i-th column of the j-th latin square L_j. We form $n^2 + n$ lines b_{jk}, $j = -1, 0, 1, 2, \ldots, n-1$; $k = 1, 2, \ldots, n$; where b_{jk} is the set of all points whose $(j+2)$-th entry is k and $b_{-1k} \equiv e_k, b_{0k} \equiv a_k$. Thus, we obtain $n+1$ sets of n parallel[6] lines. From the orthogonality of the latin squares, it follows that two non-parallel lines intersect in one and only one point.

We adjoin one point to each set of parallels and let this additional point lie on every line of the set. These additional points $E, A, B_1, B_2, \ldots, B_{n-1}$ form the *line at infinity*. Two lines then intersect in one and only one point. From this, and the fact that $n+1$ lines pass through every point, it follows that two points have at least one line in common. For, suppose on the contrary that the points Q, R have no line in common. Each of the $n+1$ lines through Q meets each of the $n+1$ lines through R in a point. This gives a total of $(n+1)^2 > n^2 + n + 1$ points all together, no two of which can coincide, otherwise two lines through Q would have a second intersection. This contradiction shows that two points have at least one line in common. It follows at once that they have exactly one line in common.

Thus, we obtain a finite projective plane with $n+1$ points on every line. □

As an example, a complete set of MOLS of order 4 and the corresponding projective plane of order 4 are shown in Figure 5.2.3. The following 21 sets of 5 collinear points are the lines of this plane:

$$
\begin{array}{ccccc}
a_1 & b_1 & e_1 & f_1 & u \\
a_2 & b_2 & e_2 & f_2 & u \\
a_3 & b_3 & e_3 & f_3 & u \\
a_1 & c_1 & e_3 & f_2 & v \\
a_2 & c_2 & e_1 & f_3 & v \\
a_3 & c_3 & e_2 & f_1 & v \\
a_1 & d_1 & e_2 & f_3 & w \\
a_2 & d_2 & e_3 & f_1 & w \\
a_3 & d_3 & e_1 & f_2 & w
\end{array}
$$

$$
\begin{array}{ccccc}
a_2 & a_3 & b_1 & c_1 & d_1 \\
a_3 & a_1 & b_2 & c_2 & d_2 \\
a_1 & a_2 & b_3 & c_3 & d_3 \\
b_1 & b_2 & b_3 & v & w \\
c_1 & c_2 & c_3 & w & u \\
d_1 & d_2 & d_3 & u & v
\end{array}
$$

$$
\begin{array}{ccccc}
b_1 & c_2 & d_3 & e_2 & e_3 \\
b_2 & c_3 & d_1 & e_3 & e_1 \\
b_3 & c_1 & d_2 & e_1 & e_2 \\
b_1 & c_3 & d_2 & f_2 & f_3 \\
b_2 & c_1 & d_3 & f_3 & f_1 \\
b_3 & c_2 & d_1 & f_1 & f_2
\end{array}
$$

---

[6]We call two lines *parallel* if they have no finite point in common.

0	1	2	3
2	3	0	1
3	2	1	0
1	0	3	2

0	1	2	3
3	2	1	0
1	0	3	2
2	3	0	1

0	1	2	3
1	0	3	2
2	3	0	1
3	2	1	0

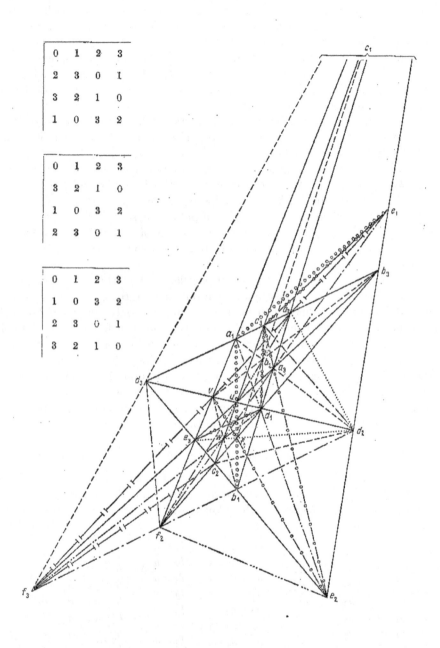

FIG. 5.2.3.

Before proceeding further, we need to say what we mean by equivalence of sets of MOLS.

DEFINITION. Two latin squares are called *equivalent* if one can be obtained from the other by re-naming the symbols and/or reordering the rows and/or the columns: that is, if they (and the quasigroups whose multiplication tables they represent) are isotopic (see Section 1.3). Two sets of MOLS of the same order are *equivalent* if the numbers of squares in the two sets are the same and if one set of squares can be obtained from the other by applying one permutation to the rows of each square, a second to the columns of each square, and then permuting the symbols in each square separately.

THEOREM 5.2.3 *For every integer n that is a power of a prime number, there exists at least one projective plane of order n (and consequently at least one complete set of MOLS of order n).*

PROOF.   It is well known that, corresponding to each integer n which is a prime power $n = p^r$, there exists a Galois field $\mathcal{F} = GF[p^r]$, unique up to isomorphism, which has p^r elements. For the points of our projective plane π, we take the totality of homogeneous co-ordinate vector triples $\mathbf{x} = (x_0, x_1, x_2)$ where x_0, x_1, x_2 are any three elements of $\mathcal{F}$, not all zero, and where the triples $\mathbf{x}$ and $\lambda\mathbf{x}$ represent the same point for all $\lambda \neq 0$ in $\mathcal{F}$. We define the line joining the points $Y(y_0, y_1, y_2)$ and $Z(z_0, z_1, z_2)$ as consisting of the set of all points whose co-ordinate vectors $\lambda\mathbf{y} + \mu\mathbf{z}$ are linear combinations of those of Y and Z, where λ and μ are in $\mathcal{F}$. It is then easy to check that if W is a point of the line YZ, the lines WZ and YZ are the same. That is to say, a line is a well-defined concept. Since each point of the line YZ except Z itself includes among its possible co-ordinate vectors one of the form $\mathbf{y} + \nu\mathbf{z}$ and since ν takes p^r values, each line contains $p^r + 1$ points.

Since each of x_0, x_1, x_2 may take any of p^r values, except that the vector $(0,0,0)$ is to be excluded, and since each point can be represented by $p^r - 1$ different vectors, corresponding to the $p^r - 1$ possible non-zero choices of λ, there are a total of

$$\frac{p^{3r} - 1}{p^r - 1} = p^{2r} + p^r + 1 = n^2 + n + 1$$

points all together.

The line YZ consists of the totality of points whose co-ordinate vectors (x_0, x_1, x_2) satisfy the relation

$$\begin{vmatrix} x_0 & x_1 & x_2 \\ y_0 & y_1 & y_2 \\ z_0 & z_1 & z_2 \end{vmatrix} = 0$$

since they are linear combinations of the co-ordinate vectors of Y and Z. We say that

$$l_0 x_0 + l_1 x_1 + l_2 x_2 = 0, \text{ where } l_i = \begin{vmatrix} y_j & y_k \\ z_j & z_k \end{vmatrix}$$

and i, j, k are a cyclic rearrangement of $0, 1, 2$ is the *equation* of the line and that $[l_0, l_1, l_2]$ are its *line co-ordinates*. The sets of line co-ordinates $[l_0, l_1, l_2]$ and $[\lambda l_0, \lambda l_1, \lambda l_2]$ represent the same line for all non-zero λ in $\mathcal{F}$ and hence, by the same argument as for points, there exists a total of $n^2 + n + 1$ lines in π.

It is an immediate consequence of the definition of a line that any two points of π are incident with exactly one line and also, since just one set of ratios $x_0 : x_1 : x_2$ satisfies the equations of two distinct lines simultaneously, each two lines of π are incident with exactly one point. Moreover, it is easy to see that, even when $\mathcal{F} = GF[2]$, there exist at least four points no three of which are on the same line. Thus, π is a projective plane of order $n = p^r$.   □

DEFINITION. A projective plane which is constructed in the manner just described is called a *Galois plane*.

For example, the plane of order 4 exhibited in Figure 5.2.3 is a Galois plane and the co-ordinates of its 21 points are as follows (where α satisfies the equation $\alpha^2 = \alpha + 1$ and addition is modulo 2):

$$
\begin{aligned}
a_1 &= (1,0,0) & c_1 &= (0,1,\alpha) & e_1 &= (\alpha,1,1) \\
a_2 &= (0,1,0) & c_2 &= (\alpha,0,1) & e_2 &= (1,\alpha,1) & u &= (1,1,1) \\
a_3 &= (0,0,1) & c_3 &= (1,\alpha,0) & e_3 &= (1,1,\alpha) & v &= (1,\alpha,\alpha^2) \\
b_1 &= (0,1,1) & d_1 &= (0,\alpha,1) & f_1 &= (\alpha^2,1,1) & w &= (1,\alpha^2,\alpha) \\
b_2 &= (1,0,1) & d_2 &= (1,0,\alpha) & f_2 &= (1,\alpha^2,1) \\
b_3 &= (1,1,0) & d_3 &= (\alpha,1,0) & f_3 &= (1,1,\alpha^2)
\end{aligned}
$$

It follows from Theorem 5.2.2 and Theorem 5.2.3 that, *if p is a prime number and r any integer, a set of $p^r - 1$ MOLS of order p^r can always be constructed.* Bose(1938) and Stevens(1939) independently gave methods of construction for any given case but it has come to light in more recent years that Moore(1896) had obtained a similar construction nearly 40 years earlier. These constructions are equivalent to the following:

THEOREM 5.2.4 *Let the elements of the Galois field $GF[p^r]$ be denoted by $\alpha_0 = 0$, $\alpha_1 = 1$, $\alpha_2 = x$, $\alpha_3 = x^2, \ldots, \alpha_{p^r-1} = x^{p^r-2}$, where x is a generating element of the multiplicative group of $GF[p^r]$. For $k = 1, 2, \ldots, p^r - 1$ define a latin square L_k in which, for $i, j = 0, 1, \ldots, p^r - 1$, the entry in row i and column j is $\alpha_i + \alpha_k \alpha_j$. Then the L_k form a complete set of MOLS of order p^r.*

PROOF. It is evident that no two elements of the j-th column of L_k can be the same. If two elements of the i-th row were the same we would have $\alpha_k \alpha_{j_1} = \alpha_k \alpha_{j_2}$ for some $j_1 \neq j_2$. Since $\alpha_k \neq 0$ it has an inverse so $\alpha_{j_1} = \alpha_{j_2}$, which is a contradiction. Thus each square L_k is latin.

Next, suppose that two distinct squares L_a and L_b are not orthogonal. There must be two places, say the (s,t)-th and (u,v)-th places, in which the same entry occurs in both squares. We have $\alpha_s + \alpha_a \alpha_t = \alpha_u + \alpha_a \alpha_v$ and $\alpha_s + \alpha_b \alpha_t =$

$\alpha_u + \alpha_b \alpha_v$. By subtraction, this leads to $(\alpha_a - \alpha_b)(\alpha_t - \alpha_v) = 0$. This implies that $t = v$ and hence also $s = u$, since by hypothesis $a \neq b$. Thus the positions (s, t) and (u, v) are not distinct and the supposition that two of the squares are not orthogonal is untenable.   □

An important property of the MOLS just constructed is stated in the next theorem.

THEOREM 5.2.5 *Let $L_1, L_2, \ldots, L_{p^r-1}$ be the MOLS constructed in Theorem 5.2.4. Then all the squares have the same first column and their remaining columns are cyclic permutations of those of L_1.*

PROOF.   The entry in row i and column 0 of L_k is $\alpha_i + \alpha_k \alpha_0 = \alpha_i$, because $\alpha_0 = 0$. Therefore, the first column of each square is in natural order 0, 1, ... , $p^r - 1$ of the suffices i. Moreover, the entry in row 0 and column j of L_1 is $\alpha_0 + \alpha_1 \alpha_j = \alpha_j$ so this too is in natural order of the suffices. Thus, the MOLS are a standardized set (relative to columns).

Also, $\alpha_a = x^{a-1}$ for $a \neq 0$, where x is a generating element of the multiplicative group of $GF[p^r]$ and so, for $j \neq 0$ and $k \neq 0$, we have $\alpha_k \alpha_j = x^{k-1+j-1} = \alpha_t$ where $t \equiv j + k - 1 \bmod p^r - 1$, and $1 \leq t \leq p^r - 1$. It follows that the j-th column of L_k, which is

$$\alpha_0 + \alpha_k \alpha_j, \ \alpha_1 + \alpha_k \alpha_j, \ \alpha_2 + \alpha_k \alpha_j, \ \ldots, \ \alpha_{p^r-1} + \alpha_k \alpha_j,$$

is the same as the t-th column

$$\alpha_0 + \alpha_t, \ \alpha_1 + \alpha_t, \ \alpha_2 + \alpha_t, \ \ldots, \ \alpha_{p^r-1} + \alpha_t$$

of L_1 In other words, the columns of $L_2, L_3, \ldots, L_{p^r-1}$ are obtained by permuting the columns of L_1 as claimed.   □

It is relevant to consider at this point whether other sets of $p^r - 1$ MOLS of order p^r exist which are mathematically distinct from the set which we have constructed above; for example, a set in which the columns of $L_2, L_3, \ldots, L_{p^r-1}$ cannot be obtained by cyclically permuting the columns of L_1.

In consequence of Theorem 5.2.2, the question of whether there exist complete sets of MOLS other than those of Theorem 5.2.4 is related to the question of whether there exist finite projective planes of order p^r which are isomorphically distinct from the Galois plane of that order. However, the former question also involves some subtleties concerning equivalent sets of MOLS which were explored by Owens and Preece(1995,1997). In the first of these papers they showed that up to equivalence, under the definition given on page 171, there are precisely 19 different complete sets of MOLS of order nine. This is despite the fact that there are, up to isomorphism, only four projective planes of order nine[7]. When we look at the construction described in the proof of Theorem 5.2.2, we see that

---

[7]For more details, see below.

the pivotal decisions are the choice of the line at infinity and of the points A and E on that line; different choices of l_∞, A or E can lead to non-equivalent sets of MOLS. Owens(1992) had earlier discovered five transformations which together are capable of converting any complete set of MOLS associated with a projective plane into any other complete set of MOLS associated with either the same plane or its dual. These transformations are described in more detail on page 367 of [DK2].

It is known that for each of the prime power orders up to and including eight, no plane other than the Galois plane of that order exists. For the orders 2, 3 and 4, this was first shown by Veblen and Wedderburn(1907); for the order 5, by MacInnes(1907) by a somewhat laborious enumeration of cases. A short proof was later given by Mann(1944).

Bose and Nair(1941) deduced the non-existence of projective planes of order 7, other than the Galois plane of that order, from an examination of the (incomplete) list of 7×7 latin squares in H.W.Norton(1939). As explained in Section 4.3, the list was later completed by Sade(1951a,b), but this did not affect the result. A properly geometrical proof was not given until 1953 (corrected in 1954) and was due to the combined efforts of Pierce(1953) and M.Hall(1953,1954).

Computers were indispensable in settling the number of different projective planes of orders 8, 9 and 10. The uniqueness of the projective plane of order 8 was shown by M.Hall, Swift and Walker(1956). The method adopted by these authors again made use of the list of 7×7 squares compiled by Norton and completed by Sade. There are exactly four isomorphically distinct planes of order 9, as was shown by Lam, Kolesova and Thiel(1991). For more details of these planes, see Section 8.2. The non-existence of a projective plane of order 10 was shown by Lam, Thiel and Swiercz(1989). A popular account of the proof can be found in Lam(1991). For more details, see pages 377-380 of [DK2].

Also, it is known that at least two isomorphically distinct planes exist for all orders p^r, p odd and $r \geq 2$, and for all orders 2^r, $r \geq 4$ [see chapter 12 of M.Hall(1967)]. Certain of these planes are represented by complete sets of latin squares in which the rows (columns) of $L_2, L_3, \ldots, L_{p^r-1}$ are not a reordering of those of L_1. That is, there exist complete sets of MOLS in which the squares are not all isotopic to each other.

We shall make further mention of these matters in later chapters of this book, but for detailed information concerning the theory of finite projective planes the reader is referred in particular to Hirschfeld(1998), Hughes and Piper(1973) and Kárteszi(1966,1972,1976).

It is relevant at this point to mention two unsolved problems. Since every known finite projective plane has order $n = p^r$, p prime, and can be represented as a set of $n - 1$ MOLS of order n based on the elementary abelian group of order n, we may ask:

(i) Is it true that there do not exist sets of $n - 1$ MOLS of order n based on a cyclic group unless n is prime?

(ii) If $m(n)$ denotes the maximum number of MOLS of order n, none of which satisfies the quadrangle criterion, what is the value of $m(n)$?

At this point in the first edition, [DK1], of this book a conjecture of Schönheim relating to (ii) was stated. The conjecture was that $m(n) < n-1$ for every $n \geq 3$. However, this is disproved by Figure 8.4.3 of [DK1] which purported to represent the eight MOLS of the Hughes plane of order 9. In fact, these MOLS represent the dual translation plane of that order. See Section 8.2.

We end the present discussion of finite planes by proving the important Theorem 5.2.6, known as the Bruck-Ryser theorem [see Bruck and Ryser(1949)].

Before doing so, we draw attention to the promised similarity between the well-known result due to Bruck(1955) that "If a finite projective plane of order n contains a projective subplane of order r, with $r < n$, then $n \geq r^2$; if $n > r^2$, then $n \geq r^2 + r$" and the following deduction from Theorem 5.1.7. "If a set of $r - 1$ MOLS of order n has a set of $r - 1$ mutually orthogonal latin subsquares of order r $(r < n)$, then $n \geq r^2$; If $n > r^2$, a necessary condition that there exists a further latin square of order n orthogonal to all those given is that $n \geq r^2 + r$." The latter result is obtained by putting $t = r - 1$ and $u = r$ in Theorem 5.1.7.

THEOREM 5.2.6 *If n is a positive integer congruent to 1 or 2 modulo 4, there cannot exist any finite projective plane of order n unless n can be expressed as a sum of two integer squares, $n = a^2 + b^2$. Equivalently, if $n \equiv 1$ or 2 modulo 4 and if the square-free part of n contains at least one prime factor p which is congruent to 3 modulo 4, there does not exist a finite projective plane of order n.*

PROOF.   We shall need two theorems from number theory concerning the representation of an integer as a sum of squares.

Theorem A. A positive integer is expressible as a sum of two integer squares if and only if each of its prime factors of the form $4k + 3$ occurs as a factor an even number of times. In particular, every prime of the form $4k + 1$ is representable as a sum of two squares.

Theorem B. Every natural number is representable as the sum of (at most) four integer squares.

Proofs of these two theorems will be found, for example, in chapter 5 of Davenport(2008). The equivalence of the two statements of Theorem 5.2.6 follows immediately from theorem A.

To prove Theorem B it suffices to prove the result for primes and then invoke the identity

$$
\begin{aligned}
(a^2 + &\; b^2 + c^2 + d^2)(A^2 + B^2 + C^2 + D^2) \\
&= (aA + bB + cC + dD)^2 + (aB - bA - cD + dC)^2 + \qquad (5.1) \\
&\quad (aC - bD - cA - dB)^2 + (aD - bC + cB - dA)^2.
\end{aligned}
$$

For the proof of Theorem 5.2.6, we observe that the number of points and lines of a projective plane π of order n are each equal to $N = n^2 + n + 1$ (as was shown in Theorem 5.2.1). For $i = 1$ to N let x_i be a variable over the rational numbers associated with the point P_i of π and let the lines of π be denoted by L_j, for $j = 1$ to N. We define *incidence numbers* a_{ij} as follows: $a_{ij} = 1$ if P_i and L_j are incident and $a_{ij} = 0$ otherwise. The $N \times N$ matrix $A = \|a_{ij}\|$ is then called the *incidence matrix* of the plane π, and it is easy to show that

$$AA^T = A^T A = nI + J \tag{5.2}$$

where the superscript T denotes transpose, I is the $N \times N$ identity matrix and J is an $N \times N$ matrix consisting entirely of 1's. However, for the present proof it is more convenient to express the content of the identity (5.2) in terms of quadratic forms.

With the line L_j we associate the linear form

$$L_j = \sum_{i=1}^{N} a_{ij} x_i,$$

which we may also denote by L_j without confusion. Here the a_{ij} and x_i are defined as above. Then

$$\sum_{i=1}^{N} L_i^2 = (n+1) \sum_{i=1}^{N} x_i^2 + 2 \sum_{r \neq s} x_r x_s = n \sum_{i=1}^{N} x_i^2 + \left(\sum_{i=1}^{N} x_i \right)^2. \tag{5.3}$$

To see this we observe that in the set of the L_j each x_r occurs with a coefficient 1 exactly $n+1$ times, since each point P_r is incident with $n+1$ lines. Also, each cross-product $2x_r x_s$ occurs in $L_1^2 + L_2^2 + \cdots + L_N^2$ exactly once, since there is exactly one line containing both P_r and P_s. We may rewrite (5.3) as

$$\sum_{i=1}^{N} L_i^2 = n \sum_{i=2}^{N} (x_i + x_1/n)^2 + \left(\sum_{i=2}^{N} x_i \right)^2 = y_1^2 + ny_2^2 + ny_3^2 + \cdots + ny_N^2$$

where $y_1 = x_2 + x_3 + \cdots + x_N$ and $y_i = x_i + x_1/n$ for $i \geq 2$. Note that after this change of variables, the y_i are rational since the x_i are.

Next suppose that $n \equiv 1$ or 2 modulo 4 and note that $N = n^2 + n + 1 \equiv 3 \bmod 4$ in both cases. We now use Theorem B to write $n = c_1^2 + c_2^2 + c_3^2 + c_4^2$, where each c_i is an integer. Next we employ (5.1) to see that for any $i \leq N - 3$

$$n(y_i^2 + y_{i+1}^2 + y_{i+2}^2 + y_{i+3}^2) = (c_1^2 + c_2^2 + c_3^2 + c_4^2)(y_i^2 + y_{i+1}^2 + y_{i+2}^2 + y_{i+3}^2)$$

$$= z_i^2 + z_{i+1}^2 + z_{i+2}^2 + z_{i+3}^2$$

for some rational variables z_i, z_{i+1}, z_{i+2} and z_{i+3}. Taking $i = 2, 6, 10, \ldots, N - 5$ successively, and remembering that $N \equiv 3 \pmod 4$, we get

$$\sum_{i=1}^{N} L_i^2 = y_1^2 + z_2^2 + z_3^2 + \cdots + z_{N-2}^2 + ny_{N-1}^2 + ny_N^2.$$

For convenience, define $z_1 = y_1, z_{N-1} = y_{N-1}$ and $z_N = y_N$, so that

$$\sum_{i=1}^{N} L_i^2 = z_1^2 + z_2^2 + z_3^2 + \cdots + z_{N-2}^2 + n(z_{N-1}^2 + z_N^2). \qquad (5.4)$$

At this point we look back through the derivation of this equation and claim that each x_j is a rational linear combination of the z_i, and hence, since $L_1 = \sum_{i=1}^{N} a_{i1}x_i$, there exist rational numbers $b_1, b_2, \ldots, b_N$ such that $L_1 = b_1 z_1 + b_2 z_2 + \cdots + b_N z_N$. Now (5.4) is an identity, so it remains true if we specialize to particular values of the z_i. If $b_1 \neq 1$ we choose z_1 so that

$$(b_1 - 1)z_1 + b_2 z_2 + \cdots + b_N z_N = 0,$$

while if $b_1 = 1$, we choose z_1 so that

$$(b_1 + 1)z_1 + b_2 z_2 + \cdots + b_N z_N = 0.$$

In the first case $L_1 = z_1$ and in the second $L_1 = -z_1$. So in either case $L_1^2 = z_1^2$ and (5.4) reduces to

$$\sum_{i=2}^{N} L_i^2 = z_2^2 + z_3^2 + \cdots + z_{N-2}^2 + n(z_{N-1}^2 + z_N^2).$$

Continuing in this way we may choose z_2 so that $L_2^2 = z_2^2$, z_3 so that $L_3^2 = z_3^2$, and so on until, for these special choices of $z_1, z_2, \ldots, z_{N-2}$, the identity (5.4) reduces to

$$L_{N-1}^2 + L_N^2 = n(z_{N-1}^2 + z_N^2). \qquad (5.5)$$

Now each of L_{N-1}, L_N, z_{N-1} and z_N is rational so (5.5) implies that there are integers $i_1, i_2, \ldots, i_8$ such that

$$n = \frac{(i_1/i_2)^2 + (i_3/i_4)^2}{(i_5/i_6)^2 + (i_7/i_8)^2} = \frac{(i_1 i_4 i_6 i_8)^2 + (i_3 i_2 i_6 i_8)^2}{(i_5 i_2 i_4 i_8)^2 + (i_7 i_2 i_4 i_6)^2} = \frac{r}{s}$$

say, where each prime factor of the form $4k + 3$ occurs an even number of times in the integer r and in the integer s by Theorem A. Hence, the integer n has the same property and so it also is the sum of two integer squares. This proves the theorem.   □

5.3   Sets of MOLS of maximum and minimum size

Many papers have been published which attempt to determine lower or upper bounds for the function $N(n)$ defined as follows:

DEFINITION. The largest possible number of $n \times n$ latin squares which can exist in a single mutually orthogonal set is denoted by $N(n)$.

For example, Theorem 5.2.3 shows that $N(p^r) = p^r - 1$ for every prime p and integer r. Also we have noted that Bose, Shrikhande and Parker(1960) showed that $N(n) \geq 2$ for all $n \geq 3$ except $n = 6$. $N(2) = 1$ is obvious and $N(6) = 1$ was proved by Tarry(1900a). Guérin(1966a,b) proved that $N(n) \geq 4$ for all $n \geq 53$, Hanani(1970) proved that $N(n) \geq 5$ for all $n \geq 63$ and R.M.Wilson(1974) proved that $N(n) \geq 6$ for all $n \geq 90$. However these results have since been improved. Another general result is that of MacNeish.

Using a product construction[8] for obtaining a latin square of order $n_1 n_2$ from latin squares of orders n_1 and n_2, MacNeish(1922) deduced from Theorem 5.2.3 that $N(p_1^{r_1} p_2^{r_2} \ldots p_s^{r_s}) \geq \min_i(p_i^{r_i} - 1)$. In view of Euler's conjecture and such other evidence as was available at that time, he further conjectured that this lower bound is also an upper bound.

One of the earlier results for a case when n is not a prime power (other than those of Tarry, MacNeish and the several people involved in the disproof of Euler's conjecture) was an explicit construction showing that $N(12) \geq 5$. We give more details of this in Section 7.1. (The reader will notice that it immediately disproves MacNeish's conjecture. See also Section 11.1.)

Many other constructions yielding results for individual values of n have been published. Probably the most important of these were $N(14) \geq 3$ obtained by Todorov(1985), $N(15) \geq 4$ obtained by Schellenberg, van Rees and Vanstone(1978) and $N(18) \geq 3$ obtained by X.Zhang and H.Zhang(1997). These results and others have enabled the values obtained by Guérin, Hanani and Wilson to be improved to $N(n) \geq 4$ for all $n \geq 23$ and $N(n) \geq 6$ for all $n \geq 76$. See Colbourn and Dinitz(1996,2006) for a table giving these values and others. It is still not known whether $N(10) > 2$ despite many attempts by computer or otherwise to resolve the problem: notably by Parker(1959c,1961,1962a,1963), Keedwell(1966,1980,1984b), Hedayat, Parker and Federer(1970), Brown and Parker (1985), Acketa and Matić-Kekić(1995) and, more recently, by Appa(2003) and McKay, Meynert, and Myrvold(2007).

A slight refinement of the problem is to ask how many latin squares, all of which are idempotent, can exist in a mutually orthogonal set. In the cases $n = 10, 14$ and 18, the answer so far obtained is the same as without this extra requirement. Again see Colbourn and Dinitz(1996,2006) for more details.

We list here results for individual values of n additional to those already mentioned. To save space, the relevant papers are not in the bibliography of this book. Instead, each result is followed by its Mathematical Reviews reference. $N(n) \geq 4$ for $n = 20$ (90c:05041); $N(n) \geq 4$ for $n = 24$ (88b:05032); $N(n) \geq 4$ for $n = 20, 30, 38, 44$ (94g:05020); $N(n) \geq 4$ for $n = 28, 52$ (92i:05047); $N(n) \geq 5$ for $n = 21$ [MR(1991)11728733]; $N(n) \geq 5$ for $n = 24$ (93d:05027); $N(n) \geq 5$ for $n = 24, 40$ (96a:05028); $N(n) \geq 5$ for $n = 36, 40, 48$ (96c:05026); $N(n) \geq 6$

---

[8]See Section 11.2 for more details.

for $n = 24, 55$ and $N(n) \geq 7$ for $n = 48$ (2000m:05045); $N(n) \geq 6$ for $n = 69$ (86d:05022); $N(n) \geq 7$ for $n = 24, 75$ and $N(n) \geq 8$ for $n = 36$ (2004m:05046); $N(n) \geq 8$ for $n = 48$ and for specified larger values (2007m:05039); $N(n) \geq 3$ idempotent squares for $n = 22, 26$ (97b:05030).

Several of these results were obtained with the aid of orthogonal orthomorphisms. A result relevant to this is that of Quinn(1999).

Further information concerning methods by which some of the above lower bounds for $N(n)$ were obtained can be found in Chapter 5 of [DK2] and in Section 11.1 and Section 11.2 of the present book.

In Wallis(1986), that author gave a fairly concise proof of the fact that $N(n) \geq 3$ for all values of n except 2, 3, 6 and possibly 10 and 14.[9] The proof is described in terms of transversal designs and is worth reading.

In Bedford(1993), that author has shown that several (and probably many) of the constructions of the past twenty years which yield improved lower bounds for $N(n)$ implicitly employ a procedure which can be described as construction by means of left neofields (see Section 4.5) as the author explains.

In Bedford and Whitaker(2000a), these authors discuss maximal sets of MOLS of order n (that is, sets of MOLS which cannot be enlarged) which are constructed from orthogonal orthomorphisms of a group. In particular, they report existence of such maximal sets containing just three squares for $n = 15, 16$ and just four squares for $n = 12, 16, 24, 28$.

Jungnickel(1996) has given a survey of results on this topic obtained prior to that date while Drake, van Rees and Wallis(1999) have proved some more general results. For example, the latter authors have shown the existence of maximal 3-sets of MOLS of order $v = 8t + 1$ whenever $6t + 1$ is a prime power and $t \neq 3$ or 5. Some other interesting results are in A.B.Evans(1991,1992). More recent papers on this topic are Jungnickel and Storme(2003), Govaerts, Jungnickel, Storme and Thas(2003) and Drake and Myrvold(2004). For a discussion of maximal 1-sets, see also Section 9.1. Also relevant to this topic is Theorem 5.1.7.

In contrast to the results on maximal sets of MOLS, Shrikhande(1961) proved the very important fact that every set of $n - 3$ MOLS can be extended to a complete set of $n - 1$ MOLS. See L-Q. Zhang(1963) for a weaker result.

A number of authors have obtained asymptotic results for $N(n)$ as $n \to \infty$. The best of these so far obtained is $N(n) \geq n^{\frac{10}{143}} - 2$ for all sufficiently large n obtained by Lu(1985) which is a marginal improvement on $N(n) \geq n^{\frac{5}{74}}$ for all sufficiently large n obtained by Beth(1983) two years earlier.

Finally, let us mention upper bounds for $N(n)$. We showed in Theorem 5.1.2 that $N(n) \leq n - 1$ for all positive integers n. For the special case of Sudoku latin squares (of order n^2), we shall show that the upper bound becomes $n^2 - n$ and that, for n^2 equal to a prime power, this upper bound can be attained.

---

[9]Todorov's result that $N(14) \geq 3$ was not known at the time Wallis wrote his paper.

$(\times a^r)$	0	1	u_2	...	u_w	...	$u_j a$	$u_j a + 1$	...	$u_j a + 2$	...	$u_j a + u_2$	...	$u_j a + u_w$	...
0															
a^r		$a^r + 1$	$u_2(a^r + 1)$	...	$u_w(a^r + 1)$	:	:	:	:	:	:	:	:	:	:
$u_2 a^r$		:	:	:	:	:	:	:	:	:	:	:	:	:	:
...		:	:	:	:	:	:	:	:	:	:	:	:	:	:
$u_w a^r$		:	:	:	:	:	:	:	:	:	:	:	:	:	:
...		:	:	:	:	:	:	:	:	:	:	:	:	:	:
$u_j a.a^r$		:	:	:	:	:	$u_j a(a^r + 1)$	:	:	:	:	:	:	:	:
$(u_j a + 1)a^r$		:	:	:	:	:	:	$(u_j a + 1)(a^r + 1)$	:	:	:	:	:	:	:
$(u_j a + u_2)a^r$		:	:	:	:	:	:	:	:	:	:	$(u_j a + u_2)(a^r + 1)$	:	:	:
...		:	:	:	:	:	:	:	:	:	:	:	:	:	:
$(u_j a + u_w)a^r$		:	:	:	:	:	:	:	:	:	:	:	:	$(u_j a + u_w)(a^r + 1)$	$\times\times$
...		:	:	:	:	:	:	:	:	:	:	:	:	:	:

Fig. 5.3.1. Square L_r for the case when the order is a prime power.

$(\times 1)$	0	1	u_2	...	u_w	...	$u_w a$	$u_w a + 1$	...	$u_w a + u_{w-1}$	$u_w a + u_w$
0											$u_w a + u_w$
1		$1 + 1$	:	:	:	:	:	:	:	$u_w a + u_w$	$u_w a + u_w$
u_2		:	$u_2(1 + 1)$	:	:	:	:	:	$u_w a + u_w$	:	:
...		:	:	:	:	:	:	:	:	:	:
u_w		:	:	:	$u_w(1 + 1)$	:	$u_w a + u_w$	:	:	:	:
...		:	:	:	:	:	:	:	:	:	:
$u_w a$		:	:	:	:	:	$u_w a(1 + 1)$	:	:	:	:
$u_w a + 1$		:	:	:	:	:	:	$(u_w a + 1)(1 + 1)$	:	:	:
...		:	:	:	:	:	:	:	:	:	:
$u_w a + u_{w-1}$		$u_w a + u_w$	:	:	:	:	:	:	:	:	:
$u_w a + u_w$	$u_w a + u_w$	:	:	:	:	:	:	:	:	:	$(u_w a + u_w)(1 + 1)$

Fig. 5.3.2. Square L_0 for the case when the elements of F have been appropriately ordered.

THEOREM 5.3.1 *Not more than* $n^2 - n$ *mutually orthogonal Sudoku latin squares of order* n^2 *exist.*

PROOF.   As in the proof of Theorem 5.1.2, each of the squares may have its symbols renamed without affecting the orthogonality of the set. By such renamings, we may arrange that the symbols which occur in the first rows of all the squares are $1, 2, \ldots, n^2$ in natural order. The symbols in the first cells of the second rows of the squares must then all be different, again as in the proof of Theorem 5.1.2. But we have the additional requirement that no symbol may occur twice in the leading $n \times n$ subsquare of any of the squares and so none of the entries $1, 2, 3, \ldots, n$ of the first row of this subsquare can occur in its second row. Thus, the maximum number of different entries which can occur in the first (or any other) cell of this second row is $n^2 - n$. Consequently, the number of mutually orthogonal squares is bounded above by this number.   □

Next, we explain how the Moore/Bose/Stevens construction (described in Theorem 5.2.4) may be modified to give a complete set of mutually orthogonal Sudoku squares. We shall require a preliminary lemma.

LEMMA. *If* c, d *are in the same coset* $h + F$ *of* F *and* $x \in K - F$, *where* F *and* K *are defined as in the following Theorem 5.3.2, then* cx, dx *are in different cosets of* F.

PROOF.   Suppose on the contrary that cx, dx are in the same coset $l + F$. Then $cx = l + f_1, dx = l + f_2$ so $cx - dx = f_1 - f_2$. That is $(c - d)x \in F$. Since c, d are in the same coset $h + F$, then $c = h + f_3, d = h + f_4$ so $c - d = f_3 - f_4 \in F$. But then $x \in F$, a contradiction.   □

THEOREM 5.3.2 *Complete sets of mutually orthogonal Sudoku latin squares exist of all prime power orders* p^{2s}. *Moreover, it is possible to construct such sets in which each of the squares is a diagonal latin square.*[10]

PROOF.   Let F be the Galois field of order $q = p^s$ (p prime) with elements $u_0 = 0$, $u_1 = 1$, $u_2, \ldots, u_w$, where $w = q - 1$ and let K be the Galois field of order q^2 with a as a generating element of its multiplicative group. Then K can be regarded as a quadratic extension of F and each non-zero element of K can be expressed either as a power of a or in the form $u_i a + u_j$, $u_i, u_j \in F$. For a fixed element u_i, the elements $u_i a + u_j$ for $j = 0, 1, \ldots, q - 1$ form a coset of F in K. Also, the elements $0, 1 = a^{(q-1)(q+1)}, a^{q+1}, a^{2(q+1)}, \ldots, a^{(q-2)(q+1)}$ are the elements of F expressed as powers of a.

We construct the Cayley table for the additive group of K with the elements of the row and column borders arranged into cosets of F as follows:

---

[10]That is, the elements along each of the main diagonals are all different. For more information about such squares, see the next chapter.

$$0, 1, u_2, \ldots, u_w | a, a+1, a+u_2, \ldots, a+u_w | \ldots$$
$$\ldots | u_i a, u_i a + 1, u_i a + u_2, \ldots, u_i a + u_w | \ldots.$$

This Cayley table L_0, which we display in Figure 5.3.2, is a latin square which can be regarded as partitioned into $q \times q$ subsquares by the cosets.

Next, we re-arrange the rows of L_0 by replacing the row labelled by $u_i a + u_j$ by the row labelled by $(u_i a + u_j) a^r$ for each $a^r \notin F$ to obtain new latin squares L_r (one for each suffix r) as shown in Figure 5.3.1.

Each pair of these latin squares, say L_r and L_s, are orthogonal. (To see this observe that the element in the (j, k)th cell of L_r is $a^r a_j + a_k$, where $a_j + a_k$ is the element in the corresponding cell of the Cayley table of the addition group of K. Then no two cells (j, k) and (l, m) of the juxtaposed pair of Sudoku squares L_r, L_s can contain the same ordered pair of elements since $a^r a_j + a_k = a^r a_l + a_m$ and $a^s a_j + a_k = a^s a_l + a_m$ imply that $j = l$ and $k = m$, whence it follows that L_r and L_s are orthogonal. (This is effectively the Moore/Bose/Stevens construction for $q^2 - 1$ mutually orthogonal latin squares.) Moreover, when $a^r \notin F$, each of the $q \times q$ subsquares into which L_r is separated contains each element of K once (as follows from the lemma above because the elements of each row of a subsquare come from the same coset but elements from different rows come from distinct cosets) so L_r is a Sudoku latin square and it has the left semi-diagonal property as is illustrated in Figure 5.3.1. Thus, we get $q^2 - q$ mutually orthogonal Sudoku latin squares all of which are left semi-diagonal.

Next, we show that we can order the elements of F in such a way that these squares all have the right semi-diagonal property as well. To do so, we need to observe that the latin square L_0 (which we exhibit in Figure 5.3.2) has the entries of its main right-to-left diagonal all equal provided that the elements of F are ordered in such a way that $u_t + u_{w-t} = u_w$ for $t = 0, 1, \ldots, w/2$ or $(w - 1)/2$ according as q is odd or even[11]. It is not a Sudoku latin square but is orthogonal to each of the $q^2 - q$ Sudoku latin squares. Consequently, provided that the elements of F have been correctly ordered before the squares L_r are constructed, each of the latter has the elements of its main right-to-left diagonal all different. Thus, all the Sudoku latin squares can be made diagonal latin squares and so, for all prime power orders p^{2s}, there do exist complete sets of mutually orthogonal diagonal Sudoku latin squares as we claimed.   □

The above modification of the Moore/Bose/Stevens construction is due to Pedersen and Vis(2009) while the observation that the squares can all be made diagonal is due to Keedwell(2010,2011a) and Hilton who suggested that looking at those of the squares L_r which are not Sudoku might lead to a proof that the squares could be made to have the right semi-diagonal property. See also Lorch(2009,2011).

---

[11]There are many such orderings, one for each choice of the element u_w.

5.4   Orthogonal quasigroups, groupoids and triple systems

In view of the fact that any bordered latin square represents the multiplication table of a quasigroup (Theorem 1.1.1), the concept of orthogonality between two latin squares leads naturally to the concept of orthogonality between two quasigroups.

DEFINITION. Two finite quasigroups $(G, \cdot)$ and $(G, *)$ defined on the same set G are said to be *orthogonal* if the pair of equations $x \cdot y = a$ and $x * y = b$ (where a and b are any two given elements of G) are satisfied by a unique pair of elements x and y from G.

It is clear that when $(G, \cdot)$ and $(G, *)$ are orthogonal quasigroups the latin squares defined by their multiplication tables are also orthogonal.

The above definition may be expressed in another way. We may say that $(G, \cdot)$ and $(G, *)$ are orthogonal if $x \cdot y = z \cdot t$ and $x * y = z * t$ together imply $x = z$ and $y = t$. The definition can then be generalized. On the one hand, we can say that two binary operations $(\cdot)$ and $(*)$ defined on the same set G are *orthogonal operations* if the equations $x \cdot y = z \cdot t$ and $x * y = z * t$ together imply $x = z$ and $y = t$, as was done by Sade(1958c) and Belousov(1965). In a later paper, Belousov(1968) discussed in detail the connections between such orthogonal operations and systems of orthogonal quasigroups. On the other hand, we can extend the above definition of orthogonality between quasigroups to the case of infinite quasigroups and also to general groupoids as was done by S.K.Stein(1957), who made the following definition:

DEFINITION. Two finite or infinite groupoids $(G, \cdot)$ and $(G, *)$ defined on the same set G are called *orthogonal* if the mapping σ of the cartesian product $G \times G$ to itself defined by $(x, y) \overset{\sigma}{\to} (x \cdot y, x * y)$ is an equivalence mapping. (If G is a finite set, this implies that σ is a one-to-one mapping of $G \times G$ onto itself.) The groupoid $(G, *)$ is said to be an *orthogonal complement* of the groupoid $(G, \cdot)$.

The following results also come from S.K.Stein(1957).

THEOREM 5.4.1 *A finite commutative groupoid $(G, \cdot)$ is a quasigroup if and only if it is orthogonal to the groupoid $(G, *)$ whose multiplication is given by $x * y = x \cdot xy$.*

PROOF.   Firstly, let $(G, \cdot)$ be a quasigroup and let a and b belong to G. Then there is an element x in G such that $xa = b$ and an element y in G such that $xy = a$. Hence $x * y = x \cdot xy = xa = b$ and so the simultaneous equations $xy = a$ and $x * y = b$ have a solution for x and y, which is necessarily unique since $(G, \cdot)$ is a quasigroup. That is, the groupoids $(G, \cdot)$ and $(G, *)$ are orthogonal.

Secondly, let $(G, \cdot)$ and $(G, *)$ be orthogonal groupoids where $x * y = x \cdot xy$. Suppose, if possible, that the equation $au = c$ has two solutions for u, say $u = y$ and $u = z$. Then $a \cdot ay = a \cdot az = ac$ and so $a * y = a * z$. But, by definition of orthogonality, the equations $ay = az$ and $a * y = a * z$ imply $y = z$. So the equation $au = b$ is uniquely soluble in $(G, \cdot)$. Since $(G, \cdot)$ is commutative, it is a quasigroup.   □

COROLLARY. *Every quasigroup possesses an orthogonal complement (which is not necessarily a quasigroup).*

PROOF.   If $(G, \cdot)$ is the given quasigroup, we define a groupoid $(G, *)$ by $x * y = x \cdot xy$. Then $(G, *)$ is orthogonal to $(G, \cdot)$ by the first part of the proof of the theorem.   □

THEOREM 5.4.2 *A quasigroup $(G, \cdot)$ which satisfies the constraint $y \cdot yx = z \cdot zx \Rightarrow y = z$ possesses an orthogonal complement which is a quasigroup.*

PROOF.   Define a groupoid $(G, *)$ by $x * y = x \cdot xy$. Then the equation $x * y = x * z$ is equivalent to $x \cdot xy = x \cdot xz$ and implies $y = z$ since $(G, \cdot)$ is a quasigroup. The equation $y * x = z * x$ is equivalent to $y \cdot yx = z \cdot zx$ and implies $y = z$ by the hypothesis of the theorem. It follows that $(G, *)$ is a quasigroup and, by the first part of the proof of Theorem 5.4.1, it is orthogonal to $(G, \cdot)$.   □

COROLLARY. *A quasigroup $(G, \cdot)$ which satisfies the constraint $y \cdot yx = xy$ has an orthogonal complement which is a quasigroup.*

PROOF.   Such a quasigroup clearly satisfies the condition $y \cdot yx = z \cdot zx \Rightarrow y = z$.   □

Let L be a latin square with transpose L^T and let $(G, \cdot)$ and $(G, *)$ be the quasigroups whose Cayley tables are obtained by bordering L and L^T with the first row and column of L. Then, by Theorem 5.4.2, $(G, \cdot)$ and $(G, *)$ are orthogonal quasigroups if $x * y = yx = x \cdot xy$. So, when the quasigroup $(G, \cdot)$ satisfies the identity $yx = x \cdot xy$, the latin square L will be orthogonal to its own transpose.

This identity was called the *Stein identity* by Belousov (see Section 2.1) and he called a quasigroup which satisfies it a *Stein quasigroup* [see Belousov(1965), page 102] while Sade(1960a) called such a quasigroup *anti-abelian*. As Stein pointed out, quasigroups which satisfy this identity are idempotent (we have $x \cdot xx = xx$ and so, by left cancellation, $xx = x$) and distinct elements do not commute. We discuss such quasigroups and their associated latin squares in the next section.

Stein hoped that the above theorems might enable him to construct counter-examples to the Euler conjecture about orthogonal latin squares. However, in this connection he was only able to prove the following:

THEOREM 5.4.3 *For all orders $n \equiv 0$, 1 or 3 (mod 4) there exist quasigroups $(G, \cdot)$ satisfying the constraint $y \cdot yx = z \cdot zx \Rightarrow y = z$.*

PROOF.   Let us denote by $A[GF(2^k), \alpha, \beta]$ the groupoid constructed from the Galois field $GF(2^k)$ by the multiplication $x \otimes y = \alpha x + \beta y$, where α and β are fixed elements of $GF(2^k)$.

    If $\alpha\beta(1 + \beta) \neq 0$ and $k \geq 2$, the groupoid $A[GF(2^k), \alpha, \beta]$ satisfies the constraint and is a quasigroup because $x \otimes y = x \otimes z$ implies $\alpha x + \beta y = \alpha x + \beta z$, whence $y = z$ if $\beta \neq 0$, and $y \otimes x = z \otimes x$ implies $\alpha y + \beta x = \alpha z + \beta x$, whence $y = z$

if $\alpha \neq 0$. As regards satisfaction of the constraint, we have $x \otimes (x \otimes z) = y \otimes (y \otimes z)$ implies $\alpha x + \beta (\alpha x + \beta z) = \alpha y + \beta (\alpha y + \beta z)$. That is, $\alpha (1 + \beta)(x - y) = 0$, giving $x = y$ if $\alpha (1 + \beta) \neq 0$. The order of this quasigroup is 2^k, which is congruent to zero modulo 4 if $k \geq 2$.

By forming the direct product of a system of this type with an abelian group of odd order, it is possible to construct systems satisfying the constraint whose orders n are congruent to 1 or 3 modulo 4. (In a group of odd order, $yy = zz$ implies $y = z$, as was proved in Theorem 1.5.3, so $y \cdot yx = z \cdot zx$ implies $y = z$.) The elements of the direct product will be ordered pairs (b_i, c_i), $b_i \in GF(2^k)$, $c_i \in (H, \cdot)$ where $(H, \cdot)$ is a group of odd order. The law of composition $(*)$ will be defined by $(b_i, c_i) * (b_j, c_j) = (b_i \otimes b_j, c_i c_j)$. $\square$

Stein gave the following examples of anti-abelian quasigroups.

(1) The quasigroup of order 4 with multiplication table as shown in Figure 5.4.1.

$$
\begin{array}{c|cccc}
 & 1 & 2 & 3 & 4 \\
\hline
1 & 1 & 3 & 4 & 2 \\
2 & 4 & 2 & 1 & 3 \\
3 & 2 & 4 & 3 & 1 \\
4 & 3 & 1 & 2 & 4 \\
\end{array}
$$

FIG. 5.4.1.

(2) The groupoid $A[GF(p^k), \alpha, \beta]$ when $5^{(p^k - 1)/2} \equiv 1 \bmod p^k$.

(3) The groupoid $A(C_n, p, q)$ constructed from the cyclic group C_n of order n by the multiplication $x \otimes y = x^p y^q$, where p and q are fixed integers, in the case when $p = q^2$, $(2q + 1)^2 \equiv 5 \bmod n$ and n is odd. (Such systems exist only when 5 is a quadratic residue of n.)

The following interesting result was proved by Belousov and Gvaramiya(1966): namely that, if a Stein quasigroup is isotopic to a group G, then the commutator subgroup of G is in the centre of G.

In a paper published several years later than that discussed in the preceding pages, S.K.Stein(1965) himself showed (i) that if a quasigroup $(Q, \cdot)$ of order n has a transitive set of n automorphisms, then there is a quasigroup $(Q, *)$ orthogonal to it (cf. Theorem 7.2.1) and (ii) that no quasigroup of order $4k+2$ has a transitive automorphism group. It follows from (ii) that no counter-examples to the Euler conjecture can be constructed by means of (i).

The properties of systems of orthogonal quasigroups have been investigated by Belousov(1962). Also, Sade(1958c) obtained a number of interesting and useful results concerning orthogonal groupoids, their parastrophes (conjugates) and their isotopes. In a later paper [Sade(1960a)], which we shall discuss in Section 11.2, the same author achieved the goal of Stein: namely, to construct counter-examples to the Euler conjecture. He used a particular type of direct

$$L_1 = \begin{array}{ccccc} 0 & 1 & 2 & 3 & 4 \\ 1 & 2 & 3 & 4 & 0 \\ 2 & 3 & 4 & 0 & 1 \\ 3 & 4 & 0 & 1 & 2 \\ 4 & 0 & 1 & 2 & 3 \end{array} \qquad \begin{array}{ccccc} 0 & 1 & 2 & 3 & 4 \\ 4 & 0 & 1 & 2 & 3 \\ 3 & 4 & 0 & 1 & 2 \\ 2 & 3 & 4 & 0 & 1 \\ 1 & 2 & 3 & 4 & 0 \end{array} \qquad \begin{array}{c|ccccc} & 0 & 4 & 3 & 2 & 1 \\ \hline 0 & 0 & 1 & 2 & 3 & 4 \\ 4 & 4 & 0 & 1 & 2 & 3 \\ 3 & 3 & 4 & 0 & 1 & 2 \\ 2 & 2 & 3 & 4 & 0 & 1 \\ 1 & 1 & 2 & 3 & 4 & 0 \end{array}$$

$$L_2 = \begin{array}{ccccc} 0 & 1 & 2 & 3 & 4 \\ 2 & 3 & 4 & 0 & 1 \\ 4 & 0 & 1 & 2 & 3 \\ 1 & 2 & 3 & 4 & 0 \\ 3 & 4 & 0 & 1 & 2 \end{array} \qquad \begin{array}{ccccc} 0 & 1 & 2 & 3 & 4 \\ 3 & 4 & 0 & 1 & 2 \\ 1 & 2 & 3 & 4 & 0 \\ 4 & 0 & 1 & 2 & 3 \\ 2 & 3 & 4 & 0 & 1 \end{array} \qquad \begin{array}{c|ccccc} & 0 & 4 & 3 & 2 & 1 \\ \hline 0 & 0 & 1 & 2 & 3 & 4 \\ 4 & 3 & 4 & 0 & 1 & 2 \\ 3 & 1 & 2 & 3 & 4 & 0 \\ 2 & 4 & 0 & 1 & 2 & 3 \\ 1 & 2 & 3 & 4 & 0 & 1 \end{array}$$

$$L_3 = \begin{array}{ccccc} 0 & 1 & 2 & 3 & 4 \\ 3 & 4 & 0 & 1 & 2 \\ 1 & 2 & 3 & 4 & 0 \\ 4 & 0 & 1 & 2 & 3 \\ 2 & 3 & 4 & 0 & 1 \end{array} \qquad \begin{array}{ccccc} 0 & 1 & 2 & 3 & 4 \\ 2 & 3 & 4 & 0 & 1 \\ 4 & 0 & 1 & 2 & 3 \\ 1 & 2 & 3 & 4 & 0 \\ 3 & 4 & 0 & 1 & 2 \end{array} \qquad \begin{array}{c|ccccc} & 0 & 4 & 3 & 2 & 1 \\ \hline 0 & 0 & 3 & 1 & 4 & 2 \\ 4 & 1 & 4 & 2 & 0 & 3 \\ 3 & 2 & 0 & 3 & 1 & 4 \\ 2 & 3 & 1 & 4 & 2 & 0 \\ 1 & 4 & 2 & 0 & 3 & 1 \end{array}$$

$$L_4 = \begin{array}{ccccc} 0 & 1 & 2 & 3 & 4 \\ 4 & 0 & 1 & 2 & 3 \\ 3 & 4 & 0 & 1 & 2 \\ 2 & 3 & 4 & 0 & 1 \\ 1 & 2 & 3 & 4 & 0 \end{array} \qquad \begin{array}{ccccc} 0 & 1 & 2 & 3 & 4 \\ 1 & 2 & 3 & 4 & 0 \\ 2 & 3 & 4 & 0 & 1 \\ 3 & 4 & 0 & 1 & 2 \\ 4 & 0 & 1 & 2 & 3 \end{array} \qquad \begin{array}{c|ccccc} & 0 & 4 & 3 & 2 & 1 \\ \hline 0 & 0 & 2 & 4 & 1 & 3 \\ 4 & 2 & 4 & 1 & 3 & 0 \\ 3 & 4 & 1 & 3 & 0 & 2 \\ 2 & 1 & 3 & 0 & 2 & 4 \\ 1 & 3 & 0 & 2 & 4 & 1 \end{array}$$

FIG. 5.4.2.

product of quasigroups which he called *"produit direct singulier"* and which preserves orthogonality.

The idea of forming new quasigroups as singular direct products came to Sade as a consequence of his investigations of the related concept of singular divisors of quasigroups introduced by him earlier in Sade(1950) and discussed further in Sade(1953a,1957). In the opinion of the authors, this concept is a fruitful one but did not at the time receive the attention it deserved.

We end this discussion of orthogonal quasigroups with a theorem of Barra(1963) which we shall use on several occasions later in connection with constructions of pairs of orthogonal latin squares.

THEOREM 5.4.4 *From a given set of* $t \leq n - 1$ *mutually orthogonal quasigroups of order* n, *a set of* t *mutually orthogonal quasigroups of the same order* n *(but usually different from those of the original set) can be constructed of which* $t - 1$ *are idempotent quasigroups.*

PROOF.   Let the multiplication tables of the quasigroups be given by the mutually orthogonal latin squares $L_1, L_2, \ldots, L_t$ all of which are bordered in the same
way. We first rearrange the rows of all the latin squares simultaneously in such
a way that the entries of the leading diagonal of one square, say the square L_1,
are all equal. Since the rearranged squares $L_2, L_3, \ldots, L_t$ are all orthogonal to
L_1, the entries of the leading diagonal of each of them must form a transversal:
that is, be all different. A relabelling of all the entries in any one of the squares
does not affect its orthogonality to the remainder, so we may suppose that the
squares $L_3, L_4, \ldots, L_t$ are relabelled in such a way that the entries of the leading
diagonal of each become the same as those of L_2. If now each of the squares
is bordered by its elements in the order in which these elements appear in the
leading diagonal of L_2, the resulting Cayley tables define mutually orthogonal
quasigroups $Q_1, Q_2, \ldots, Q_t$ of which all but the first are idempotent.   □

An example of the process described in the proof of Theorem 5.4.4 is given
in Figure 5.4.2.

A similar theorem was proved by N.S.Mendelsohn(1971c) who pointed out
that, when n is a prime power, it is always possible to construct sets of $n - 2$
mutually orthogonal idempotent latin squares.

We mention here a curiousity: In Norton and Stein(1956), with each idempotent latin square of order n an integer N associated with that square has been
introduced whose value is dependent on the disposition of the off-diagonal elements of the square (that is, those not on the main left-to-right diagonal). It has
been proved that the relation $N \equiv n(n - 1)/2$ modulo 2 always holds whatever
the explicit value of N.

ORTHOGONAL TRIPLE SYSTEMS

We showed in Section 2.3 that Steiner and Mendelsohn triple systems can
both be represented as quasigroups. If the quasigroups which correspond to two
triple systems of the same kind and size are orthogonal [or, more correctly, *perpendicular* in the case of Steiner triple systems (see later)], we say that the triple
systems are orthogonal. In the case of Steiner triple systems, the quasigroups are
totally symmetric and idempotent and so an equivalent statement is that two
Steiner triple systems of the same order and defined on the same set of elements
are *orthogonal* if (i) the systems have no triples in common and if (ii) when two
pairs of elements appear with the same third element in one system, then they
appear with distinct third elements in the other system. As an illustration, we
give the following two orthogonal Steiner triple systems of order 7:

$$S_1 = \big\{(1\ 2\ 6),\ (2\ 3\ 7),\ (3\ 4\ 1),\ (4\ 5\ 2),\ (5\ 6\ 3),\ (6\ 7\ 4),\ (7\ 1\ 5)\big\}$$
$$S_2 = \big\{(1\ 2\ 4),\ (2\ 3\ 5),\ (3\ 4\ 6),\ (4\ 5\ 7),\ (5\ 6\ 1),\ (6\ 7\ 2),\ (7\ 1\ 3)\big\}$$

In the case of Mendelsohn triple systems, the corresponding quasigroups are
semi-symmetric and idempotent: that is, each satisfies the identities $x(yx) =
y$ and $xx = x$. It turns out that the conditions for two such systems to be

orthogonal are the same as those for Steiner triple systems. We remind the reader that we showed in Section 2.1 that $x(yx) = y \Leftrightarrow (xy)x = y$.

At the time of writing, the question of whether and when directed triple systems are orthogonal had not been considered. We recall from Section 2.3 that when such a system can be represented by a quasigroup, it is called a *latin directed triple system*.

In O'Shaughnessy(1968), that author gave a construction for Room squares[12] which employed a pair of orthogonal Steiner triple systems. For this purpose, he successfully constructed orthogonal pairs of orders 7, 13 and 19 but was not able to construct a pair of order 9. This led him to conjecture that orthogonal pairs exist of all orders $v \equiv 1 \bmod 6$ but that they do not exist for orders $v \equiv 3 \bmod 6$. Mullin and Németh(1969b) proved that orthogonal pairs exist when $v \equiv 1 \bmod 6$ is a prime power and in Mullin and Németh(1970a) that they do not exist when $v = 9$. See also N.S.Mendelsohn(1970) for some earlier results. Then Rosa(1974) constructed an orthogonal pair of order $v = 27$, thus disproving the O'Shaughnessy conjecture. After many partially successful attempts to resolve the existence question, it was finally proved in a joint paper by Colbourne, Gibbons, Mathon, Mullin and Rosa(1994) that orthogonal pairs of Steiner triple systems exist for all $v \equiv 1$ or $3 \bmod 6$ and $v \geq 7$ except $v = 9$. The method used by these authors in their construction requires a generalization of the notion of orthogonality to group divisible designs and also the concept of parastrophic orthogonal quasigroups which we introduce in the next section.

As regards the corresponding question for pairs of Mendelsohn triple systems, it was shown by Bennett and Zhu(1992), who again made use of parastrophic orthogonal quasigroups, that self-orthogonal (which we define below) Mendelsohn triple systems exist of all orders $v \equiv 1 \bmod 3$ except $v = 10$. Later, in Bennett, Zhang and Zhu(1996), those authors showed that such systems also exist of all orders $v \geq 15$ and $v \equiv 0 \bmod 3$ except possibly $v = 18$. This answers the question of existence in the affirmative for all $v \equiv 0$ or 1 except $v = 3, 6, 10$ (which had earlier been ruled out) and possibly $v = 12$ and 18.

5.5   Self-orthogonal and other parastrophic orthogonal latin squares and quasigroups

We showed in Section 1.4 that every latin square is associated with six parastrophes (also called conjugates) including itself of which 1, 2, 3 or all 6 may be distinct. For particular latin squares, it may happen that some or all of these parstrophes are orthogonal. A considerable number of papers have been published which investigate this situation. In particular, every latin square which is not symmetric has a transpose distinct from itself. A latin square which is orthogonal to its own transpose has been mis-called *self-orthogonal*. It is believed

---

[12]We define Room squares and explain O'Shaughnessy's construction in Section 6.4.

that Németh was the first to use this name in his Ph.D. thesis. Shortly afterwards, the name was adopted by N.S.Mendelsohn(1969,1971b). The name has stuck and so we shall follow convention and adopt it.

The first to write about self-orthogonal latin squares were S.K.Stein(1957) and Sade(1960a). However, both writers considered instead the closely related topic of anti-abelian quasigroups, pointing out that an anti-abelian quasigroup has a multiplication table whose body is a self-orthogonal latin square. We have already described some of their results in the preceding section. Both authors and later also Mendelsohn(1971b) obtained results equivalent to the following:

THEOREM 5.5.1 *If $a, b, a + b$ and $a - b$ are integers prime to h in the ring Z_h of residue classes modulo h, then the law of composition $x \cdot y = ax + by + c$ defines an anti-abelian quasigroup $(Z_h, \cdot)$ on the set Z.*

PROOF. The quasigroup $(Z_h, \cdot)$ will be orthogonal to its own transpose provided that, for all x, y, z, t, $xy = zt \Rightarrow yx \neq tz$. So suppose on the contrary that $xy = zt$ and $yx = tz$. These conditions become $ax + by \equiv az + bt$ (mod h) and $ay + bx \equiv at + bz$ (mod h). Therefore, $a(x - z) \equiv b(t - y)$ (mod h) and $b(x - z) \equiv a(t - y)$. From the first equality, $ab(x - z) \equiv b^2(t - y)$ and, from the second, $ab(x - z) \equiv a^2(t - y)$. Thence, $(a - b)(a + b)(t - y) \equiv 0$ (mod h), and so $y = t$ whence also $x = z$.   □

Since $a, b, a + b$ and $a - b$ are all to be prime to h, we find that h has to be prime to 6 since, if a and b are both odd, then $a + b$ is even and so h must be prime to 2; if neither a nor b is a multiple of 3 then either $a + b$ or $a - b$ is a multiple of 3 whence h must also be prime to 3. For all such integers h, anti-abeian quasigroups and consequently pairs of orthogonal latin squares of order h such that one is the transpose of the other, actually exist. Sade takes as an example, $a = 2$ and $b = -1$, from which we see that the operation $x \cdot y = 2x - y$ gives an anti-abeian (and idempotent) quasigroup whenever h is an integer prime to 6.

THEOREM 5.5.2 *In any Galois field $GF(p^n)$, every quasigroup defined on the elements of the field by a law of composition of the form $x \cdot y = ax + by + c$, where a and b are non-zero elements of the field such that $a^2 - b^2 \neq 0$, is anti-abelian. Moreover, such quasigroups exist in all finite fields except $GF(2)$ and $GF(3)$.*

PROOF. Exactly as in Theorem 5.5.1, the conditions $xy = zt$ and $yx = tz$ together imply $(a^2 - b^2)(t - y) = 0$ if a and b are non-zero. Then, provided that $a^2 - b^2 \neq 0$, we have $y = t$ and $x = z$, so $xy = zt \Rightarrow yx \neq tz$ as required for an anti-abelian quasigroup. In any field except $GF(2)$ and $GF(3)$ non-zero elements a and b with distinct squares exist. For example, we can take $a = 1$ and b any element of the field distinct from $0, 1, -1$.   □

In the early 1970s, several papers were devoted to providing constructions for self-orthogonal latin squares of order n for isolated values of n and for various infinite classes of n. These are listed in the bibliography of Brayton, Coppersmith and Hoffman(1976). The problems of existence and construction were brought to the attention of the latter authors by a question concerning a particular type of mixed-doubles tennis tournament (which was christened "spouse-avoiding") because these authors discovered that such tournaments involving n married couples can be constructed when and only when a self-orthogonal latin square of order n exists. Full details can be found in the aforementioned paper, in a summary paper of Brayton, Coppersmith and Hoffman(1974), and also in Keedwell(2000). Brayton *et al* were able to prove that self-orthogonal latin squares exist of all orders n except $n = 2, 3$ and 6.

In T.Evans(1973), that author considered a generalization of self-orthogonality to latin cubes (which latter we define later in this chapter).

Next, the more general question of existence of latin squares which are orthogonal to any one or more of their parastrophes was considered. Phelps(1978) showed that, if there exists a latin square of order n which is orthogonal to its $(2, 3, 1)$-parastrophe, then we can construct from it a latin square of the same order which is orthogonal to its $(3, 1, 2)$-parastrophe and conversely. An exactly similar statement can be made about the $(3, 2, 1)$- and $(1, 3, 2)$-parastrophes. Phelps proved existence of, and gave constructions for, latin squares orthogonal to their $(2, 3, 1)$-parastrophe for all orders n except 2 and 6. He also obtained a similar result for latin squares orthogonal to their $(3, 2, 1)$-parastrophe except that, in the latter case, he was not able to guarantee existence for the orders $n = 14$ and 26.

Belousov(1983b,2005) and, independently, Bennett and Zhu(1992) considered the related question of which quasigroup identities would ensure that the quasigroups defined by such identities would be orthogonal to one or more of their parastrophes. This latter question had earlier been considered by T.Evans(1975) and by Lindner and N.S.Mendelsohn(1973).

Belousov began by considering identities in an algebra $Q(\Sigma)$, where Σ is some system of quasigroup operations (quasigroups) defined on a set Q (cf. Section 2.2). He showed that a non-trivial quasigroup identity must be of minimum length five and must involve two different free variables, one appearing twice and the other three times (as for example in the Stein identity $x \cdot xy = yx$). It is convenient for the following discussion if we use A, B, C, etc. to denote quasigroup operations so that, for example, we write $A(x, B(x, y)) = y$ instead of $x \cdot (x \circ y) = y$, where A, B are respectively the binary operations $(\cdot)$ and $(\circ)$ operating on a set Q of elements.[13] Belousov proved that any non-trivial minimal identity defined in (Q, Σ) can be written in the form $A(x, B(x, C(x, y))) = y$, where A, B, C represent three operations possibly all different. For example, he showed that the identity $A(B(x, y), C(x, y)) = x$ can be re-written as $A^{(13)}(x, C(x, y)) = B(x, y)$ and thence as $A^{(13)}(x, C(x, B^{(23)}(x, z))) = z$, where $z = B(x, y)$. He defined a special binary operation E (not a quasigroup operation) by $E(x, y) = y$.

---

[13]See Section 2.1 for an earlier discussion of notation for parastrophes.

Then, for brevity, he abbreviated the minimal identity $A(x, B(x, C(x, y))) = y$ to $ABC = E$ and remarked that it is easy to see that a quasigroup which satisfies this minimal identity also satisfies the minimal identities $BCA = E$, $CAB = E$, $C^r B^r A^r = E$, $B^r A^r C^r = E$ and $A^r C^r B^r = E$, where r denotes the permutation (2 3). Next, he proved

BELOUSOV LEMMA. *Let A, B be quasigroup operations. Then A, B are orthogonal operations if and only if there is a quasigroup operation K such that $K(x, B(x, (A^{(23)}(x, y))) = E(x, y)$.*

T.Evans(1975) had earlier and independently proved an equivalent lemma for the special case of parastrophic operations: namely,

EVANS LEMMA. *Let A, B be parastrophic (or conjugate) operations on a quasi-group. Then A, B are orthogonal operations if and only if there is a further operation L such that $L(A(x, y), B(x, y)) = x$.*

Evans called an identity of the type just described a *short conjugate-orthogonal identity*. Belousov, on the other hand, called a quasigroup $Q(A)$ which satisfies an identity of the form $A^\alpha(x, A^\beta(x, A^\gamma(x, y))) = y$, where $A^\alpha, A^\beta, A^\gamma$ are parastrophic operations (that is, operations from the set $P = \{A, A^{(12)}, A^{(13)}, A^{(23)}, A^{(123)}, A^{(132)}\}$ as defined in Section 2.1), a Π-*quasigroup of type* $[\alpha, \beta, \gamma]$.

By writing the minimal identity $ABC = E$ in the form $ABC^{rr} = E$, it follows directly from Belousov's lemma that the quasigroup operation B is orthogonal to the operation $C^{(2\ 3)}$ (which we shall write as $B \perp C^r$) and so also $C \perp A^r$ and $A \perp B^r$.

It follows from this that, if $Q(A)$ is of type $[\alpha, \beta, \gamma]$, then $A^\beta \perp A^{\gamma r}$, $A^\gamma \perp A^{\alpha r}$ and $A^\alpha \perp A^{\beta r}$.

Since (as we remarked above) a quasigroup which satisfies the minimal identity $ABC = E$ also satisfies other minimal identities of this canonical form, we may expect that a Π-quasigroup $Q(A)$ of type $[\alpha, \beta, \gamma]$ will also be of other types as well. (For example, it will be of type $[\beta, \gamma, \alpha]$.) The types $[\alpha, \beta, \gamma]$ and $[\beta, \gamma, \alpha]$ are said to be *parastrophically equivalent*.

Belousov showed that there are just seven parastrophically inequivalent types of minimal identity such that a quasigroup which satisfies one of these identities is orthogonal to one or more of its parastrophes.

We list these in Figure 5.5.1 which is taken from Table 1 in Belousov(2005). In that table, r represents the permutation (2 3) as before and l represents the permutation (1 3). However, we treat permutations as right-hand mappings whereas Belousov treats them as left-hand mappings, so the table below differs from that in Belousov(2005). We give identities (3) and (6) from the table as examples.

Consider the identity No. 3. $x(x(y/x)) = y$.

Let $y/x = z$. Then $y = zx$. So $A(z, x) = y$, whence $A^{(13)}(y, x) = z$.

Then $A^{(13)(12)}(x, y) = z$.

No.	Type	Identity	Derived form	Note
1.	$T_1 = [1,1,1]$	$x(x \cdot xy) = y$	$x(x \cdot xy) = y$	
2.	$T_2 = [1,1,l]$	$x(x(x/y)) = y$	$x(y \cdot yx) = y$	
3.	$T_4 = [1,1,rl]]$	$x(x(y/x)) = y$	$x \cdot xy = yx$	Stein's 1st law
4.	$T_6 = [1,l,rl]$	$x(x/(y/x)) = y$	$xy \cdot x = y \cdot xy$	Stein's 2nd law
5.	$T_{10} = [1,lr,l]$	$(x/xy)/x = y$	$xy \cdot yx = y$	Stein's 3rd law
6.	$T_8 = [1,lr,rl]$	$x((y/x)\backslash x) = y$	$xy \cdot y = x \cdot xy$	Schröder's 1st law
7.	$T_{11} = [1,rl,lr]$	$x((y\backslash x)/x) = y$	$yx \cdot xy = y$	Schröder's 2nd law

Fig. 5.5.1.

Now, $(1\ 3)(1\ 2) = (1\ 3\ 2) = (2\ 3)(1\ 3) = rl$. So, $A^{rl}(x,y) = z$.

Therefore, $y/x = z \Leftrightarrow A^{rl}(x,y) = z$ and so the above identity can be written as $A(x, A(x, A^{rl}(x,y))) = y$ or as $[1,1,rl]$.

Consider the identity No. 6. $x((y/x)\backslash x) = y$.

Let $y/x = z$ as before [so that $A^{rl}(x,y) = z$] and let $v = z\backslash x$.

Then $zv = x$ or, equivalently, $A^{(123)}(x,z) = v$ where the identity (6) is $xv = y$ or $A(x, A^{(123)}(x,z)) = y$. Since $(1\ 2\ 3) = (1\ 3)(2\ 3) = lr$, the identity (6) is $A(x, A^{lr}(x, A^{rl}(x,y))) = y$ or $[1, lr, rl]$.

As an example of how the table may be used, consider the Π-quasigroup which satisfies the identity $[1,1,rl]$. Since $rl = (2\ 3)(1\ 3) = (2\ 1\ 3)$, this is the identity
$$A(x, A(x, A^{(2\ 1\ 3)}(x,y))) = y.$$
Let $A^{(2\ 1\ 3)}(x,y) = z$.[14] Then $z \cdot x = y$ so the identity becomes $x \cdot xz = zx$ which is Stein's first law. From above, we find that $A \perp A^{rlr}$, $A^{rl} \perp A^r$ and $A \perp A^r$. Thus, in particular, $Q(A)$ is orthogonal to its parastrophes $Q(A^{(1\ 2)})$ and $Q(A^{(2\ 3)})$. The first of these implies that $Q(A)$ is self-orthogonal.

A full list of orthogonalities between parastrophic operations in Π-quasigroups is given in Belousov's paper.

Working independently, Bennett and Zhu(1992) obtained the same result. However, the list of seven inequivalent quasigroup identities obtained by the latter authors differed slightly from that obtained by Belousov in that the identities T_1 and T_6 were replaced by their duals and T_2 was replaced by the dual of Belousov's $T_5 = [1,1,s]$, where $s = (1\ 2)$, which is parastrophically equivalent to T_2.

For further details of this topic, the reader should refer to the very extensive and detailed papers of Belousov(1983b,2005) and Bennett and Zhu(1992).

[14]If $A(u,v) = w$, then $A^r(u,w) = v$ and so $A^{rl}(v,w) = u$. Thus, $A^{rl}(x,y) = z \Rightarrow A(z,x) = y$.

In some more recent work on self-orthogonal latin squares by Graham and Roberts(1991,2002), these authors have considered maximal sets of pairwise orthogonal self-orthogonal latin squares, say $\{A_1, A_1^T, A_2, A_2^T, \ldots, A_m, A_m^T\}$ of order n. They have shown that, when $n = p^k$, p prime, this maximal number is $(2^k - 2)/2$ when $p = 2$ and is $(p^k - 3)/2$ otherwise. [cf. Theorem 5.1.2 and Theorem 5.3.1.] They have given constructions for such sets using an affine plane and/or a left nearfield. In a later paper, Graham and Roberts(2007), the same authors have established a relationship between complete sets of orthogonal self-orthogonal latin squares and projective planes analogous to that of Theorem 5.2.2.

5.6   Orthogonality in other structures related to latin squares

In this section we consider how the orthogonality concept may be generalized to apply to a number of structures related to latin squares. We consider in turn latin rectangles, permutation cubes, latin cubes and hypercubes and orthogonal arrays.

Latin rectangles were defined in Section 3.1. We say that two latin rectangles of the same size are *orthogonal* if, when one is superimposed on the other, each ordered pair of symbols (r, s) occurs in at most one cell of the superimposed pair. Also, a set of $n - 1$ mutually orthogonal latin rectangles of size $m \times n$, with $m \leq n$ is a *complete set*.

It is easy to see that the definition of orthogonality for latin squares is included as a special case of this more general definition and, by the method of Theorem 5.1.2, that one cannot have more than $n - 1$ mutually orthogonal $m \times n$ latin rectangles if $2 \leq m \leq n$.

The following result was proved by Quattrocchi(1968).

THEOREM 5.6.1 *For every prime p and integer q such that q has no prime divisor less than p there exists at least one complete set of mutually orthogonal $p \times pq$ latin rectangles.*

PROOF.   Throughout this proof equivalences will be modulo pq. For each $k = 1, 2, \ldots, pq - 1$, we define a $p \times pq$ matrix $R_k = \|\alpha_{ij}\|$, $i = 0, 1, \ldots, p - 1$; $j = 0, 1, \ldots, pq - 1$, by $\alpha_{ij} \equiv ik + j$ and $0 \leq \alpha_{ij} \leq pq - 1$.

It is immediate that each of the integers $0, 1, \ldots, pq - 1$ occurs exactly once in each row of R_k and at most once in each column. Consequently, R_k is a $p \times pq$ latin rectangle.

Let us consider two rectangles R_{k_1} and R_{k_2}. Suppose that when they are placed in juxtaposition the ordered pair (s, t) appears both in the cell in row i_1 and column j_1 and in the cell in row i_2 and column j_2. Without loss of generality we may assume that $i_1 > i_2$. Then by the definition of the R_k we have

$$i_1 k_1 + j_1 \equiv s, \quad i_1 k_2 + j_1 \equiv t,$$

and

$$i_2 k_1 + j_2 \equiv s, \quad i_2 k_2 + j_2 \equiv t.$$

These relations imply that

$$(i_1 - i_2)k_1 + (j_1 - j_2) \equiv 0 \equiv (i_1 - i_2)k_2 + (j_1 - j_2) \tag{5.6}$$

and so

$$(i_1 - i_2)(k_1 - k_2) \equiv 0. \tag{5.7}$$

Now recall that q has no prime divisor less than p and $0 \leq i_1 < i_2 \leq p-1$. Hence $i_1 - i_2$ is relatively prime to pq, so (5.7) implies $k_1 \equiv k_2$ which implies that we did not start with distinct rectangles. This is sufficient to show that the $pq - 1$ rectangles are pairwise orthogonal.   □

Quattrocchi made use of this theorem in a construction of generalized affine spaces (equivalent to a certain type of balanced incomplete block design) from similar spaces of smaller order. The latin rectangles were used to define the incidence relation between point and line in the synthesized structure. Much later, Mullen and Shiue(1991) used Quattrocchi's construction to build orthogonal latin rectangles of more general sizes than those constructed by Theorem 5.6.1.

Horák, Rosa and Širáň(1997) considered pairs of what they called *maximal orthogonal rectangles*, which are orthogonal $r \times n$ latin rectangles which cannot be extended to $(r+1) \times n$ latin rectangles. They conjectured that for sufficiently large n such pairs exist for precisely those r which satisfy $n/3 < r \leq n$, and they proved some results in that direction.

We remark here that Wanless(2001) has given a construction for four mutually orthogonal 9×10 latin rectangles and, moreover, these form a latin power set as defined in Section 10.2.[15]

Finally, Asplund and Keranen(2011) have introduced what they call *equitable latin rectangles* and have shown how to construct mutually orthogonal sets of these.

A latin square is a two-dimensional object and the latin rectangle is a generalization of it in the sense that the "size" of one of these dimensions is allowed to be different from the other. A different generalization is obtained if, while retaining a fixed size, we allow the number of dimensions to be increased. If we increase the number of dimensions to three, we obtain what should properly be called a latin cube; an object having n^2 rows, n^2 columns and n^2 files such that each of a set of n elements occurs once in each row, once in each column and once in each file.

An illustrative example for the case $n = 3$ is given in Figure 5.6.1. If the number of dimensions increases still further to m say, we obtain an object which could reasonably be called an m-dimensional latin hypercube. Unfortunately, the terms latin cube and latin hypercube have been used by statisticians to denote

---

[15]It is interesting to compare this result with Brouwer's construction of four almost-orthogonal 10×10 latin squares mentioned on pages 147, 149 and elsewhere in [DK2].

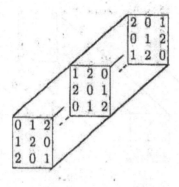

FIG. 5.6.1.

another kind of combinatorial object which we shall describe later in this section. In view of this we shall follow the lead given by Heppes and Révész(1956) and call the objects just introduced *permutation cubes.*

We make the following formal definition:

DEFINITION. An m-dimensional $n \times n \times \cdots \times n$ matrix the elements of which are the integers $0, 1, 2, \ldots, n-1$ will be called an *m-dimensional permutation cube* of order n if every column (that is, every sequence of elements parallel to an edge of the cube) of the matrix contains a permutation of the integers $0, 1, 2, \ldots, n-1$. In particular, a two-dimensional permutation cube is simply a latin square of order n.

The appropriate generalization of a pair of orthogonal latin squares is an m-tuple of "orthogonal" m-dimensional permutation cubes which Heppes and Révész have called a *variational cube.* They have made the following definition.[16]

DEFINITION. The m-dimensional permutation cubes $c_1, c_2, \ldots, c_m$ constitute a variational m-tuple of cubes or, more briefly, a *variational cube* if, among the n^m m-tuples of elements chosen from corresponding cells of the m cubes, every distinct ordered m-tuple involving the integers $0, 1, 2, \ldots, n-1$ occurs exactly once.

As an example, the three 3-dimensional permutation cubes of order 3 shown in Figure 5.6.2 form a variational cube. Each ordered triple of the integers 0,1,2 occurs exactly once. The triple $(0, 0, 1)$, for example, is given by the cell in the second place of the file of the second row and first column. The triple $(2, 1, 0)$ is given by the cell in the third place from the front of the same file.

A set of MOLS is a set such that each pair of its squares forms a graeco-latin square. In just the same way we define a *variational set* (mutually orthogonal set) of k (where $k \geq m$) m-dimensional permutation cubes of order n to be

---

[16] A quite different definition of orthogonality for permutation cubes has been introduced in Höhler(1970) and yet another in Warrington(1973).

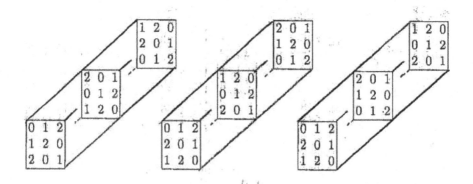

FIG. 5.6.2.

a set having the property that each subset of m cubes is a variational cube. The four 3-dimensional permutation cubes of order 3 shown in Figure 5.6.1 and Figure 5.6.2 together form such a variational set.

In Heppes and Révész(1956), the following two theorems giving a lower bound for k were proved.

THEOREM 5.6.2 *If p is a prime and $m \leq p - 1$ then a system of $p - 1$ m-dimensional permutation cubes on p elements $0, 1, \ldots, p - 1$ can be constructed each m of which form a variational m-tuple of cubes.*

PROOF.   We prove our result by an actual construction. Let the k-th m-dimensional cube of the system be denoted by C_k, $k = 1, 2, \ldots, p - 1$. Let the element of C_k whose co-ordinates are $(x_1, x_2, \ldots, x_m)$, $0 \leq x_i \leq p - 1$, be defined as the residue modulo p of the integer $\sum_{j=1}^{m} x_j k^{j-1}$. Then the difference between the elements in the h-th and the i-th places of that column of C_k which is given by varying x_l and keeping all the other co-ordinates fixed is $(h - i)k^{l-1}$ if $x_l = h$ in the h-th place and $x_l = i$ in the i-th place. Since p is a prime and $h - i$ and k are both integers less than p we know that $(h - i)k^{l-1} \not\equiv 0$ mod p. Consequently, the elements of this typical column are all different, so C_k is a permutation cube.

Next, let m cubes $C_{k_1}, C_{k_2}, \ldots, C_{k_m}$ be chosen arbitrarily from the system. We require to show that these cubes form a variational m-tuple of cubes. That is, we require to show that the p^m m-tuples formed by cells of $C_{k_1}, C_{k_2}, \ldots, C_{k_m}$ respectively having the same co-ordinates are all distinct. Suppose, on the contrary, that those corresponding to the sets of cells having co-ordinates $(x_1, x_2, \ldots, x_m)$ and $(x'_1, x'_2, \ldots, x'_m)$ were the same. Then we would have
$$\sum_{j=1}^{m} x_j k_i^{j-1} \equiv \sum_{j=1}^{m} x'_j k_i^{j-1} \text{ mod } p, \text{ for } i = 1, 2, \ldots, m.$$
That is,
$$\sum_{j=1}^{m} (x_j - x'_j) k_i^{j-1} \equiv 0 \text{ mod } p, \text{ for } i = 1, 2, \ldots, m.$$
We may regard these equations as being m homogeneous linear equations in m quantities $(x_j - x'_j)$ not all of which are simultaneously zero by hypothesis. The equations are consistent only if

$$\begin{vmatrix} 1 & k_1 & k_1^2 & \cdots & k_1^{m-1} \\ 1 & k_2 & k_2^2 & \cdots & k_2^{m-1} \\ \vdots & \vdots & \vdots & \ddots & \vdots \\ 1 & k_m & k_m^2 & \cdots & k_m^{m-1} \end{vmatrix} = 0.$$

The left hand side of this equation is an alternant and can only be zero if two of the k_i are equal. Since the m-tuple of cubes under consideration are distinct by hypothesis, the equations above cannot be simultaneously satisfied and so the cubes form a variational set, as required. □

THEOREM 5.6.3 *If* $n = p_1^{\alpha_1} p_2^{\alpha_2} \cdots p_r^{\alpha_r}$, *where the* p_i *are primes and the* α_i *are positive integers, then a system of* $u = \min_i(p_i^{\alpha_i} - 1)$ m-*dimensional permutation cubes on* n *elements can be constructed each* m *of which form a variational* m-*tuple of cubes provided that* $m \leq u$.

PROOF. As before, we prove the result by giving a construction for the set of cubes. Let $F_i \equiv GF[p_i^{\alpha_i}]$, $i = 1, 2, \ldots, r$, denote the Galois field with $p_i^{\alpha_i}$ elements. Let $\gamma_h = (f_{h1}, f_{h2}, \ldots, f_{hr})$ where $f_{hi} \in F_i$ for $i = 1, 2, \ldots, r$ and $h = 0, 1, \ldots, n-1$. Under the operations $\gamma_g + \gamma_h = (f_{g1} + f_{h1}, f_{g2} + f_{h2}, \ldots, f_{gr} + f_{hr})$ and $\gamma_g \gamma_h = (f_{g1} f_{h1}, f_{g2} f_{h2}, \ldots, f_{gr} f_{hr})$, the r-tuples γ_h form a ring R with a unit element $\gamma_1 = (f_{11}, f_{12}, \ldots, f_{1r})$ where f_{1i} is the unit element of F_i. R contains divisors of zero, but every divisor of zero in R is a vector γ_h which has at least one zero component. The remaining elements of R have inverses in the ring. We take the n elements $\gamma_0, \gamma_1, \ldots, \gamma_{n-1}$ as the elements of our permutation cubes. We may also label the cells of each cube by means of co-ordinates $(x_1, x_2, \ldots, x_m)$ with $x_i \in R$ for $i = 1, 2, \ldots, m$.

We may suppose without loss of generality that $u = \min_i(p_i^{\alpha_i} - 1) = p_1^{\alpha_1} - 1$. Assuming this, we select a set U containing u elements of R as follows:

$$\tilde{\gamma}_1 = (f_{11}, f_{12}, \ldots, f_{1r})$$
$$\tilde{\gamma}_2 = (f_{21}, f_{22}, \ldots, f_{2r})$$
$$\vdots$$
$$\tilde{\gamma}_u = (f_{u1}, f_{u2}, \ldots, f_{ur})$$

where $f_{11}, f_{21}, \ldots, f_{u1}$ are the complete set of non-zero elements of the Galois field F_1 and where $f_{1i}, f_{2i}, \ldots, f_{ui}$ are u distinct non-zero elements of the Galois field F_i, $i = 2, 3, \ldots, r$. Then none of the elements of U is a divisor of zero and neither is any difference $\tilde{\gamma}_l - \tilde{\gamma}_m$ if $\tilde{\gamma}_l$ and $\tilde{\gamma}_m$ are both elements of U.

We are now able to construct the u m-dimensional permutation cubes $C_1, C_2, \ldots, C_u$ whose existence is claimed in the theorem by defining the element of the cube C_k whose co-ordinates are $(x_1, x_2, \ldots, x_m)$, with $x_i \in R$, to be the element $\sum_{j=1}^{m} x_j \tilde{\gamma}_k^{j-1}$ for $k = 1, 2, \ldots, u$ and $\tilde{\gamma}_k \in U$.

As in the previous theorem, we show easily from the fact that $(\tilde{\gamma}_h - \tilde{\gamma}_i)\tilde{\gamma}_k = 0$ implies $h = i$ that each of our cubes has the entries in each of its columns all different and so is a permutation cube. Also, as in the previous theorem, the set of m simultaneous equations

$$\sum_{j=1}^{m}(x_j - x_j')\tilde{\gamma}_k^{j-1} = 0, \ \ k = k_1, k_2, \ldots, k_m,$$

where $\tilde{\gamma}_{k_1}, \tilde{\gamma}_{k_2}, \ldots, \tilde{\gamma}_{k_m}$ are m distinct elements of the set U, cannot be consistent if the set of m quantities $x_j - x_j'$ are not simultaneously all zero unless the determinant

$$\begin{vmatrix} 1 & \tilde{\gamma}_{k_1} & \tilde{\gamma}_{k_1}^2 & \cdots & \tilde{\gamma}_{k_1}^{m-1} \\ 1 & \tilde{\gamma}_{k_2} & \tilde{\gamma}_{k_2}^2 & \cdots & \tilde{\gamma}_{k_2}^{m-1} \\ \vdots & \vdots & \vdots & \ddots & \vdots \\ 1 & \tilde{\gamma}_{k_m} & \tilde{\gamma}_{k_m}^2 & \cdots & \tilde{\gamma}_{k_m}^{m-1} \end{vmatrix}$$

is zero.

This is not the case because none of the quantities $\tilde{\gamma}_{k_l} - \tilde{\gamma}_{k_m}$ is either equal to zero or is a divisor of zero in R. Consequently every subset of m of the u permutation cubes forms a variational set. □

The reader should compare Theorem 5.6.3 with MacNeish's theorem mentioned in Section 5.3. As in the case of MacNeish's theorem, the lower bounds given by Theorems 5.6.2 and 5.6.3 can be exceeded. Our four cubes given in Figures 5.6.1 and 5.6.2 show that the bound given in Theorem 5.6.2 can be exceeded. As regards Theorem 5.6.3, Arkin(1973a) has constructed a variational set of three 3-dimensional permutation cubes on ten elements (that is a variational cube of order 10). Here $m = 2 \times 5$ and $u = \min_i(p_i^{\alpha_i} - 1) = 1$ so the lower bound of 1 is certainly exceeded. Arkin's method makes use of the existence of pairs of orthogonal latin squares of order 10. See also Arkin(1973b,1974) and Arkin and Strauss(1974).

An unsolved problem of considerable difficulty is that of finding, for given values of m and n, the maximum number $\pi_m(n)$ of permutation cubes in a variational set. The corresponding problem for latin squares $(m = 2)$ has already been mentioned in Section 5.2 and is discussed in greater detail in later chapters of this book and also in Chapter 5 of [DK2]. However, as in the case of latin squares, it is not difficult to find an upper bound for $\pi_m(n)$. We have the following theorem, which was first proved by Humblot(1971).

THEOREM 5.6.4 If $\pi_m(n)$ denotes the maximum number of m-dimensional permutation cubes of order n in a variational set, then $\pi_m(n) \le (m - 1)(n - 1)$.

PROOF. The argument is similar to that of Theorem 5.1.2 (which proves the result for the case $m = 2$) and could be used to provide an alternative proof of that theorem. We shall denote the cells of each m-dimensional permutation cube by means of m-tuples of co-ordinates $(x_1, x_2, \ldots, x_m)$, where each x_i takes the values $1, 2, \ldots, n$.

We may rename the symbols in all the cells of any one permutation cube simultaneously without affecting the orthogonality. By such renamings we may arrange that the symbol occurring in the cell with co-ordinates $(i, 1, 1, \ldots, 1)$ is the symbol $i - 1$ and is the same in all the permutation cubes, for $i = 1, 2, \ldots, n$. Assuming this has been done, it now follows that the symbol 0 can occur in the cell $(x_1, 2, 1, 1, \ldots, 1)$ for the same value of $x_1(x_1 \neq 1)$ in at most $m - 1$ of the permutation cubes in the variational set: for if it occurred in the same place in m of the cubes, these m cubes would not form a variational cube. This is true for each of the $n - 1$ possible values of x_1 (namely, $x_1 = 2, 3, \ldots, n$). However, the symbol 0 must occur in some cell of the row $x_2 = 2$ in every permutation cube of the variational set so the total number of permutation cubes in the set cannot exceed $(m - 1)(n - 1)$. □

Next let us introduce the concepts of latin cube and latin hypercube as used by statisticians.

DEFINITION. An $n \times n \times n$ three-dimensional matrix comprising n layers each having n rows and n columns is called a *latin cube* if it has n distinct elements each repeated n^2 times and so arranged that, in each slice parallel to a face of the cube, all the n distinct elements appear and each is repeated exactly n times. [See Fisher(1966), page 85 and Kishen(1950), page 21.] A *latin hypercube* is the analogous concept in more than three dimensions.

This concept differs from that of a permutation cube in that, in the statistician's latin cube, there may exist rows, columns or files in which some of the elements do not occur and others are repeated. Preece, Pearce and Kerr(1973) distinguished the various possibilities by calling a latin cube regular (or 3-regular) if every row, column and file contains each element exactly once (implying that a regular latin cube is the same concept as a permutation cube). Latin cubes can then be called 0-regular, 1-regular or 2-regular according to how many of the three directions (row, column and file) in which they are regular. So for example, the first latin cube in Figure 5.6.3 has each element exactly once in every row and column, though not in the files, so it is 2-regular.

Latin cubes and hypercubes were first introduced by Kishen(1942) and independently by Fisher(1945). Latin hypercubes of a slightly different kind were also introduced in the following year by Rao(1946). Both concepts find a use in connection with the design of statistical experiments. For a description of such applications, see Kerr, Pearce and Preece(1973).

Both Kishen and Fisher also defined orthogonality of latin cubes, as follows.

DEFINITION. Two latin cubes of order n are *orthogonal* if, among the n^3 pairs of elements chosen from corresponding cells of the two cubes, each distinct ordered pair of elements occurs exactly n times.

The maximum number of latin cubes in a set of pairwise orthogonal ones is $n^2 + n - 2$. This was first proved by Plackett and Burman(1943-1946), who actually obtained a more general result of which this is a special case, see page 203.

A construction for orthogonal sets of latin cubes and hypercubes of prime order was described by Brownlee and Loraine(1948). Their method employs the elementary abelian group of type $(p, p, \ldots, p)$. Later, a more comprehensive paper on the subject was written by Kishen(1950) of which we shall give more details below, but first we give some examples to illustrate the concepts just introduced.

In Figure 5.6.3 we exhibit a pair of orthogonal latin cubes of order 3. A set of ten (which is the maximum possible number) pairwise orthogonal ones of the same order is given in Figure 5.6.4.

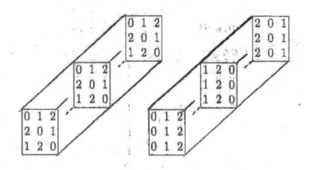

FIG. 5.6.3.

Kishen's paper [Kishen(1950)] on latin cubes and hypercubes was a very detailed one. In the first place, Kishen gave a general formula for the maximum number of m-dimensional latin hypercubes of order n in a set of pairwise orthogonal ones. This number is

$$\frac{n^m - 1}{n - 1} - m.$$

He also showed how a complete set of pairwise orthogonal latin cubes or hypercubes can be constructed from a finite projective space of the appropriate dimension. His method is a generalization of that given in Theorem 5.2.2 for the construction of a complete set of MOLS from a finite projective plane. In the

0 1 2	0 1 2	0 1 2	0 1 2	0 0 0	0 0 0	0 1 2	0 2 1	0 2 1	0 2 1
1 2 0	2 0 1	0 1 2	0 1 2	1 1 1	1 1 1	1 2 0	1 0 2	2 0 1	1 2 0
2 0 1	1 2 0	0 1 2	0 1 2	2 2 2	2 2 2	2 0 1	2 1 0	1 2 0	2 0 1
0 1 2	0 1 2	1 2 0	2 0 1	1 1 1	2 2 2	1 2 0	1 0 2	1 2 0	2 0 1
1 2 0	2 0 1	1 2 0	2 0 1	2 2 2	0 0 0	2 0 1	2 1 0	0 1 2	0 1 2
2 0 1	1 2 0	1 2 0	2 0 1	0 0 0	1 1 1	0 1 2	0 1 2	2 0 1	1 2 0
0 1 2	0 1 2	2 0 1	1 2 0	2 2 2	1 1 1	2 0 1	2 1 0	2 0 1	1 2 0
1 2 0	2 0 1	2 0 1	1 2 0	0 0 0	2 2 2	0 1 2	0 2 1	1 2 0	2 0 1
2 0 1	1 2 0	2 0 1	1 2 0	1 1 1	0 0 0	1 2 0	1 0 2	0 1 2	0 1 2

FIG. 5.6.4.

second place, Kishen made further generalizations of the concepts both of latin hypercube and of orthogonality between pairs of these objects.

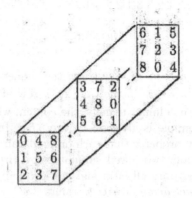

FIG. 5.6.5.

He defined an m-dimensional latin hypercube of order n and *of the r-th class* to be an $n \times n \times \cdots \times n$ m-dimensional matrix having n^r distinct elements, each repeated n^{m-r} times, and such that each element occurs exactly n^{m-r-1} times in each of its m sets of n parallel $(m-1)$-dimensional linear subspaces (or "layers"). Two such latin hypercubes of the same order n and class r with the property that, when one is superimposed on the other, every element of the one occurs exactly n^{m-2r} times with every element of the other, are said to be *orthogonal*. Thus the latin cubes which we have discussed above are all of the first class. An example of a $3 \times 3 \times 3$ latin cube of the second class is exhibited in Figure 5.6.5. Latin cubes of the second class were discussed by Saxena(1960) who gave some methods for their construction additional to those of Kishen.

In Laywine, Mullen and Whittle(1995), these authors proved that the maximum number of mutually orthogonal hypercubes of order $n \geq 2$, dimension $m \geq 2$, and fixed class j with $0 \leq j \leq m-1$ is bounded above by

$$\frac{1}{n-1}\left(n^m - 1 - \binom{m}{1}(n-1) - \binom{m}{2}(n-1)^2 - \ldots - \binom{m}{j}(n-1)^j \right).$$

They also proved that the m-dimensional versions of both Euler's and MacNeish's conjectures (see Section 5.1 and Section 5.3) are false for $m > 2$ whenever they are false for $m = 2$.

The significance of latin cubes and hypercubes is more easily understood if we regard them as special cases of another design which arises by generalizing the concept of a set of orthogonal latin squares in a different way.

Let $L_1, L_2, \ldots, L_{k-2}$ be a set of $k-2$ MOLS of order n whose elements are the integers $1, 2, \ldots, n$ and let a $k \times n^2$ matrix M be formed in the following way. The first row of M has as its elements the integer 1 repeated n times, the integer 2 repeated n times, and so on until finally the integer n is repeated

$$M = \begin{pmatrix} 1\,1\,1\,2\,2\,2\,3\,3\,3 \\ 1\,2\,3\,1\,2\,3\,1\,2\,3 \\ 1\,2\,3\,2\,3\,1\,3\,1\,2 \\ 1\,2\,3\,3\,1\,2\,2\,3\,1 \end{pmatrix} \qquad \begin{array}{|ccc|} \hline 1 & 2 & 3 \\ 2 & 3 & 1 \\ 3 & 1 & 2 \\ \hline \end{array} \qquad \begin{array}{|ccc|} \hline 1 & 2 & 3 \\ 3 & 1 & 2 \\ 2 & 3 & 1 \\ \hline \end{array}$$

FIG. 5.6.6.

n times. The second row of M comprises the sequence of integers $1, 2, \ldots, n$ repeated n times. If $L_h = (a_{ij}^{(h)})$, the $(h + 2)$-th row of M has $a_{ij}^{(h)}$ as the entry in the $[(i - 1)n + j]$-th column: namely, the column whose first two entries are i and j. A simple example is given in Figure 5.6.6. In general, the matrix M so constructed has the property that each ordered pair of elements appears just once as a column in any two-rowed submatrix by virtue of the facts that the squares $L_1, L_2, \ldots, L_{k-2}$ are all latin and are also pairwise orthogonal. We say that it is an *orthogonal array* of k constraints and n levels having strength 2 and index 1. (See also Section 11.1.) More generally, we may make the following definition.

DEFINITION. An *orthogonal array of size N with k constraints, n levels, strength t and index λ* is a $k \times N$ matrix M having n different elements and with the property that each different ordered t-tuple of elements occurs exactly λ times as a column in any t-rowed submatrix of M.

It is immediate from the definition that $N = \lambda n^t$. It is also clear from the construction given above that any orthogonal array of k constraints, n levels, strength 2 and index 1 is equivalent to a set of $k - 2$ MOLS of order n. We shall defer a formal proof of this fact until Section 11.1. From Theorem 5.1.2, we may deduce that, if M is an orthogonal array with $t = 2$ and $\lambda = 1$ then $k \leq n+1$. [See also Rao(1949).] The investigation of upper bounds for the number of constraints in other cases has been the subject of numerous papers. See, in particular, Rao(1946,1947), Bose and Bush(1952), Bush(1952a,b) and Seiden and Zemach(1966). Also, we refer the reader to a more recent monograph by Hedayat, Sloane and Stufken(1999), and to the bibliography therein, for a fairly comprehensive treatment of the subject which, however, puts its main emphasis on the application to coding and to statistical design of experiments.

The connection between these orthogonal arrays and latin cubes and hypercubes is given by the statements: (i) a latin cube of order n is equivalent to an orthogonal array of 4 constraints and n levels having strength 2 and index n; (ii) a set of $k - 3$ mutually orthogonal latin cubes of order n is equivalent to an orthogonal array of k constraints and n levels having strength 2 and index n; (iii) a set of $k - m$ mutually orthogonal m-dimensional latin hypercubes of order n is equivalent to an orthogonal array of k constraints and n levels having strength 2 and index n^{m-2}.

As an illustration of statement (ii), we give in Figure 5.6.7 the orthogonal array M corresponding to the two orthogonal latin cubes shown in Figure 5.6.3.

$$M = \begin{bmatrix} 0\,0\,0\,0\,0\,0\,0\,0\,0\,1\,1\,1\,1\,1\,1\,1\,1\,2\,2\,2\,2\,2\,2\,2\,2\,2 \\ 0\,0\,0\,1\,1\,1\,2\,2\,2\,0\,0\,0\,1\,1\,1\,2\,2\,2\,0\,0\,0\,1\,1\,1\,2\,2\,2 \\ 0\,1\,2\,0\,1\,2\,0\,1\,2\,0\,1\,2\,0\,1\,2\,0\,1\,2\,0\,1\,2\,0\,1\,2\,0\,1\,2 \\ 0\,0\,0\,1\,1\,1\,2\,2\,2\,2\,2\,2\,0\,0\,0\,1\,1\,1\,1\,1\,1\,2\,2\,2\,0\,0\,0 \\ 0\,1\,2\,1\,2\,0\,2\,0\,1\,0\,1\,2\,1\,2\,0\,2\,0\,1\,0\,1\,2\,1\,2\,0\,2\,0\,1 \end{bmatrix}$$

FIG. 5.6.7.

Here the first, second and third rows of M give the row, column and file respectively of the cell of the latin cube whose entry appears below them in the fourth or fifth row of M according to which of the two latin cubes is being considered.

Plackett and Burman(1943-1946) proved that the maximum number of constraints k for an orthogonal array of n levels having strength 2 and index λ satisfies the inequality $k \leq (\lambda n^2 - 1)/(n - 1)$. An alternative proof of the same result was later given by Bose and Bush(1952). If we put $\lambda = n$, we get $k \leq n^2 + n + 1$ for this case and it follows that the maximum number of latin cubes in a pairwise orthogonal set is $n^2 + n - 2$, as stated earlier.

CHAPTER 6

Connections between latin squares and magic squares

The close link that exists between magic squares and latin squares has been recognized by mathematicians ever since the by-now famous investigation of magic squares by Euler(1776,1779) more than two centuries ago. Consequently, no further justification for devoting the first three sections of the present chapter to this subject seems necessary.

In the fourth section of the chapter, we give a historical account of research into the construction of Room squares, since these designs may be regarded as magic squares of a sort and because investigations have shown that they also are inter-connected with latin squares.

6.1 Diagonal (or magic) latin squares

A *magic square* of order n is an arrangement of n^2 integers (usually, but not necessarily, consecutive and usually, but not necessarily, distinct) in a square such that the sums of each row, each column and each of the main diagonals are the same.

Magic squares were known in ancient times in China and India and were often engraved on metal or stone. Even to this day they are worn as amulets. For more details, see L.D.Andersen(2013). The first known European writer on the subject was Moschopoulos, a Greek of Constantinople (c. A.D. 1300). Later, Agrippa (1486-1535) constructed magic squares of orders 3 to 9. A magic square of order 4 is shown in Dürer's picture "Melancholy". Further general information and historical details on the subject of magic squares may be found in Ahrens(1901), Andrews(1917), Armstrong(1955), Bachet(1612), Ball(1939), Dénes(1961), van Driel(1936,1939), Dudeney(1917), Euler(1776), Frolov(1884), Gunther(1876b), Kraitchik(1930,1942), Lucas(1882), Maillet(1906), Postnikov(1964), Schubert(1898) and on the website of C. Boyer among other places. Also, a number of results on the connection between magic squares and latin squares are given in the above books and expository papers.

In this first section we shall discuss the more unusual type of magic square in which only n of the n^2 integers are distinct and each is repeated n times. Among the magic squares of this type are the so-called "diagonal latin squares".

DEFINITION. In this chapter, the diagonal formed by the elements $a_{11}, a_{22}, \ldots, a_{nn}$ of a latin square $A = ||a_{ij}||$ will be called the *main left-to-right diagonal* while the diagonal formed by the elements $a_{1,n}, a_{2,n-1}, \ldots, a_{n,1}$ will be called the *main*

Latin Squares and their Applications. http://dx.doi.org/10.1016/B978-0-444-63555-6.50006-4

right-to-left diagonal. A latin square is called *left semi-diagonal*[1] if the elements on its main left-to-right diagonal are distinct. Similarly, if the elements on the main right-to-left diagonal are distinct, the latin square is called *right semi-diagonal.* A latin square is called *diagonal* if it is both left and right semi-diagonal simultaneously and, of course, every such latin square is also a magic square according to our definition.

THEOREM 6.1.1 *For each positive integer $n \neq 2$ there exists at least one left semi-diagonal latin square of order n and at least one right semi-diagonal latin square of order n.*

PROOF.   For any given $n \neq 2$ there exists an idempotent quasigroup of order n (see Theorem 1.5.6) and the multiplication table of an idempotent quasigroup is a left semi-diagonal latin square. By reversing the order of the columns of a left semi-diagonal latin square one will obtain a right semi-diagonal latin square.   □

$$
\begin{array}{ccccc}
0 & \alpha & 2\alpha & \cdots & (n-1)\alpha \\
\beta & \beta+\alpha & \beta+2\alpha & \cdots & \beta+(n-1)\alpha \\
2\beta & 2\beta+\alpha & 2\beta+2\alpha & \cdots & 2\beta+(n-1)\alpha \\
\vdots & \vdots & \vdots & \ddots & \vdots \\
(n-1)\beta & (n-1)\beta+\alpha & (n-1)\beta+2\alpha & \cdots & (n-1)\beta+(n-1)\alpha
\end{array}
$$

FIG. 6.1.1.

THEOREM 6.1.2 *If n is any odd integer which is not a multiple of three there exists at least one diagonal latin square of order n.*

PROOF.   We first show that if α and β are positive integers such that α, β, $\alpha + \beta$ and $\alpha - \beta$ are all relatively prime to n then the latin square exhibited in Figure 6.1.1 (where the elements are taken modulo n) is a diagonal latin square. To see this, we observe that the main left-to-right diagonal contains the elements

$$0, \beta + \alpha, 2(\beta+\alpha), \ldots, (n-1)(\beta+\alpha)$$

and the other main diagonal contains the elements

$$(n-1)\alpha, \beta + (n-2)\alpha, 2\beta + (n-3)\alpha, \ldots, (n-1)\beta.$$

The latter elements are respectively equal to

$$(n-1)\alpha, (n-1)\alpha - (\alpha - \beta), (n-1)\alpha - 2(\alpha - \beta), \ldots, (n-1)\alpha - (n-1)(\alpha - \beta).$$

Since $\alpha + \beta$ and $\alpha - \beta$ are both relatively prime to n, the elements $i(\alpha + \beta)$ for $i = 0, 1, \ldots, n - 1$, are all distinct and so also are the elements $i(\alpha - \beta)$. Consequently, the displayed latin square is diagonal.

[1]These definitions are given in Margossian(1931) but probably date from much earlier. See, for example, Tarry(1904) where diagonal latin squares are mentioned.

Now let us choose $\alpha = 2$ and $\beta = 1$. Then $\alpha - \beta = 1$ and $\alpha + \beta = 3$, so it is immediately obvious that if $n > 1$ is odd and not a multiple of three, the integers $\alpha, \beta, \alpha - \beta$ and $\alpha + \beta$ are all relatively prime to n. The existence of a diagonal latin square of order n follows.   □

The above construction fails when n is even, but our next two results will deal with specific subcases of this situation.

THEOREM 6.1.3 *If $n = 2^k$ for an integer $k \geq 0$ then there exists at least one diagonal latin square of order n if and only if $k \neq 1$.*

PROOF. Existence for $k = 0$ and non-existence for $k = 1$ are trivial to see and the cases $k = 2$ and $k = 3$ are settled by Figure 6.1.2, which shows that diagonal latin squares of orders 4 and 8 exist.

1	2	3	4
3	4	1	2
4	3	2	1
2	1	4	3

1	2	3	4	5	6	7	8
2	3	5	6	7	8	4	1
8	6	4	1	2	5	3	7
6	7	8	5	3	2	1	4
4	8	7	2	6	1	5	3
3	4	6	8	1	7	2	5
5	1	2	7	4	3	8	6
7	5	1	3	8	4	6	2

FIG. 6.1.2.

For $k > 3$ we proceed by an induction argument. Let $k = r$, so that $n = 2^r$, and take as induction hypothesis that there exists at least one diagonal latin square of every order $n = 2^k$ for which $1 < k < r$.

Since $2^r = 4 \cdot 2^{r-2}$, it follows from the induction hypothesis that 2^r can be written as the product $n_1 n_2$ of two integers n_1, n_2 such that diagonal latin squares exist both of order n_1 and of order n_2. Let us suppose now that A is a diagonal latin square of order n_1 and that $B_0, B_1, \ldots, B_{n_1-1}$ are isomorphic diagonal latin squares of order n_2 such that B_i is defined on the set $\{in_2, in_2+1, \ldots, in_2+n_2-1\}$ for $i = 0, 1, \ldots, n_1 - 1$. If the elements $0, 1, \ldots, n_1 - 1$ of A are replaced by $B_0, B_1, \ldots, B_{n_1-1}$ respectively, it is easy to see that we shall obtain a diagonal latin square of order $n_1 n_2 = 2^r$. The proof can now be completed by induction on the integer k.   □

THEOREM 6.1.4 *If n is any integral multiple of 4, there exists at least one diagonal latin square of order n.*

PROOF. We show first that if n is even and $n = n_1 n_2$, where n_1, n_2 are integers such that there exists at least one left semi-diagonal latin square of order n_1, and also at least one diagonal latin square of order n_2, and if, further, n_2 is even, then a diagonal latin square of order n can be constructed. The construction is similar to that used in the proof of Theorem 6.1.3 and is as follows.

Let A be a diagonal latin square of order n_2. Let the elements $0, 1, \ldots, n_2 - 1$ of A, excluding those which appear in the main right-to-left diagonal, be replaced respectively by $B_0, B_1, \ldots, B_{n_2-1}$ where B_i, for $i = 0, 1, 2, \ldots, n_2 - 1$, is a left semi-diagonal latin square of order n_1 defined on the set $\{in_1, in_1 + 1, \ldots, in_1 + n_1 - 1\}$. Further, let those elements of A which are contained in the main right-to-left diagonal be replaced successively by $B_0^*, B_1^*, \ldots, B_{n_2-1}^*$ where, for each i, B_i^* is defined on the same set as B_i and has the same order, but is right instead of left semi-diagonal. Then, it is easy to see that, provided n_2 is even, the resulting square of order $n_1 n_2$ is a diagonal latin square.

We now take the special case when $n_2 = 4$, corresponding to the case when $n = 4n_1$ is an integral multiple of 4. By Theorem 6.1.3, there exists a diagonal latin square of order 4 and, by Theorem 6.1.1, there exist both left and right semi-diagonal latin squares of order n_1 so the conditions required for our construction hold. This proves the theorem.   □

It is easy to see that the constructions of Theorem 6.1.2, Theorem 6.1.3 and Theorem 6.1.4 do not provide the only ways of producing diagonal latin squares since none of the above methods is applicable if $n = 6$, 9 or 10 and yet diagonal squares of these orders certainly exist, as is shown by Figure 6.1.3. However, it is easy to prove that no diagonal latin square of order 3 can exist.

THEOREM 6.1.5 *Diagonal latin squares of order 3 do not exist.*

PROOF.   Suppose that $\|a_{ij}\|$ is a diagonal latin square of order 3. By the diagonal property, the elements a_{11}, a_{22}, a_{33} are all different and the same is true for the elements a_{13}, a_{22}, a_{31}. This necessarily implies that either $a_{11} = a_{13}$ or $a_{11} = a_{31}$. However, neither equality is compatible with $\|a_{ij}\|$ being a latin square.   □

Some examples of diagonal latin squares of orders 5 and 7 as well as 6, 9, and 10, are given in Figure 6.1.3. Examples of orders 4 and 8 have already been given in Figure 6.1.2.

In a lecture at the University of Surrey, one of the authors of the present book (first edition) conjectured that diagonal latin squares exist for all orders $n > 3$. The truth of this conjecture has since been shown independently by a number of different people. First to give a proof was Hilton(1973), who made use of so-called *cross latin squares*: that is, latin squares such that all the elements of the main left-to-right diagonal are equal and all the elements of the main right-to-left diagonal are equal. (The definition has to be modified slightly for squares of odd order.) Shortly afterwards, Lindner(1974) found a somewhat simpler proof which made use of prolongations.[2] Later, a proof more elementary than either of those just mentioned was obtained by Gergely(1974a). We shall now explain Gergely's proof, which depends essentially on the following result.

---

[2]Both Hilton and Lindner used the term *diagonal latin square* to mean a left semi-diagonal latin square and the term *doubly diagonal latin square* to mean a diagonal latin square.

1	2	3	4	5
4	5	1	2	3
2	3	4	5	1
5	1	2	3	4
3	4	5	1	2

1	2	3	4	5	6
6	5	4	3	2	1
2	4	6	1	3	5
3	1	5	2	6	4
5	3	1	6	4	2
4	6	2	5	1	3

1	2	3	4	5	6	7
7	3	4	5	6	1	2
6	1	2	7	3	4	5
5	7	1	6	4	2	3
4	6	5	2	7	3	1
3	4	7	1	2	5	6
2	5	6	3	1	7	4

1	2	3	4	5	6	7	8	9
6	3	2	8	7	5	9	1	4
7	9	8	5	6	4	3	2	1
5	7	6	9	1	8	4	3	2
9	8	5	7	4	1	2	6	3
8	1	9	6	3	2	5	4	7
3	4	7	1	2	9	6	5	8
4	5	1	2	9	3	8	7	6
2	6	4	3	8	7	1	9	5

1	2	3	4	5	6	7	8	9	10
4	10	8	7	6	9	5	3	1	2
10	8	2	6	7	5	1	4	3	9
9	4	10	8	1	3	2	7	6	5
2	9	6	1	3	8	4	5	10	7
3	5	1	10	9	7	6	2	4	8
5	6	7	3	2	4	9	10	8	1
7	1	5	9	4	10	8	6	2	3
8	7	4	2	10	1	3	9	5	6
6	3	9	5	8	2	10	1	7	4

FIG. 6.1.3.

THEOREM 6.1.6 *For every $n \geq 3$, there exist left semi-diagonal latin squares of order n which possess at least one transversal disjoint from the main left-to-right diagonal.*

PROOF. Gergely's original proof of Theorem 6.1.6 made use of prolongations of cyclic group tables, but we can deduce the result very simply as follows. It follows from the results of Bose, Shrikhande and Parker(1960) described in Chapter 5 that, for all orders n except 2 and 6, there exist latin squares of order n which have n disjoint transversals. By rearranging the rows, we may easily arrange that one of the transversals lies along the main left-to-right diagonal. For the case $n = 6$, the truth of the theorem is shown by the example in Figure 6.1.4 which actually has three transversals disjoint from each other and from the main left-to-right diagonal. [Four disjoint transversals is the maximum number possible for a latin square of order 6. See Wanless(2007).] □

1	2	3_b	4_c	5	6_d
4	6	5_c	1	3_d	2_b
5_d	3_c	2	6	1_b	4
6_b	5	4_d	3	2	1_c
2_c	1_d	6	5_b	4	3
3	4_b	1	2_d	6_c	5

FIG. 6.1.4.

We are now able to prove our main result.

THEOREM 6.1.7 *For $n \geq 4$, there exists at least one diagonal latin square of order n.*

PROOF. The cases $n = 4$ and $n = 5$ are settled by the examples given in Figure 6.1.2 and Figure 6.1.3 respectively, so we may assume that $n \geq 6$.

We suppose first that n is even. We first construct a left semi-diagonal latin square D_1 of order $k = n/2$ on the symbols $0, 1, 2, \ldots, k-1$ having the property described in Theorem 6.1.6. We construct a second latin square D_2 of order k by reversing the order of the columns of D_1 and then adding k to each of its elements.

With the aid of the squares D_1 and D_2, we construct a latin square L of order n of the form

$$L = \left| \begin{array}{cc} D_1 & D_2 \\ D_4 & D_3 \end{array} \right.$$

where the subsquares D_3 and D_4 are yet to be defined.

Let τ_1 denote the off-diagonal transversal of D_1 whose existence is guaranteed by Theorem 6.1.6, and let τ_2 denote the corresponding transversal of D_2. We choose the elements of the main left-to-right diagonal of D_3 to be the elements of the transversal τ_2 which occur in the corresponding columns of D_2. Similarly, we choose the elements of the main right-to-left diagonal of D_4 to be the elements of the transversal τ_1 which occur in the corresponding columns of D_1. Now, for $i \in \{1, 3\}$ let $d_{11}^{(i)} d_{22}^{(i)} \cdots d_{kk}^{(i)}$ denote the elements of the main left-to-right diagonal of D_i.

We define a permutation π_L of the symbols of D_1 by

$$\pi_L = \left(\begin{array}{ccccc} d_{11}^{(1)} & \cdots & d_{jj}^{(1)} & \cdots & d_{kk}^{(1)} \\ d_{11}^{(3)} - k & \cdots & d_{jj}^{(3)} - k & \cdots & d_{kk}^{(3)} - k \end{array} \right)$$

Similarly, we define a permutation π_R of the symbols of D_2 by

$$\pi_R = \left(\begin{array}{ccccc} d_{1k}^{(2)} & \cdots & d_{j,k+1-j}^{(2)} & \cdots & d_{k1}^{(2)} \\ d_{1k}^{(4)} + k & \cdots & d_{j,k+1-j}^{(4)} + k & \cdots & d_{k1}^{(4)} + k \end{array} \right)$$

where in this case the permutation is expressed in terms of the elements of the main right-to-left diagonals of D_2 and D_4.

Each element of D_3 not on the main left-to-right diagonal is defined to be the transform of the corresponding element of D_1 by the permutation π_L. Each element of D_4 not on the main right-to-left diagonal is defined to be the transform of the corresponding element of D_2 by the permutation π_R. Finally, to make the square L into a latin square, we increase each of the elements of the transversal τ_1 of D_1 by k and at the same time reduce each of the elements of the transversal τ_2 of D_2 by k. Then L is of order n and is the diagonal latin square whose existence was to be shown. We note further that the elements in the cells of the transversal

τ_1 in D_1 together with the elements in the corresponding positions in D_3 form an off-diagonal transversal of L.

Using this fact, it is easy to construct from L a diagonal latin square L^* of order $n+1$, where $n+1$ is odd. (We illustrate the procedure below.) We can thereby construct squares of the odd orders required to complete the proof. □

To illustrate the construction of L^* from L, we shall give it in detail for the special case $n = 10$.

$$D_1 = \begin{array}{ccccc} 0 & \boxed{1} & 2 & 3 & 4 \\ 1 & 2 & \boxed{3} & 4 & 0 \\ 2 & 3 & 4 & \boxed{0} & 1 \\ 3 & 4 & 0 & 1 & \boxed{2} \\ \boxed{4} & 0 & 1 & 2 & 3 \end{array} \qquad D_2 = \begin{array}{ccccc} 9 & 8 & 7 & \boxed{6} & 5 \\ 5 & 9 & \boxed{8} & 7 & 6 \\ 6 & \boxed{5} & 9 & 8 & 7 \\ \boxed{7} & 6 & 5 & 9 & 8 \\ 8 & 7 & 6 & 5 & \boxed{9} \end{array}$$

$$\pi_L = \begin{pmatrix} 0 & 2 & 4 & 1 & 3 \\ 7-5 & 5-5 & 8-5 & 6-5 & 9-5 \end{pmatrix} = \begin{pmatrix} 0 & 1 & 2 & 3 & 4 \\ 2 & 1 & 0 & 4 & 3 \end{pmatrix}$$

$$\pi_R = \begin{pmatrix} 5 & 7 & 9 & 6 & 8 \\ 2+5 & 0+5 & 3+5 & 1+5 & 4+5 \end{pmatrix} = \begin{pmatrix} 5 & 6 & 7 & 8 & 9 \\ 7 & 6 & 5 & 9 & 8 \end{pmatrix}$$

$$L = \begin{array}{ccccc|ccccc} 0 & \boxed{6} & 2 & 3 & 4 & 9 & 8 & 7 & 1 & 5 \\ 1 & 2 & \boxed{8} & 4 & 0 & 5 & 9 & 3 & 7 & 6 \\ 2 & 3 & 4 & \boxed{5} & 1 & 6 & 0 & 9 & 8 & 7 \\ 3 & 4 & 0 & 1 & \boxed{7} & 2 & 6 & 5 & 9 & 8 \\ \boxed{9} & 0 & 1 & 2 & 3 & 8 & 7 & 6 & 5 & 4 \\ 8 & 9 & 5 & 6 & 2 & 7 & \boxed{1} & 0 & 4 & 3 \\ 7 & 8 & 9 & 0 & 6 & 1 & 5 & \boxed{4} & 3 & 2 \\ 6 & 7 & 3 & 9 & 5 & 0 & 4 & 8 & \boxed{2} & 1 \\ 5 & 1 & 7 & 8 & 9 & 4 & 3 & 2 & 6 & \boxed{0} \\ 4 & 5 & 6 & 7 & 8 & \boxed{3} & 2 & 1 & 0 & 9 \end{array}$$

$$L^* = \begin{array}{ccccc|cccccc} 0 & \alpha & 2 & 3 & 4 & 6 & 9 & 8 & 7 & 1 & 5 \\ 1 & 2 & \alpha & 4 & 0 & 8 & 5 & 9 & 3 & 7 & 6 \\ 2 & 3 & 4 & \alpha & 1 & 5 & 6 & 0 & 9 & 8 & 7 \\ 3 & 4 & 0 & 1 & \alpha & 7 & 2 & 6 & 5 & 9 & 8 \\ \alpha & 0 & 1 & 2 & 3 & 9 & 8 & 7 & 6 & 5 & 4 \\ 9 & 6 & 8 & 5 & 7 & \alpha & 3 & 1 & 4 & 2 & 0 \\ 8 & 9 & 5 & 6 & 2 & 1 & 7 & \alpha & 0 & 4 & 3 \\ 7 & 8 & 9 & 0 & 6 & 4 & 1 & 5 & \alpha & 3 & 2 \\ 6 & 7 & 3 & 9 & 5 & 2 & 0 & 4 & 8 & \alpha & 1 \\ 5 & 1 & 7 & 8 & 9 & 0 & 4 & 3 & 2 & 6 & \alpha \\ 4 & 5 & 6 & 7 & 8 & 3 & \alpha & 2 & 1 & 0 & 9 \end{array}$$

W.Taylor(1972) discussed the analogue of a diagonal latin square in higher dimensions. He also remarked that he and Faber had devised yet another proof of Theorem 6.1.7.

As we remarked at the beginning of this section, all diagonal latin squares are magic squares. However, the more usual type of magic square contains n^2 different integers, and usually these are required to be consecutive integers. In the next section, we explain how to construct the latter type of magic square with the aid of pairs of orthogonal latin squares.

6.2   Construction of magic squares with the aid of orthogonal latin squares.

For our first construction, we shall need to consider only pairs of orthogonal latin squares which are isotopic to the square which represents the multiplication table of the cyclic group C_n. The method which we shall describe is effective for all odd integers n. Our procedure is substantially equivalent to that of De la Hire[3] but the technique of our proof involves the use of latin squares.

$$
\begin{array}{|cccccccc|}
\hline
0 & 1 & 2 & \cdots & \frac{1}{2}(n-3) & \frac{1}{2}(n-1) & \cdots & n-2 & n-1 \\
1 & 2 & 3 & \cdots & \frac{1}{2}(n-1) & \frac{1}{2}(n+1) & \cdots & n-1 & 0 \\
\vdots & \vdots & \vdots & \ddots & \vdots & \vdots & \ddots & \vdots & \vdots \\
n-1 & 0 & 1 & \cdots & \frac{1}{2}(n-5) & \frac{1}{2}(n-3) & \cdots & n-3 & n-2 \\
\hline
\end{array}
$$

FIG. 6.2.1.

Let L^* denote the latin square exhibited in Figure 6.2.1, which is the un-bordered Cayley table of the cyclic group C_n, for odd n, when represented as an additive group. By interchanging the elements $n - 1$ and $\frac{1}{2}(n - 1)$ in each of the rows of L^*, we can transform it into a latin square L whose row, column and diagonal sums are each equal to $\frac{1}{2}n(n - 1)$ and which can be characterized as follows. The main left-to-right diagonal forms a transversal of L (so L is left semi-diagonal) and each broken diagonal parallel to the main left-to-right diagonal also forms a transversal. The main right-to-left diagonal contains the element $\frac{1}{2}(n - 1)$ duplicated n times and each broken diagonal parallel to the main right-to-left diagonal has all its elements equal.

A latin square L' orthogonal to L can be obtained from L by reversing the order of its columns. The orthogonality follows from the fact that all the elements of each broken diagonal parallel to the main left-to-right diagonal of L' are the same, whereas in L they are all different. Also, the elements of each broken diagonal parallel to the main right-to-left diagonal of L' form a transversal, whereas in L they are all the same.

We illustrate the above construction in Figure 6.2.2 by exhibiting the squares L^*, L and L' for the case $n = 5$.

We can express our result in the form of a theorem as follows:

THEOREM 6.2.1 *Let C_n denote the cyclic group of an odd order n, with elements represented by the integers $0, 1, \ldots, n - 1$ under addition modulo n and let its*

---

[3]The work of the French mathematician, astronomer, physicist, naturalist and painter Philippe de la Hire (1640–1718) is frequently quoted, without giving an explicit reference, in the literature on magic squares.

$$L^* = \begin{vmatrix} 0 & 1 & 2 & 3 & 4 \\ 1 & 2 & 3 & 4 & 0 \\ 2 & 3 & 4 & 0 & 1 \\ 3 & 4 & 0 & 1 & 2 \\ 4 & 0 & 1 & 2 & 3 \end{vmatrix} \quad L = \begin{vmatrix} 0 & 1 & 4 & 3 & 2 \\ 1 & 4 & 3 & 2 & 0 \\ 4 & 3 & 2 & 0 & 1 \\ 3 & 2 & 0 & 1 & 4 \\ 2 & 0 & 1 & 4 & 3 \end{vmatrix} \quad L' = \begin{vmatrix} 2 & 3 & 4 & 1 & 0 \\ 0 & 2 & 3 & 4 & 1 \\ 1 & 0 & 2 & 3 & 4 \\ 4 & 1 & 0 & 2 & 3 \\ 3 & 4 & 1 & 0 & 2 \end{vmatrix}$$

Fig. 6.2.2.

isotope under the isotopism $\rho = (\alpha, \beta, \gamma)$ *be denoted by* $\rho(C_n)$. *Let* $\rho_1 = (\epsilon, \epsilon, \gamma)$ *and* $\rho_2 = (\epsilon, \beta, \gamma)$ *where* ϵ *is the identity permutation,*

$$\beta = \begin{pmatrix} 0 & 1 & \cdots & \frac{1}{2}(n-3) & \frac{1}{2}(n-1) & \cdots & n-1 \\ n-1 & n-2 & \cdots & \frac{1}{2}(n+1) & \frac{1}{2}(n-1) & \cdots & 0 \end{pmatrix}$$

and γ *simply transposes* $\frac{1}{2}(n-1)$ *and* $n-1$. *Then the latin squares which represent the unbordered multiplication tables of* $\rho_1(C_n)$ *and* $\rho_2(C_n)$ *are orthogonal and have the structure described above.*

We can use this result to build magic squares, as follows:

THEOREM 6.2.2 *Let* $L = \|a_{ij}\|$ *and* $L' = \|b_{ij}\|$ *be two orthogonal latin squares of odd order* n *formed in the manner described in Theorem 6.2.1 and having as their elements the integers* $0, 1, \ldots, n-1$. *Let a square matrix* $M = \|c_{ij}\|$ *be constructed from* L *and* L' *by putting* $c_{ij} = na_{ij} + b_{ij}$. *Then the sum of the elements of each row, column, and diagonal of* M *is equal to* $\frac{1}{2}n(n^2-1)$ *and the elements of* M *are the consecutive integers* $0, 1, \ldots, n^2-1$.

PROOF. In order to show that the row, column, and diagonal sums are each equal to $\frac{1}{2}n(n^2-1)$, we derive the following equalities directly from the structure of the latin squares L and L' as described in Theorem 6.2.1. For each fixed i,

$$\sum_j c_{ij} = \sum_j (na_{ij} + b_{ij}) = n\sum_j a_{ij} + \sum_j b_{ij} = (n+1)\left[\tfrac{1}{2}n(n-1)\right] = \tfrac{1}{2}n(n^2-1),$$

and, for each fixed j,

$$\sum_i c_{ij} = n\sum_i a_{ij} + \sum_i b_{ij} = (n+1)\left[\tfrac{1}{2}n(n-1)\right] = \tfrac{1}{2}n(n^2-1).$$

Similar calculations applied to the main diagonals show that

$$\sum_i c_{ii} = \tfrac{1}{2}n(n^2-1) = \sum_i c_{i,n+1-i}.$$

To complete the proof, we have to show that no two elements of M are equal and that $0 \le c_{ij} \le n^2 - 1$ for all values of i, j. Suppose first that two distinct elements c_{ij} and c_{kl} of M were equal. This would imply that $n(a_{ij} - a_{kl}) + (b_{ij} - b_{kl}) = 0$. Since the elements of L' are all less than n, the latter equation could only hold if $a_{ij} = a_{kl}$ and then, since L is a latin square, we would necessarily have $i \neq k$ and $j \neq l$. It would follow that a_{ij} and a_{kl} were in the same broken diagonal parallel to the main right-to-left diagonal. However, in L' the elements of each such diagonal are all different, so $b_{ij} \neq b_{kl}$. It follows that $c_{ij} \neq c_{kl}$. Finally, since $0 \leq a_{ij} \leq n - 1$ and $0 \leq b_{ij} \leq n - 1$, the largest value that c_{ij} can take is $n(n - 1) + n - 1 = n^2 - 1$, so $0 \leq c_{ij} \leq n^2 - 1$ as required.   □

We demonstrate the above procedure in Figure 6.2.3, where we give the final step of the construction for the example exhibited in Figure 6.2.2.

$$
nL = \begin{vmatrix}
0 & 5 & 20 & 15 & 10 \\
5 & 20 & 15 & 10 & 0 \\
20 & 15 & 10 & 0 & 5 \\
15 & 10 & 0 & 5 & 20 \\
10 & 0 & 5 & 20 & 15
\end{vmatrix}
\qquad
L' = \begin{vmatrix}
2 & 3 & 4 & 1 & 0 \\
0 & 2 & 3 & 4 & 1 \\
1 & 0 & 2 & 3 & 4 \\
4 & 1 & 0 & 2 & 3 \\
3 & 4 & 1 & 0 & 2
\end{vmatrix}
$$

$$
M = nL + L' = \begin{vmatrix}
2 & 8 & 24 & 16 & 10 \\
5 & 22 & 18 & 14 & 1 \\
21 & 15 & 12 & 3 & 9 \\
19 & 11 & 0 & 7 & 23 \\
13 & 4 & 6 & 20 & 17
\end{vmatrix}
$$

FIG. 6.2.3.

In the following theorem, we give another construction which uses orthogonal latin squares.

THEOREM 6.2.3 *If n is any integer for which an orthogonal pair of diagonal latin squares of order n exists, then an $n \times n$ magic square whose entries are the consecutive integers 0 to $n^2 - 1$ can be constructed.*

PROOF. We first write the two squares in juxtaposed form, as for example in Figure 6.2.6. Since the sums of the elements of each row, column, and main diagonal are all equal for each of the two squares, the same is true in the juxtaposed form. Moreover, this is true regardless of the number base selected. If we take the number base as the integer n we get a magic square M whose entries are the integers 0 to $n^2 - 1$ as required. Thus, to get the matrix M exhibited in Figure 6.2.6 the number base 4 has been selected.

We may express the construction in another way by saying that M is related to L_1 and L_2 by the matrix equation $M = nL_1 + L_2$. (Compare the proof of Theorem 6.2.2.)   □

The construction of magic squares by means of orthogonal pairs of diagonal latin squares or by means of orthogonal pairs of latin squares of the type given by

Theorem 6.2.1 have been known and used for more than two centuries. See, for example, Euler(1779), Maillet(1894a) and Maillet(1894c) or (1896) [Maillet(1894c) and Maillet(1896) are two publications of the same paper], Barbette(1896), Tarry(1904,1905), Kraichik(1942,1953) and Ball(1939). These constructions are very useful but they are not the only methods available. For example, neither of these constructions can be used to obtain a magic square of order 6 and yet magic squares of this order certainly exist, as is demonstrated by Figure 6.2.4.

35	1	6	26	19	24
3	32	7	21	23	25
31	9	2	22	27	20
8	28	33	17	10	15
30	5	24	12	14	16
4	36	29	13	18	11

FIG. 6.2.4.

Because of their importance in connection with the construction of magic squares, many authors have tried to answer the question: "For which orders n distinct from 2, 3 and 6 do there exist orthogonal pairs of diagonal latin squares?"

That such pairs exist for the case $n = 4$ has been known at least since 1723, as has been pointed out by Ball(1939), page 190. The fact that a solution exists whenever n is odd and not a multiple of three (see Theorem 6.2.4) has also been known at least since the nineteenth century. Using both these results, Tarry(1905) proved at the beginning of the twentieth century that orthogonal pairs of diagonal latin squares exist for every order n which is a multiple of 4. Moreover, he proved in Tarry(1904) that if $n = 8m$, where m is not a multiple of two or three, then the method of Theorem 6.2.3 can be used to obtain a magic square with the additional property that the sums of the squares of the elements in each row, column and main diagonal are equal. A magic square with the latter property is called *bimagic*. We discuss such squares further in the next section.

The complete resolution of the existence question was not achieved until 1993. We shall now explain the main steps.

THEOREM 6.2.4 *There exist orthogonal pairs of diagonal latin squares of every odd order which is not a multiple of* 3.

PROOF. It follows directly from Theorem 6.1.2 that, if $\alpha = 2$ and $\beta = 1$, the latin square exhibited in Figure 6.1.1 is a diagonal latin square provided that n is odd and not divisible by 3. Its transpose is also a diagonal latin square and we shall show that the two squares so obtained are orthogonal, again under the condition that n is not divisible by 3.

Let us denote the square obtained from Figure 6.1.1 when $\alpha = 2$ and $\beta = 1$ by L_1 and its transpose by L_2 (the squares L_1 and L_2 for the special case when $n = 5$ are exhibited in Figure 6.2.5). We shall show first that the contents of the cells of L_1 which correspond to the cells containing the element 0 in L_2 are all different.

$$L_1 = \begin{vmatrix} 0 & 2 & 4 & 1 & 3 \\ 1 & 3 & 0 & 2 & 4 \\ 2 & 4 & 1 & 3 & 0 \\ 3 & 0 & 2 & 4 & 1 \\ 4 & 1 & 3 & 0 & 2 \end{vmatrix} \qquad L_2 = \begin{vmatrix} 0 & 1 & 2 & 3 & 4 \\ 2 & 3 & 4 & 0 & 1 \\ 4 & 0 & 1 & 2 & 3 \\ 1 & 2 & 3 & 4 & 0 \\ 3 & 4 & 0 & 1 & 2 \end{vmatrix}$$

FIG. 6.2.5.

Rows and columns will be indexed with the set $\{0, 1, \ldots, n-1\}$ and all calculations will be performed modulo n. In the r-th row of L_2, the element 0 appears in the $(n - 2r)$-th column. In the cell of the r-th row and $(n - 2r)$-th column of L_1, the element $r + 2(n - 2r)$ appears. To see this, observe that the entry in the cell of the r-th row and 0-th column of the square L_1 is r and that for each step taken to the right along this row, the cell entry is increased by 2. Since $r + 2(n - 2r) \equiv -3r \bmod n$ and since -3 is relatively prime to n, the elements $-3r$ are all different modulo n as r varies through the set $\{0, 1, 2, \ldots, n-1\}$. This proves the result.

By a similar argument, we could show that the contents of the cells of L_1 which correspond to the cells containing the element i in L_2 are all different and that this is true for each choice of i in the range $0 \le i \le n-1$. It follows that L_1 and L_2 are orthogonal diagonal latin squares.   □

It is an obvious consequence of Theorem 6.1.5 that orthogonal pairs of diagonal latin squares of order 3 do not exist. However, our next theorem, which comes from Ball(1939), gives a further set of values of n for which orthogonal pairs of diagonal latin squares exist.

THEOREM 6.2.5 *Orthogonal pairs of diagonal latin squares of order n can be constructed whenever n is odd or a multiple of 4 except, possibly, when n is a multiple of 3 but not of 9.*

PROOF.   We first point out that there exist orthogonal pairs of diagonal latin squares of orders 4, 8, 9 and 27. Examples of such pairs corresponding to the first three of these values of n are exhibited in Figure 6.2.6, Figure 6.2.7 and Figure 6.2.8 respectively and a construction which gives a pair of order 27 is described by Ball(1939), page 192. (In Figure 6.2.6 the diagonal squares L_1 and L_2 are first shown separately and then in juxtaposition. In the remaining cases, only the juxtaposed form is shown.)

Then, making use of Theorem 6.2.4 and of the construction described in the proof of Theorem 6.1.3, it is easy to deduce the truth of the present theorem which comes from Ball(1939), page 192.   □

$$L_1 = \begin{vmatrix} 0 & 1 & 2 & 3 \\ 2 & 3 & 0 & 1 \\ 3 & 2 & 1 & 0 \\ 1 & 0 & 3 & 2 \end{vmatrix} \qquad L_2 = \begin{vmatrix} 0 & 3 & 1 & 2 \\ 2 & 1 & 3 & 0 \\ 3 & 0 & 2 & 1 \\ 1 & 2 & 0 & 3 \end{vmatrix}$$

$$\begin{vmatrix} 00 & 13 & 21 & 32 \\ 22 & 31 & 03 & 10 \\ 33 & 20 & 12 & 01 \\ 11 & 02 & 30 & 23 \end{vmatrix} \qquad M = \begin{vmatrix} 0 & 7 & 9 & 14 \\ 10 & 13 & 3 & 4 \\ 15 & 8 & 6 & 1 \\ 5 & 2 & 12 & 11 \end{vmatrix}$$

FIG. 6.2.6.

$$\begin{vmatrix} 17 & 50 & 43 & 04 & 32 & 75 & 66 & 21 \\ 31 & 76 & 65 & 22 & 14 & 53 & 40 & 07 \\ 00 & 47 & 54 & 13 & 25 & 62 & 71 & 36 \\ 26 & 61 & 72 & 35 & 03 & 44 & 57 & 10 \\ 45 & 02 & 11 & 56 & 60 & 27 & 34 & 73 \\ 63 & 24 & 37 & 70 & 46 & 01 & 12 & 55 \\ 52 & 15 & 06 & 41 & 77 & 30 & 23 & 64 \\ 74 & 33 & 20 & 67 & 51 & 16 & 05 & 42 \end{vmatrix}$$

FIG. 6.2.7.

The preceding two theorems left the cases when n is an odd multiple of two or three unresolved. Proof of existence for these cases was finally completed by the combined efforts of six authors after a lapse of more than three-quarters of a century. For full details, see Brown, Cherry, Most, Most, Parker and Wallis(1993) but we shall now describe some of the steps along the way.

Lindner(1973) made an attempt to solve the problem by means of a construction which used the singular direct product of Sade. (The latter concept is defined in Section 11.2.) His main result is as follows: "If there are t mutually

$$\begin{vmatrix} 76 & 82 & 64 & 15 & 27 & 00 & 41 & 53 & 38 \\ 11 & 23 & 08 & 46 & 52 & 34 & 75 & 87 & 60 \\ 45 & 57 & 30 & 71 & 83 & 68 & 16 & 22 & 04 \\ 62 & 74 & 86 & 07 & 10 & 25 & 33 & 48 & 51 \\ 03 & 18 & 21 & 32 & 44 & 56 & 67 & 70 & 85 \\ 37 & 40 & 55 & 63 & 78 & 81 & 02 & 14 & 26 \\ 84 & 66 & 72 & 20 & 05 & 17 & 58 & 31 & 43 \\ 28 & 01 & 13 & 54 & 36 & 42 & 80 & 65 & 77 \\ 50 & 35 & 47 & 88 & 61 & 73 & 24 & 06 & 12 \end{vmatrix}$$

FIG. 6.2.8.

orthogonal diagonal latin squares of order $v = 2m$, t mutually orthogonal left semi-diagonal latin squares of order q each containing a diagonal latin square of order p in its top left-hand corner, and t mutually orthogonal right semi-diagonal latin squares of order $q - p$, then there are t mutually orthogonal diagonal latin squares of order $v(q - p) + p$." Taking $t = 2$, $v = 8$, $p = 1$ and $q = 5$, Lindner deduced that there is a pair of orthogonal diagonal latin squares of order 33, which number is a multiple of 3 but not of 9. Similarly, taking $t = 6$, $v = 8$, $p = 1$ and $q = 17$, he deduced that there are at least six mutually orthogonal diagonal latin squares of order 129. By means of the direct product construction, it is easy to show from these results that pairs of mutually orthogonal diagonal latin squares exist for infinitely many orders which are multiples of 3 but not of 9. (Contrast Theorem 6.2.5.)

By methods quite similar to those used by Lindner, Hilton and Scott(1974) showed that pairs of orthogonal diagonal latin squares exist for some orders n which are multiples of 2 but not of 4. The smallest such order for which their construction works is $n = 50$. They also exhibited a pair of orthogonal diagonal latin squares of order 21. Later, Hilton(1975b) showed that it is possible to obtain a set of at least four mutually orthogonal diagonal latin squares of order 50.

Wallis and Zhu(1981,1982,1983) obtained a set of four mutually orthogonal diagonal latin squares of order 12 and disposed of some of the remaining outstanding cases. See also Du(1991) who considered the existence of triples. Meanwhile, Heinrich and Hilton(1983) used various constructions to show existence of pairs for all remaining orders (at the time their paper was submitted) except 10, 12, 14, 15, 18 and 26. Wallis(1984) obtained pairs of orders 15, 18 and 26 (which are actually self-orthogonal). Zhu(1984a) dealt with order 14. The last case of all to be resolved was order 10 in the paper of Brown, Cherry et al (1993).

With regard to the existence of self-orthogonal diagonal latin squares, two further papers should be mentioned: namely, Danhof, Phillips and Wallis(1990) and Du and Wallis(1999)

As regards the maximum possible number $N_D(n)$ of diagonal latin squares in a pairwise orthogonal set of squares of order $n = p_1^{\alpha_1} p_2^{\alpha_2} \ldots p_r^{\alpha_r}$ (where the p_i are distinct primes), Gergely(1974b) proved that

$$N_D(n) \geq \alpha(n) - 3$$

if n is odd, and

$$N_D(n) \geq \alpha(n) - 2$$

if n is even, where $\alpha(n) = \min_{1 \leq i \leq r} p_i^{\alpha_i}$ (cf. MacNeish's theorem, page 178). Gergely also pointed out that $N_D(n) \leq n - 3$ if $n \geq 3$ is odd and $N_D(n) \leq n - 2$ if n is even (cf. Theorem 5.1.2).

Hilton(1974,1975a) showed that $N_D(n) \to \infty$ as $n \to \infty$ [cf. Chowla, Erdös and Straus(1960) who proved the same result for $N(n)$].

Let d_r be the least integer such that for all $n > d_r$ there exist r pairwise orthogonal diagonal latin squares of order n. Wallis and Zhu(1984a) and

Du(1993) have given bounds on some of the d_r. Also, considerable work has been done on constructing parastrophic orthogonal diagonal latin squares. See Du(1996a,1996b,1998) and Bennett, Du and Zhang(1997, 1998, 2001).

6.3 Additional results on magic squares

A magic square is called *pandiagonal* if the sets of elements in each of its broken diagonals have the same sum and this sum is the same as the sum of the elements of each row, column, and the two main diagonals.[4]

McClintock(1897) investigated pandiagonal squares, combining some of the results of De la Hire and extending them further. It is well known and easy to check that, if α and β are integers which satisfy the conditions of Theorem 6.1.2, then the latin square exhibited in Figure 6.1.1 is a pandiagonal latin (magic) square.[5] [See also Ball(1939) and McClintock(1897).] In fact, this square has the property that the elements in each of its broken diagonals (parallel to the main left-to-right and right-to-left diagonals) are all different. Such a square is called a *totally diagonal latin square* or a *Knut Vik design*. The latter name arose because such squares have been used in the statistical design of experiments. (For details, see [DK2].) Consequently, Theorem 6.1.2 can be reformulated as follows.

THEOREM 6.3.1 *For every odd integer n which is not a multiple of 3 there exists at least one totally diagonal latin square of size $n \times n$.*

Hedayat and Federer(1975) and Hedayat(1977) have proved that $n \times n$ Knut Vik designs (totally diagonal latin squares) exist if and only if n is not a multiple of 2 or 3.[6] Hedayat has also proved the following theorem.

THEOREM 6.3.2 *If n is not a multiple of 2 or 3, then orthogonal pairs of totally diagonal latin squares exist.*

PROOF. This result follows from Theorem 6.3.3 below. □

By exactly the same argument and construction that we used in Theorem 6.2.2, it is easy to see that we can use such a pair of totally diagonal latin squares to build a pandiagonal magic square. This fact is illustrated in Figure 6.3.1. (To obtain a pandiagonal magic square using the integers 1 to 49, it is only necessary to increase each of the entries in the square shown by one. If this is done, the sum of the elements of each of the rows, columns and diagonals becomes 175.) In Xu and Lu(1995), these authors have shown that, for many orders n, it is possible to use self-orthogonal pandiagonal latin squares for the same purpose.

An interesting question to ask is "What is the largest number of totally diagonal latin squares of order n that can exist in a pairwise orthogonal set?"

[4]For a detailed account of the history of such squares, see Problem 6.2 of [DK1] on page 346 of this book.

[5]See further down the page for construction of a pandiagonal magic square on consecutive integers.

[6]A much earlier proof of this is attributed to Euler(1779).

$$
\begin{vmatrix}
0 & 1 & 2 & 3 & 4 & 5 & 6 \\
3 & 4 & 5 & 6 & 0 & 1 & 2 \\
6 & 0 & 1 & 2 & 3 & 4 & 5 \\
2 & 3 & 4 & 5 & 6 & 0 & 1 \\
5 & 6 & 0 & 1 & 2 & 3 & 4 \\
1 & 2 & 3 & 4 & 5 & 6 & 0 \\
4 & 5 & 6 & 0 & 1 & 2 & 3
\end{vmatrix}
\quad
\begin{vmatrix}
0 & 1 & 2 & 3 & 4 & 5 & 6 \\
2 & 3 & 4 & 5 & 6 & 0 & 1 \\
4 & 5 & 6 & 0 & 1 & 2 & 3 \\
6 & 0 & 1 & 2 & 3 & 4 & 5 \\
1 & 2 & 3 & 4 & 5 & 6 & 0 \\
3 & 4 & 5 & 6 & 0 & 1 & 2 \\
5 & 6 & 0 & 1 & 2 & 3 & 4
\end{vmatrix}
\quad
\begin{vmatrix}
0 & 8 & 16 & 24 & 32 & 40 & 48 \\
23 & 31 & 39 & 47 & 6 & 7 & 15 \\
46 & 5 & 13 & 14 & 22 & 30 & 38 \\
20 & 21 & 29 & 37 & 45 & 4 & 12 \\
36 & 44 & 3 & 11 & 19 & 27 & 28 \\
10 & 18 & 26 & 34 & 35 & 43 & 2 \\
33 & 41 & 42 & 1 & 9 & 17 & 25
\end{vmatrix}
$$

FIG. 6.3.1.

Hedayat(1977) showed that such a set can contain at most $t = n - 3$ squares by the following simple argument. Suppose that the set contains t squares. Each member of the set is orthogonal to each of the latin squares $\|a_{ij}\|$ and $\|b_{ij}\|$, where $a_{ij} = i + j \bmod n$ and $b_{ij} = i - j \bmod n$. Moreover, the latter two squares are themselves orthogonal when n is odd. But, as we showed in Theorem 5.1.2, there are at most $n - 1$ latin squares in a complete set of mutually orthogonal squares so $t + 2 \leq n - 1$.

In the same paper, Hedayat also proved

THEOREM 6.3.3 *If n is a prime, there are $n - 3$ pairwise orthogonal Knut Vik designs. If n is not a prime and is not divisible by 2 or 3, there is at least a pair of orthogonal Knut Vik designs.*

PROOF.   The proof is carried out in two steps. (i) Let $A = \|a_{ij}\|$, where $a_{ij} = \lambda i + j \bmod n$. Then A is a Knut Vik design provided that $\lambda - 1, \lambda$ and $\lambda + 1$ are all relatively prime to n; and (ii) $B = \|b_{ij}\|$ and $C = \|c_{ij}\|$, where $b_{ij} = \lambda_1 i + j \bmod n$ and $c_{ij} = \lambda_2 i + j \bmod n$ are orthogonal latin squares provided that $\lambda_1 - \lambda_2$ is prime to n.

Proof (i). It is easy to see that $a_{ij} = a_{ik} \Rightarrow j = k$. Also, $a_{hj} = a_{ij} \Rightarrow \lambda(h - i) \equiv 0 \bmod n$ so $h = i$ if λ is prime to n. Thus A is a latin square if λ is prime to n. Next, $a_{ij} = a_{i+r,j+r} \Rightarrow \lambda i + j \equiv \lambda(i + r) + (j + r) \Rightarrow (\lambda + 1)r \equiv 0 \bmod n$ so the elements of each broken left-to-right diagonal are distinct provided that $\lambda + 1$ is prime to n. Similarly, $a_{ij} = a_{i+r,j-r} \Rightarrow (\lambda - 1)r \equiv 0 \bmod n$ so the elements of each broken right-to-left diagonal are distinct provided that $\lambda - 1$ is prime to n.

Proof (ii). If B and C were not orthogonal, there would exist distinct cells (i, j) and (u, v) such that $a_{ij} = a_{uv}$ and $b_{ij} = b_{uv}$. By subtraction of the expressions for these quantities, this would imply that $(\lambda_1 - \lambda_2)i = (\lambda_1 - \lambda_2)u$. If $\lambda_1 - \lambda_2$ is prime to n, the latter equality is impossible unles $i = u$ and then $j = v$, so the squares are orthogonal when $\lambda_1 - \lambda_2$ is prime to n.

Finally, we observe that $\lambda - 1, \lambda$ and $\lambda + 1$ are all relatively prime to n for all non-zero choices of λ except 1 and $n - 1$ when n is prime and that, in that case, $\lambda_1 - \lambda_2$ is prime to n for all choices of λ_1 and λ_2. When n is not prime but is not divisible by 2 or 3, the choices $\lambda_1 = 3$ and $\lambda_2 = 2$ ensure that $\lambda_1 - 1, \lambda_1, \lambda_1 + 1, \lambda_2 - 1, \lambda_2, \lambda_2 + 1$ and $\lambda_1 - \lambda_2$ are all prime to n so there is at least one pair of orthogonal Knut Vik designs for all permissible orders n.   □

The squares used in Figure 6.3.1 were constructed by Hedayat's method.

The existence question regarding orthogonal sets has been further investigated in Stoffel(1976), Afsarinejad(1986,1987) and Ishihara(2006).

It is a property of pandiagonal magic squares, long known and easily recognized, that a row or column which forms one of the four edges of the square can be moved to the opposite side of the square without destroying the magic and pandiagonal attributes of the square[7]. By reason of this remarkable property, pandiagonal squares have been given by some authors, beginning with De la Hire, the name *perfect* and by others, beginning with Lucas, the name *diabolic*.

As regards the construction and enumeration of Knut Vik designs, two further papers Atkin, Hay and Larson(1977,1983) deserve mention.

ADDITION-MULTIPLICATION MAGIC SQUARES.

Horner(1952,1955) investigated a method of constructing magic squares in which not only the sum but also the product of the elements in each row, column or main diagonal is a constant. Such a square is called an *addition-multiplication magic square*. In Horner(1952), that author showed how to construct addition-multiplication magic squares of any odd order and in Horner(1955) he obtained addition-multiplication magic squares of orders 8 and 16. Both papers make use of latin squares. In Figure 6.3.2 and Figure 6.3.3 we give examples of addition-multiplication magic squares of orders 8 and 9 respectively, obtained by Horner's methods.

162	207	51	26	133	120	116	25
105	152	100	29	138	243	39	34
92	27	91	136	45	38	150	261
57	30	174	225	108	23	119	104
58	75	171	90	17	52	216	161
13	68	184	189	50	87	135	114
200	203	15	76	117	102	46	81
153	78	54	69	232	175	19	60

FIG. 6.3.2.

The magic sum (that is, the sum of the elements of each row, column or diagonal) of the addition-multiplication magic square exhibited in Figure 6.3.2 is 840 and its magic product is 2 058 068 231 856 000. In our second example (Figure 6.3.3) the magic sum is 848 and the magic product 5 804 807 833 440 000.

For details of recent work on this topic, see Boyer(2012).

MULTIMAGIC SQUARES.

As we remarked earlier (see page 215), it is possible to construct magic squares of certain orders for which not only are the sums of the elements in each row,

---

[7]Hence, any cyclic permutation of the rows and/or columns also preserves these properties.

200	87	95	42	99	1	46	108	170
14	44	10	184	81	85	150	261	19
138	243	17	50	116	190	56	33	5
57	125	232	9	7	66	68	230	54
4	70	22	51	115	216	171	25	174
153	23	162	76	250	58	3	35	88
145	152	75	11	6	63	270	34	92
110	2	28	135	136	69	29	114	225
27	102	207	290	38	100	55	8	21

FIG. 6.3.3.

column and main diagonal all equal but also the squares of those elements have the same properties. Such squares are *bimagic*. More generally, we may make the following definition.

DEFINITION. An $n \times n$ matrix whose entries are the integers $0, 1, \ldots, n^2 - 1$ (or, alternatively, $1, 2, \ldots, n^2$) is called m-*multimagic* if the sums of the elements in each row, column and main diagonal are all equal and if the same is true when the elements are replaced by their squares, or their cubes, ... , or their mth powers.

In Tarry(1906), that author gave a method of constructing bimagic squares of all sizes $p^2 \times p^2$ where p is a prime. He also gave a similar method for constructing trimagic squares of size $p^3 \times p^3$ and claimed, without giving a proof, that m-multimagic squares exist for all positive integers m. The truth of the latter claim has been shown by Derkson, Eggermont and van den Essen(2007).

Tarry's method for constructing bimagic and trimagic squares is too lengthy to be given here but is described in detail in Cazalas((1934)) and, for the case of bimagic squares only, in an expository paper by the present author [Keedwell(2011b)].

Just as orthogonal diagonal latin squares can be used to construct magic squares (see Theorem 6.2.3) so orthogonal diagonal Sudoku latin squares can be used to construct bimagic squares. (Existence of the former was shown in Section 5.6.) The latter observation was first made by Boyer(2006) when studying Tarry's construction methods (which do not use Sudoku squares explictly). It has been exploited by the present author to give a construction of $p^2 \times p^2$ bimagic squares valid for all primes p except $p = 5$. [See Keedwell(2011c).] It is likely that, by using a structurally different pair of orthogonal diagonal Sudoku squares, the restriction $p \neq 5$ can be removed or replaced by a similar but different restriction.

The constructions used in the proofs of Theorem 6.1.2, Theorem 6.2.1 and Theorem 6.2.3 suggest the importance of orthogonal pairs of diagonal latin squares which satisfy the quadrangle criterion. Since such squares play an important role in the construction of magic squares it is interesting to ask the question: "What is the largest number of elements which can be deleted from the two members of an orthogonal pair of diagonal latin squares both of which satisfy the quadrangle criterion, in such a way that the pair can be reconstructed uniquely from the remainder?" This problem is solved in Theorem 6.3.4 which does not require the two squares to be diagonal.

THEOREM 6.3.4 *Let $A = ||a_{ij}||$ and $B = ||b_{ij}||$ be two orthogonal latin squares of order $n \neq 4$, both of which satisfy the quadrangle criterion, and let the matrix of ordered pairs (a_{ij}, b_{ij}) be denoted by C. Let $U(C)$ be the maximum number of entries which can be deleted from C in such a way that C can always be uniquely constructed from those left, regardless of which entries are removed. Then $U(C) = 2n - 1$.*

PROOF.   It follows from Theorem 3.3.2 that $2n - 1$ arbitrary elements can be deleted from C without jeopardising the unique reconstructibility. Hence $U(C) \geq 2n - 1$, so it suffices to show that $U(C) < 2n$.

Choose any two distinct symbols k and l from A and form a new latin square A' by interchanging these two symbols throughout A. Note that A' satisfies the quadrangle criterion and is orthogonal to B. Let C' be obtained by juxtaposing A' and B. Then C' differs from C in exactly $2n$ places, which shows that $U(C) < 2n$.   □

As an illustration of the above argument, take the orthogonal pair of latin squares in Figure 6.2.5 to be C and let $k = 0$ and $l = 1$. The resulting matrices C and C' are exhibited in Figure 6.3.4.

$$
\begin{array}{|lllll}
00 & 21 & 42 & 13 & 34 \\
12 & 33 & 04 & 20 & 41 \\
24 & 40 & 11 & 32 & 03 \\
31 & 02 & 23 & 44 & 10 \\
43 & 14 & 30 & 01 & 22
\end{array}
\qquad
\begin{array}{|lllll}
10 & 21 & 42 & 03 & 34 \\
02 & 33 & 14 & 20 & 41 \\
24 & 40 & 01 & 32 & 13 \\
31 & 12 & 23 & 44 & 00 \\
43 & 04 & 30 & 11 & 22
\end{array}
$$

FIG. 6.3.4.

We note, as Dénes did in Dénes(1961), that the proof of Theorem 6.3.4 shows that $U(C) < 2n$ for any matrix C formed by the juxtaposition of two orthogonal latin squares of order n. However, without the restriction of the quadrangle criterion, $U(C)$ can be very much smaller than $2n$ as the next theorem shows.

THEOREM 6.3.5 *For every $n \geq 9$ there is a matrix C formed by the juxtaposition of two orthogonal latin squares of order n for which $U(C) \leq 5$.*

PROOF.   Wallis and Zhu(1984b) have shown that for all $n \geq 9$ there exists a pair of orthogonal squares of order n containing orthogonal latin subsquares of order 3. We can simply apply Theorem 6.3.4 to the subsquares, since all latin squares of order 3 satisfy the quadrangle criterion.   □

In Keedwell(2006a), the present author has initiated an investigation of the sizes of uniquely completable partial magic squares (cf. Section 3.2). He has made the following dfinition.

DEFINITION. A *defining set* for a magic square is a set of entries of an $n \times n$ matrix, fixed in orientation, which determines the remaining entries so as to form a unique $n \times n$ magic square.

For a 3×3 magic square, the size of a minimal defining set is two. Since, up to equivalence, there is only one such square, this is easy to show. However, in the case of 4×4 magic squares, even when variations of a square obtained by rotation and reflection are not counted, there are 880 such squares and, in the case of 5×5 magic squares, there are more than 275 million. In Berlekamp, Conway and Guy(2004), the 4×4 squares have been separated into 12 classes, of which 304 are of type CH (central horizontal). Keedwell has shown that there exist squares of the latter type which have critical sets (cf. Section 3.2) of at most four elements. This result does not rule out the possibility that there exist 4×4 magic squares with critical sets of smaller size either in this class or in one of the others. The author conjectures that no 4×4 magic square can have a critical set of less than three entries but he has no convincing proof that such is the case. Many questions regarding this topic remain to be answered.

Another question considered by Gridgeman(1972), by the present author [in Keedwell(2006b)] and, later also by Lorch(2012), is that of constructing $n^2 \times n^2$ Sudoku latin squares (see Section 3.2) each of whose $n \times n$ subsquares is a magic square. The present author has shown that it is easy to construct a 16×16 Sudoku latin square all of whose 4×4 subsquares are magic (to be called *magic Sudoku latin squares*) and indeed that they can all be versions of the same pandiagonal magic square. [See also Gridgeman(1972) for another example.] He has raised the question whether it is possible to construct an $n^2 \times n^2$ Sudoku latin square ($n \geq 4$) all of whose $n \times n$ subsquares are magic squares of which no two are equivalent. He conjectures that the answer is in the affirmative for all sufficiently large n and that a construction may be possible for the case $n = 4$.

Lorch(2012) has given a construction for orthogonal magic Sudoku latin squares.

We end this section by pointing out that magic squares should not be regarded solely as a mathematical amusement because they also play an important role in practical applications such as the design of experiments and the construction of error detecting and correcting codes.

Samoilenko(1965) illustrated how generalized magic squares can be utilized for error correction. He used latin squares to obtain his generalized magic squares and then constructed variable parameter codes with the aid of the latter. For more information about error correcting codes see Section 11.5. A typical application of variable parameter codes is in the transmission of data between computers.

Phillips(1964) showed how magic squares can be applied to statistics. The magic squares which he used were obtained by means of latin squares and orthogonal pairs of latin squares.

6.4   Room squares: their construction and uses

Under the title "A new type of magic square", Room(1955) introduced what was believed at the time to be a new kind of combinatorial structure and which has subsequently become known as a *Room square* or *Room design*. As in the case of magic squares, many of these designs can be constructed with the aid of latin squares.

A *Room design* of order $2n$ comprises a square array having $2n-1$ cells in each row and column and such that each cell is either empty or contains an unordered pair of symbols chosen from a set of $2n$ elements. Without loss of generality, we can take these elements as the numbers $1, 2, \ldots, 2n-1, \infty$. Each row and column of the design contains $n-1$ empty cells and n cells each of which contains a pair of symbols. Each row and column contains each of the $2n$ symbols exactly once, and, further, each of the $n(2n-1)$ possible distinct pairs of symbols is required to occur exactly once in a cell of the square. As illustrations of the concept, we exhibit two Room designs of order 8 in Figure 6.4.1.

$\infty,1$			$6,2$		$5,7$	$3,4$
$4,5$	$\infty,2$			$7,3$		$6,1$
$7,2$	$5,6$	$\infty,3$			$1,4$	
	$1,3$	$6,7$	$\infty,4$			$2,5$
$3,6$		$2,4$	$7,1$	$\infty,5$		
	$4,7$		$3,5$	$1,2$	$\infty,6$	
		$5,1$		$4,6$	$2,3$	$\infty,7$

$\infty,1$	$5,6$	$2,4$		$3,7$		
	$\infty,2$	$6,7$	$3,5$		$4,1$	
		$\infty,3$	$7,1$	$4,6$		$5,2$
$6,3$			$\infty,4$	$1,2$	$5,7$	
	$7,4$			$\infty,5$	$2,3$	$6,1$
$7,2$		$1,5$			$\infty,6$	$3,4$
$4,5$	$1,3$		$2,6$			$\infty,7$

FIG. 6.4.1.

A Room square of side $2n - 1$ is synonymous with a Room design of order $2n$.

Room came across structures of this kind in connection with a study of Clifford matrices, a concept from algebraic geometry. However, what was not realised until 1970 is that such designs had begun to be studied some fifty years earlier in connection with the design of tournaments for the card game known as "Bridge".

The purpose of a Duplicate Bridge tournament is to establish comparisons between every pair of players taking part. In each separate game, two pairs of players compete. We may designate each pair of players by a single symbol and, if there are $2n$ pairs, we may take the symbols denoting these to be the numbers $1, 2, \ldots, 2n - 1, \infty$ as above. The tournament consists of $2n - 1$ rounds and all the players take part in every round. During the course of the tournament each pair of players is required to play each other pair exactly once and also each pair of players is required to play at each of a number of different tables exactly once. The arrangements of the cards at a particular table are the same for each set of players who play at that table (but are not disclosed to any pair of players until they reach that table). In this way, the desired comparisons between the play of the different pairs is effected. All these requirements can be met if and only if there exists a Room design of order $2n$, where $2n$ is the number of Bridge pairs.

To see this, let us regard the rows of the Room square as giving the rounds and the columns as giving the tables. There are $2n - 1$ of the latter, but in any particular round only n of them are in use. If the cell which lies at the intersection of the r-th row and s-th column of the Room square is occupied, the two numbers which appear in it indicate the two pairs of players who should play the s-th table in the r-th round of the tournament.

The existence of such designs for use in Bridge tournaments was first investigated in 1897 by Howell, then Professor of Mathematics at the Massachusetts Institute of Technology. According to N.S. Mendelsohn, to whom the authors of [DK1] are indebted for pointing out the foregoing application of Room designs to Bridge Tournaments, Howell constructed designs for all values of n from 4 up to and including 15. In books giving instructions for the organization of Bridge tournaments, these designs are known as *Howell master sheets*. In Beynon(1943), for example, master sheets for $n = 4, 5$ and 7 are listed and, in Beynon(1944), master sheets for $n = 6$ and 8 as well. In Gruenther(1933), master sheets for all values of n from 4 to 15 inclusive are given. In Figure 6.4.1, we exhibit two master sheets (Room designs) of order 8 (that is, $n = 4$) attributed by Beynon to Ach and Kennedy of Cincinnati and to McKennedy and Baldwin (of whom further details are lacking) respectively.

It will be noted that each of the designs shown in Figure 6.4.1 is completely determined by its first row in the sense that each successive pair along a broken left-to-right diagonal of the square is obtained from the preceding pair in that diagonal by addition of 1 modulo 7 (or modulo $2n - 1$ in the general case) to each member of that pair. Such a Room design is called *cyclic*. Non-cyclic Room designs also exist. An example of order 8 is given in Figure 6.4.2.

1, 2		3, 4		5, 6		7, 8	
	3, 7	2, 5				4, 8	1, 6
4, 7	1, 5			3, 8	2, 6		
			6, 8	1, 4	5, 7	2, 3	
5, 8		6, 7	2, 4			1, 3	
	4, 6	1, 8	3, 5	2, 7			
3, 6	2, 8		1, 7				4, 5

FIG. 6.4.2.

We return to the post-1950 history of Room designs. In his paper Room(1955), the author pointed out the non-existence of Room designs for $n = 2$ and 3 and gave an example of a non-cyclic design for $n = 4$ which we reproduce in Figure 6.4.2. Here, the digits $1, 2, \ldots, 7$ and 8 are used, as the eighth symbol is no longer specially treated.

The next authors to write on this subject were Archbold and Johnson(1958), who gave a geometrical construction of cyclic Room designs for all values of n of the form 4^m and made use of Singer's theorem [Singer(1938)] to enable them to express the designs they obtained in a canonical form. These authors also showed how Room designs might be used as statistical designs for a suitable kind of experiment. (More information on the subject of statistical designs is given in Section 11.5.) Later, Archbold(1960) published a further paper in which he gave another construction for Room squares, based on difference sets, which yielded designs of orders 8, 12, 20 and 24 ($n = 4$, 6, 10 and 12). Both these kinds of design were cyclic and it is interesting to note that the design obtained for $n = 6$ is exactly the same as that published in Beynon(1944) sixteen years earlier. A more detailed investigation of the effectiveness of Room squares for use in statistical designs has been carried out by Shah(1970).

In Bruck(1963b), that author showed an interesting connection between Room designs and idempotent quasigroups, as follows:

THEOREM 6.4.1 *A Room design of order $2n$ is equivalent to a pair of commutative idempotent quasigroups, say $(Q, \mathbf{r})$ and $(Q, \mathbf{c})$, each of order $2n - 1$ and satisfying the following two orthogonality conditions:*

(i) *if $a, x, y \in Q$ are such that $xry = a = xcy$ then $x = y = a$; and*

(ii) *if a and b are distinct elements of Q, then there exists at most one unordered pair of elements x, y of Q such that $xry = a$ and $xcy = b$.*

PROOF. To see the equivalence, we suppose that the given Room design has symbols $1, 2, \ldots, 2n - 1, \infty$ and we permute its rows and then its columns in such a way that the ordered pair (∞, i) occurs at the intersection of the i-th row and the i-th column. The quasigroups $(Q, \mathbf{r})$ and $(Q, \mathbf{c})$ are now defined by the statements that they are idempotent and that, for $x \neq y$, $x\mathbf{r}y$ and $x\mathbf{c}y$ are respectively equal to the numbers of the row and column in which the cell containing the unordered pair x, y appears in the normalized Room square.   □

As an example, the square due to Room exhibited in Figure 6.4.2 takes the form shown in Figure 6.4.3 if we carry out the above rearrangements of rows and columns after replacing the symbol 8 by the symbol ∞.

$\infty, 1$	$4, 6$	$2, 7$			$3, 5$	
	$\infty, 2$			$3, 6$	$1, 7$	$4, 5$
	$1, 5$	$\infty, 3$	$2, 6$	$4, 7$		
$2, 5$	$3, 7$		$\infty, 4$			$1, 6$
$6, 7$			$1, 3$	$\infty, 5$	$2, 4$	
		$1, 4$	$5, 7$		$\infty, 6$	$2, 3$
$3, 4$		$5, 6$		$1, 2$		$\infty, 7$

$(\mathbf{r})$	1	2	3	4	5	6	7
1	1	7	5	6	3	4	2
2	7	2	6	5	4	3	1
3	5	6	3	7	1	2	4
4	6	5	7	4	2	1	3
5	3	4	1	2	5	7	6
6	4	3	2	1	7	6	5
7	2	1	4	3	6	5	7

$(\mathbf{c})$	1	2	3	4	5	6	7
1	1	5	4	3	2	7	6
2	5	2	7	6	1	4	3
3	4	7	3	1	6	5	2
4	3	6	1	4	7	2	5
5	2	1	6	7	5	3	4
6	7	4	5	2	3	6	1
7	6	3	2	5	4	1	7

FIG. 6.4.3.

Following Bruck, we shall call this the *normalized form* of the design.

Also shown in Figure 6.4.3 are the multiplication tables of the quasigroups $(Q, \mathbf{r})$ and $(Q, \mathbf{c})$ which are thus defined by this square.

The reader should note that the orthogonality conditions given in Theorem 6.4.1 do not imply that the quasigroups are orthogonal in the sense defined in Section 5.4. (Since both quasigroups are commutative, the equations $x\mathbf{r}y = a$ and $x\mathbf{c}y = b$ are not soluble simultaneously for all choices of a and b.) They provide our second example of a pair of perpendicular commutative quasigroups, further details of which are given in connection with quasi-orthogonal latin squares in Section 10.1.

Using Theorem 6.4.1, Bruck(1963b) simplified the construction of Archbold and Johnson for a Room design of order 2^{2m+1}. Let Q comprise the $2^{2m+1} - 1$

non-zero elements of the Galois field $GF[2^{2m+1}]$ and define two quasigroups $(Q, \mathbf{r})$ and $(Q, \mathbf{c})$ on the set Q by the statements $x\mathbf{r}x = x = x\mathbf{c}x$, $x\mathbf{r}y = x + y$ and $x\mathbf{c}y = (x^{-1} + y^{-1})^{-1}$ for all $x, y \in Q$. Then, using the properties of the finite field, it is easy to check that the orthogonality conditions described above are satisfied for these quasigroups and so a Room design of order 2^{2m+1} can be constructed. (That is, $n = 4^m$.)

DEFINITION.   A pair of idempotent quasigroups $(Q, \mathbf{r})$ and $(Q, \mathbf{c})$ which are commutative and satisfy the orthogonality conditions of Theorem 6.4.1 are called a *Room pair* of quasigroups.

Bruck asserted that it was easy to see that the direct product of two Room pairs of quasigroups was itself a Room pair of quasigroups. Let $(Q_1, \mathbf{r_1})$, $(Q_1, \mathbf{c_1})$ be one Room pair and $(Q_2, \mathbf{r_2})$, $(Q_2, \mathbf{c_2})$ another. The direct product is defined as the pair $(Q, \mathbf{r})$, $(Q, \mathbf{c})$ where $Q = Q_1 \times Q_2$ and $(q_1, q_2)\mathbf{r}(q'_1, q'_2) = (q_1\mathbf{r_1}\, q'_1, q_2\mathbf{r_2}\, q'_2)$, $(q_1, q_2)\mathbf{c}(q'_1, q'_2) = (q_1\mathbf{c_1}\, q'_1, q_2\mathbf{c_2}\, q'_2)$. Later, Mullin and Németh(1969a) pointed out that such a direct product need not satisfy the second orthogonality condition and used Bruck's own construction of Room pairs of quasigroups of order $2^{2m+1} - 1$ to give an explicit counter-example. Had Bruck's assertion been true, it would have implied that from two Room designs of orders $2m$ and $2n$ respectively, a Room design of order $(2m - 1)(2n - 1) + 1$ could be constructed. Stanton and Horton(1970,1972) have shown that, although Bruck's proof of it was fallacious, the statement just made is true. We shall now give their proof.

THEOREM 6.4.2   *If Room squares of sides $2m - 1$ and $2n - 1$ exist, then one can construct a Room square of side $(2m - 1)(2n - 1)$.*

PROOF.   Let R and S be Room squares of sides $r = 2m - 1$ and $s = 2n - 1$ whose entries are the symbols $0, 1, 2, \ldots, r$ and $0, 1, 2, \ldots, s$ respectively. Let L_1 and L_2 be a pair of (arbitrarily chosen) orthogonal latin squares of order $r = 2m - 1$ whose entries are the symbols $1, 2, \ldots, r$. To construct our Room square T of side $rs = (2m - 1)(2n - 1)$, we regard T as an $s \times s$ square each of whose cells is an $r \times r$ subsquare, and we prescribe these subsquares by the following rules:

(a) If the cell (i, j) of the Room square S is empty, then the $r \times r$ subsquare t_{ij} in the corresponding cell of T is to consist entirely of empty cells.

(b) If the cell (i, j) of S is occupied by the pair $(0, k)$, then the $r \times r$ subsquare t_{ij} in the corresponding cell of T is to be the Room square obtained from R by adding kr to each of its non-zero symbols. (The zero symbol is to be left unchanged.)

(c) If the cell (i, j) of S is occupied by the pair (h, k), with $h \neq 0$ and $k \neq 0$, then the $r \times r$ subsquare t_{ij} in the corresponding cell of T is to be the square with every cell occupied by an ordered pair of symbols and which is constructed from the latin squares L_1 and L_2 in the following manner. First add hr to each of the symbols of L_1 to form a new latin square L_1^*. Similarly form a new latin square L_2^* by adding kr to each of the symbols in L_2. Finally, juxtapose L_1^* and L_2^* so as to form a square $t_{ij} = (L_1^*, L_2^*)$

whose entries are ordered pairs of symbols (l_1, l_2) with $l_1 \in \{1 + hr, 2 + hr, \ldots, r + hr\}$ and $l_2 \in \{1 + kr, 2 + kr, \ldots, r + kr\}$.

The square T so constructed has

$$\{0, 1 + r, 2 + r, \ldots, r + r, 1 + 2r, 2 + 2r, \ldots, r + sr\}$$

as its set of symbols. Also, by the method of construction, each of these symbols occurs just once in each row and once in each column of T. For, we have that each of the symbols $1, 2, \ldots, r$ occurs just once in a row of R and once in a row of L_1 and L_2. Since each of the symbols $1, 2, \ldots, s$ occurs just once in a row of S, each of the symbols $x + yr$, $1 \leq x \leq r$, $1 \leq y \leq s$ occurs just once in a row of T. Since the symbol 0 occurs just once in a row of S, it occurs just once in a row of T. The preceding statements are also valid with "row" replaced by "column" throughout.

Moreover, in each subsquare t_{ij} no unordered pair of symbols occurs more than once (an immediate consequence of the mode of formation of these subsquares), and no two subsquares t_{ij} and t_{uv} have any pair in common. For, suppose that the pair $(x_1 + y_1 r, x_2 + y_2 r)$ with $x_1 \neq 0$ and $x_2 \neq 0$ were common to t_{ij} and t_{uv}. If $y_1 \neq y_2$, it would follow that (y_1, y_2) occurred in each of the cells (i, j) and (u, v) of S, a contradiction. If $y_1 = y_2 = y$, it would follow that $(0, y)$ occurred in each of the cells (i, j) and (u, v) of S. Finally, if $(0, x + yr)$ were common to t_{ij} and t_{uv} it would again follow that $(0, y)$ occurred in each of the cells (i, j) and (u, v) of S.

We conclude that T is a Room square, as desired. $\square$

Stanton and Mullin(1968), with the aid of a computer, investigated the possibility of the existence of cyclic Room squares of side $2m + 1$ whose first row contains the unordered pairs

$$(\infty, 0), (1, 2m), (2, 2m - 1), \ldots, (m, m + 1),$$

not necessarily in this order. They called such squares *patterned Room squares*, and the unordered pairs just listed were said to form a *starter* for such a square. They were able to construct such patterned Room squares of all odd orders $2m + 1$ from 7 to 49 except 9. For the latter order, patterned Room squares do not exist. However, a cyclic Room square of this order ($n = 5$ in our previous notation) had previously been obtained by Weisner(1964) and, of course, Room designs of this order had also been constructed much earlier for use as Howell master sheets. The construction by Stanton and Mullin was later generalized by Mullin and Németh(1969b). Also, Byleen(1970) proved that patterned Room squares of side p exist for all primes p not of the form $1 + 2^s$.

Next it was shown that there exist Room designs, not necessarily all cyclic, for all values of n except those for which $2n - 1$ has a Fermat prime of the form $2^{2^r} + 1$ as an unrepeated factor. This result is a consequence of Theorem 6.4.2 and a construction described by Mullin and Németh(1969c) which uses a generalized form of patterned Room squares and which gives Room squares of side $2n - 1$, where $2n - 1$ is any odd prime power which is not a Fermat prime of the form $2^{2^r} + 1$.

In the early 1970s, the existence problem for Room designs was finally settled, the conclusion being that these designs exist for every even order $2n$ except when $n = 2$ or 3. That is to say, there do not exist Room squares of side 3 or 5 but, for every other odd integer $2n - 1$, Room squares of side $2n - 1$ do exist. A good survey of the results which led to this conclusion was given by Mullin and Wallis(1975) who with the benefit of hindsight were also able to condense the proof.

We mentioned at the beginning of this section that Room designs can be constructed with the aid of latin squares. The following theorem is due to Byleen and Crowe(1971). See also Mullin and Németh(1970b), Mullin and Stanton(1971).

THEOREM 6.4.3 *Let L be a latin square of odd order $2n-1$ which is orthogonal to its transpose L^T, which has elements $1, 2, \ldots, 2n - 1$, and which is standardized in such a way that these elements occur along its main left-to-right diagonal in natural order. Let the entries of the main left-to-right diagonal of L^T be all replaced by the symbol ∞ and let M denote the matrix of ordered pairs formed when L and the modified square L^T are juxtaposed. Then a Room design may be obtained from M by deletion of a selected set of n of its left-to-right broken diagonals provided that the n diagonals to be deleted can be chosen so that*

(i) in each row of M, every element of L appears exactly once among the remaining pairs, and

(ii) if the diagonal which contains the cell m_{1j} of M is deleted then the diagonal which contains the cell m_{j1} is not deleted.

PROOF. It is an immediate consequence of the fact that L and L^T are transposes that if the cell m_{ik} of M, for $i \neq k$, contains the ordered pair (a, b), then the cell m_{ki} of M contains the ordered pair (b, a). Also, since L and L^T are orthogonal, every ordered pair of distinct elements a, b occurs just once in a cell of M. Hence, if n diagonals of M are deleted, which are chosen so that (ii) is satisfied, the remaining cells of M will contain each unordered pair of distinct elements a, b chosen from the set $\{1, 2, \ldots, 2n-1\}$ just once. Moreover, the pairs $(\infty, 1), (\infty, 2), \ldots, (\infty, 2n - 1)$ will occur along the main left-to-right diagonal of M.

It follows easily that if in addition, condition (i) is satisfied, then the equivalent condition for columns will also hold, so that the structure will form a Room square. □

Byleen and Crowe showed how to construct a latin square L and corresponding matrix M for which the requirements of Theorem 6.4.3 are satisfied whenever $2n - 1$ is an odd prime power not of the form $1 + 2^s$. We give an example for the case $2n - 1 = 7$ in Figure 6.4.4.

We discuss the general problem of constructing latin squares which are orthogonal to their own transposes in Section 5.5.

$$L = \begin{vmatrix} 1 & 7 & 6 & 5 & 4 & 3 & 2 \\ 3 & 2 & 1 & 7 & 6 & 5 & 4 \\ 5 & 4 & 3 & 2 & 1 & 7 & 6 \\ 7 & 6 & 5 & 4 & 3 & 2 & 1 \\ 2 & 1 & 7 & 6 & 5 & 4 & 3 \\ 4 & 3 & 2 & 1 & 7 & 6 & 5 \\ 6 & 5 & 4 & 3 & 2 & 1 & 7 \end{vmatrix}$$

$1,\infty$	$7,3$	$6,5$		$4,2$		
	$2,\infty$	$1,4$	$7,6$		$5,3$	
		$3,\infty$	$2,5$	$1,7$		$6,4$
$7,5$			$4,\infty$	$3,6$	$2,1$	
	$1,6$			$5,\infty$	$4,7$	$3,2$
$4,3$		$2,7$			$6,\infty$	$5,1$
$6,2$	$5,4$		$3,1$			$7,\infty$

Fig. 6.4.4.

Finally, we should like to mention a construction of Room designs with the aid of a pair of orthogonal Steiner triple systems which was first given by O'Shaughnessy(1968). We remind the reader that orthogonal Steiner triple systems were defined in Section 5.4. O'Shaughnessy's theorem is as follows:

THEOREM 6.4.4 *Let S and S' be two orthogonal Steiner triple systems of the same order v (necessarily congruent to 1 or 3 modulo 6, as shown in Section 2.3) and defined on the same set $\{1, 2, \ldots, v\}$. Then, a Room square of side v may be constructed by means of S and S' by putting the unordered pair of elements (i, j) in the cell of the k-th row and k'-th column of the square, where k is the third element of the triple of S which contains i and j, while k' is similarly defined by S'. The square is completed by putting the ordered pair (∞, h) in the cell of the h-th row and h-th column, for $h = 1, 2, \ldots, v$.*

PROOF.   The square constructed by the method just described clearly contains all unordered pairs of distinct elements obtainable from the set $\{1, 2, \ldots, v, \infty\}$. Since i occurs with k in exactly one triple of S, i occurs exactly once in the k-th row of the square. This is true for every i and every k. Similarly, i occurs exactly once in the k'-th column of the square, and again this is true for every i and every k'. Hence, the proof is complete.   □

In O'Shaughnessy(1968), the author used his method to construct Room designs of order 14 ($v = 13$) and order 20 ($v = 19$). For the interest of the reader, we give the two Steiner triple systems, S_1 and S_2 which generate the first of these designs and also the first row of the design itself (which is cyclic).

S_1 has triples $(1+i, 4+i, 5+i)$ and $(1+i, 6+i, 12+i)$, for $i = 0, 1, 2, \ldots, 12$, all addition being modulo 13. S_2 has triples $(1+i, 2+i, 5+i)$ and $(1+i, 7+i, 12+i)$. The Room design has first row:

$$(\infty, 1), (7, 9), -, (6, 12), -, -, -, (4, 5), (10, 0), (3, 8), -, (2, 11), -.$$

We note that the pair of Room quasigroups which correspond to a Room square constructed by O'Shaughnessy's method are the Steiner quasigroups produced by the two triple systems and so they are totally symmetric as well as perpendicular.

In Keedwell(1978), the present author showed that O'Shaughnessy's construction can be generalized to give a similar construction using perpendicular uniform P-circuit designs. Such a design separates the edges of the complete undirected graph K_v into circuits of length h. The special case $h = 3$ is the case of Steiner triple systems. Examples when $h = 5$, $v = 31$ and when $h = 7$, $v = 29$ are given in the paper. We discuss P-circuit designs in more detail in Section 8.3.

We end this section with a question. Two Room designs R_1 and R_2 of the same order $2n$ and defined on the same set S of symbols are *isomorphic* if R_2 can be obtained from R_1 by any sequence of permuting rows, permuting columns and permuting the symbols of S. [See Lindner(1972c).] They are *equivalent* if they are isomorphic or if R_2 is isomorphic to the transpose of R_1. We may ask "How many non-isomorphic and non-equivalent Room designs of order $2n$ exist?" So far as the authors are aware, the answers are only known for $n = 4$ and 5. We give them in the table below.

$n =$	2	3	4	5
Non-isomorphic	0	0	10	511562
Inequivalent	0	0	6	257630

For a comprehensive account of earlier results concerning Room squares, the reader should consult Mullin and Wallis(1975) and also part two of Wallis, Street and Wallis(1972) and the bibliographies therein.

In the early 1970s, Horton introduced an analogue of Room designs for more than two dimensions and he also pointed out an interesting connection between Room designs of order $2n$ and one-factorizations of the complete graph on $2n$ vertices which he attributed to Németh.

A fairly recent survey on the topic of Room squares and the above generalization to higher dimensions is in Dinitz and Stinson(1992). Also, a summary of results up to the publication date(s) of their handbook is in Colbourn and Dinitz(1996,2006). The topic of skew Room squares is mentioned in Section 10.1 of this book in connection with quasi-orthogonal latin squares.

Another, more recent, design related to Room squares is the so-called *referee square* which first arose in an attempt to solve a problem concerning the scheduling of the game of rugby.

For an account of this and of the application of latin squares to designing tournaments for further games and sports other than the card game of Bridge (such as whist, tennis and football), see Section 11.5 and also Keedwell(2000) and the references therein.

CHAPTER 7

Constructions of orthogonal latin squares which involve rearrangement of rows and columns

The many known methods of constructing two or more mutually orthogonal latin squares of an assigned order n can all be put into one of two categories. On the one hand, we have methods which involve obtaining all the squares by rearrangements of the rows or columns of a single one of the set, the square in question being usually referred to as the *basis square*[1]; and, on the other hand, we have methods which entail the use of previously determined sets of mutually orthogonal latin squares of smaller order, the squares of these sets being then modified or adjoined one to another in various ways to form squares of the order required. In the present chapter, we shall give an account of all those constructions which can be assigned to the first category, reserving our discussion of the second kind until Section 11.1 and Section 11.2. We may remark at this point that the construction of Bose, Shrikhande and Parker(1960) by means of which the Euler conjecture was disproved is of the second kind.

Although not strictly relevant to its title, we end the chapter by showing the close connection which exists between complete mappings of groups and left neofields.

7.1 Generalized Bose construction: constructions based on abelian groups

It has been shown in Keedwell(1966,1967) that all the known constructions of the first category can be regarded as special cases of a generalization of the construction which was described in Theorem 5.2.4. We may formulate this generalization as follows:

THEOREM 7.1.1 *Let $S_0 = I, S_1, S_2, \ldots, S_{r-1}$ be the permutations representing the rows of an $r \times r$ latin square L_1 as permutations of its first row and $M_1 \equiv I, M_2, M_3, \ldots, M_h$, $h \leq r-1$, be permutations keeping one symbol of L_1 fixed. Then the squares L_i^* whose rows are represented by the permutations $M_i S_0, M_i S_1, M_i S_2, \ldots, M_i S_{r-1}$ for $i = 1, 2, \ldots, h$ are certainly all latin and will be mutually orthogonal if, for every choice of $i, j \leq h$, the set of permutations*

$$S_0^{-1} M_i^{-1} M_j S_0, S_1^{-1} M_i^{-1} M_j S_1, \ldots, S_{r-1}^{-1} M_i^{-1} M_j S_{r-1}$$

is exactly simply transitive (sharply transitive) on the symbols of L_1.

[1]Each of the squares of the set is isotopic to the basis square.

Latin Squares and their Applications. http://dx.doi.org/10.1016/B978-0-444-63555-6.50007-6

PROOF. Let us remark first that, since each column of L_1 contains each symbol exactly once, the permutations $S_0, S_1, S_2, \ldots, S_{r-1}$ must form a sharply transitive set and that then the set of permutations $M_i S_0, M_i S_1, \ldots M_i S_{r-1}$ will also be sharply transitive. Consequently, the columns (and, of course, the rows) of L_i^* will contain each symbol exactly once, so L_i^* will be latin.

Secondly, if $U_0, U_1, \ldots, U_{r-1}$ are permutations representing the rows of one latin square L_i^* as permutations of $1, 2, \ldots, n$ and if $V_0, V_1, \ldots, V_{r-1}$ are the similarly defined permutations representing the rows of another latin square L_j^*, then the permutations $U_0^{-1}V_0, U_1^{-1}V_1, \ldots, U_{r-1}^{-1}V_{r-1}$ map the first, second,..., rth rows of L_i^* respectively onto the first second,..., rth rows of L_j^*. When, and only when, these squares are orthogonal, each symbol of the square L_i^* must map exactly once onto each symbol of the square L_j^* since each symbol of L_i^* occurs in positions corresponding to those of a transversal of L_j^*. Thus, when and only when L_i^* and L_j^* are orthogonal, the permutations $U_0^{-1}V_0, U_1^{-1}V_1, \ldots, U_{r-1}^{-1}V_{r-1}$ are a sharply transitive set. $\square$

The representation of a latin square by means of permutations was introduced originally by Schönhardt(1930), but the above two properties seem to have been observed first by Mann(1942).

The requirement in the above construction that the permutations $M_1, M_2, \ldots, M_h$ be permutations keeping one symbol of L_1 fixed is equivalent to requiring that the mutually orthogonal latin squares $L_1, L_2^*, L_3^*, \ldots, L_h^*$ be standardized in such a way that one column is the same for all the squares and, as we have shown in Section 5.1, such a requirement does not lead to any loss of generality. [This fact was first pointed out by Mann(1942)]. Notice also that the columns of any square L_i^* will always be a rearrangement of the columns of the basis square L_1 and this rearrangement will be that defined by the corresponding permutation M_i. (M_i reorders the symbols before the permutations $S_0, S_1, S_2, \ldots, S_{r-1}$ act.)

Now let us take the special case when the square L_1 is the addition table of an abelian group G. In this case, the S_i are the permutations of the Cayley representation of G and the M_i are one-to-one mappings of G onto itself. The entry in the cell of the xth row and yth column of the square L_i^* will be $xM_iS_y = xM_i + y$, where x and y belong to G and G is written in additive notation. If G is the additive group of a Galois field $\mathcal{F}$ and the M_i effect the multiplications of $\mathcal{F}$ so that $xM_i = xx_i$ for every x in G, then the construction of Theorem 7.1.1 becomes precisely the same as that described in Theorem 5.2.4.

We shall consider a number of other possibilities.

First we mention two other constructions which are applicable to the case when L_1 is the addition table of an abelian group.

(i) The construction of D.M. Johnson, A.G. Dulmage and N.S. Mendelsohn.

If we again take the case when the square L_1 is the addition table of an abelian group G and the S_i are the permutations of the Cayley representation of G, then the square L_i^* will be orthogonal to the square L_1 if the permutations

$S_y^{-1}M_iS_y$, where y ranges through G, form a sharply transitive set. That is, if and only if
$$wS_y^{-1}M_iS_y = wS_z^{-1}M_iS_z$$
implies $y = z$ for any w in G. That is, if and only if
$$(w - y)M_i + y = (w - z)M_i + z$$
implies y = z. Subtracting w from each side and writing $w - y = u$, $w - z = v$, we have that L_i^* will be orthogonal to L_1 if and only if $uM_i - u = vM_i - v$ implies $u = v$.

A mapping M_i of the abelian group G onto itself which has the latter property was called an *orthomorphism* by Johnson, Dulmage and N.S.Mendelsohn(1961) who were the first authors to use this term.[2]

Moreover, repetition of the argument leads at once to the fact that squares L_i^* and L_j^* will be orthogonal if $M_i^{-1}M_j$ is also an orthomorphism, as the above authors have shown. They have pointed out further that a one-to-one correspondence between orthomorphisms of G and transversals of the latin square representing the Cayley table of G can be established (cf. Section 1.5).

The entries in the cells $(x_1, y_1), (x_2, y_2), ..., (x_r, y_r)$, where (x_k, y_k) denotes the cell of the x_kth row and y_kth column, will form a transversal if and only if the mapping M_i defined by $x_kM_i = -y_k$ for $k = 1, 2, \ldots, r - 1$ is an orthomorphism of G. For suppose that we define $-y_k = x_kM_i$ for each k so that the entry in the (x_k, y_k)th cell is $x_k - x_kM_i$. Then these entries will be all distinct and form a transversal if and only if $x_h - x_hM_i = x_k - x_kM_i$ implies $x_h = x_k$; that is, if and only if M_i is an orthomorphism.

In their paper already referred to above, Johnson and her co-authors devised an algorithm for constructing orthomorphisms which is suitable for a computer search and with its aid they found a set of four non-identity orthomorphisms of the group $C_6 \times C_2$ of order 12 suitable for the construction of five mutually orthogonal latin squares of that order. They thus established that $N(12) \geq 5$, a result which has not been bettered up to the present.

(ii) The construction of R.C. Bose, I.M. Chakravarti and D.E. Knuth.

The necessary and sufficient condition
$$uM_i^{-1}M_j - u = vM_i^{-1}M_j - v \Rightarrow u = v$$
that the squares L_i^* and L_j^* defined above be orthogonal may be re written in the form
$$wM_j - wM_i = xM_j - xM_i \Rightarrow w = x,$$
where $w = uM_i^{-1}$ and $x = vM_i^{-1}$. In other words, the squares L_i^* and L_j^* will be orthogonal if and only if the equation $xM_j - xM_i = t$ is uniquely soluble for x. In Bose, Chakravarti and Knuth(1960,1961,1978), these authors have shown how mappings M_i having this property may be computed for abelian groups G of order $4t$ (with $4t - 1$ a prime power) and have thus obtained further sets of five mutually orthogonal latin squares of order 12. These authors called such mappings M_i *orthogonal mappings.*

---

[2]Their usage is consistent with the definition we gave in Section 1.5.

7.2   The automorphism method of H.B. Mann

The latin squares $L_1, L_2^*, L_3^*, \ldots, L_h^*$ of the construction described in Theorem 7.1.1 can be modified by the definition $L_i = L_i^* M_i^{-1}$ for $i = 2, 3, \ldots, h$. That is to say, the xth row of the latin square L_i will be represented by the permutation $M_i S_x M_i^{-1}$. This permutation, being conjugate to the permutation S_x, is very easy to calculate when the permutation S_x is known. We note also that the squares $L_1, L_2, \ldots, L_h$ will be mutually orthogonal whenever the squares $L_1, L_2^*, L_3^*, \ldots, L_h^*$ are so, and that each of the squares $L_1, L_2, \ldots, L_h$ has the identity permutation as first row. Thus, these squares[3] are a *standardized set* as defined in Section 5.2.

This modified form of the construction described in Theorem 7.1.1 we shall call the *K-construction* and we shall refer to it several times in the present chapter. Let the square L_1 be the addition table of a group (written in additive notation, but not necessarily abelian) and let the mappings M_i^{-1}, for $i = 1, 2, \ldots, h$, represent automorphisms τ_i of G. Let the elements of G be denoted by $a, b, c, \ldots$. Then the rows of the square L_1 are represented by the permutations $S_0 \equiv I, S_a, S_b, S_c$ and so on. The sth row of the square L_i is represented by the permutation

$$
M_i S_s M_i^{-1} = \begin{pmatrix} a\tau_i & b\tau_i & \cdots \\ a & b & \cdots \end{pmatrix} \begin{pmatrix} a & b & \cdots \\ a+s & b+s & \cdots \end{pmatrix} \begin{pmatrix} a & \cdots & a+s & \cdots \\ a\tau_i & \cdots & (a+s)\tau_i & \cdots \end{pmatrix}
$$

$$
= \begin{pmatrix} \cdots & a\tau_i & \cdots \\ \cdots & (a+s)\tau_i & \cdots \end{pmatrix} = \begin{pmatrix} \cdots & t & \cdots \\ \cdots & t+s\tau_i & \cdots \end{pmatrix} = S_{s\tau_i}
$$

since τ_i is an automorphism of G. The squares L_i and L_j will be orthogonal if $I, S_{a\tau_i}^{-1} S_{a\tau_j}, S_{b\tau_i}^{-1} S_{b\tau_j}, \ldots$ is a sharply transitive set of permutations. Since G is a group, and τ_i is an automorphism,

$$
S_{a\tau_i}^{-1} S_{a\tau_j} = S_{-a\tau_i} S_{a\tau_j} = S_{-a\tau_i + a\tau_j}.
$$

Thus, the squares will be orthogonal provided that $-s\tau_i + s\tau_j \neq -t\tau_i + t\tau_j$ for distinct elements s and t of G. That is, provided that $t\tau_i - s\tau_i \neq t\tau_j - s\tau_j$.

On writing $t - s = u$, we have that the squares L_i, and L_j will be orthogonal provided that the automorphisms τ_i and τ_j have the property $u\tau_i \neq u\tau_j$ for any element u other than the identity in G. Hence we may state:

THEOREM 7.2.1 *Let G be a group and suppose that there exist w automorphisms $\tau_1, \tau_2, \ldots, \tau_w$ of G every pair of which possesses the property that $u\tau_i \neq u\tau_j$ for any element $u \in G$ except the identity element. Then we shall be able to construct w mutually orthogonal latin squares based on the group G.*

---

[3]The squares $L_1, L_2^*, L_3^*, \ldots, L_h^*$ also form a standardized set, all of them having the same first column but differing first rows.

Theorem 7.2.1 was first proved by Mann(1942) and, in the same paper, the author obtained an upper bound for w in terms of the number of conjugacy classes of G. See the next theorem.

THEOREM 7.2.2 *Let L be a latin square based on a group G (that is, L represents the multiplication table of the group G) and let c_q be the number of conjugacy classes of elements with the property that all the elements of each class are elements of order q in the given group G. Let $v = \min c_q$ when q ranges through the factors of the order of G. Then not more than v mutually orthogonal latin squares containing L can be constructed from G by the automorphism method.*

PROOF. We have just shown that the squares L_i and L_j will be orthogonal provided that the automorphisms τ_i and τ_j have the property $u\tau_i \neq u\tau_j$ for any element u other than the identity in G, so suppose that w is the largest number of automorphisms of G each pair of which has this property and denote the members of such a set by $\tau_1, \tau_2, \ldots, \tau_w$. This requires that $u\tau_i\tau_j^{-1} \neq u$ for $u \neq e$; in other words, that each of the automorphisms $\sigma_{ij} = \tau_i\tau_j^{-1}$, i and $j = 1, 2, \ldots, w$, $i \neq j$, leaves no element other than the identity e of G fixed. We shall show that such an automorphism σ_{ij} of G maps each element of G into an element of a different conjugacy class. That is, for each i and j, $g\sigma_{ij}$ and g are in different conjugacy classes, where $g \in G$. It follows that $g\tau_i = g\sigma_{ij}\tau_j$ and $g\tau_j$ are in different conjugacy classes for each two automorphisms τ_i and τ_j of the set $\tau_1, \tau_2, \ldots, \tau_w$. Consequently, w cannot exceed the number of conjugacy classes whose elements have orders equal to that of g. Since this is true for each element $g \in G$, the result of the theorem will follow.

Let σ be an automorphism of G which leaves no element other than the identity e of G fixed. If g_i and g_j are distinct elements of G, then $g_i^{-1}(g_i\sigma) \neq g_j^{-1}(g_j\sigma)$ since equality would imply $(g_i\sigma)(g_j\sigma)^{-1} = g_ig_j^{-1}$ and so $(g_ig_j^{-1})\sigma = g_ig_j^{-1}$, whence $g_ig_j^{-1} = e$ and g_i, g_j would not be distinct. It follows that, as g_i ranges through the elements of G, so does $g_i^{-1}(g_i\sigma)$. Now let a be an element of G of order q. Then the element $a\sigma$ is likewise of order q. Suppose, if possible, that a and $a\sigma$ are in the same conjugacy class so that $a\sigma = h^{-1}ah$ for some element h in G. We can represent h in the form $g^{-1}(g\sigma)$ for some g in G. Then $gh = g\sigma$ and so
$$(gag^{-1})\sigma = g\sigma.a\sigma.(g\sigma)^{-1} = gh.h^{-1}ah.h^{-1}g^{-1} = gag^{-1}$$
implying $gag^{-1} = e$ and $a = e$. Thus, a and $a\sigma$ must be in different conjugacy classes as claimed above. This completes the proof. □

COROLLARY. *If $n = p_1^{r_1}p_2^{r_2}\ldots p_s^{r_s}$, where $p_1, p_2, \ldots, p_s$ are distinct primes, then not more than $t = \min(p_i^{r_i} - 1)$ MOLS of order n can be constructed from any group G by the automorphism method.*

PROOF. Since the Sylow p_i-subgroups of G are all conjugate (for fixed i), each conjugacy class of elements whose orders are powers of p_i has a representative element in each Sylow p_i-subgroup. Since the number of elements in a Sylow p_i-subgroup is $p_i^{r_i}$, c_q cannot exceed $p_i^{r_i} - 1$ when q is equal to p_i. (At least one element of G has order p_i by Cauchy's theorem.) Hence, $v = \min c_q$ cannot exceed $\min(p_i^{r_i} - 1)$.   □

7.3   The construction of pairs of orthogonal latin squares of order ten

The original construction by Parker(1959b) of a pair of orthogonal latin squares of order 10 involved the use of orthogonal latin squares of order three. However, Lyamzin (1963) and Weisner (1963) subsequently produced a pair in which the columns of one square are a rearrangement of the columns of the other. Although the two authors worked independently, the pairs of squares which they obtained are equivalent. In both pairs, one of the two squares is symmetric. Unfortunately, neither author has given a proper account of the means by which his squares were obtained.

Using the K-construction described above, the present author more recently tried unsuccessfully to extend the Ljamzin-Weisner squares to a set of three mutually orthogonal squares. [For the details, see Keedwell(1966).]

It is appropriate to point out at this point that, although the squares of the generalized set $L_1, L_2^*, L_3^*, \ldots, L_h^*$ have the property that the columns of any square L_i^* are a rearrangement of the columns of the square L_1, this is no longer necessarily true of either the rows or the columns of the set $L_1, L_2, \ldots, L_h$. We have the following theorem.

THEOREM 7.3.1   *The necessary and sufficient condition that the squares $L_1, L_2,$ $\ldots, L_h$ of the K-construction have the property that the rows of any one square L_i are the same as those of any other square L_j of the set, except that they occur in a different order, is that the operation $(\cdot)$ defined by the relation $aM_x = ax$ for each of $M_1, M_2, \ldots, M_h$, be right distributive over the operation $(+)$ defined by $aS_x = a + x$.*

PROOF.   It is necessary and sufficient to show that the permutations representing the rows of each square L_k are a reordering of the permutations representing the rows of the square L_1. That is, it is necessary and sufficient to have $M_k S_p M_k^{-1} = S_q$ for some q, or $M_k^{-1} S_q M_k = S_p$. Since $S_0 \equiv I$ and $M_1 \equiv I$, 0 and 1 are respective identities for $(+)$ and $(\cdot)$. Thus

$$M_k^{-1} S_q M_k = S_p = \begin{pmatrix} 0 \ldots k \ldots xk \ldots \\ 0 \ldots 1 \ldots x \ldots \end{pmatrix} \begin{pmatrix} 0 \ldots & x & \ldots \\ q \ldots & x+q \ldots \end{pmatrix} \times$$

$$\times \begin{pmatrix} 0 & 1 \ldots & q & \ldots & x+q & \ldots \\ 0 & k \ldots & qk & \ldots & (x+q)k & \ldots \end{pmatrix} = \begin{pmatrix} 0 \ldots & xk & \ldots \\ qk \ldots & (x+q)k \ldots \end{pmatrix}.$$

Therefore, $M_k^{-1} S_q M_k = S_p$ if and only if $p = qk$, and then

$$S_p = \begin{pmatrix} 0 \ldots & xk & \ldots \\ qk \ldots & xk+qk \ldots \end{pmatrix}; \text{ so } (x+q)k = xk + qk$$

for all x (and evidently also for all q and k); $x, q = 1, 2, \ldots, r-1$; $k = 1, 2, \ldots, h$.   □

COROLLARY.   *When the conditions of the theorem are fulfilled, the permutation*

M_k^{-1} *represents the rearrangement of the rows of L_1 which is required to turn it into the square L_k.*

PROOF.   Suppose that the pth row of the square L_k is the same as the qth row of the square L_1. Then $M_k S_p M_k^{-1} = S_q$ and so $p = qk$. That is, M_k maps q into p. Thus the mapping $M_k^{-1} = \begin{pmatrix} 0 \ldots p \ldots \\ 0 \ldots q \ldots \end{pmatrix}$ represents replacement of the pth row of L_1 by its qth row; that is, it rearranges the rows of L_1 in such a way that they become the rows of L_k.   □

For the squares of Ljamzin mentioned above, the row permutations are as follows:

$$L_1 = \{S_0, S_1, S_2, \ldots, S_8, S_9\}$$
$$L_2 = \{S_0, S_2, S_3, \ldots, S_9, S_1\}$$

where

$$S_0 = I$$
$$S_1 = (0\ 1)(2\ 5)(3\ 8\ 6\ 7\ 9\ 4)$$
$$S_2 = (0\ 2)(3\ 6)(4\ 9\ 7\ 8\ 1\ 5)$$
$$S_3 = (0\ 3)(4\ 7)(5\ 1\ 8\ 9\ 2\ 6)$$
$$S_4 = (0\ 4)(5\ 8)(6\ 2\ 9\ 1\ 3\ 7)$$
$$S_5 = (0\ 5)(6\ 9)(7\ 3\ 1\ 2\ 4\ 8)$$
$$S_6 = (0\ 6)(7\ 1)(8\ 4\ 2\ 3\ 5\ 9)$$
$$S_7 = (0\ 7)(8\ 2)(9\ 5\ 3\ 4\ 6\ 1)$$
$$S_8 = (0\ 8)(9\ 3)(1\ 6\ 4\ 5\ 7\ 2)$$
$$S_9 = (0\ 9)(1\ 4)(2\ 7\ 5\ 6\ 8\ 3)$$

The mapping $M_2 \equiv M_x$ is the permutation $(0)(9\ 8\ 7\ 6\ 5\ 4\ 3\ 2\ 1)$ and we have, for example, $(7+5)x = 3x = 2 = 6+4 = 7x + 5x$. That is, $7S_5 M_x = 7M_x S_{5M_x}$. So, the right-distributive law holds. Moreover, the rows of the square L_2 are obtained from the rows of the square L_1 by carrying out the permutation M_2^{-1} on those rows.

For the squares constructed by the automorphism method the row permutations are as follows:

$$L_1 = \{S_a, S_b, S_c, \ldots\}; \qquad L_i = \{S_{a\tau_i}, S_{b\tau_i}, S_{c\tau_i}, \ldots\}$$

for $i = 2, 3, \ldots, h$; and the mapping M_i is such that $xM_i = x\tau_i^{-1}$. Since τ_i is an automorphism, it is clear that the right-distributive law $(x+y)\tau_i^{-1} = x\tau_i^{-1} + y\tau_i^{-1}$ holds; and, moreover, the xth row of the square L_i is the $x\tau_i$th row of the square L_1, so the permutation M_i^{-1} rearranges the rows of L_1 in such a way that they become the rows of L_i.

Another illustration of the theorem is provided, for example, by the complete sets of mutually orthogonal latin squares which correspond to the Veblen-Wedderburn-Hall translation planes. (See Section 8.2.)

We notice further that, when the square L_1 is the addition table of a group G and the conditions of Theorem 7.3.1 are satisfied, each permutation M_x defines an automorphism of G: for the validity of the right-distributive law $(a+b)x = ax+bx$ implies that the mapping $a \to ax$ is an automorphism of G.

Although it is not strictly relevant to the subject matter of this chapter, we end this section by remarking that, as well as the orthogonal pair of 10×10 latin squares described above, Weisner(1963) constructed two other pairs of orthogonal latin squares of order ten. The second of these pairs which we display in Figure 7.3.1 [Figure 3 in Weisner(1963)] has the following two remarkable properties: if x,y is the entry in the cell of the ith row and jth column, then (1) i,j is the entry in the cell of the xth row and yth column; and (2) the pair of squares obtained by putting y,j in the xth row and ith column consists of a square and its (orthogonal) transpose: that is, each of the latter squares is self-orthogonal according to the definition given in Section 5.5.

	0	1	2	3	4	5	6	7	8	9
0	0 0	6 5	1 8	5 2	9 6	2 9	4 7	8 1	3 4	7 3
1	4 5	1 1	7 6	2 0	6 3	9 7	3 9	5 8	0 2	8 4
2	1 3	5 6	2 2	8 7	3 1	7 4	9 8	4 9	6 0	0 5
3	7 1	2 4	6 7	3 3	0 8	4 2	8 5	9 0	5 9	1 6
4	6 9	8 2	3 5	7 8	4 4	1 0	5 3	0 6	9 1	2 7
5	9 2	7 9	0 3	4 6	8 0	5 5	2 1	6 4	1 7	3 8
6	2 8	9 3	8 9	1 4	5 7	0 1	6 6	3 2	7 5	4 0
7	8 6	3 0	9 4	0 9	2 5	6 8	1 2	7 7	4 3	5 1
8	5 4	0 7	4 1	9 5	1 9	3 6	7 0	2 3	8 8	6 2
9	3 7	4 8	5 0	6 1	7 2	8 3	0 4	1 5	2 6	9 9

Fig. 7.3.1.

Weisner called the first property *the involutary property*. More recently, Mendelsohn called it *the Weisner property*. In N.S.Mendelsohn(1979), the latter author proved that pairs of orthogonal latin squares exist which are both self-orthogonal and have the involutary property for all orders $n \equiv 0$ or $1 \pmod 4$ except 5 and possibly also 12 and 21.

Let A be a latin square which is orthogonal to its transpose A^T. If the pair A, A^T has the involutary property, then Bennett, Du and Zhang(1998) have called the square A *self-conjugate self-orthogonal* thus giving yet another meaning to the overworked word "conjugate" and adding to the confusion that multiple meanings for the same word cause.

For the connection between the 10×10 orthogonal latin squares of Weisner and Ljamzin and two of the cyclic neofields of order 10, see page 36 of Bedford(1993).

7.4   The column method

This method, which is described in Keedwell(1966) and in [DK1], is another specialization of the K-construction in which once again the latin square L_1 is taken to be the multiplication table of a (not-necessarily-abelian) group. The method was originally used by its author to construct (for the first time) a triad of mutually orthogonal latin squares of order fifteen and also a pair of orthogonal latin squares based on the dihedral group of order 8. However, it has subsequently been shown that $N(15) \geq 4$. See Section 5.3.

7.5   The diagonal method

For this construction it is not necessary that the latin square L_1 be the multiplication table of a group and, mainly for this reason, the method is effective in obtaining either two or a complete set of mutually orthogonal latin squares (according as r is not or is a prime power) for every order r except one, two and six up to at least the value $r = 20$.[4] It is therefore of considerable interest to see why the method fails when $r = 6$ so as to obtain some explanation for the peculiarity of that integer in regard to the theory. In order to motivate the construction, let us look at the set-up given by the K-construction in the case when we have a set of mutually orthogonal latin squares based on a Galois field $GF(r)$. In that case, the multiplications are effected by the elements $1, x, x^2, \ldots, x^{r-2}$ of a cyclic group of order $r - 1$ and the corresponding multiplication permutations are $M_1 \equiv I$, $M_x = (0)(1\ x\ x^2\ \cdots\ x^{r-2})$, M_x^2, $M_x^3, \ldots, M_x^{r-2}$. The addition permutations are $S_0 \equiv I$, S_1, $S_x, \ldots, S_{x^{r-2}}$ and, because the latin squares L_1 and L_x are orthogonal, the permutations

$$
\begin{aligned}
M_x &= & (0)(1 & \quad x & \quad x^2 & \quad \cdots & \quad x^{r-2}) \\
S_{x^{r-2}}^{-1} M_x S_{x^{r-2}} &= (x^{r-2})(1 + x^{r-2} & \quad x + x^{r-2} & \quad \cdots & \quad x^{r-2} + x^{r-2}) \\
S_{x^{r-3}}^{-1} M_x S_{x^{r-3}} &= (x^{r-3})(1 + x^{r-3} & \quad x + x^{r-3} & \quad \cdots & \quad x^{r-2} + x^{r-3}) \\
\cdots\ \cdots &= & \cdots & \quad \cdots & \quad \cdots & \quad \cdots\ \cdots\ \cdots \\
\cdots\ \cdots &= & \cdots & \quad \cdots & \quad \cdots & \quad \cdots\ \cdots\ \cdots \\
S_x^{-1} M_x S_x &= & (x)(1 + x & \quad x + x & \quad \cdots & \quad x^{r-2} + x) \\
S_1^{-1} M_x S_1 &= & (1)(1 + 1 & \quad x + 1 & \quad \cdots & \quad x^{r-2} + 1)
\end{aligned}
$$

are a sharply transitive set.

We note that, if the first row and column are disregarded, the quotients of the elements in corresponding places of any two adjacent secondary[5] diagonals

---

[4]The author conjectures that the foregoing statement is true for all positive integers $r > 6$ but this is still unproved.

[5]By a secondary diagonal of an $m \times m$ matrix $A = \|a_{rs}\|$, where $r, s = 0, 1, \ldots, (m - 1)$, is meant a set of elements $a_{0\,p}, a_{1\,p-1}, \ldots, a_{p\,0}, a_{p+1\,m-1}, a_{p+2\,m-2}, \ldots a_{m-1\,p+1}$. The secondary diagonal given by $p = m - 1$ will be called the *main secondary diagonal*.

are constant. Moreover, in each such diagonal, the element of the pth row and qth column is x times the element of the $(p+1)$th row and $(q-1)$th column. In consequence of this fact and the Galois field relationship between addition and multiplication, it is clear that each of the elements $1, x, x^2, \ldots, x^{r-2}$ occurs exactly once in each secondary diagonal. These properties will be exploited in the method of construction we are about to explain.

We shall suppose that all elements except 0 are expressed as powers of x and write indices only. We shall write $r-1$ in place of the element 0. Let the indices (natural numbers) $0, 1, 2, \ldots, r-3$ be ordered in such a way that the differences between adjacent numbers are all different, taken modulo $r-1$, and so that no difference is equal to 1. As an example, take the case $r=8$. Then a solution is 4 0 2 1 5 3, the differences being 3, 2, 6, 4, 5. We set up an array whose main secondary diagonal consists entirely of 7s and such that all other secondary diagonals consist of the indices 0, 1, 2, ... , 6 in descending order, columns being taken cyclically, column 0 = column 7, and so on. The result is shown in Figure 7.5.1

$$\begin{vmatrix} 4 & 0 & 2 & 1 & 5 & 3 & 7 \\ 6 & 1 & 0 & 4 & 2 & 7 & 3 \\ 0 & 6 & 3 & 1 & 7 & 2 & 5 \\ 5 & 2 & 0 & 7 & 1 & 4 & 6 \\ 1 & 6 & 7 & 0 & 3 & 5 & 4 \\ 5 & 7 & 6 & 2 & 4 & 3 & 0 \\ 7 & 5 & 1 & 3 & 2 & 6 & 4 \end{vmatrix}$$

FIG. 7.5.1.

It is clear from the method of construction that, if the entries of the first column of this array are all different, so are the entries of each other column. We seek arrays A_8^* such that this is the case. Figure 7.5.2, with first row and column deleted, provides an example of such an array. In fact, Figure 7.5.2 is derived from the array A_8^* by bordering the array with a 0th row and 0th column whose entries are those missing from the appropriate column or row of the array A_8^*. Thus, the complete square shown in Figure 7.5.2 is a latin square.

Since the differences between adjacent entries of the first row of the array A_8^* are all different, the same is true of each other row provided that each such row is regarded as starting and ending at the entry 7. Moreover, the disposition of the 7s is such that every other integer follows and precedes an entry 7 exactly once. Thus, Figure 7.5.2 provides a sharply transitive set of permutations from which we can derive permutations

$$S_o = I, \ S_1 = S_{x^o}, S_{x^1}, S_{x^2}, \ldots, S_{x^6}$$

representing the rows of a latin square L_1 and with the property that the latin square L_x derived from L_1 by means of the multiplication permutation M_x will be orthogonal to L_1.

$$
\begin{aligned}
M &= (7)(0\,1\,2\,3\,4\,5\,6) \\
S_{x^6}^{-1} M_x S_{x^6} &= (6)(2\,5\,0\,4\,3\,1\,7) \\
S_{x^5}^{-1} M_x S_{x^5} &= (5)(4\,6\,3\,2\,0\,7\,1) \\
S_{x^4}^{-1} M_x S_{x^4} &= (4)(5\,2\,1\,6\,7\,0\,3) \\
S_{x^4}^{-1} M_x S_{x^4} &= (3)(1\,0\,5\,7\,6\,2\,4) \\
S_{x^4}^{-1} M_x S_{x^4} &= (2)(6\,4\,7\,5\,1\,3\,0) \\
S_{x^4}^{-1} M_x S_{x^4} &= (1)(3\,7\,4\,0\,2\,6\,5) \\
S_{x^4}^{-1} M_x S_{x^4} &= (0)(7\,3\,6\,1\,5\,4\,2)
\end{aligned}
$$

FIG. 7.5.2.

In fact, the sharply transitive set of permutations shown in Figure 7.5.2 arises from the Galois field GF(8): for, in this field, $x^8 - x = 0$ and a primitive root x satisfies $x^3 + x + 1 = 0$ [or $x^3 = x + 1$, since $-1 = 1$ in GF(8)]. So, if $M_x = (0)(1\,x\,x^2\,x^3\,x^4\,x^5\,x^6)$, we have

$$
\begin{aligned}
S_{x^6}^{-1} M_x S_{x^6} &= (x^6)(1 + x^6\ x + x^6\ \ldots\ x^5 + x^6\ x^6 + x^6) \\
&= (x^6)(x^2\ x^5\ 1\ x^4\ x^3\ x\ 0),
\end{aligned}
$$

which is equivalent to the expression given in Figure 7.5.2. However, arrays of type A_r^* exist when r is not a prime power. For example, when $r = 10$, we get just two possible arrays A_{10}^*. One of these is shown (bordered by the appropriate 0th row and 0th column) in Figure 7.5.3 and leads to one of the pairs of orthogonal latin squares of order 10 obtained by Weisner and displayed in Fig. 2 of his paper Weisner(1963).

$$
\begin{aligned}
&(9)(0\ 1\ 2\ 3\ 4\ 5\ 6\ 7\ 8) \\
&(8)(5\ 3\ 0\ 2\ 7\ 6\ 1\ 4\ 9) \\
&(7)(2\ 8\ 1\ 6\ 5\ 0\ 3\ 9\ 4) \\
&(6)(7\ 0\ 5\ 4\ 8\ 2\ 9\ 3\ 1) \\
&(5)(8\ 4\ 3\ 7\ 1\ 9\ 2\ 0\ 6) \\
&(4)(3\ 2\ 6\ 0\ 9\ 1\ 8\ 5\ 7) \\
&(3)(1\ 5\ 8\ 9\ 0\ 7\ 4\ 6\ 2) \\
&(2)(4\ 7\ 9\ 8\ 6\ 3\ 5\ 1\ 0) \\
&(1)(6\ 9\ 7\ 5\ 2\ 4\ 0\ 8\ 3) \\
&(0)(9\ 6\ 4\ 1\ 3\ 8\ 7\ 2\ 5)
\end{aligned}
$$

FIG. 7.5.3.

The discussion above may be summarized into the following theorem:

THEOREM 7.5.1 *If r is an integer for which an array A_r^* exists, then there exist at least two mutually orthogonal latin squares of order r. When r is a power of a prime, there exists at least one array A_r^* which can be used to generate a complete set of mutually orthogonal latin squares of order r, representing the desarguesian projective plane of that order.*

Following Keedwell(1966), we shall now give two simple criteria for the existence of an array A_r^* corresponding to a given integer r.

THEOREM 7.5.2 *A necessary and sufficient condition that an array A_r^* exists for a given integer r is that the residues $2, 3, \ldots, (r-2)$, modulo $(r-1)$, can be arranged in a row array P_r in such a way that the partial sums of the first one, two $, \ldots, (r-3)$ are all distinct and non-zero modulo $(r-1)$ and so that, in addition, when each element of the array is reduced by 1, the new array P_r' has the same property.*

PROOF.   Suppose firstly that an array A_r^*, such as is given in Figure 7.5.2, exists. The differences between successive entries of the first row of A_r^*, excluding the last element $(r-1)$, form an array P_r of the type specified in the theorem, since, if this were not the case, the entries of that first row would not be all distinct. Moreover, if we write

$$A_r^* = \begin{vmatrix} a_{11} \ a_{12} \ a_{13} \cdots \\ a_{21} \ a_{22} \ a_{23} \cdots \\ a_{31} \ a_{32} \ a_{33} \cdots \\ \cdots \ \cdots \ \cdots \cdots \end{vmatrix}$$

we have $a_{ij} = a_{i-1,j+1} - 1$. Consequently,

$a_{21} - a_{11} = (a_{12} - 1) - a_{11} = (a_{12} - a_{11}) - 1$,

$a_{31} - a_{21} = (a_{22} - 1) - a_{21} = (a_{22} - a_{21}) - 1 = (a_{13} - a_{12}) - 1$, and generally,

$a_{i+1,1} - a_{i1} = (a_{i2} - 1) - a_{i1} = (a_{i2} - a_{i1}) - 1 = (a_{1,i+1} - a_{1i}) - 1$,

so that the differences between successive entries of the first column of A_r^*, excluding the last element $(r-1)$, form an array of the type specified in the theorem, with each element reduced by one from the corresponding element of P_r: for, if this were not the case, the entries of the first column of A_r^* would not be all distinct.

Conversely, suppose that the residues $2, 3, \ldots, (r-2)$, modulo $(r-1)$, can be arranged in the manner described in the theorem. Then an array A_r^* exists. We shall find it easiest to illustrate this by means of an example. We take the case $r = 8$.

$$P_8 = 3, 2, 4, 6, 5; \ P_8' = 2, 1, 3, 5, 4.$$

Here, 3=3, 3+2=5, 3+2+4=2, 3+2+4+6= 1, 3+2+4+6+5 = 6, so the entries of the first row of A_8^* are

$$x, x + 3, x + 5, x + 2, x + 1, x + 6, r - 1 = 7$$

Since the entry $r - 2 = 6$ is not to appear in the first row, we must have $x + 4 = 6$: that is, $x = 2$. Then A_8^* is as shown in Figure 7.5.2, the entries in the first row and column being all distinct in virtue of the properties of the row array P_8. □

COROLLARY 1. *With each array A_r^* occurs a dual array $A_r^{(d)}$ obtained from A_r^* by replacing each entry q in the corresponding row arrays P_r, P_r', by its complement $(r-1) - q$ taken modulo $(r-1)$ to obtain a dual row array $P_r^{(d)}$ and hence a dual array $A_r^{(d)}$.*

PROOF.   If P_r and P'_r are transformed in the manner specified, the entries of the tranform of P'_r become one greater than the corresponding entries of the transform of P_r. It is easy to see, therefore, that the transform of P'_r must have the same properties as P_r.   □

For example, if P_8 is as given in the theorem above, we have
$$P_8^{(d)} = 5, 6, 4, 2, 3 \text{ and } P_8^{'(d)} = 4, 5, 3, 1, 2.$$

COROLLARY 2. *With each array A_r^* occurs a mirror image array $A_r^{(m)}$ (which may coincide with $A_r^{(d)}$), obtained from A_r^* by first constructing P_r, reversing the order of its entries to obtain a row array $P_r^{(m)}$, and then constructing the corresponding array $A_r^{(m)}$ in the manner described in the theorem.*

PROOF.   We have only to show that the row array obtained by reversing the order of the entries of the row array P_r has the same properties as P_r. Let
$$P_r = d_1, \ldots, d_{r-3}; \ P'_r = d_1 - 1, d_2 - 1, \ldots, d_{r-3} - 1.$$
The property possessed by P_r is that
$$e_1 = d_1, e_2 = d_1 + d_2, \ldots, e_{r-3} = \sum_{i=1}^{i=r-3} d_i$$
are all distinct and non-zero modulo $r - 1$. Writing
$$P_r^{(m)} = d_{r-3}, d_{r-4}, \ldots, d_2, d_1,$$
we have
$$d_{r-3} = e_{r-3} - e_{r-4}, d_{r-3} + d_{r-4} = e_{r-3} - e_{r-5}, \ldots,$$
$$\sum_{i=r-3}^{i=2} d_i = e_{r-3} - e_1, \sum_{i=r-3}^{i=1} d_i = e_{r-3} - 0,$$
and these are evidently all distinct and non-zero, since $e_{r-3}, e_{r-4}, \ldots, e_2, e_1, 0$ are all distinct, and $e_{r-3} \neq 0$.

Consequently, $P_r^{(m)}$ has the same property as P_r. The same argument applied to P'_r shows that $P_r^{'(m)}$ has the same property. Thus, $P_r^{(m)}$ has all the same properties as P_r, as required.

For our second criterion for the existence of an array A_r^* of the type described above, we shall need the concept of a neofield.

DEFINITION. A set $J = \{a, b, c, \ldots\}$ on which are defined two binary operations $(+)$ and $(\cdot)$ such that J is a loop with respect to the operation $(+)$ with identity element 0 say, $J - 0$ is a group with respect to the operation $(\cdot)$ and the distributive laws $a(b + c) = ab + ac$ and $(b + c)a = ba + ca$ hold, is called a *neofield*. The neofield is *commutative* if the loop $(J, +)$ is commutative.

Neofields were first introduced by Paige(1949), and he derived a number of their principal properties in that paper. They had not subsequently been used until the following theorem was proved by the present author in Keedwell(1966):

THEOREM 7.5.3 *A necessary and sufficient condition that an array A_r^* exists for a given integer r is that there exists a neofield of r elements whose multiplicative group is cyclic of order $r - 1$ and which possesses the property that*

$$(1 + x^t)/(1 + x^{t-1}) = (1 + x^u)/(1 + x^{u-1}) \Rightarrow t = u,$$
where x is any generating element of the multiplicative group.

PROOF.   Suppose first that such a neofield of order r exists. There exists a unique element x^s of the additive loop such that $1 + x^s = 0$.[6]

Then $x^t + x^{s+t} = 0$ for all integers t. The addition table of the neofield will consequently be of the form shown in Figure 7.5.4. Here, if the 0th row and 0th column be disregarded, the main secondary diagonal consists entirely of zeros. The remaining secondary diagonals comprise elements which can be represented as powers of x in descending natural order.

Moreover, the differences between the indices of adjacent elements of the first row of the addition table are all distinct and non-zero modulo $r - 1$: for
$$(x^{r-2} + x^{s+p})/(x^{r-2} + x^{s+p-1}) = (x^{r-2} + x^{s+q})/(x^{r-2} + x^{s+q-1})$$
would imply
$$(1 + x^{s+p+1})/(1 + x^{s+p}) = (1 + x^{s+q+1})/(1 + x^{s+q})$$
with $p \neq q$, contrary to hypothesis.[7] Likewise, the differences between the indices of adjacent elements of the first column of the addition table are all distinct and non-zero modulo $r - 1$: for
$$(x^{r-p} + x^s)/(x^{r-p+1} + x^s) = (x^{r-q} + x^s)/(x^{r-q+1} + x^s)$$
would imply
$$(1 + x^{s+p-1})/(1 + x^{s+p-2}) = (1 + x^{s+q-1})/(1 + x^{s+q-2})$$
with $p \neq q$, contrary to hypothesis. Consequently, if we replace the nonzero elements of the addition table by their indices when represented as powers of x, and the zeros by $r - 1$, we shall obtain an array A_r^*.

$(+)$	x^s	x^{s+1}	$\cdots$	x^{r-2}	1	$\cdots$	x^{s-2}	x^{s-1}
x^{r-2}	$x^{r-2} + x^s$	$x^{r-2} + x^{s+1}$	$\cdots$	$x^{r-2} + x^{r-2}$	$x^{r-2} + 1$	$\cdots$	$x^{r-2} + x^{s-2}$	0
x^{r-3}	$x^{r-3} + x^s$	$x^{r-3} + x^{s+1}$	$\cdots$	$x^{r-3} + x^{r-2}$	$x^{r-3} + 1$	$\cdots$	0	$x^{r-3} + x^{s-1}$
$\star\star$	$\star\star\star$	$\star\star\star$	$\cdots$	$\star\star\star$	$\star\star\star$	$\cdots$	$\star\star\star$	$\star\star\star$
$\star\star$	$\star\star\star$	$\star\star\star$	$\cdots$	$\star\star\star$	$\star\star\star$	$\cdots$	$\star\star\star$	$\star\star\star$
x	$x + x^s$	0	$\cdots$	$x + x^{r-2}$	$x + 1$	$\cdots$	$x + x^{s-2}$	$x + x^{s-1}$
1	0	$1 + x^{s+1}$	$\cdots$	$1 + x^{r-2}$	$1 + 1$	$\cdots$	$1 + x^{s-2}$	$1 + x^{s-1}$

FIG. 7.5.4.

Conversely, suppose that an array A_r^* is given. We border the array with a 0th row and 0th column in such a way that the bordered array forms an $r \times r$ latin square. Upon replacing each element $t \neq r - 1$ by x^t and each element $r - 1$ by zero, and then identifying the square with that given in Figure 7.5.4 for a

---

[6]Note that the property $(1 + x^t)/(1 + x^{t-1}) = (1 + x^u)/(1 + x^{u-1}) \Rightarrow t = u$ certainly holds if one of $t, t - 1, u, u - 1$ is equal to s.

[7]Moreover, no index is 1 since $(1 + x^t)/(1 + x^{t-1}) = x$ would imply $1 + x^t = x + x^t$ which is impossible for a loop.

suitable choice of s (determined by the position of the 1 in the 0th row), we shall define the addition table of a neofield of the type specified in the theorem. $\square$

The property that $(1 + x^t)/(1 + x^{t-1}) = (1 + x^u)/(1 + x^{u-1})$ holds if and only if $t = u$ in a neofield with a cyclic multiplicative group has been called the *divisibility property* and a neofield of the kind specified in Theorem 7.5.3 has been called a neofield with property D or, more briefly, a *D-neofield*.

In Keedwell(1966), the author has shown by a consideration of the addition table that a D-neofield is commutative if and only if the row arrays $P_r^{(d)}$ and $P_r^{(m)}$ associated with it, and defined as in the corollaries to Theorem 7.5.2, coincide.

A quite different proof of this result will be found in Keedwell(1967). In the latter paper, the conditions under which two D-neofields of a given order are isomorphic or anti-isomorphic have been analysed and, with the aid of a computer, all isomorphically distinct D-neofields of orders less than or equal to 17 have been catalogued. Some examples of such neofields have also been given for the orders 18, 19 and 20. It appears clear from the results that the number of D-neofields of an assigned order r increases rapidly with r and it is conjectured by the author that D-neofields exist for all orders r except 6. (We note at this point that, in particular, every Galois field is a D-neofield.)

As regards the explanation for the non-existence of a D-neofield of order 6 (and consequent non-existence of a pair of orthogonal latin squares of that order constructible by the above method), we easily see that, when $r = 6$, a row array P_r having the properties of Theorem 7.5.2 cannot exist: for the integers 2, 3, 4 and $2 - 1, 3 - 1, 4 - 1$ cannot be simultaneously reordered so that their partial sums taken modulo 5 are all distinct and non-zero. Essentially, this is due to the fact that the integers 2, 3 occur in each triad and must occur consecutively in one or the other. This observation suggests that the non-existence of orthogonal latin squares of order 6 may be a consequence of nothing more profound than the paucity of combinatorial rearrangements of the integers 0 to 4.

A number of unsolved problems concerning D-neofields have been listed in Keedwell(1970) and these also appear as Problems 7.3, 7.4 and 7.5 in [DK1].

7.6   Left neofields and orthomorphisms of groups

The reader will recall from Section 5.1 that, if a group has an orthomorphism (or, equivalently, a complete mapping) then its Cayley table has an orthogonal mate. There is a very close connection between orthomorphisms and left neofields. A *left neofield* $(N, \oplus, \cdot)$ differs from a neofield as defined in the preceding section in having only one distributive law: namely, $a(b \oplus c) = ab \oplus ac$ so multiplication may not distribute over addition from the right. A left neofield whose multiplicative group is $(G, \cdot)$ is said to be *based on* that group.

We recall that an orthomorphism ϕ of a finite group $(G, \cdot)$ is a one-to-one mapping $g \to \phi(g)$ of G onto itself such that the mapping $g \to \theta(g)$, where $\theta(g) = g^{-1}\phi(g)$ is again a one-to-one mapping of G onto itself. θ is the *complete mapping* of G onto itself which corresponds to the orthomorphism ϕ. ϕ is in *canonical form* if $\phi(1) = 1$, where 1 is the identity element of G.

An orthomorphism in canonical form may be regarded as a permutation
$$\phi = (1)(g_{11}\ g_{12}\ \cdots\ g_{1k_1})(g_{21}\ g_{22}\ \cdots\ g_{2k_2})\cdots(g_{s1}\ g_{s2}\cdots\ g_{sk_s})$$
of G such that the elements $g_{ij}^{-1}g_{i,j+1}$ (where $i = 1, 2, \ldots, s$ and the second suffix j is added modulo k_i) comprise the non-identity elements of G each counted once. Then $\phi(g_{ij}) = g_{i,j+1}$ and $\theta(g_{ij}) = g_{ij}^{-1}g_{i,j+1}$. The mapping θ is the *complete mapping* associated with the orthomorphism ϕ.

If we express an orthomorphism in cycle form as above, we see that $\prod_{g \in G} g = \prod \theta(g_{ij}) = \prod(g_{ij}^{-1}g_{i,j+1}) = 1$. That is, the product of all the elements of G in some appropriate order is equal to the identity element (as already proved in Theorem 2.5.1).

The concept of an orthomorphism and also that of a near orthomorphism (defined below) is important in the study of neofields because, in particular, the mapping defined by $1 \to 1$ and $w \to 1 \oplus w$ (for $w \neq 1$) is an orthomorphism of the multiplicative group of every left neofield for which $1 \oplus 1 = 0$.

Suppose now that the elements of the finite group $(G, \cdot)$ can be arranged in the form of a sequence $[g_1'\ g_2'\ g_3'\ \cdots\ g_h']$ followed by s cyclic sequences $(g_{11}\ g_{12}\ \cdots\ g_{1k_1}), (g_{21}\ g_{22}\ \cdots\ g_{2k_2}), \ldots, (g_{s1}\ g_{s2}\cdots\ g_{sk_s})$ such that the elements $g_j'^{-1}g_{j+1}'$ and $g_{ij}^{-1}g_{i,j+1}$ together with the elements $g_{ik_i}^{-1}g_{i1}$ comprise the non-identity elements of G each counted once. Then the mapping θ of $G - \{g_h'\}$ onto $G - \{1\}$ given by $\theta(g_j') = g_j'^{-1}g_{j+1}'$ for $j = 1, 2, \ldots, h - 1$ and $\theta(g_{ij}) = g_{ij}^{-1}g_{i,j+1}$ (where arithmetic of second suffices is modulo k_i) is called a *near complete mapping* of G. The associated mapping $\phi : g \to g\theta(g)$ of $G - \{g_h'\}$ onto $G - \{g_1'\}$ is called a *near orthomorphism* of G. It is said to be in *canonical form* if $g_1' = 1$, where 1 is the identity element of G.

We shall represent a near orthomorphism in the following way:
$$\phi = [g_1'\ g_2'\ g_3'\ \cdots\ g_h'](g_{11}\ g_{12}\ \cdots\ g_{1k_1})(g_{21}\ g_{22}\ \cdots\ g_{2k_2})\cdots(g_{s1}\ g_{s2}\cdots\ g_{sk_s})$$
When the near orthomorphism is in canonical form so that $g_1' = 1$, we shall denote the element which has no image under the mapping by η and call it the *ex-domain element*.

It is immediate to see from the definition of θ that
$$\eta = (\textstyle\prod_{j=1}^{j=h-1} \theta(g_j'))(\textstyle\prod_{i=1}^{i=s} \textstyle\prod_{j=1}^{j=k_s} \theta(g_{ij}))$$

That is, η is the product of all the elements of G in some appropriate order.

A special case of this arises when the group $(G, \cdot)$ is sequenceable. In that case, $\phi = [e\ b_1\ b_2\ \cdots\ b_{n-1}]$ is a near orthomorphism of the group, where $b_i = ea_1a_2\ldots a_i$ for $i = 1, 2, \ldots, n-1$ if the elements of G are $e, a_1, a_2, \ldots, a_{n-1}$. See Chapter 3 of [DK2] and Section 2.6 of this book.

The connection between left neofields and orthomorphisms and near orthomorphisms of a group is given by the following theorem:

THEOREM 7.6.1 *Let $(N, \oplus, \cdot)$ be a finite left neofield with multiplicative group $(G, \cdot)$, where $G = N - \{0\}$. Then, if $1 \oplus 1 = 0$ in N, N defines an orthomorphism* *(and corresponding complete mapping) of $(G, \cdot)$, which is in canonical form. If $1 \oplus 1 \neq 0$ but $1 \oplus \eta = 0$, N defines a near orthomorphism of G in canonical form and with η as ex-domain element.*

Conversely, let $(G, \cdot)$ be a finite group with identity element 1 which possesses an orthomorphism ϕ (in canonical form). Let 0 be a symbol not in the set G and define $N = G \cup \{0\}$. Then $(N, \oplus, \cdot)$ is a left neofield, where we define $\psi(w) = 1 \oplus w = \phi(w)$ for all $w \neq 0, 1$ and $\psi(O) = 1 \oplus 0 = 1$, $\psi(1) = 1 \oplus 1 = 0$. Also, we define $x \oplus y = x(1 \oplus x^{-1}y)$ for $x \neq 0$, $0 \oplus y = y$ and $0 \cdot x = 0 = x \cdot 0$ for all $x \in N$.

Alternatively, let $(G, \cdot)$ possess a near orthomorphism ϕ in canonical form. Then, with N defined as before, $(N, \oplus, \cdot)$ is a left neofield, where we define $\psi(w) = 1 \oplus w = \phi(w)$ for all $w \neq 0, \eta$, where η is the ex-domain element of ϕ and $\psi(0) = 1 \oplus 0 = 1$, $\psi(\eta) = 1 \oplus \eta = 0$. Also, as before, $x \oplus y = x(1 \oplus x^{-1}y)$ for $x \neq 0$, $0 \oplus y = y$ and $0 \cdot x = 0 = x \cdot 0$ for all $x \in N$.

PROOF. *Part 1: neofield given.*

Case when $1 \oplus 1 = 0$.

In $N - \{0\}$, define $\phi(w) = 1 \oplus w$ for $w \neq 1$ and $\phi(1) = 1$. Since $(N, \oplus)$ is a loop, the equation $\phi(w) = 1 \oplus w = z$ has a unique solution for w and $w \neq 1$ if $z \neq 0$, so ϕ is a one-to-one mapping of $G = N - \{0\}$ onto itself.

Then $\theta(w) = w^{-1}\phi(w) = w^{-1} \oplus 1$ if $w \neq 1$, and $\theta(1) = 1$. Since $(N, \oplus)$ is a loop, the equation $\theta(w) = w^{-1} \oplus 1 = z$ has a unique solution for w and $w \neq 1$ if $z \neq 0$, so θ is a one-to-one mapping of $G = N - \{0\}$ onto itself.

It follows that ϕ is an orthomorphism in canonical form of $G = N - \{0\}$ with θ as corresponding complete mapping.

Case when $1 \oplus 1 \neq 0$ and $1 \oplus \eta = 0$.

In $N - \{0\}$, define $\phi(w) = 1 \oplus w$ for $w \neq \eta$. Since $(N, \oplus)$ is a loop, the equation $\phi(w) = 1 \oplus w = z$ has a unique solution for w with $w \neq 0, \eta$ when $z \in G - \{1\}$, where $G = N - \{0\}$. Thus, ϕ is a one-to-one mapping from $G - \{\eta\}$ to $G - \{1\}$.

Then $\theta(w) = w^{-1}\phi(w) = w^{-1} \oplus 1$ if $w \neq 0, \eta$. Since $(N, \oplus)$ is a loop, the equation $\theta(w) = w^{-1} \oplus 1 = z$ has a unique solution for w with $w \neq 0, \eta$ when $z \in G - \{1\}$. So θ is a one-to-one mapping from $G - \{\eta\}$ to $G - \{1\}$. [Note that $\eta^{-1} \oplus 1 = 0$ in $(N, \oplus, \cdot)$.]

It follows that ϕ is a near orthomorphism in canonical form mapping $G - \{\eta\}$ onto $G - \{1\}$ with θ as corresponding near complete mapping. Since the identity element 1 of G has no pre-image, the near orthomorphism is in canonical form.

PROOF. *Part 2: mapping of group given.*

We require to show that $(N, \oplus)$ is a loop with identity element 0 and that multiplication is left-distributive over addition.

We have $tu \oplus tv = tu(1 \oplus (tu)^{-1}tv) = tu(1 \oplus u^{-1}v) = t(u \oplus v)$ from the definition $x \oplus y = x(1 \oplus x^{-1}y)$, so the left distributive law holds.

Also, in the case when ϕ is an orthomorphism,

$$\psi(w) = 1 \oplus w = \phi(w) \text{ for } w \neq 0, 1;$$
$$\psi(1) = 1 \oplus 1 = 0;$$

$\psi(0) = 1 \oplus 0 = 1$. [Note that $\phi(1) = 1$ since ϕ is in canonical form.]
Therefore, the elements $a \oplus y = a(1 \oplus a^{-1}y) = a\psi(a^{-1}y)$ are distinct as y varies.

When $x \neq 0, a$; $x \oplus a = x(1 \oplus x^{-1}a) = x\phi(x^{-1}a) = x \cdot x^{-1}a\theta(x^{-1}a) = a\theta(x^{-1}a)$. These are all different as x varies and none is equal to 0 or a. Also, $a \oplus a = a(1 \oplus 1) = 0$ and $0 \oplus a = a$. Therefore, the elements $x \oplus a$ are all distinct as x varies.

In the case when ϕ is a near orthomorphism,

$\psi(w) = 1 \oplus w = \phi(w)$ when $w \neq 0, \eta$;

$\psi(\eta) = 1 \oplus \eta = 0$;

$\psi(0) = 1 \oplus 0 = 1$. [Note that $\phi(\eta)$ is not defined since η is ex-domain element and that $\phi(w) \neq 1$ for $w \in G$ since ϕ is in canonical form.]
Therefore, the elements $a \oplus y = a(1 \oplus a^{-1}y) = a\psi(a^{-1}y)$ are distinct as y varies.

When $x^{-1}a \neq 0, \eta$; $x \oplus a = x(1 \oplus x^{-1}a) = x\phi(x^{-1}a) = x \cdot x^{-1}a\theta(x^{-1}a) = a\theta(x^{-1}a)$. These are all different as x varies and none is equal to 0 or a. Furthermore, $x^{-1}a = \eta \Rightarrow x = a\eta^{-1}$ and $a\eta^{-1} \oplus a = a\eta^{-1}(1 \oplus \eta) = 0$. Also, $0 \oplus a = a$. Therefore, the elements $x \oplus a$ are all distinct as x varies.

We conclude that $(N, \oplus)$ is a loop with 0 as two-sided identity. $\square$

Note. The mapping $\psi : z \Rightarrow 1 \oplus z$ is called the *presentation function* of the left neofield because it determines the complete addition table of the neofield by virtue of the fact that $x \oplus y = x(1 \oplus x^{-1}y)$.

A left neofield becomes a neofield if the right distributive law holds. It was shown in Hsu and Keedwell(1984) that necessary and sufficient conditions for this are that the complete mapping (in the case when $1 \oplus 1 = 0$) or the near complete mapping (in the case when $1 \oplus \eta = 0$, $\eta \neq 1$) of the group G which defines the neofield maps conjugacy classes to conjugacy classes and additionally in the latter case that η is in the centre of G.

Most useful in practice are cyclic neofields: that is, neofields whose multiplication group is cyclic. Such neofields have been investigated in considerable detail in a book by Hsu(1980).

In Keedwell(2001), the present author has investigated the properties that an orthomorphism or near orthomorphism of a cyclic group must have to enable it to define a finite field rather than just a cyclic neofield. (As remarked earlier in this book, for abelian groups the existence of orthomorphisms or near orthomorphisms are mutually exclusive according as the group has not or has a unique element of order two.)

CHAPTER 8

Connections with geometry and graph theory

In this chapter we shall show firstly that there is a very intimate connection between latin squares and geometric nets and, as already noted in Section 5.2, between complete sets of MOLS and projective planes. As a consequence, a number of problems concerning the former may more easily be dealt with in the guise of geometry.

In the first section, we show that, with each bordered latin square or quasigroup, there is associated a corresponding geometric 3-net and that the geometrical properties of the 3-net reflect the algebraic properties of the quasigroup; while, as shown in Chapter 2, these in turn influence the structure of the latin square. Conversely, with each given 3-net, there is associated a class of isotopic quasigroups and related latin squares.

We go on to show that a k-net, for $k > 3$, is correspondingly associated with a set (or sets) of $k-2$ MOLS. In particular, when $k = n+1$, the k-net becomes a projective plane and the $n-1$ MOLS are then a complete set of MOLS. We point out that non-isomorphic projective planes exist for some orders n and explain how this leads to the existence of structurally distinct complete sets of MOLS of those orders.

In the third section, we discuss various connections between latin squares and graphs and indicate how these connections help in the solution of latin square problems and vice versa.

8.1 Quasigroups and 3-nets

The concept of a 3-net arises naturally in connection with the problem of assigning co-ordinates to the points and lines of an affine or projective plane. Historically, the concept arose also in the study of certain topological problems of differential geometry. Among early papers on the subject are Baer(1939,1940), Blaschke(1928), Blaschke and Bol(1938), Bol(1937), Reidemeister(1929) and Thomsen(1930). Later papers concerning the connections between quasigroups, geometric nets and projective planes are Bruck(1951), Ostrom(1968) and Pickert(1954). Extensive bibliographies of papers on the subject will be found in Aczél, Pickert and Radó(1960), in Aczél's survey paper (1964) or (1965), in Bruck(1958), Pickert(1955) and Chapter 11 of Belousov(1967b). Two books devoted to the connection between nets and quasigroups are Belousov (1971,1972).

More recently, the subject has been treated in Chapter 2 of Pflugfelder(1990) (where it is shown in particular that the representation of a quasigroup by

Latin Squares and their Applications. http://dx.doi.org/10.1016/B978-0-444-63555-6.50008-8

a geometric net leads naturally to the concepts of parastrophy and isostrophy(paratopy). Also, several chapters of Chien, Pflugfelder and Smith(1990) treat different aspects of the subject. (In the latter books, the word *web* is used in place of net.) The subject is also treated in various sections of the CRC Handbook of Combinatorial Designs [Colbourn and Dinitz(1996,2006)] where, in particular, the fact that a geometric net and a transversal design are geometrically dual concepts is pointed out.

Let us begin by giving a general definition of a net as used in geometry. We should mention here that the "lines" of our definition may be curves of the real plane in the applications to differential geometry or to nomograms.

DEFINITION.   A *geometric net* is a set of objects called "points" together with certain designated subsets called "lines". The lines occur in classes called "parallel classes" such that (a) each point belongs to exactly one line of each parallel class; (b) if l_1 and l_2 are lines of different parallel classes, then l_1 and l_2 have exactly one point in common; (c) there are at least three parallel classes and at least two points on a line. A net possessing k parallel classes is called a *k-net*. (If the number of parallel classes is one or two and the remaining conditions are fulfilled, the system may be called a *trivial net*.)

If the net is finite, then it is characterized by a parameter n, called the *order* of the net, such that (i) each line contains exactly n points; (ii) each parallel class consists of exactly n lines; and (iii) the total number of points is n^2.

Statements (i) and (ii) follow at once from conditions (a) and (b) of the definition of a net as soon as we postulate that one line of one parallel class has n points or that one parallel class has n lines. Then, since the lines of any one of the parallel classes contain all the points and since each of the n lines of this class has n points, statement (iii) follows. (Note that the number of parallel classes may be more or less than n.)

DEFINITION.   An *affine plane* π^* comprises a set of objects called "points" together with certain distinguished subsets called "lines" such that (a) two distinct points belong to (are incident with) exactly one line; (b) every point exterior to a given line l of π^* is incident with exactly one line which has no point in common with l; and (c) there are at least three points not all incident with one line.

The connection between affine planes and geometric nets follows immediately from the statement that:

THEOREM 8.1.1 *An affine plane is a geometric net in which each pair of points is incident with a line. (Equivalently, we may say that an affine plane of order n is a net of order n with $n + 1$ parallel classes.)*

PROOF.   Let l be a line of the net containing n points and let P be a point not on l. Then there are exactly $n + 1$ lines through P if each pair of points of the net is connected by a line: for the joins of P to the points of l give n lines and there is also one line through P belonging to the parallel class of l. These $n + 1$

lines through P necessarily belong to distinct parallel classes, so there are $n+1$ parallel classes. □

For the purpose of introducing co-ordinates for the points of a geometric net or an affine plane, we may assign arbitrary symbols to the lines of just two of its parallel classes, C_1 and C_2, and then assign the co-ordinate pair (x,y) to the point through which pass the line x of class C_1 and the line y of class C_2. There will be no loss of generality in using symbols from the same set Q (of cardinal n) for each of the two parallel classes C_1 and C_2. If the same set Q of symbols is used to label the lines of a third parallel class C_3 and if the line w of this class is incident with the point (x,y), then we can define a binary operation $(*)$ on the set Q by the statement $x*y = w$. The properties of the geometric net then ensure that each equation $x*y = w$ is uniquely soluble for x, y or w in Q when the other two variables are specified, and so $(Q, *)$ is a quasigroup.

Conversely, we may prove:

THEOREM 8.1.2 *If a bordered latin square of order n is given representing the multiplication table of a quasigroup $(Q, *)$, then there can be associated with it a geometric net of order n having exactly three parallel classes and such that the lines x, y of the parallel classes C_1 and C_2 are incident with the line w of the parallel class C_3 if and only if $x*y = w$.*

PROOF. As the n^2 points of our 3-net, we take the n^2 ordered pairs (x, y) of our quasigroup $(Q, *)$. For each fixed choice of x, the n points (X, y), $y \in Q$, form a line $l_X^{(1)}$ of the parallel class C_1. For each fixed choice of y, the n points (x, Y), $x \in Q$, form a line $l_Y^{(2)}$ of the parallel class C_2. For each fixed choice of w, the set of all points (x, y) such that $x*y = W$ form a line $l_W^{(3)}$ of the parallel class C_3. Since the multiplication table of $(Q, *)$ is a latin square, each line of C_3 has exactly n points.

It is immediate from the definitions that no two lines of C_1 or C_2 or C_3 have a point in common and that each point belongs to exactly one line of each parallel class. Also, two lines belonging to distinct parallel classes have exactly one point in common. The lines $l_X^{(1)}$ and $l_Y^{(2)}$ have the point (X, Y) in common. The lines $l_X^{(1)}$ and $l_W^{(3)}$ have the point $(X, \bar{y})$ in common. where $\bar{y}$ is the unique solution of the equation $X*\bar{y} = W$. The lines $l_Y^{(2)}$ and $l_W^{(3)}$ have the point $(\bar{x}, Y)$ in common. where $\bar{x}$ is the unique solution of the equation $\bar{x}*Y = W$. This completes the proof. □

As an example, we give in Figure 8.1.1 below a quasigroup of order 4 and its associated 3-net N.

We shall show shortly that the algebraic properties of the quasigroup $(Q.*)$ are reflected in the geometrical properties of its associated 3-net.

There is a close connection between an affine plane and a projective plane which latter we defined in Section 5.2. Indeed, the former may be regarded simply

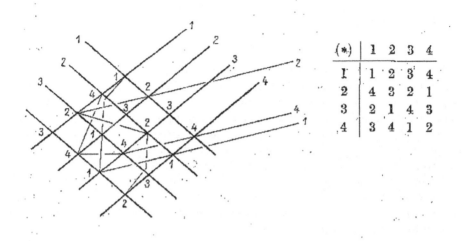

(∗)	1	2	3	4
1	1	2	3	4
2	4	3	2	1
3	2	1	4	3
4	3	4	1	2

FIG. 8.1.1.

as a projective plane from which one line has been deleted. Precisely, we may make the following statement:

A *projective plane* of order n is an affine plane to which one extra line l_∞ has been adjoined, each of the $n+1$ points of l_∞ being a point of concurrence (point at infinity) for the $n+1$ lines of a parallel class. (Such a point on l_∞ is sometimes called the *vertex* of the parallel class.) Thus, a projective plane has a total of n^2+n+1 lines and n^2+n+1 points. When one line and the points on it are deleted, we get a geometric net of order n having $n+1$ parallel classes. That is, we get an affine plane.

In co-ordinatizing an affine (or projective) plane, it is usual to introduce two binary operations denoted by $(+)$ and $(\cdot)$ respectively, the first being associated with a 3-net whose three "parallel" classes have their vertices A, B, E collinear on the special line l_∞ (that is, the lines of each parallel class are "parallel" in the sense used with reference to an affine plane), and the second being associated with a 3-net such that two of its "parallel" classes coincide with two of those of the addition net and have vertices on l_∞ while the third one comprises the pencil of lines through a finite point O say. The finite point (vertex of the parallel class) is regarded as having been deleted when treating the system as a net. More details are given in the next section of this chapter.

In the case of the real affine plane, these binary operations coincide with addition and multiplication of real numbers if the four parallel classes in question comprise the lines parallel to the x-axis, the lines parallel to the y-axis, the lines of gradient 1 and the lines through the origin O of cartesian co-ordinates.

Let us consider further the relation between latin squares and geometric 3-nets. In the first place, let us observe that if a 3-net is given, the choice of co-ordinatizing set Q and the procedure for assigning its elements one-to-one to the lines of each parallel class are not unique. Indeed, it is easy to see that change of the co-ordinatizing set of a 3-net N corresponds to a change of symbols in the associated latin square L, while changes of the order in assigning the elements of a given set of symbols to the lines of the parallel classes correspond to re-orderings of (i) the row labels, (ii) the column labels, and (iii) the symbols in the body of the associated (bordered) latin square.

Equivalently, we may state:

THEOREM 8.1.3 *Isotopic quasigroups[1] are associated with the same geometric 3-net and correspond to changes in the labelling set.*

PROOF.   To see this, let $(G, \cdot)$ and $(H, *)$ be two quasigroups. These quasigroups are isotopic if there exists an ordered triple (θ, ϕ, ψ) of one-to-one mappings θ, ϕ, ψ of G onto H such that $(x\theta) * (y\phi) = (x \cdot y)\psi$ for all x, y in G. Geometrically, the mapping θ effects a replacement of the symbols of the set G which are assigned to the lines of one parallel class C_1 of a geometric net by symbols of the set H. The mappings ϕ and ψ respectively effect similar re-labellings of the lines of the parallel classes C_2 and C_3. In the re-labelled set, the lines of the classes C_1 and C_2 labelled $x\theta$ and $y\phi$ are incident with the line $(x \cdot y)\psi$ when $(x\theta) * (y\phi) = (x \cdot y)\psi$: that is, when the lines of the classes C_1 and C_2 which were originally labelled x and y are incident with the line of C_3 which was originally labelled $x \cdot y$. The latter lines are the same as the former.   □

COROLLARY. *Isotopisms of a quasigroup $(G, \cdot)$ onto itself correspond to changes of order in the labelling set G of the associated 3-net.*

PROOF.   The mappings θ, ϕ, ψ are now mappings of the set G onto itself. That is, they represent permutations of the labelling set.   □

In Chapter 1, we proved that every isotope $(H, *)$ of a quasigroup $(G, \cdot)$ is isomorphic to a principal isotope of the quasigroup (Theorem 1.3.2) and that, among the principal isotopes of a quasigroup $(G, \cdot)$, there always exist loops (Theorem 1.3.3). These facts imply the truth of the following theorem:

THEOREM 8.1.4 *If $(G, \cdot)$ is a quasigroup associated with a given 3-net, then there are as many isomorphically distinct quasigroups associated with that net as there are principal isotopes of $(G, \cdot)$. Moreover, among these quasigroups there always exist loops. Consequently any 3-net has loops among its co-ordinate systems.*

We can interpret Theorem 1.3.3 of Chapter 1 geometrically as follows:
In the 3-net whose parallel classes are the lines through the points X, Y, W, we may select the particular lines of the (X) and (Y) parallel classes which carry the labels u, v of $(G, \cdot)$ as the ones to be re-labelled as the identity lines e for the principal isotope $(G, \otimes)$. The point of intersection U of these particular lines will

---

[1]The concept of isotopy between quasigroups was introduced in Section 1.3.

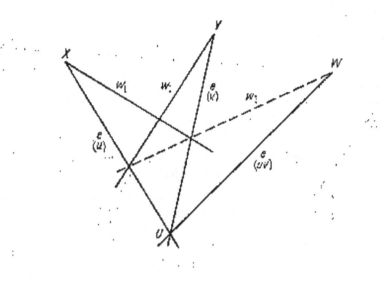

FIG. 8.1.2.

be called unit point. We now re-label the lines of the parallel classes (X) and (Y) in such a way that if w_i is any line of the parallel class (W) other than the line WU, then the line of the parallel class (X) through the point $w_i \cap YU$ will be labelled w_i and the line of the parallel class (Y) through the point $w_i \cap XU$ will also be labelled w_i. In the quasigroup $(G, \otimes)$ thus defined, $e \otimes w_i = w_i = w_i \otimes e$. Moreover, if we now consider how we should re-label the line WU, we see that it should carry the same label e as XU and YU. So, the identity element e of $(G, \otimes)$ is the element uv of $(G, \cdot)$ as in Theorem 1.3.3. (See Figure 8.1.2.)

Let us now show as promised how the two most important properties of a quasigroup are reflected in the geometrical properties of its associated 3-net.

Before doing so, it is desirable to mention again that, in plane projective geometry, two distinct types of 3-net arise, *affine* nets and *triangular* nets. In the first, the points A, B, E of l_∞ take the roles of X, Y, W and the parallel classes are the pencils with the collinear vertices A, B, E. For such nets, the quasigroup is usually written additively. Lines through E have "equations" of the form $y = x + a$. For triangular nets on the other hand, the vertices A, B, O of a proper triangle are the vertices of three pencils of lines representing the "parallel classes" (the lines of the pencil with vertex O being no longer parallel in the geometrical sense), and, for such nets, the associated quasigroup is usually written multiplicatively. Lines through O have "equations" of the form $y = xm$. (See Figure 8.1.3.) If the elements of either the additive or the multiplicative quasigroup satisfy some algebraic identity, this corresponds to closure of a certain geometrical configuration.

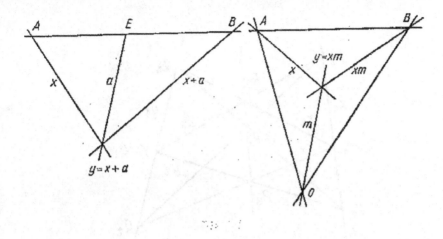

FIG. 8.1.3.

Let us consider first the algebraic effect of requiring closure of the Pappus configuration in a triangular or affine 3-net. The assertion that the Pappus configuration has incidence closure is usually called *Pappus' theorem*. This may be stated as follows: if P_1, P_2, P_3 and Q_1, Q_2, Q_3 are two triads of collinear points, then the triad of points R_1, R_2, R_3, where $R_i = P_j Q_k \cap P_k Q_j (i, j, k = 1, 2, 3; \; i \neq j \neq k)$, are collinear. We have

THEOREM 8.1.5 *If the parallel classes of the net comprise the lines through the points X, Y, W and if the lines through X are denoted by $x_1, x_2, \ldots$ and the lines through Y are denoted by $y_1, y_2, \ldots$, then Pappus' theorem implies the algebraic relation*

$$x_1 y_2 = x_2 y_1 \text{ and } x_1 y_3 = x_3 y_1 \Rightarrow x_2 y_3 = x_3 y_2$$

for the quasigroup $(G, \cdot)$, where the x_i and y_i are the symbols of G and where we identify the points P_i, Q_i, R_i with the points shown in Figure 8.1.4.

PROOF. If we label the various lines of the configuration as in Figure 8.1.4, the result becomes obvious since $x_1 y_2 = x_2 y_1$ implies that the points P_2, Q_3 are collinear with W and $x_1 y_3 = x_3 y_1$ implies that the points P_3, Q_2 are collinear with W. ☐

To interpret the above result in terms of latin squares, let $(G, \cdot)$ be a quasigroup associated with a 3-net in which the Pappus configuration has incidence closure and let a and a' be fixed elements of the set $G = \{a, b_1, b_2, \ldots\}$. Then, for $b_i \in G$, $b_i a' = h_i$ for some $h_i \in G$ and there exists an element $b_i' \in G$ such that $a b_i' = h_i$. Thus, to each $b_i \in G$, there corresponds another element $b_i' \in G$ such that $b_i a' = a b_i'$. Moreover, when $b_i = a$, then $b_i' = a'$ since $aa' = aa'$ is a tautological statement. Therefore, $G = \{a', b_1', b_2', \ldots\}$. Then, if the Pappus configuration closes in the associated net, we shall have $b_i b_j' = b_j b_i'$ for $i, j = 1, 2, 3, \ldots$; $i \neq j$. This follows from the fact that $a b_i' = b_i a'$ and $a b_j' = b_j a'$ together

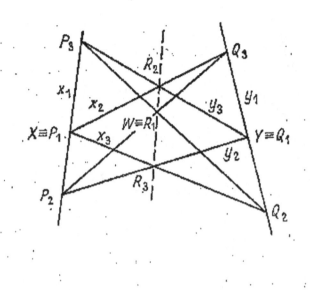

FIG. 8.1.4.

imply $b_i b_j' = b_j b_i'$. Thus, the multiplication table of $(G, \cdot)$ will be a bordered latin square of the form shown in Figure 8.1.5.

$(\cdot)$	a'	b_1'	b_2'	b_3'	$\ldots$
a	$\cdot$	u	v	w	$\ldots$
b_1	u	$\cdot$	x	y	$\ldots$
b_2	v	x	$\cdot$	z	$\ldots$
b_3	w	y	z	$\cdot$	$\ldots$
$\cdot$	$\cdot$	$\cdot$	$\cdot$	$\cdot$	$\cdot$
$\cdot$	$\cdot$	$\cdot$	$\cdot$	$\cdot$	$\cdot$

FIG. 8.1.5.

If, instead of a triangular net with vertices X, Y, W (or say A, B, O in the application to the co-ordinatization of a projective plane discussed later), we have to deal with an affine net whose parallel classes have collinear vertices A, B and E, then the whole of the above discussion remains valid if the Pappus configuration is replaced by the Thomsen configuration [see Thomsen(1930)] which is illustrated in Figure 8.1.6. This may be regarded as the configuration obtained from the Pappus configuration when $O \to E$ on l_∞. The assertion that the Thomsen configuration has incidence closure is often called the *axial minor theorem of Pappus*. This may be stated as follows: If P_1, P_2, P_3, Q_2, Q_3 are five given points such that P_1, P_2, P_3 are collinear and if R_1 is the point

$P_2Q_3 \cap P_3Q_2$ and P_1R_1 meets Q_2Q_3 at Q_1, then the points R_1, $R_2 \equiv P_1Q_3 \cap P_3Q_1$ and $R_3 \equiv P_1Q_2 \cap P_2Q_1$ are collinear.

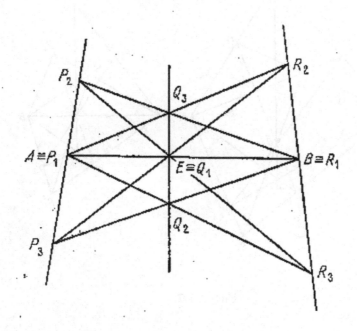

FIG. 8.1.6.

We consider next the algebraic consequence for a triangular net of requiring closure of the large Reidemeister configuration [see Reidemeister(1929)]. This asserts that, if the quadrangles $P_1P_2P_3P_4$ and $P_1'P_2'P_3'P_4'$ have the joins of vertices P_1P_2, P_3P_4, $P_1'P_2'$, $P_3'P_4'$ concurrent in a point B, P_2P_3, P_1P_4, $P_2'P_3'$, $P_1'P_4'$ concurrent in a point A and P_1P_1', P_2P_2', P_3P_3', concurrent in a point O, then the join P_4P_4' passes through the same point O (see the left-hand diagram in Figure 8.1.7). In Argunov(1950), this configurational proposition has been denoted by the symbol $mA(11, 11, 12)$.

   We have the following important result.

THEOREM 8.1.6 *If the lines through the point A have symbols x_1, x_2, ... , and the lines through the point B have symbols y_1, y_2, ... , then closure of the large Reidemeister configuration implies satisfaction of the algebraic relation*

$$x_1y_2 = x_2y_1, \ x_1y_4 = x_2y_3 \ and \ x_3y_2 = x_4y_1 \Rightarrow x_3y_4 = x_4y_3$$

in the quasigroup $(G, \cdot)$ associated with the net whose parallel classes are the pencils of lines through the points A, B and O.

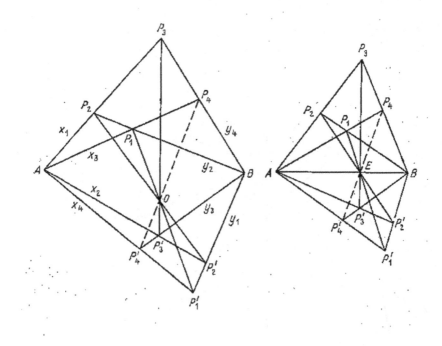

FIG. 8.1.7.

PROOF.   If we label the various lines of the configuration as in the left-hand diagram of Figure 8.1.7, the result becomes obvious since the first three equalities in the implication statement respectively imply that the pairs of points P_2, P_2'; P_3, P_3'; and P_1, P_1' are collinear with the point O. $\square$

Let us observe that the algebraic relation given in Theorem 8.1.6 is precisely the quadrangle criterion which was first introduced in Theorem 1.2.1. This observation makes it clear that our next result is only to be expected.

THEOREM 8.1.7 *Closure of the large Reidemeister configuration in the triangular 3-net associated with the quasigroup $(G, \cdot)$ implies that all loops isotopic to $(G, \cdot)$ are groups.*

PROOF.   Let $x \otimes y = (x\sigma) \cdot (y\tau)$, where $x\sigma^{-1} = xy_1$ and $y\tau^{-1} = x_1 y$. Then $(G, \otimes)$ is a loop-principal isotope (LP-isotope) of $(G, \cdot)$ with identity element $x_1 y_1$ by Theorem 1.3.3. Let p, q, r be arbitrarily chosen in G. Then, for fixed choices of x_1 and y_1, there exist elements x_2, y_2, x_3 and y_3 of G such that $p = x_3 y_1$, $q = x_1 y_2 = x_2 y_1$ and $r = x_1 y_3$. For x_2, y_2, x_3 and y_3 so defined, there are elements x_4 and y_4 in G such that $x_1 y_4 = x_2 y_3$ and $x_4 y_1 = x_3 y_2$. By the hypothesis of the theorem, we then have $x_3 y_4 = x_4 y_3$. Also,

$$(p \otimes q) \otimes r = [(x_3 y_1) \otimes (x_1 y_2)] \otimes r = (x_3 \sigma^{-1} \otimes y_2 \tau^{-1}) \otimes r = (x_3 y_2) \otimes r$$
$$= (x_4 y_1) \otimes (x_1 y_3) = x_4 \sigma^{-1} \otimes y_3 \tau^{-1} = x_4 y_3;$$

and

$$p \otimes (q \otimes r) = p \otimes [(x_2 y_1) \otimes (x_1 y_3)] = p \otimes (x_2 \sigma^{-1} \otimes y_3 \tau^{-1}) = p \otimes (x_2 y_3)$$
$$= (x_3 y_1) \otimes (x_1 y_4) = x_3 \sigma^{-1} \otimes y_4 \tau^{-1} = x_3 y_4.$$

Therefore, $(p \otimes q) \otimes r = p \otimes (q \otimes r)$ and every LP-isotope of $(G, \cdot)$ is associative. Since every loop isotopic to $(G, \cdot)$ is isomorphic to a principal isotope, the statement of the theorem follows. $\qquad \square$

The following result is both interesting and important.

THEOREM 8.1.8 *If all the loop-principal isotopes of the quasigroup* $(G, \cdot)$ *are commutative, they are all associative too.*

PROOF. Let a, b, c be arbitrary elements of the quasigroup $(G, \cdot)$ and let u, v be fixed elements of $(G, \cdot)$. Then there exist elements s, t such that $a = us$ and $b = ut$. Also, there exists an element w such that $c = wv$.

Let $x \otimes y = (x\sigma) \cdot (y\tau)$, where $x\sigma^{-1} = xv$ and $y\tau^{-1} = uy$ (so that $s\tau^{-1} = a$ and $t\tau^{-1} = b$). Then $(G, \otimes)$ is an LP-isotope of $(G, \cdot)$ with identity element uv as in the previous theorem. Also, let $x \oplus y = (x\sigma) \cdot (y\theta)$, where $y\theta^{-1} = wy$ (whence $v\theta^{-1} = wv = c$). Then $(G, \oplus)$ is another loop-principal isotope of $(G, \cdot)$ with identity element $wv = c$.

Thence, $h \otimes b = (h\sigma) \cdot (b\tau) = (h\sigma) \cdot t = h \oplus (t\theta^{-1}) = h \oplus (wt) = h \oplus (w\sigma^{-1} \otimes t\tau^{-1}) = h \oplus [(wv) \otimes (ut)] = h \oplus (c \otimes b)$.

In particular, putting $h = (c \otimes a)$, we get

$$(c \otimes a) \otimes b = (c \otimes a) \oplus (c \otimes b) \text{ for arbitrary elements } a, b, c \in G.$$

Similarly, $h \otimes a = (h\sigma) \cdot (a\tau) = (h\sigma) \cdot s = h \oplus s\theta^{-1}) = h \oplus (ws) = h \oplus (w\sigma^{-1} \otimes s\tau^{-1}) = h \oplus [(wv) \otimes (us)] = h \oplus (c \otimes a)$. In particular, putting $h = (c \otimes b)$, we get

$$(c \otimes b) \otimes a = (c \otimes b) \oplus (c \otimes a)$$

Since $\oplus$ is commutative by hypothesis[2], we deduce that $(c \otimes a) \otimes b = (c \otimes b) \otimes a$ and thence, since $\otimes$ also is commutative, $(a \otimes c) \otimes b = a \otimes (c \otimes b)$ and so $(G, \otimes)$ satisfies the associative law, as claimed. $\qquad \square$

COROLLARY 1. *Under the hypotheses of the theorem (that is, in a net for which Pappus' theorem is satisfied relative to the parallel classes with vertices A, B, O), every loop isotopic to $(G, \cdot)$ is an abelian group.*

PROOF. As every isotope of a quasigroup is isomorphic to a principal isotope by Theorem 1.3.2, it follows that every loop isotopic to $(G, \cdot)$ is isomorphic to an LP-isotope. The latter are commutative and associative, so they are abelian groups. $\qquad \square$

COROLLARY 2. *If Pappus' theorem holds relative to the vertices A, B, O for all choices of lines with one line fixed through O, then it holds for all choices of lines through O.*

---

[2] If c is kept fixed as here, then w is as before and so the loop $(G, \oplus)$ is unaltered.

PROOF.   In the argument above, the mapping σ is kept fixed as the LP-isotope varies. Geometrically, this corresponds to using the same line $y = xv$ through O throughout.   □

In the first corollary to Theorem 8.1.8 above, we showed that, in a triangular net for which Pappus' theorem is satisfied relative to the parallel classes with vertices A, B and O, every co-ordinatizing loop is an abelian group. This means, a fortiori, that every such loop is associative and so the large Reidemeister configuration closes (since $(p \otimes q) \otimes r = p \otimes (q \otimes r)$ implies that $x_4y_3 = x_3y_4$ in Theorem 8.1.8). Thus, the geometric equivalent of Theorem 8.1.8 is the following:

THEOREM 8.1.9 *Closure of the Pappus configuration relative to the parallel classes with vertices A, B and O in a projective plane implies closure of the large Reidemeister configuration relative to these same parallel classes.*

A wholly geometrical proof of this result is possible. See Hessenburg(1905) and Klingenberg(1952,1955).

When $O \to E$ (a point on AB), our triangular net is replaced by an affine net and the large Reidemeister configuration is replaced by the small Reidemeister configuration, denoted by the symbol $aA(10; 11, 13)$ in Argunov(1950)[3] and illustrated in the right-hand diagram of Figure 8.1.7.

In the application to co-ordinatizing a projective plane, closure of the latter configuration is the geometrical condition that the additive loop of the co-ordinate system of the projective plane is associative. The geometrical equivalent of Theorem 8.1.8 for such an affine net is

THEOREM 8.1.10 *Closure of the Thomsen configuration (or validity of the axial minor theorem of Pappus) relative to the parallel classes with vertices A, E and B in a projective plane implies closure of the configuration $aA(10; 11, 13)$ relative to those same parallel classes.*

A purely geometrical proof of this result has been given in Keedwell(1964).

Among the further infinite number of configurational propositions whose algebraic interpretations relative to a given 3-net we might consider, the so-called Bol configurations [first introduced by Bol(1937)], the central minor theorem of Pappus and the hexagon configuration have shown themselves to be of most importance.[4]

---

[3]In Argunov's notation, which makes some attempt to be systematic, the leading lower case letter m or a stands for "multiplication" or "addition", the capital letter A stands for "associativity" and the three numbers following, say $(R; P, L)$ are respectively the rank, the number of points, and the number of lines of the configuration. The *rank* R is defined by the formula $R = 2(P + L) - I$, where I is the number of incidences in the unclosed configuration.

[4]The configuration of Desargues is of crucial importance in the theory of affine and projective planes but not so much so in the theory of 3-nets.

However, it would be inappropriate in the present text to discuss the algebraic implications of each of these configurations in full detail and so we shall content ourselves with a summary of the most important results[5].

The Bol configurations arise from the Reidemeister configurations when P_1 lies on P_3P_3' (see Figure 8.1.8) and are denoted by $mA(10; 11, 11)$ and $aA(9; 11, 12)$ respectively in Argunov's notation.

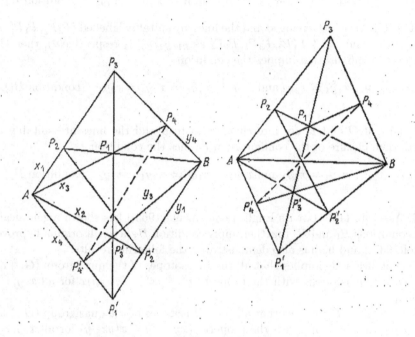

FIG. 8.1.8.

From the manner of the derivation of these configurations from the Reidemeister configurations, it is immediately clear that the roles of the vertices A, B, O are no longer interchangeable. (It can be shown that closure of the Pappus and Reidemeister configurations relative to the parallel classes with vertices A, B, O taken in any one order implies their closure relative to these parallel classes taken in any other order. In particular, the following re-labelling of the points in the left-hand diagram of Figure 8.1.7 shows that the roles of O and B are interchangeable for the large Reidemeister configuration. Let

$$P_2'P_2P_3P_3' \to Q_1Q_2Q_3Q_4 \text{ and } P_1'P_1P_4P_4' \to Q_1'Q_2'Q_3'Q_4'.$$

[5]For a detailed and comprehensive account of the subject of the inter-relations between geometrical configurations and quasigroup identities treated from the geometrical point of view, the reader is referred to Argunov's paper(1950) and to Pickert's book(1955) on projective planes. Among other sources of information are Aczél's paper(1964) [or (1965)], Belousov(1971) and chapter 8 of Belousov(1967b).

This shows us that, when the Reidemeister configuration has incidence closure with respect to the quadrangles $P_1P_2P_3P_4$ and $P_1'P_2'P_3'P_4'$, it also has incidence closure with respect to the quadrangles $Q_1Q_2Q_3Q_4$ and $Q_1'Q_2'Q_3'Q_4'$ and, for the latter quadrangles, the roles of the vertices O and B are interchanged)

If the vertices A, O, B are as in Figure 8.1.8, closure of the configuration $mA(10; 11, 11)$ implies the condition

$$x_1y_2 = x_2y_1 \text{ and } x_1y_4 = x_2y_3 = x_3y_2 = x_4y_1 \Rightarrow x_3y_4 = x_4y_3 \quad (\text{condition } B_3).$$

If O and A are interchanged and the lines are suitably labelled (P_2P_2', P_1P_1', P_4P_4' as x_1, x_2, x_4 and $P_3'P_4', P_1'P_2', P_3P_4, P_1P_2$ as y_1, y_2, y_3, y_4 respectively), then closure of the configuration implies the condition

$$x_1y_2 = x_2y_1, \ x_1y_4 = x_2y_3 \text{ and } x_3y_2 = x_4y_1 \Rightarrow x_2y_4 = x_4y_3 \quad (\text{condition } B_1).$$

Finally, if O and B are interchanged instead and the lines are suitably labelled, then closure of the configuration implies the condition

$$x_1y_2 = x_2y_1, \ x_1y_4 = x_2y_2 \text{ and } x_3y_2 = x_4y_1 \Rightarrow x_3y_4 = x_4y_2 \quad (\text{condition } B_2).$$

J. Aczél [in (1964) or its English equivalent (1965)] has shown algebraically that conditions B_1 and B_2 together imply condition B_3 [a result originally proved in Bol(1937)] and he has also demonstrated the following results:

(1) condition B_1 implies that all the LP-isotopes of the quasigroup $(G, \cdot)$ are M_1-loops: that is, loops with the property $y(z \cdot yx) = (y \cdot zy)x$ for all x, y, z in the loop;

(2) condition B_2 implies that all the LP-isotopes of the quasigroup $(G, \cdot)$ are M_2-loops: that is, loops with the property $(xy \cdot z)y = x(yz \cdot y)$ for all x, y, z in the loop;

(3) if a loop is an M_1-loop and an M_2-loop, then it is a Moufang loop: that is, it has the property $(xy \cdot z)y = x(y \cdot zy)$;

(4) if a loop is an M_1-loop, its associated net satisfies condition B_1; if a loop is an M_2-loop, its associated net satisfies condition B_2; if a loop is Moufang, its associated net satisfies all of the conditions B_1, B_2, B_3.

An equivalent set of results has also been given by Bruck(1963b). One implication of these results is that the multiplicative loop of any co-ordinate system of a projective plane that is based on the co-ordinatizing points A, B, O, E will be Moufang if and only if the configuration $mA(10; 11, 11)$ has incidence closure relative to the "parallel" classes with vertices A, B and O.

The *central minor theorem of Pappus* is the geometric dual of the axial minor theorem of Pappus discussed earlier. It is illustrated in the left-hand diagram of Figure 8.1.9. For a triangular net, its closure implies the condition

$$x_1y_2 = x_2y_1 \text{ and } x_1y_3 = x_2y_2 = x_3y_1 \Rightarrow x_2y_3 = x_3y_2 \quad (\text{condition } H).$$

For an affine net, the same condition is implied by the *third minor theorem of Pappus*: that is, by closure of the *hexagon configuration* which arises from the central minor theorem when $O \rightarrow E$ on AB. The third minor theorem of Pappus may be stated as follows: if P_1, P_2, Q_1, Q_2 are four points which form a proper quadrangle and if R_3 is the point $P_1Q_2 \cap P_2Q_1$, S is the point $P_1P_2 \cap Q_1Q_2$, R_3S intersects P_1Q_1 at R_1 and $P_2R_1 \cap Q_1Q_2 \equiv Q_3$, $Q_2R_1 \cap P_1P_2 \equiv P_3$, then the point $R_2 \equiv P_1Q_3 \cap P_3Q_1$ lies on R_1R_3 (see the right-hand diagram of Figure 8.1.9).

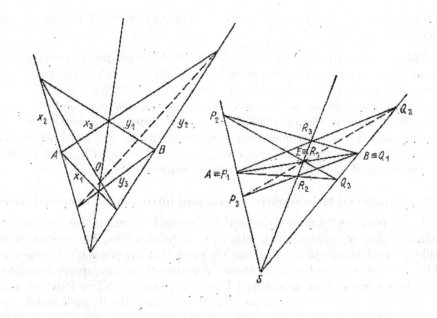

FIG. 8.1.9.

The following facts are well-known:

(5) Condition H implies and is implied by the statement that all loops isotopic to $(G, \cdot)$ are power associative.

(6) Condition H implies and is implied by the statement that, in each loop isotopic to $(G, \cdot)$, each element has a two-sided inverse.

With the aid of the results (5) and (6), it is quite easy to prove the following important result:

THEOREM 8.1.11 *If all loops isotopic to the quasigroup $(G, \cdot)$ are power associative, then they are all strongly power associative too. That is, if in any loop isotope $(G, \otimes)$ of $(G, \cdot)$ we define x^n by $x^0 = e$ and $x^n = x \otimes x^{n-1}$ for $n > 0$, x^{-1} by $x^{-1} \otimes x = e$ and $x^n = (x^{-1})^{-n}$ for $n < 0$, then $x^m \otimes x^n = x^{m+n}$ for all integers m and n.*

The interested reader will find proofs of all these results in Aczél(1964,1965).

We may summarize the results of this section by saying that every quasigroup $(G, \cdot)$ (that is, every bordered latin square) determines a 3-net of a particular geometrical type which it co-ordinatizes. Conversely, to each such particular 3-net there corresponds a family of isotopic quasigroups any one of which co-ordinatizes it and this family includes loops. An identity such as $(p \otimes p) \otimes p = p \otimes (p \otimes p)$ (validity of power associativity) which is valid in all loops isotopic to the quasigroup $(G, \cdot)$ corresponds to closure of a certain geometrical configuration in the associated 3-net. This leads to the following definition:

DEFINITION. An identity in a loop $(G, \otimes)$ is called *universal* if it is satisfied in every loop isotopic to $(G, \otimes)$.

Belousov and Ryzkov(1966) have given an algorithm for constructing the closure figure corresponding to a given universal identity and, as an illustration, have applied it to the Moufang identity $x \otimes [y \otimes (x \otimes z)] = [(x \otimes y) \otimes x] \otimes z$.

Some further discussion of connections between quasigroups and geometric nets will be found in Neumann(1960,1962), Sade(1958b) and Belousov(1967d).

This topic has recently undergone a renascence by the study of 3-nets whose points and lines are those of a projective plane. See Urzúa(2010), Blokhuis, Korchmáros and Mazzocca(2011). Korchmáros, Nagy and Pace(2014).

8.2   Orthogonal latin squares, k-nets and introduction of co-ordinates

In the previous Section, we defined a k-net and showed that every 3-net N defines a class of corresponding latin squares (representing the multiplication tables of a set of isotopic quasigroups Q_N) and that the geometrical properties of the 3-net are reflected in the algebraic properties of the quasigroups associated with these squares. Also, in Section 5.2, we showed that each $(n+1)$-net of order n (that is to say, each affine plane) defines a set of mutually orthogonal latin squares and conversely. Both of these results are special cases of the general statement that each geometric k-net N of order n defines and is defined by a corresponding set of $k - 2$ MOLS of order n.

Moreover, the discussion of the previous section makes it evident that the algebraic properties of the quasigroups whose multiplication tables are represented by these latin squares reflect geometrical properties of the various corresponding sub-3-nets of the net N.

A slight modification of the argument of Theorem 8.1.2 is all that is necessary to show that our general statement is true.

In view of its importance, we formulate it as a theorem:

THEOREM 8.2.1 *Each geometric k-net N of order n defines, and is defined by, a corresponding set of $k - 2$ MOLS of order n.*

PROOF. We designate the various parallel classes of the k-net N by calligraphic letters $\mathcal{A}, \mathcal{E}, \mathcal{B}_1, \mathcal{B}_2, \ldots, \mathcal{B}_{k-2}$. Let the lines of these parallel classes be labelled as follows: $a_1, a_2, \ldots, a_n$ are the lines of the class $\mathcal{A}$; $e_1, e_2, \ldots, e_n$ are the lines of the class $\mathcal{E}$; $b_{j1}, b_{j2}, \ldots, b_{jn}$ are the lines of the class $\mathcal{B}_j$. Every point $P(h, i)$ of

N can then be identified with a set of k numbers $(h, i, l_1, l_2, \dots, l_{k-2})$ describing the k lines $e_h, a_i, b_{1l_1}, b_{2l_2}, \dots, b_{k-2\,l_{k-2}}$ with which it is incident, one from each of the k parallel classes, and a set of $k-2$ MOLS can be formed in the following way: In the jth square, put l_j in the (h, i)th place. Each square is latin since, as h varies with i fixed, so does l_j; and, as i varies with h fixed, so does l_j. Each two squares L_p and L_q are orthogonal: for, if not, we would have two lines belonging to distinct parallel classes with more than one point in common.

Conversely, from a given set of $k-2$ MOLS, we may construct a k-net N. We define a set of n^2 points (h, i), $h = 1, 2, \dots, n$; $i = 1, 2, \dots, n$; where the point (h, i) is to be identified with the k-tuple of numbers $(h, i, l_1, l_2, \dots, l_{k-2})$, l_j being the entry in the hth row and ith column of the jth latin square L_j. We form kn lines b_{jl}, $j = -1, 0, 1, 2, \dots, k-2$; $l = 1, 2, \dots, n$; where b_{jl} is the set of all points whose $(j+2)$th entry is l and $b_{-1l} \equiv e_l$, $b_{0l} \equiv a_l$. Thus, we obtain k sets of n parallel lines. (Two lines are *parallel* if they have no point in common.) Also, from the orthogonality of the latin squares, it follows that any two lines from distinct parallel classes intersect in one and only one point, so we have a k-net. $\qquad\qquad\qquad\qquad\qquad\qquad\qquad\qquad\qquad\qquad\qquad\qquad$ □

Bruck(1951) has used the representation of a set of $k-2$ MOLS by means of a k-net to obtain an interesting criterion for such a set of squares to have a common transversal and hence has obtained a simple necessary (but not sufficient) condition for such a set of squares to be extendible to a larger set (of the same order). In a subsequent paper, Bruck(1963a), and again using the net representation, he has obtained a sufficient condition in terms of the relative sizes of k and n for a set of MOLS to be extendible to a complete set. We gave a detailed account of these results in Chapter 9 of [DK1]. However, Bruck's results have more recently been improved by Metsch(1991). See the next section.

In his well-known paper M.Hall(1943), that author has shown that co-ordinates may be introduced into an arbitrary projective plane (or k-net) in the following way:

We select any two points A, B in the given projective plane π. If from π we remove the line l_∞ joining A, B and the points on l_∞, the remaining plane π^* is an affine plane. We use A and B as centres of perspectivities to introduce co-ordinates for the points of π^*. No ambiguity will arise if, as we shall sometimes find convenient, we speak of π and π^* as the same plane as long as l_∞ is fixed.

Let the lines of the pencil through A in π^* be denoted by $x = x_0$, $x = x_1, \dots,$ $x = x_{r-1}$, where r is their cardinal number and $x_0, x_1, \dots, x_{r-1}$ are r different symbols. Similarly, denote the lines of the pencil through B by $y = y_0$, $y = y_1$, $\dots, y = y_{r-1}$. (The cardinal number of the y's is necessarily the same as that of the x's.) Through every point P of π^*, there is exactly one line $x = x_i$ and one line $y = y_j$. Denote P by (x_i, y_j). On an arbitrary line L of π^*, not through A or B, there are points (x_i, y_j) where each x and each y occurs exactly once. Henceforward, suppose that the symbols $x_0, x_1, \dots, x_{r-1}$ are the same as $y_0, y_1,$ $\dots, y_{r-1}$, though not necessarily in the same order. Then, an arbitrary line L, not through A or B, is associated with the permutation $\begin{pmatrix} \cdots & x_i & \cdots \\ \cdots & y_i & \cdots \end{pmatrix}$, where the (x_i, y_i) are the points of L. This expresses the fact that L determines a one-to-one correspondence between the lines of the pencil through A and the lines of the pencil through B. Without confusion we may write $L = \begin{pmatrix} \cdots & x_i & \cdots \\ \cdots & y_i & \cdots \end{pmatrix}$ since distinct lines contain at most one point in common and hence are associated with distinct permutations.

It is evident that the same method of co-ordinatization may be used for a k-net of order r.

We now choose two particular points O and I which are joined by a line but are not collinear with A or B and assign them the co-ordinates $(0,0)$ and $(1,1)$ respectively. (In the case of a projective plane, every two points are joined by a line and so the choice of O and I is entirely arbitrary.) This assignment of co-ordinates is equivalent to assigning the labels $0, 1$ to two of the symbols $x_0, x_1, \ldots, x_{r-1}$, say $x_0 = 0$, $x_1 = 1$.

Let OI meet AB at E and let the remaining points of AB be denoted by $B_2, B_3, \ldots, B_{k-2}, k \leq r + 1$. If the line OB_m meets the line AI at the point $(1, m)$, denote the point B_m by the symbol (m). In particular, denote the point $E(\equiv B_1)$ by the symbol (1). We are now able to define operations $(+)$ and $(\star)$ on the symbols $x_0, x_1, \ldots, x_{r-1}$ by means of which the lines of the pencils with vertices E and O may be assigned equations. Each line through E has a permutation representation of the form $\begin{pmatrix} 0 & \cdots & x_i & \cdots \\ a & \cdots & y_i & \cdots \end{pmatrix}$. We define $x_i + a = y_i$ for $i = 0, 1, 2, \ldots, r - 1$ and say that the line has equation $y = x + a$. This defines the result of the operation $(+)$ on every ordered pair of the r symbols. Each line through O has a permutation representation of the form $\begin{pmatrix} 0 & 1 & \cdots & x_i & \cdots \\ 0 & m & \cdots & y_i & \cdots \end{pmatrix}$. We define $x_i \star m = y_i$ for $i = 0, 1, 2, \ldots, r - 1$ and say that the line has equation $y = x \star m$. This defines the result of the operation $(\star)$ for all choices of x and for $k - 1$ finite values of m.

We can provide equations for the remaining lines of the plane with the aid of a ternary operation defined on the set of symbols $x_0, x_1, \ldots, x_{r-1}$ in the following way. If x, m, a are any three of the symbols, but with m restricted to $k - 1$ finite values in the case of a k-net, we define $T(x, m, a) = y$, where y is the second co-ordinate of the point on the line joining the points (m) and $(0, a)$ whose first co-ordinate is x. We can then say that $y = T(x, m, a)$ is the "equation" of the line joining the points (m) and $(0, a)$. The connection between this ternary operation on the symbols $x_0, x_1, \ldots, x_{r-1}$ and the binary operations $(+)$ and $(\star)$ previously introduced is given by the relations $a + b = T(a, 1, b)$ and $a \star b = T(a, b, 0)$. We may also observe that, both in the case of a complete projective plane and in the case of a k-net with $k < r + 1$, the lines which pass through the point E are those which are represented in the permutation representation by permutations which displace all symbols and by the identity permutation. We shall denote this subset of the set S of permutations representing the lines by the symbol $\bar{S}$.

Since each finite projective plane of order n defines, and is defined by, a complete set of $n-1$ mutually orthogonal latin squares of order n (see Theorem 5.2.2), it is evident that the answer to the question "How many non-equivalent complete sets of mutually orthogonal latin squares of order n exist?" is closely related to the question[6] "How many geometrically distinct projective planes of order n exist?". We have already shown that there exists a Galois or desarguesian[7] plane of every positive integral order n that is a power of a prime number (Theorem 5.2.3) and we wish now to demonstrate the existence of non-desarguesian planes.

The simplest type of non-desarguesian projective planes to describe are the so-called translation planes.

We may specify such a plane by the nature of its co-ordinate system. We suppose co-ordinates to have been introduced into the plane in the manner described above, so that there is a special line $AB = l_\infty$ whose points are represented by symbols (m), there are two special points $(0,0)$ and $(1,1)$, and so that all other points are represented by co-ordinate pairs (x, y), where m, x, y belong to a co-ordinate set Σ. Also lines through A have equations of the type $x = x_r$, lines through B have equations of the form $y = y_r$ and all other lines have equations of the form $y = T(x, m, a)$, where T is a ternary operation on the set Σ. Further, by means of T, binary operations $(+)$ and $(\star)$ may be defined on Σ. If then the algebraic system $(\Sigma, +, \star)$ has the properties (i) $(\Sigma, +)$ is an abelian group with 0 as its identity element, (ii) $(\Sigma - 0, \star)$ is a loop with 1 as its identity element, (iii) $(a + b) \star m = (a \star m) + (b \star m)$ for all a, b, m in Σ, and (iv) if $r \neq s$, the equation $x \star r = (x \star s) + t$ has a unique solution x in Σ when r, s, t are in Σ, we say that it is a *Veblen Wedderburn system* or a *right quasifield*. If, further, the ternary operation T on Σ satisfies the condition $T(a, m, b) = a \star m + b$, then we may easily verify as below that the axioms for a projective plane are satisfied and we call the resulting plane a *translation plane*. (It is easy to check that the condition $T(a, m, b) = a \star m + b$ is consistent with the relations describing the binary operations $(+)$ and $(\star)$ in terms of T previously given.)

PROOF.   In the first place, we have to check that there is a unique line joining two given points (x_r, y_r) and (x_s, y_s). If $x_r = x_s$, the required unique line is that with equation $x = x_r$. Tf $y_r = y_s$ the required line is $y = y_s$. Jf $x_r \neq x_s$ and $y_r \neq y_s$, the required line is that with equation $y = x \star m + b$ where m and b satisfy the equations $x_r \star m + b = y_r$ and $x_s \star m + b = y_s$: that is, m is the unique solution of the equation $(x_r - x_s) \star m = y_r - y_s$ and b is then determined uniquely by the requirement that $x_r \star m + b = y_r$. In the second place, there is a unique point common to each two lines. The lines $x = x_r$, $x = x_s$ have the point A in common. Similarly, the lines $y = y_r$, $y = y_s$ have the point B in common. The lines $x = x_r$, $y = y_s$ have the point (x_r, y_s) in common. The lines $x = x_r$

---

[6] The questions are not the same because different choices of l_∞ in a particular plane may yield inequivalent sets of latin squares. See later in this section.

[7] A Galois plane is called a desarguesian plane because in such a plane the well-known configurational theorem of Desargues concerning perspective triangles is universally valid. (It is conjectured that every plane of prime order is desarguesian.)

and $y = x \star m + b$ have the point $(x_r, x_r \star m + b)$ in common. The lines $y = y_s$, $y = x \star m + b$ have the point (x_s, y_s) in common, where x_s is the unique solution of the equation $x_s \star m = y_s - b$.

Finally, the lines $y = x \star m_1 + b_1$ and $y = x \star m_2 + b_2$ intersect in the point whose x co-ordinate is the unique solution of the equation $x \star m_1 = x \star m_2 + (b_2 - b_1)$ and whose y co-ordinate can then be obtained from either one of the equations of the two lines. $\qquad \Box$

It is easy to see that, in particular, every Galois field is a right quasifield. Moreover, in the case when the right quasifield is a Galois field, homogeneous co-ordinates can be introduced in a manner analogous to that used in elementary geometry and it is then quite simple to show that the plane is isomorphic to the Galois plane of the same order. For the details, the reader is referred to books on projective planes. [See, for example, Pickert(1955).] We deduce that the existence of finite translation planes which are geometrically distinct from the Galois planes is dependent upon the existence of finite right quasifields which are not fields.

M.Hall(1943) gave one method for constructing such quasifields but many methods for obtaining translation planes are now known and many papers on the subject have been published.

When the complete set of mutually orthogonal latin squares which are defined by a desarguesian plane, by a translation plane, or by the dual of a translation plane, are constructed and put into standardized form (see Section 5.1), it is found that the rows of any one square L_k of the set are the same as those of any other square L_h of the set, except that they occur in a different order. That this is necessarily so was proved for the case of desarguesian planes by Bose and Nair(1941). However, there exist other types of projective plane for which the squares do not have this property as the same authors pointed out, among them being a class of planes constructed by D.R.Hughes(1957) and called the *Hughes planes*.[8]

From the point of view of the theory of latin squares it is therefore important to have a criterion for distinguishing the two cases. Such a necessary and sufficient condition has been given by Hughes himself [see Hughes(1955)], who has shown that linearity of the ternary operation is the required criterion, and in a more geometrical form by the present author, who has given a direct proof of the equivalent geometrical criterion: namely,

THEOREM 8.2.2 *A necessary and sufficient condition that, in a standardized complete set of mutually orthogonal latin squares, the rows of the square L_k be the same as those of the square L_h, except that they occur in a different order is that the squares represent the incidence structure of a projective plane*

---

[8] In fact, the Hughes plane of order 9 was first constructed by Veblen and Wedderburn(1907).

in which the first minor theorem of Desargues[9] holds affinely with E as a vertex of perspective and A, B_h, B_k as meets of corresponding sides of the two triangles.[10] (The notation is that used earlier in this section.)

In Figure 8.4.3 of [DK1], a complete set of MOLS of order 9 was displayed [taken from Paige and Wexler(1953)] which it was claimed represented the Hughes plane of that order but in fact they represent the dual of the unique translation plane of order 9 as has been proved by Owens(1992). For more details of how this came to light, see Section 6 of Chapter 11 in [DK2]. Since that book was published, Owens and Preece(1995,1997) have carried out a complete analysis of the possible non-equivalent sets[11] of MOLS of order 9 and, for each, which plane it represents. They found that there are 19 non-equivalent sets. We also draw the reader's attention to the facts that it is now known that there are just four different planes of order 9 (namely, the Desarguesian plane, one translation plane, its dual, and the Hughes plane) and that no projective plane of order 10 exists. It has not, so far as the author is aware been decided whether a triple of MOLS of order 10 exists though recent work, in particular that of McKay, Meynert and Myrvold(2007), makes it extremely unlikely.

For more recent work on "Latin Squares and Geometry" which supercedes and includes much of that in [DK1] other than that included here, the reader is recommended to read Chapter 11 of [DK2].

We end this section by re-presenting a problem mentioned in [DK1] but which remains unsolved.

A finite projective plane Π is said to be *with characteristic h* when a positive integer h exists such that, in any co-ordinatization of the plane by means of a Hall ternary ring (described earlier in this section), all elements of the loop formed by the co-ordinate symbols under the operation $(+)$ have the same order h. In the case when h is a prime this is equivalent to requiring that each proper quadrangle (set of four points no three of which are collinear) of the plane generates a subplane of order h. Thus, for example, every desarguesian plane of order p^n, p prime, has characteristic p.

Two questions arise:

Firstly, as to what can be said about the orders of such planes and, secondly, whether non-desarguesian planes with characteristic can exist. As regards the case when $h = 2$, it has been shown by Gleason(1956) that every finite projective plane of characteristic two is desarguesian and consequently that it has order

---

[9]The first minor theorem of Desargues is the special case of Desargues' theorem in which the vertex of perspectivity of the two triangles lies on the axis of perspectivity. It is said to be satisfied affinely when this axis is the special line l_∞.

[10]Theorem 6.3 on page 368 of [DK2] is a strengthened form of this theorem due to Owens(1992).

[11]Two complete sets of latin squares are said to be *equivalent* if one set can be obtained from the other by a combination of the following operations: (i) simultaneously permuting the rows of all the squares; (ii) simultaneously permuting the columns of all the squares; and (iii) (for each independently) permuting the symbols of any of the squares.

equal to a power of two. For the case $h = 3$, it has been shown in Keedwell(1963) that, under an additional restriction, the order must be a power of three. More recently, the same author has shown that, if $h = p$, p prime, then, under the same restriction, the order of the plane is a power of p. [See Keedwell(1971).] In the author's attempt to deal with the second question for the case $h = 3$, it proved effective to use the latin square representation [see Keedwell(1965)] and two orthogonal latin squares suitable for the construction of a non-desarguesian projective plane in which affine quadrangles would generate subplanes of order three were quite easily constructed. However, the question as to the existence or non-existence of such planes remains unanswered to this day.

8.3   Latin squares and graphs

There are several ways of representing a latin square as an edge (or vertex) coloured graph. See Keedwell(1996)[12] for three of these. See also Shee(1970). Such representations enable latin square problems to be translated into graph theory questions or vice versa. One which has proved particulaly useful is as follows:

Let L be a latin square of order n. We denote the complete (undirected) bipartite graph with $2n$ vertices by $K_{n,n}$. Let one partite set $\{c_1, c_2, \ldots, c_n\}$ denote the columns of L and let the second partite set $\{s_1, s_2, \ldots, s_n\}$ denote the symbols. If, in row r_i, symbol s_k occurs in column c_j, colour the edge $[c_j, s_k]$ with colour r_i. This defines a proper n-colouring of the edges of $K_{n,n}$ in which the edges coloured with a particular colour form a 1-factor of $K_{n,n}$.

If we regard L as a collection of n^2 triples (see page 14), we easily see that the roles of row, column and symbol can be permuted in this representation and so each latin square L defines six ways of colouring $K_{n,n}$ of which up to three may be distinct.

We may use this representation to re-interpret the problem of finding partial latin squares which are uniquely completable to L (see Section 3.2) into that of finding partial edge-colourings of $K_{n,n}$ which can be completed uniquely to a proper colouring of all the edges of $K_{n,n}$. For more details of this and of the general concepts of uniquely completable and critical partial colourings of the vertices or edges of a graph, see Keedwell(1994,1996); also Mahmoodian, Naserasr and Zaker(1997), who re-introduced the same idea without acknowledgement, Mahmoodian and Mahdian(1997) and Hajiabolhassan, Mehrabadi, Tusserkani and Zaker(1999), Burgess and Keedwell(2001),

Harary(1960) has pointed out that, in the above representation, a 1-factor of $K_{n,n}$ whose edges have distinct colours corresponds to existence of a transversal in L. So, if no such 1-factor exists, L is without transversals and is consequently a bachelor square as defined in Section 9.1.

---

[12]Cayley(1878b) was the first to propose a way of representing a group both graphically and as a latin square.

The above representation has also been used by Wanless in his investigation of so-called atomic latin squares (see Section 9.4). For that purpose, he needed 1-factorizations of $K_{n,n}$ with the property that the union of every pair of 1-factors defines a Hamiltonian circuit of $K_{n,n}$. Such 1-factorizations are called *perfect*.

The similar property for 1-factorizations of the complete directed and undirected graphs K_n^* and K_n is closely connected to the existence of latin squares and rectangles which are row complete. (See Section 2.6 for the definition.) We explained in Section 3.1 the connections between row complete latin rectangles and decompositions of K_n into Hamiltonian paths and cycles and also decompositions into Eulerian cycles. There are analogous connections between latin squares which are row complete and decompositions of K_n^*. We have:

THEOREM 8.3.1 *If a row complete latin square of order n exists, then (i) the complete directed graph on n vertices can be separated into disjoint Hamiltonian paths, and (ii) the complete directed graph on $n+1$ vertices can be separated into n disjoint Hamiltonian circuits.*

PROOF.   For (i), we associate a directed graph with the given row complete latin square in such a way that the vertices of the graph correspond to the n distinct elements of the latin square and that an edge of the graph directed from the vertex x to the vertex y exists when and only when the elements x and y appear as an ordered pair of adjacent elements in some row of the latin square. Then, because the latin square is row complete, the graph obtained will be the complete directed graph on n vertices. Moreover, it is immediate to see that the rows of the latin square define n disjoint Hamiltonian paths into which the graph can be decomposed.

For (ii) we adjoin an extra column to the given row complete latin square L, the elements of which are equal but distinct from the n elements of the latin square. We associate a directed graph with the $n \times (n + 1)$ matrix so formed in the same way as before but this time treating each row cyclically so that if the rth row ends with the element x and begins with the element y then the associated graph has an edge directed from the vertex labelled x to the vertex labelled y. Because each element of L appears just once in its last column and just once in its first column, the directed graph on $n+1$ vertices which we obtain is complete. Also, the rows of the $n \times (n + 1)$ matrix define n disjoint circuits into which it can be decomposed.   □

The relationship described in Theorem 8.3.1(i) has been pointed out in Dénes and Török(1970) and also in Mendelsohn(1968). That described in Theorem 8.3.1 (ii) does not seem to have been noticed until it was mentioned in [DK1].

Illustrative examples of these decompositions are given in Figure 8.3.1 and Figure 8.3.2. Figure 8.3.1 shows the decomposition of the complete directed graph with four vertices into disjoint Hamiltonian paths with the aid of the 4×4 complete latin square displayed earlier in Figure 2.6.1. Figure 8.3.2 shows the decomposition of the complete directed graph with five vertices into disjoint circuits of length five with the aid of the 4 × 5 matrix obtained by augmenting this same 4 × 4 complete latin square in the way described in Theorem 8.3.1.

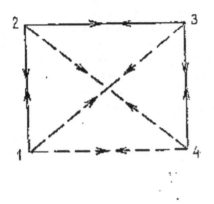

$$\begin{array}{cccc} 1 & 2 & 3 & 4 \\ 2 & 4 & 1 & 3 \\ 3 & 1 & 4 & 2 \\ 4 & 3 & 2 & 1 \end{array}$$

FIG. 8.3.1.

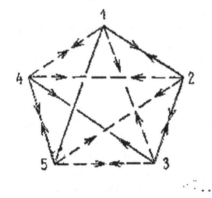

$$\begin{array}{ccccc} 1 & 2 & 3 & 4 & 5 \\ 2 & 4 & 1 & 3 & 5 \\ 3 & 1 & 4 & 2 & 5 \\ 4 & 3 & 2 & 1 & 5 \end{array}$$

FIG. 8.3.2.

By virtue of Theorem 2.6.1, decompositions of the above kind always exist when n is even. Also, as noted in Section 3.1, when the decomposition is effected with the aid of a row complete latin square formed in the manner described in Theorem 2.6.1, one half of the paths obtained are the same as the other half but described in the opposite direction.

K.O. Strauss posed the question whether a complete directed graph with an odd number n of vertices can likewise be separated into n disjoint Hamiltonian paths. By virtue of the fact that row complete latin squares of every composite order exist [see Higham(1998) and Section 2.6], the question is answered in the affirmative for non-prime odd orders but it remains unanswered for prime values of n.

Another graph problem of a similar kind to that just discussed and which has been shown by Kotzig to have connections with a particular type of latin square concerns the more general problem of the decomposition of a complete undirected graph into a set of disjoint circuits of arbitrary lengths.

In Kotzig(1970), that author gave the name P-*groupoid* (*partition groupoid*) to a groupoid $(V, \cdot)$ which has the following properties: (i) $a \cdot a = a$ for all $a \in V$; (ii) $a \neq b$ implies $a \neq a \cdot b \neq b$ for all $a, b \in V$; (iii) $a \cdot b = c$ implies and is implied by $c \cdot b = a$ for all $a, b, c \in V$.

	1	2	3	4	5
1	1	5	5	5	3
2	4	2	4	3	4
3	5	4	3	2	1
4	2	3	2	4	2
5	3	1	1	1	5

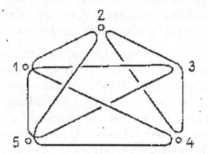

FIG. 8.3.3.

He showed that there exists a one-to-one correspondence between P-groupoids of n elements and decompositions of complete undirected graphs of n vertices into disjoint circuits. This correspondence is established by labelling the vertices of the graph with the elements of the P-groupoid and prescribing that the edges $[a, b]$ and $[b, c]$ shall belong to the same closed path of the graph if and only if $a \cdot b = c$, $a \neq b$. We illustrate this relationship in Figure 8.3.3 and from it we easily deduce:

THEOREM 8.3.2 *In any P-groupoid $(V, \cdot)$ we have (i) the number of elements is necessarily odd, and (ii) the equation $x \cdot b = c$ is uniquely soluble for x.*

PROOF. The result (i) is deduced by using the correspondence between P-groupoids and graphs just described. Since for a complete undirected graph which separates into disjoint circuits the number of edges which pass through each vertex must clearly be even, any such complete undirected graph must have an odd number of vertices all together. This is because each vertex has to be joined to an even number of others. The number of elements of a P-groupoid is equal to the number of vertices in its associated graph.

The result (ii) is a consequence of the definition of a groupoid and the fact that $x \cdot b = c$ implies $c \cdot b = x$. □.

COROLLARY. *The multiplication table of a P-groupoid is a column latin square.*

(See Section 3.1 for the definition of the latter concept.)

The example given in Figure 8.3.3 illustrates the fact that there exist *P*-groupoids which are not quasigroups. This observation leads us to make the following definition:

DEFINITION. A *P*-groupoid which is also a quasigroup will be called a *P-quasigroup.*

THEOREM 8.3.3 *Let $(V, \cdot)$ be a P-quasigroup and let a groupoid $(V, \star)$ be defined by the statement that $a \cdot (a \star b) = b$ holds for all $a, b \in V$. Then $(V, \star)$ is an idempotent and commutative quasigroup. Moreover, with any given idempotent commutative quasigroup $(V, \star)$ there is associated a P-quasigroup $(V, \cdot)$ related to $(V, \star)$ by the correspondence $a \cdot b = c \Leftrightarrow a \star c = b$.*

PROOF. In a *P*-quasigroup, the equation $a \cdot x = b$ is uniquely soluble for x, so the binary operation $(\star)$ is well-defined. Also, $a \cdot x = b$ implies $b \cdot x = a$ so $b \star a = a \star b$; that is $(V, \star)$ is commutative. The equation $a \cdot x = a$ has the solution $x = a$, so $a \star a = a$ and $(V, \star)$ is idempotent. If the equation $a \star y = c$ or the equation $y \star a = c$ had two solutions for y, the groupoid property of $(V, \cdot)$ would be contradicted[13]. Hence, $(V, \star)$ is a quasigroup.

The second statement of the theorem may be justified similarly by defining the operation $(\cdot)$ in terms of the operation $(\star)$ by the statement that $a \star (a \cdot b) = b$ for all $a, b \in V$.   □

The multiplication table of the quasigroup $(V, \star)$ is an idempotent symmetric latin square. Thus, a consequence of Theorem 8.3.2 and Theorem 8.3.3 above is another proof that idempotent symmetric latin squares exist only for odd orders n, a result which we may contrast with the fact that unipotent symmetric latin squares exist only when n is even. (See also Theorem 1.5.4 and Theorem 2.1.1.)

It also follows from Theorem 8.3.3 that each idempotent symmetric latin square of (necessarily odd) order n defines a *P*-quasigroup of order n and hence a decomposition of the complete undirected graph on n vertices into disjoint circuits.

Kotzig has pointed out that each idempotent symmetric latin square of order n also defines a partition of the complete undirected graph on n vertices into n nearly linear factors. By a *nearly linear factor* is meant a set F of $(n-1)/2$ edges such that each vertex of the graph is incident with at most one edge of F. It is immediately clear that exactly one vertex of the graph is isolated relative to a given nearly linear factor F.

As an illustration of this concept, let us point out that the three edges [1, 2], [3, 7], [4, 6] of the complete undirected graph on seven vertices $1, 2, \ldots, 7$ form a nearly linear factor F_5. Likewise, the three edges [1, 3], [4, 7], [5, 6] form another nearly linear factor F_6 of the same graph. (See also Figure 8.3.4.) We shall formulate this result of Kotzig's as a theorem:

---

[13]$a \star y_1 = c$ and $a \star y_2 = c \Rightarrow a \cdot c = y_i$ for $i = 1, 2$.

THEOREM 8.3.4 *To each idempotent symmetric latin square of order n there corresponds a partition of the complete undirected graph on n vertices into n nearly linear factors, and conversely.*

PROOF.   The correspondence is established as follows. Let k be a fixed element of the idempotent commutative quasigroup $(V, \star)$ defined by the given symmetric latin square. Then there exist $(n - 1)/2$ unordered pairs (a_i, b_i) of elements of V such that $a_i \star b_i = k(= b_i \star a_i)$. These $(n - 1)/2$ pairs define the $(n - 1)/2$ edges $[a_i, b_i]$ of a nearly linear factor of the complete undirected graph K_n whose n vertices are labelled by the elements of V. The truth of this statement follows immediately from the fact that $(V, \star)$ is an idempotent and commutative quasigroup and that the element k consequently occurs $(n - 1)/2$ times in that part of its multiplication table which lies above the main left-to-right diagonal and at most once in each row and at most once in each column.

The converse is established by defining a quasigroup $(V, \star)$ by means of a given decomposition of K_n into nearly linear factors according to the following rules: (i) if $[a_i, b_i]$ is an edge of the nearly linear factor whose isolated vertex is labelled k, then $a_i \star b_i = k = b_i \star a_i$ and (ii) for each symbol $k \in V$, $k \star k = k$. □

Kotzig has asserted further that if $n = 2k - 1$ and either n or k is prime then there exists a partition of the corresponding complete graph on n vertices into n nearly linear factors with the property that the union of every two of them is a Hamiltonian path of the graph and has suggested that a partition of the same kind may exist for all odd values of n. [See Kotzig(1970).] For more recent results, see the comment on Problem 9.1 of [DK1] later on in this book.

As an example of the above situation we give in Figure 8.3.4 an idempotent and commutative quasigroup $(V, \star)$ of order 7 which defines a decomposition of the complete graph K_7 on seven vertices into nearly linear factors whose unions in pairs are Hamiltonian paths of K_7.

(∗)	1	2	3	4	5	6	7
1	1	5	2	6	3	7	4
2	5	2	6	3	7	4	1
3	2	6	3	7	4	1	5
4	6	3	7	4	1	5	2
5	3	7	4	1	5	2	6
6	7	4	1	5	2	6	3
7	4	1	5	2	6	3	7

F_1 (2, 7), (3, 6), (4, 5)

F_2 (1, 3), (4, 7), (5, 6) · · · ·

F_3 (1, 5), (2, 4), (6, 7)

F_4 (1, 7), (2, 6), (3, 4) –·–·–

F_5 (1, 2), (3, 7), (4, 6) ——

F_6 (1, 4), (2, 3), (5, 7) – – – –

F_7 (1, 6), (2, 5), (3, 4)

FIG. 8.3.4.

In Figure 8.3.5 we give the P-quasigroup $(V, \cdot)$ which defines and is defined by the idempotent commutative quasigroup $(V, \star)$ of Figure 8.3.4 and show the associated partitions of the graph K_7 into circuits.

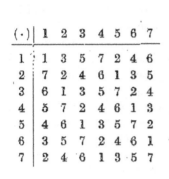

$(\cdot)$	1	2	3	4	5	6	7
1	1	3	5	7	2	4	6
2	7	2	4	6	1	3	5
3	6	1	3	5	7	2	4
4	5	7	2	4	6	1	3
5	4	6	1	3	5	7	2
6	3	5	7	2	4	6	1
7	2	4	6	1	3	5	7

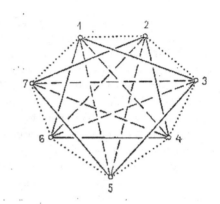

FIG. 8.3.5.

It is also easy to see that each idempotent symmetric latin square (and consequently each corresponding P-quasigroup) of order $2m - 1$ defines a 1-factorization of K_{2m}. Let an additional vertex 0 be adjoined to K_{2m-1} in Theorem 8.3.4. Then each nearly linear factor F_h of K_{2m-1} defines a 1-factor of K_{2m} whose additional edge is $[0, h]$. However, as shown by a counter-example in Keedwell(1978), the correspondence between P-quasigroups of order $2m - 1$ and 1-factorizations of K_{2m} is certainly not always one-to-one.

We showed in Theorem 6.4.1 that a Room design of order $2m$ is equivalent to a pair of perpendicular commutative idempotent quasigroups each of order $2m - 1$. It follows that such a design is also equivalent to a pair of perpendicular P-quasigroups where

DEFINITION. Two P-quasigroups $(Q, \odot)$ and $(Q, \otimes)$ are *perpendicular* if (i) for $a, x, y \in Q$, the equations $x \odot a = y$ and $x \otimes a = y$ both hold if and only if $x = y = a$; and (ii) for $a, b \in Q$, there is at most one pair of elements $x, y \in Q$ such that $x \odot a = y$ and $x \otimes b = y$.

In fact, given a Room square R of side $2m - 1$, we may define the (not necessarily unique) associated pair of perpendicular P-quasigroups of order $2m - 1$ directly as follows: We may assume that the Room square is defined on the symbols $\{\infty, 1, 2, \ldots, 2m - 1\}$. We first permute the rows and subsequently the columns of R so as to obtain it in standard form $R^\star$ with the pair (∞, i) in the cell (i, i). If $(b_1, b_1'), (b_2, b_2'), \ldots, (b_{m-1}, b_{m-1}')$ are the pairs distinct from (∞, i) which are to be found in the ith row [column] of $R^\star$ then the ith column in the multiplication table of the P-quasigroup $(Q, \odot)$ [$(Q, \otimes)$] which is associated with the rows [columns] of $R^\star$ is obtained by applying the permutation whose cycle form is $(i)(b_1, b'_1)(b_2, b'_2) \ldots (b_{m-1}, b'_{m-1})$ to the column border of $(Q, \odot)$ $[(Q, \otimes)]$, where Q is the set $\{1, 2, \ldots, 2m - 1\}$.

In Keedwell(1978), the set of edge-disjoint circuits into which a P-quasigroup of order $2m - 1$ separates the edges of the complete graph K_{2m-1} were said to form a P-*circuit design* and the separate circuits were called *linked blocks* of that design. The P-circuit design was called *uniform* if all its blocks were of the same size and none contained repeated vertices (elements).

Concepts of perpendicularity, skew-perpendicularity, and of being cyclic were introduced and it was shown that P-quasigroups with these special properties could be used to construct Room squares with the corresponding properties.

Steiner triple systems provide particular examples of uniform P-circuit designs and so the above results generalize Theorem 6.4.4.

In his paper of 1970, Kotzig raised the further question: "For what orders n does a P-quasigroup exist which defines a partition of G into an Eulerian circuit?"

Subsequently, the present author proved in Keedwell(1975) that P-quasigroups of the above kind exist whenever n is an integer of the form $4r + 3$ such that $r \not\equiv 2 \bmod 5$ and $r \not\equiv 1 \bmod 6$. They also exist when $n = 4r + 3$ and $0 \le r \le 7$. In a subsequent paper with Hilton, the two authors together proved that such P-quasigroups exist for all $n = 4r + 3$ except possibly when $r \equiv 127 \bmod 595$ and they also showed existence for many values of $n = 4r + 1$. [See Hilton and Keedwell(1976).] The problem was solved completely by Korovima(1984) who showed existence for all odd $n \ne 5$.

Some further connections between special kinds of latin square, P-quasigroups and graph decompositions are mentioned in Keedwell(1976a,1976b,1982) and in [DK1], pages 308-311.

As already mentioned earlier in this chapter, Bruck(1963a) used the concepts of both k-nets and graphs to obtain conditions under which a given set of MOLS may be extended to a larger set. His main result was as follows:

Let S be a set of $k - 2$ MOLS of order n and let $d = (n - 1) - (k - 2)$, which is the *deficiency* from being a complete set. Then, (1) whenever $n > (d - 1)^2$, if S can be completed to a complete set of MOLS at all, it can be so completed in only one way; (2) a sufficient (but not necessary) condition that such a set S can be augmented to a complete set of MOLS in at least one way is that $n > \frac{1}{2}(d - 1)^4 + (d - 1)^3 + (d - 1)^2 + \frac{3}{2}(d - 1)$.

However, more recently this result has been improved and superceded by one due to Metsch(1991):

Let S be a set of $k - 2$ MOLS of order n. and let $d = n - k + 1$, be the deficiency from being a complete set as before. Then, S can be extended to a complete set if $n > \frac{8}{3}d^3 - 6d^2 + \frac{8}{3}d + \frac{4}{3}$.

In Jungnickel(1996), that author gave a survey of results on this topic so we refer the reader to that paper for further information. However, a paper by Bose(1963) introducing the concept of a *partial geometry* (or partial geometric net) and relating it to strongly regular graphs deserves special mention because it generalizes some of the results of Bruck mentioned above.

We end this section by mentioning a connection between latin squares and graphs which is somewhat indirect. In a series of papers, Choudhuri(1948,1949,1957), that author has devised a means of defining an ordering relation on any quasi-group Q and with the aid of this definition has associated a directed graph Γ with Q. The properties of Γ serve to illuminate those of Q.

CHAPTER 9

Latin squares with particular properties

In Section 2.1, we discussed quasigroups of some special types and their associated latin squares. Also, in Section 2.6, we introduced complete latin squares. Here, we discuss latin squares with various other special properties most of which had not been thought of when [DK1] was published.

9.1 Bachelor squares

The name *bachelor square* for a latin square which has no orthogonal mate was introduced by Van Rees(1990) who drew attention to Mann's theorem of 1944 on this subject (which applies to squares of orders $4m + 1$ and $4m + 2$ [see Theorem 5.1.6]) and who proved for the case of latin squares of order $4m + 3$ that (i) such a square can be in a set of at most $2m + 2$ MOLS if it contains a latin subsquare of order $2m + 1$; and (ii) that there exist latin squares of such orders which cannot be in a set of three MOLS. Statement (i) had earlier been proved by Drake(1977). See Theorem 5.2 of Chapter 2 in [DK2].

Clearly, a latin square which has no transversals is a bachelor square so, in particular, cyclic squares of even order and latin squares of q-step type are bachelor squares. (See Section 4.5.)

It follows as a corollary to Mann's theorem that if a latin square L is of order $4m + 2$ and contains a latin subsquare of order $2m + 1$ or if it is of order $4m + 1$ and contains a latin subsquare of order $2m$ then L is a bachelor square. However, this leaves completely open the question for which orders $n \equiv 3 \mod 4$ (other than 3 itself) bachelor squares exist.

The general question "For which orders do bachelor squares exist?" was solved almost simultaneously by Wanless and Webb(2006) and by A.B.Evans(2006) using a similar method: namely, to modify a certain latin square by exchanging the elements in particular cells. However, the proof given by the first two authors has the virtue of elegance and of being applicable to latin squares of all orders, not just those of form $4m + 3$. These authors showed that, for every order $n > 3$, there exists a latin square which contains a cell that is not included in any transversal. We gave their proof in Section 3.5. Since a latin square which has an orthogonal mate has a decomposition into disjoint transversals of which precisely one must pass through every cell, it follows at once that bachelor squares exist of every order except three.

Latin Squares and their Applications. http://dx.doi.org/10.1016/B978-0-444-63555-6.50009-X

9.2   Homogeneous latin squares

DEFINITION. A latin square is said to be *h-homogeneous* if each of its cells lies in exactly h intercalates.

For example, a cyclic latin square of order 4 is 1-homogeneous whereas a latin square of order 4 based on Klein's Viergruppe $\mathbf{Z}_2 \times \mathbf{Z}_2$ is 3-homogeneous.

A latin square which is 0-homogeneous has also been called an N_2-*square*. (In this notation, a latin square which has no latin subsquares of any size is an N_∞-*square*. We discuss these in Section 9.4.) By combining the results obtained in a series of papers, it was proved in the late 1970s that N_2-squares exist for all orders n except $n = 2$ and 4. Details of the proof are given on pages 113-116 of [DK2].

To return to h-homogeneous latin squares, it is easy to see that 1-homogeneous latin squares exist only of even orders and that $(n-2)$-homogeneous latin squares of order n do not exist since any such square must in fact be $(n-1)$-homogeneous. Also, $(n-1)$-homogeneous latin squares exist if and only if n is a power of 2. See Heinrich and Wallis(1981). It was shown in Hobbs, Kotzig and Zacs(1982) that $(n-3)$-homogeneous latin squares exist if and only if $n \in \{3, 4, 6, 8, 12, 16\}$. Killgrove(2001) called a 1-homogeneous latin square *two-tiled* and he proved:

KILLGROVE'S LEMMA. *A group-based latin square is two-tiled if and only if the group has a unique element of order two.*

PROOF. Suppose without loss of generality that the latin square is in reduced form and is bordered by its own first row and column so as to form the Cayley table of a group.

If the square is 2-tiled, the identity element e in the cell (e, e) belongs to a unique intercalate whose remaining cells are (h, h), (e, h) and (h, e) for some element h. But then $h^2 = e$, since the cell (h, h) must contain e. Thus, the group has an element of order two. If there were a second element k of order two, then the cells (e, e), (k, k), (e, k) and (k, e) would form an intercalate overlapping the first, which is a contradiction. Therefore, the element h of order two must be unique.

Conversely, suppose that there is a unique element h of order two. Any inner automorphism fixes h since $g^{-1}hg$ also has order two, so h is in the centre of the group. Consider any cell (i, j). This cell, and also the cell (ih, hj), contains the element ij. Similarly, the cells (ih, j) and (i, hj) both contain the element ihj so the three cells (ih, hj), (ih, j) and (i, hj) form an intercalate with the cell (i, j). Suppose, if possible, that the cells (r, s), (r, j) and (i, s) also form an intercalate with the cell (i, j). Then $rs = ij$ and $is = rj$. Thus, $r^{-1}i = sj^{-1}$ and $r^{-1}i = js^{-1}$ whence $sj^{-1} = js^{-1}$ or $(sj^{-1})^2 = e$. That is, $sj^{-1} = h$. Therefore, $s = hj$ and $r = ih$. This proves that the cell (i, j) lies in only one intercalate, so the square is 2-tiled.   □

In two unpublished papers by D'Angelo and Turgeon and by Abrham and Kotzig, these authors obtained some further results; in particular, the latter authors show that 4-homogeneous latin squares do not exist if $n < 8$ but that they do exist of all orders $n = 8k$. (This leaves open the question whether and/or for which other orders such squares exist.) The former authors mention and/or prove several results: (i) There are 2-homogeneous latin squares of all orders divisible by 4 except 4 itself; (ii) If $n = 4k$ and $k \geq 1$, there is an $(n - k)$- or $(n - k + 1)$-homogeneous latin square of order n according as k is odd or even; (iii) If A is an h-homogeneous latin square of order m and B is a k-homogeneous latin square of order n, then $A \times B$ is an $(hk + h + k)$-homogeneous latin square of order mn (where $A \times B$ denotes the direct product of the quasigroups whose Cayley tables are defined by A and B); (iv) The Cayley table of a group G is an h-homogeneous latin square, where h is the number of elements of order 2 in G (cf. Killgrove's lemma above). From (iv) it follows that, for each positive integer $h \geq 2$, there is an h-homogeneous or $(h + 1)$-homogeneous latin square of order $2h$ according as h is odd or even. This last result makes use of the dihedral groups and had already been proved earlier in Heinrich and Wallis(1981) and in Hobbs, Kotzig and Zaks(1982).

So far as the present author is aware, the complete spectrum of integers h for which h-homogeneous latin squares exist is still an open question. But Hobbs and Kotzig(1983) contains a summary of results which were known at the time that paper was written regarding squares of orders up to 26. Another interesting question is: For which integers h do orthogonal h-homogeneous latin squares exist or, alternatively, can it be proved that they do not exist for some values of h? This question is partly answered by remark (iv) above. We notice, for example, that the three mutually orthogonal latin squares which are based on Klein's Viergruppe are all 3-homogeneous.

9.3 Diagonally cyclic latin squares and Parker squares

A latin square of order n is said to be *diagonally cyclic* (or sometimes, and more accurately, *left diagonally cyclic*) if each of its left-to-right broken diagonals contains each of its n symbols and each is a cyclic permutation of the main left-to-right diagonal. If the same statement is true for the right-to-left diagonals also, the square is called *totally diagonal* or to be a *Knut Vik design*, see page 219. Two examples of the latter type of square are given in Figure 6.3.1.

The term "diagonally cyclic" is due to Franklin(1984a,b) who was probably the first to discuss construction of such squares explicitly. He also used what he called "bordered cyclic latin squares" (obtained by a prolongation of a diagonally cyclic latin square) to construct orthogonal latin squares and, in particular, pairs of MOLS of order 10. However, Franklin was certainly not the first to conceive and make use of the latter idea.

In the papers already referred to, Franklin frequently abbreviated "diagonally cyclic" to "cyclic". However, not all left diagonally cyclic latin squares are cyclic in the sense used elsewhere in this book: that is, isotopic to a multiplication group of the cyclic group of the same order. A smallest counter-example, kindly supplied by Ian Wanless, is given in Figure 9.3.1. Wanless remarks that this is the multiplication table of a Steiner quasigroup of order 7 and that it cannot be isotopic to $(\mathbf{Z}_7, +)$ because it has too many subsquares, also that it is smallest because (see Theorem 9.3.1) all orthomorphisms of $(\mathbf{Z}_n, +)$ for $n \leq 5$ are linear[1] and so define cyclic groups.

$$
\begin{array}{|ccccccc}
0 & 3 & 6 & 1 & 5 & 4 & 2 \\
3 & 1 & 4 & 0 & 2 & 6 & 5 \\
6 & 4 & 2 & 5 & 1 & 3 & 0 \\
1 & 0 & 5 & 3 & 6 & 2 & 4 \\
5 & 2 & 1 & 6 & 4 & 0 & 3 \\
4 & 6 & 3 & 2 & 0 & 5 & 1 \\
2 & 5 & 0 & 4 & 3 & 1 & 6 \\
\end{array}
$$

FIG. 9.3.1.

A left diagonally cyclic latin square $A = \|a_{ij}\|$ is completely determined by its first row (or column) and by the order of the symbols in the main left-to-right diagonal. To see this, suppose that the latter order is $a_o, a_1, \ldots, a_{n-1}$. We regard the square as a collection of ordered triples, say (i, j, a_{ij}), as we did on page 14. Then, if (i, j, a_k) is a triple, so is $(i + 1, j + 1, a_{k+1})$ where arithmetic is modulo n.

In Bedford(1995), that author showed that:

THEOREM 9.3.1 *A diagonally cyclic latin square of order n exists if and only if the group $(\mathbf{Z}_n, +)$ has an orthomorphism.*

PROOF.   Suppose that $A = \|a_{ij}\|$ is diagonally cyclic and based on the integers $0, 1, \ldots, n - 1$. Define a bijection ϕ of $\mathbf{Z}_n$ by the statement that $\phi(j) = a_{0j}$ for $j = 0, 1, \ldots, n - 1$. Since A is diagonally cyclic, $a_{i,t+i} = a_{0t} + i$, where arithmetic is modulo n, so $a_{u-t,u} = a_{0t} + (u - t) = \phi(t) + (u - t)$. But $a_{u-t,u} \neq a_{0u}$ since A is a latin square. That is, $\phi(t) + (u - t) \neq \phi(u)$ or $\phi(t) - t \neq \phi(u) - u$ for all $t, u \in \mathbf{Z}_n$. Thus, the mapping ϕ is an orthomorphism of $(\mathbf{Z}_n, +)$.

Conversely, let ϕ be an orthomorphism of $(\mathbf{Z}_n, +)$. We construct an $n \times n$ matrix $A = \|a_{ij}\|$ in which $a_{0j} = \phi(j)$ for $j = 0, 1, \ldots, n - 1$ and put $a_{i,t+i} = a_{0t} + i = \phi(t) + i \bmod n$ so that each left-to-right diagonal is cyclic. Then, $a_{ij} = \phi(j-i)+i$ so $a_{ij} \neq a_{ik}$ because ϕ is a permutation of $\mathbf{Z}_n$ and so $\phi(j-i) \neq \phi(k-i)$.

Also, $a_{hj} \neq a_{ij}$ otherwise $\phi(j - h) + h = \phi(j - i) + i$ which would imply that $\phi(j - h) - (j - h) = \phi(j - i) - (j - i)$ and contradict the fact that ϕ is an orthomorphism. Thus, A is a latin square as well as being diagonally cyclic.   □

COROLLARY. *A left diagonally cyclic latin square of order n exists if and only if n is odd.*

---

[1]A linear orthomorphism is a map that sends x to $ax + b$ for some constants a, b. All diagonally cyclic squares based on linear orthomorphisms are isotopic to cyclic groups.

PROOF.   Since cyclic groups of even order have no orthomorphisms, this follows immediately from the theorem.   □

Later, Bedford [see Bedford(1998b)] used the fact that a diagonally cyclic latin square A can be constructed by putting $a_{ij} = \phi(j - i) + i$ to obtain his example of a non-cyclic latin square of order 15 with property P as defined in Section 9.4. The orthomorphism of $(\mathbf{Z}_{15}, +)$ which he used was
$$\phi = (0)(1\ 2)(3\ 9\ 13)(4\ 11\ 7)(5\ 8\ 10)(6\ 14\ 12).$$
Since each broken diagonal of a left diagonally cyclic latin square of order n is a transversal, such a square can be n times prolonged and this fact has been exploited by very many authors including Parker(1959b) who used it (though not explicitly as he described his construction in the language of orthogonal arrays) to disprove Euler's conjecture by constructing for the first time a pair of MOLS of order 10 and, more generally, a pair of MOLS of order $(3q-2)/2$, where q is any prime power congruent to 3 modulo 4. Yamamoto(1961) and, later, Hedayat and Seiden(1971,1974) under the title of *sum composition* used the idea of multiple prolongation[2] to construct t MOLS of order $n + m$ from $t + 1$ MOLS of order n and t MOLS of order m $(n > 2m)$. The required procedure is described in detail on pages 436-442 of [DK1]. The method has been generalized and also re-presented in a more easily understood form by Guérin(1966b) and this author has christened it "Yamamoto's method". For some related constructions and for a proof that, as $t \to \infty$, the number of non-equivalent pairs of MOLS of order t also tends to infinity, the reader is referred to chapter 6 of Guérin(1966b) and also to Barra and Guérin(1963a,b) and Guérin(1963a,b,1964,1966a).

In Wanless(2004b), that author has made the following definition: We adjoin b infinity symbols to the cyclic group $\mathbf{Z}_m$ to obtain the set $\mathbf{Z}_{m,b} = \mathbf{Z}_m \cup \{\infty_1, \infty_2, \ldots \infty_b\}$ and we define z^+ for $z \in \mathbf{Z}_{m,b}$ by $z^+ = z + 1 \bmod m$ if $z \in \mathbf{Z}_m$ and $z^+ = z$ otherwise. Then a *bordered diagonally cyclic latin square* (henceforward to be called a *Parker square*, see below) with rows, columns and symbols indexed by $\mathbf{Z}_{m,b}$ is one for which the presence of any triple (i, j, a_{ij}) implies that the triple (i^+, j^+, a_{ij}^+) also occurs in the square. We say that this Parker square is of B_b-type if it is based on $\mathbf{Z}_{m,b}$ in the above way for $m \geq 1$ or if its symbols can be mapped bijectively to $\mathbf{Z}_{m,b}$ in such a way that the resulting square has the above properties.

It was because of the important role which squares of B_b-type played in the disproof of Euler's conjecture that Wanless christened them *Parker squares*. In fact, the pair of MOLS of order 10 obtained by Parker and illustrated in Figure 9.3.2 is of B_3-type. Here, the integers 7, 8, 9 play the roles of $\infty_1, \infty_2, \infty_3$. This pair of squares could have been constructed directly by sum composition from a triple of MOLS of order 7 of which the first has constant left-to-right

---

[2]Yamamoto used the term *extension* instead of *prolongation* and he called the inverse operation *contraction* as already mentioned in Section 1.5. The term *prolongation* is due to Belousov, as stated in the same section. These two authors worked independently neither being aware of the other's work.

diagonals and the remaining two are diagonally cyclic (shown in Figure 9.3.3) and a pair of MOLS of order 3.[3]

0	4	1	7	2	9	8		3	6	5	0	7	8	6	9	3	5		4	1	2

(The figure consists of two latin squares of order 9 displayed side by side.)

Left square:

```
0 4 1 7 2 9 8   3 6 5
8 1 5 2 7 3 9   4 0 6
9 8 2 6 3 7 4   5 1 0
5 9 8 3 0 4 7   6 2 1
7 6 9 8 4 1 5   0 3 2
6 7 0 9 8 5 2   1 4 3
3 0 7 1 9 8 6   2 5 4

1 2 3 4 5 6 0   7 8 9
4 5 6 0 1 2 3   8 9 7
2 3 4 5 6 0 1   9 7 8
```

Right square:

```
0 7 8 6 9 3 5   4 1 2
6 1 7 8 0 9 4   5 2 3
5 0 2 7 8 1 9   6 3 4
9 6 1 3 7 8 2   0 4 5
3 9 0 2 4 7 8   1 5 6
8 4 9 1 3 5 7   2 6 0
7 8 5 9 2 4 6   3 0 1

4 5 6 0 1 2 3   7 8 9
2 3 4 5 6 0 1   9 7 8
1 2 3 4 5 6 0   8 9 7
```

FIG. 9.3.2.

```
0 1 2 3 4 5 6     0 4 1 5 2 6 3     0 2 4 6 1 3 5
6 0 1 2 3 4 5     4 1 5 2 6 3 0     6 1 3 5 0 2 4
5 6 0 1 2 3 4     1 5 2 6 3 0 4     5 0 2 4 6 1 3
4 5 6 0 1 2 3     5 2 6 3 0 4 1     4 6 1 3 5 0 2
3 4 5 6 0 1 2     2 6 3 0 4 1 5     3 5 0 2 4 6 1
2 3 4 5 6 0 1     6 3 0 4 1 5 2     2 4 6 1 3 5 0
1 2 3 4 5 6 0     3 0 4 1 5 2 6     1 3 5 0 2 4 6
```

FIG. 9.3.3.

In his paper of 2004, Wanless has extended Theorem 9.3.1 to provide necessary and sufficient conditions for a Parker square of any given B_i-type to exist by introducing the idea of a partial orthomorphism of $(Z_m, +)$. (In the case of a B_1-type Parker square, this is the same as a near orthomorphism. See Section 7.6.)

In the same paper, he has given an example of a B_1-type square and a B_2-type square which are *quasigroup isomorphic*: that is, there exists an isotopism which transforms one square to the other and which re-labels the rows, columns and symbols in the same way. Thus, Parker squares of different types may lie in the same isomorphism class and, a fortiori, in the same isotopism or main class of latin squares. Also, the B_2-type Parker square given by Wanless illustrates the fact that not all Parker squares can be contracted to diagonally cyclic latin

---

[3]But, in order to achieve orthogonality, the order of the last three rows and columns is important so the procedure described by Yamamoto(1961) must be followed.

squares.

We note that the diagonally cyclic latin squares in Figure 9.3.3 are orthogonal. A number of authors (including Franklin) have discussed the conditions for a diagonally cyclic latin square to have one or more orthogonal mates which are also diagonally cyclic. Since each such square is orthogonal to the latin square with the property that all the cells of each broken left-to-right diagonal contain the same fixed element, the maximum number of left diagonally cyclic squares of order n which are mutually orthogonal cannot exceed $n-2$ by Theorem 5.1.2. Wanless(2004b) has shown more generally that, given arbitrary integers n and b, there cannot be more than $n-1$ if $b = 1$ or $\min[N(b),(n-b/b]$ if $b > 1$ mutually orthogonal B_b-type latin squares of order n. He has further shown that both these bounds can be attained.

The Wanless paper of 2004 contains many further results concerning Parker squares and also it lists a surprisingly large number of papers whose authors had made use of such squares for their constructions up to the time when that paper was published. For example, Owens and Preece(1996) constructed their non-cyclic latin squares of prime order $p \geq 11$ with property P (see Section 9.4) by modifying a Parker square of B_1-type. In particular, Wanless himself, in collaboration with Bryant and Maenhaut [see Bryant, Maenhaut and Wanless(2006)] has used such squares to construct infinite families of atomic latin squares (again see Section 9.4 for the definition) which are of B_1-type.

A number of the papers cited by Wanless had also been mentioned earlier in Bedford(1993) and in the latter paper the close connection between left diagonally cyclic latin squares and neofields had been shown: namely, the addition table for a cyclic neofield of even order can always be obtained by the prolongation of an appropriate left diagonally cyclic latin square taking its main diagonal as the transversal.

Parker squares continue to be valuable as a means of obtaining new results.

Finally, let us mention a paper [Cavenagh and Wanless(2010)] in which the authors show that the number of transversals in Cayley tables of cyclic groups of odd order increases exponentially with the order and they remark that, by virtue of Theorem 9.3.1, the number of diagonally cyclic latin squares likewise increases exponentially with the order of the square.

9.4   Non-cyclic latin squares with cyclic properties

One of the rare occasions when the authors of the first edition of the present book were in serious disagreement was when the question arose in the late 1980s (in connection with their research) as to whether a given latin square of prime order p with the property that each of its rows was a cyclic permutation of the first row would necessarily be isotopic to the cyclic group of order p. Dénes was strongly of the opinion that it would while Keedwell was equally strongly of the opinion that it would not. The issue was resolved when the two authors constructed counter-examples in Dénes and Keedwell(1988) for latin squares not only of prime orders but for all orders $n \geq 7$. Shortly afterwards, a simpler construction of counter-examples was given in Keedwell(1991).

Next, in the development of the idea of constructing non-cyclic latin squares with cyclic properties, Owens and Preece(1996) constructed, for all prime orders $p \geq 11$, a latin square with the stronger property P that the permutation from row i to row j is cyclic for all i and j, $j \neq i$. The earlier construction only gave a square which had property P for the permutation from row 0 to row j, $1 \leq j \leq n - 1$ (in the case of a latin square of order n). Let us remark here that every cyclic latin square of prime order has property P and also has the similar property for columns. On the other hand, no latin square of composite order based on a group has property P. (See below.) Despite this, Bedford(1998b), in an unpublished paper (see Section 9.3), gave an example of a latin square of order 15 which has property P.

The next step came when Wanless(1999) observed that every latin square L of order n defines a 1-factorization of the complete bipartite graph $K_{n,n}$ (and in several ways, see Section 8.3). He also observed that a latin square L having property P would not contain any proper latin subsquare M because, if rows i and j of L both contained rows of such a subsquare M, then the permutation from row i to row j of L would include one or more sub-cycles involving only the elements of M. That is, any latin square with property P is an N_∞-square and conversely.

Let L be as above and let one partite set $\{c_1, c_2, \ldots, c_n\}$ of vertices of $K_{n,n}$ denote the columns of L and let the second partite set $\{s_1, s_2, \ldots, s_n\}$ denote the symbols. Then each row of L defines a 1-factor as follows: if the symbol s_k occurs in column c_j of row r_h, join c_j to s_k. Since no two entries in r_h are the same, these edges form a 1-factor which corresponds to the permutation

$$\sigma_h = \begin{pmatrix} c_1 & c_2 & \cdots & c_n \\ s_{1h} & s_{2h} & \cdots & s_{nh} \end{pmatrix}$$

defining row r_h as a permutation from natural order. Also, the 1-factors associated with different rows are disjoint since each column contains each symbol only once. The permutation which maps row r_h to row r_i is then $\sigma_h^{-1}\sigma_i$. Each of σ_h, σ_i defines a 1-factor of $K_{n,n}$ and $\sigma_h^{-1}\sigma_i$ defines a means of traversing the edges of the graph: $s_{lh} \to c_l \to s_{li} \to$ etc. If property P holds, then $\sigma_h^{-1}\sigma_i$ is a full length cycle and all edges of the graph are traversed by the union of the 1-factors corresponding to σ_h, σ_i. This led Wanless to say that a latin square (or rectangle) with property P is *pan-Hamiltonian* because of the connection with Hamiltonian decompositions of the complete bipartite graph just described.

For example, in Figure 9.4.1 the 1-factors which define the first three rows of L_4 are F_1, F_2, F_3. The unions $F_1 \cup F_2$ and $F_1 \cup F_3$ form Hamiltonian circuits in $K_{n,n}$ while $F_2 \cup F_3$ does not. As illustration, the union $F_1 \cup F_3$ gives $s_3 \to c_1 \to s_1 \to c_2 \to s_4 \to c_3 \to s_2 \to c_4 \to s_3$. The latin rectangles which are given by the first two rows or the first and third rows are pan-Hamiltonian and so contain no sub-rectangles.

$$L_4 = \begin{vmatrix} 3 & 1 & 4 & 2 \\ 2 & 3 & 1 & 4 \\ 1 & 4 & 2 & 3 \\ 4 & 2 & 3 & 1 \end{vmatrix}$$

F_1 $\quad ((1,3),\ (2,1),\ (3,4),\ (4,2)$

F_2 $\quad (1,2),\ (2,3),\ (3,1),\ (4,4)$

F_3 $\quad (1,1),\ (2,4),\ (3,2),\ (4,3)$

FIG. 9.4.1.

Later, Wanless refined the name to row-panHamiltonian since the roles of rows, columns and symbols can be permuted to give parastrophes of L which may be column-panHamiltonian or symbol-panHamiltonian.

In his paper of 1999, Wanless also proved the following three lemmas:

LEMMA A. *If a $k \times n$ latin rectangle R is pan-Hamiltonian, then either n is odd or $k = 2$.*

PROOF. Suppose that n is even and that r_h and r_i are two rows of R. Then $\sigma_h^{-1}\sigma_i$ is a cycle of length n on an even number of symbols so it is an odd permutation. Consequently, σ_h and σ_i must be of opposite parity. As this must be true for any two rows of R, there cannot be more than two rows. $\square$

This lemma and with the same proof had earlier been given as a closing remark in the paper of Owens and Preece(1996).

LEMMA B. *If a latin square L is pan-Hamiltonian, then so is any square isotopic to L and so is the row inverse $L^{(2\ 3)}$ of L.*

PROOF. Isotopies involve re-arranging rows, symbols and/or columns. The first of these does not affect the row cycles $\sigma_1, \sigma_2, \ldots, \sigma_n$ but only their order so, if $\sigma_h^{-1}\sigma_i$ was a single cycle, that property is not affected. Also, that property is not affected if the symbols are permuted. Thirdly, if the entries in two rows of L are permuted in the same way by a re-arrangement of columns, this does not affect the permutation which transforms one to the other.

The latin square $L = (a_{hk})$ can be represented as a set of n^2 triples (h, k, a_{hk}) so the permutation σ_h which defines the hth row of L is $\begin{pmatrix} \cdots & k & \cdots \\ \cdots & a_{hk} & \cdots \end{pmatrix}$. The triples of the row inverse $L^{(2\ 3)}$ take the form (h, a_{hk}, k) so the permutation which defines the hth row is $\sigma_h^{-1} = \begin{pmatrix} \cdots & a_{hk} & \cdots \\ \cdots & k & \cdots \end{pmatrix}$. If $\sigma_h^{-1}\sigma_i$ is a full length cycle for all choices of h and i, $h \neq i$, then so is $(\sigma_h^{-1})^{-1}\sigma_i^{-1} = \sigma_h\sigma_i^{-1}$ because $\sigma_h\sigma_i^{-1} = \sigma_i(\sigma_i^{-1}\sigma_h)\sigma_i^{-1} = \sigma_i(\sigma_h^{-1}\sigma_i)^{-1}\sigma_i^{-1}$ and both conjugate and inverse permutations have the same cycle structure.[4] $\square$

---

[4]The proof of the last result given in Wanless(1999) is incorrect.

LEMMA C. *If a latin square L is pan-Hamiltonian, let $\nu(L)$ denote the number of its parastrophes (including itself) which are also pan-Hamiltonian. If L is isotopic to its transpose $L^{(1\ 2)}$ then $\nu(L) = 4$ or 6. If L is isotopic to its row inverse $L^{(2\ 3)}$ then $\nu(L) = 2$ or 6.*

PROOF.   Let $A \sim B$ denote that both A and B are pan-Hamiltonian or that neither is. Then, by Lemma B, $L \sim L^{(2\ 3)}$, $L^{(1\ 2)} \sim L^{(1\ 2)(2\ 3)} = L^{(1\ 3\ 2)}$ and $L^{(1\ 3)} \sim L^{(1\ 3)(2\ 3)} = L^{(1\ 2\ 3)}$.

If L is isotopic to $L^{(1\ 2)}$ then, from Lemma B again, it follows that $L \sim L^{(1\ 2)}$ and so $L \sim L^{(2\ 3)} \sim L^{(1\ 2)} \sim L^{(1\ 3\ 2)}$. If also one of $L^{(1\ 3)}$ or $L^{(1\ 2\ 3)}$ is pan-Hamiltonian, then both are. Thus, $\nu(L) \in \{4, 6\}$.

If L is isotopic to $L^{(2\ 3)}$, then their transposes $L^{(1\ 2)}$ and $L^{(2\ 3)(1\ 2)} = L^{(1\ 2\ 3)}$ also are isotopic and so are $L^{(1\ 3)}$ and $L^{(2\ 3)(1\ 3)} = L^{(1\ 3\ 2)}$ so $L^{(1\ 2)} \sim L^{(1\ 2\ 3)}$ and $L^{(1\ 3)} \sim L^{(1\ 3\ 2)}$. But $L \sim L^{(2\ 3)}$, $L^{(1\ 2)} \sim L^{(1\ 3\ 2)}$, $L^{(1\ 3)} \sim L^{(1\ 2\ 3)}$ so $L^{(1\ 2\ 3)} \sim L^{(1\ 2)} \sim L^{(1\ 3\ 2)} \sim L^{(1\ 3)}$. If any one of these is pan-Hamiltonian, they all are and so $\nu(L) \in \{2, 6\}$.   □

Wanless called a latin square *atomic* if all six of its parastrophes are pan-Hamiltonian. A latin square is atomic if and only if none of its parastrophes has a proper latin sub-rectangle.

To test whether a given latin square L is atomic, it suffices to check that L, its transpose $L^{(1\ 2)}$ and the transpose of its row inverse $L^{(2\ 3)(1\ 2)} = L^{(1\ 2\ 3)}$ are atomic. This follows from the first statement in the proof of Lemma C.

By Theorem 4.2.2, if a latin square L is the Cayley table of a group or of a loop with the inverse property, then all five of its parastrophes are isotopic to it. It follows from Lemma B above that in that case, if L is pan-Hamiltonian then it is in fact atomic. In the Cayley table of a group of order n (which we may assume to be in reduced form), the rows are represented by regular permutations with cycle lengths which divide n and so when and only when[5] n is prime and the group is cyclic, it is pan-Hamiltonian. Thus, the only group-based square which is atomic is that of the cyclic group C_p.

Wanless(1999) used a connection between perfect 1-factorizations of the complete graph K_{n+1} and perfect 1-factorizations of $K_{n,n}$ which we next describe to obtain two non-isomorphic perfect 1-factorizations of $K_{p,p}$ and hence the following result:

"Let $p \geq 11$ be a prime. If 2 is a primitive root modulo p, then there exists an atomic square of order p outside the main class of latin squares which contains C_p."

In a subsequent paper [see Bryant, Maenhaut and Wanless(2006)], these authors used similar ideas to construct five different main classes M_1, M_2, M_3, M_4, M_5 of latin squares of prime order which sometimes or always contain atomic latin

---

[5] If n is not prime and p is a prime which divides n, then, by Cauchy's theorem, at least one row is represented by a permutation whose cycles have length $p < n$.

squares. The squares in M_1 include C_p and so are atomic always. The squares in M_3 include those constructed by Owens and Preece(1996) who showed that they too are atomic. The squares in M_2 include that constructed by Wanless and referred to in the preceding paragraph. These squares are atomic when p is such that it has 2 as primitive root and so also are the squares in M_4 and M_5.

In Maenhaut and Wanless(2004), a complete enumeration of atomic latin squares of order 11 (which is the smallest order for which there are examples distinct from isotopes of C_p) is given. They belong to seven main classes.

Suppose that the complete graph K_{n+1} has a perfect 1-factorization $F_1, F_2, \ldots,$ F_n. We label the vertices of K_{n+1} by $v_1, v_2, \ldots, v_n, v_\infty$. We denote the 1-factor which contains the edge (v_h, v_∞) by F_h. Let the remaining edges of F_h be $\ldots, (v_x, v_y), \ldots$, etc. Let $K_{n,n}$ have partite sets $\{c_1, c_2, \ldots, c_n\}$ and $\{s_1, s_2, \ldots, s_n\}$. We use F_h to construct a 1-factor F_h' of $K_{n,n}$ whose edges are $\ldots, (c_h, s_h), (c_x, s_y),$ $(c_y, s_x), \ldots$, etc. and thence construct a latin square L whose hth row is given by the involutary permutation $\sigma_h = (h)(x\ y) \ldots$. If the union of F_h and F_i is a Hamiltonian circuit in K_{n+1}, then the union of F_h' and F_i' is a Hamiltonian circuit of $K_{n,n}$, since the sequence of vertices $\to v_a \to v_b \to v_d \to v_e \to v_f \to$, where $(v_a, v_b), (v_d, v_e), \ldots \in F_h$ and $(v_b, v_d), (v_e, v_f), \ldots \in F_i$, is replaced by the sequences $\to c_a \to s_b \to c_d \to s_e \to c_f \to$ and $\to s_a \to c_b \to s_d \to c_e \to s_f \to$ in $K_{n,n}$ and the sequence $\to v_h \to v_\infty \to v_i \to$ by the sequences $\to c_h \to s_h \to$ and $\to s_i \to c_i \to$. This construction had earlier been noted and used for various purposes by several authors, in particular see Keedwell(1978), Laufer(1980), Wanless and Ihrig(2005) and page 116 of [DK2]. Note that $n + 1$ must be even for such a 1-factorization to exist.

F_1	$((v_1, v_\infty),$	$(v_2, v_3),$	(v_4, v_5)		F_1'	(c_1, s_1)	$(c_2, s_3),$	$(c_3, s_2),$	$(c_4, s_5),$	(c_5, c_4)
F_2	$((v_2, v_\infty),$	$(v_3, v_5),$	(v_1, v_4)		F_2'	(c_2, s_2)	$(c_3, s_5),$	$(c_5, s_3),$	$(c_1, s_4),$	(c_4, c_1)
F_3	$((v_3, v_\infty),$	$(v_2, v_4),$	(v_1, v_5)		F_3'	(c_3, s_3)	$(c_2, s_4),$	$(c_4, s_2),$	$(c_1, s_5),$	(c_5, c_1)
F_4	$((v_4, v_\infty),$	$(v_1, v_3),$	(v_2, v_5)		F_4'	(c_4, s_4)	$(c_1, s_3),$	$(c_3, s_1),$	$(c_2, s_5),$	(c_5, c_2)
F_5	$((v_5, v_\infty),$	$(v_1, v_2),$	(v_3, v_4)		F_5'	(c_5, s_5)	$(c_1, s_2),$	$(c_2, s_1),$	$(c_3, s_4),$	(c_4, c_3)

$F_1 \cup F_2 \quad v_3 \to v_5 \to v_4 \to v_1 \to v_\infty \to v_2 \to v_3$

$F_1' \cup F_2' \quad s_3 \to c_5 \to s_4 \to c_1 \to s_1 \to c_4 \to s_5 \to c_3 \to s_2 \to c_2 \to s_3$

FIG. 9.4.2.

As illustration, Figure 9.4.2 gives a perfect 1-factorization of K_6 and the corresponding perfect 1-factorization of $K_{5,5}$ and Figure 9.4.3 shows the pan-Hamiltonian latin square L_5 thereby defined. The method of construction shows that this square will always be idempotent and involutary. That is, each row permutation σ_h is equal to its own inverse and so $L = L^{(2\ 3)}$. By Lemma C, it follows that $\nu(L) = 2$ or 6 for such a square. If rows and symbols are interchanged to obtain the parastroph $L_5^* = L^{(1\ 3)}$, the latter will be symmetric, idempotent

$$
L_5 = \begin{vmatrix} 1 & 3 & 2 & 5 & 4 \\ 4 & 2 & 5 & 1 & 3 \\ 5 & 4 & 3 & 2 & 1 \\ 3 & 5 & 1 & 4 & 2 \\ 2 & 1 & 4 & 3 & 5 \end{vmatrix} \qquad L_5^* = \begin{vmatrix} 1 & 5 & 4 & 2 & 3 \\ 5 & 2 & 1 & 3 & 4 \\ 4 & 1 & 3 & 5 & 2 \\ 2 & 3 & 5 & 4 & 1 \\ 3 & 4 & 2 & 1 & 5 \end{vmatrix}
$$

FIG. 9.4.3.

and symbol-panHamiltonian.

Using the above ideas, Wanless(1999) was able to construct an atomic latin square not in the main class which contains C_p for each prime $p \geq 11$ for which 2 is a primitive root (as we stated earlier). Wanless also claimed that, by applying his construction procedure to a particular perfect 1-factorization of K_{28}, he had obtained an atomic latin square of order 27.

In Wanless(2005), the same author used cyclotomic orthomorphisms in finite fields to construct atomic latin squares of other prime power orders but, so far as the present author is aware, no atomic latin square of composite but non-prime-power order has yet been found.

We end this section with a brief discussion of N_∞-squares. Such squares need not be atomic. Andersen and E.Mendelsohn(1982) showed that N_∞-squares exist of all orders $n \neq 2^a 3^b$. We outlined their proof in Chapter 4, Section 2 of [DK2]. A further construction of N_∞-squares was devised by Elliott and Gibbons(1992) who obtained subsquare-free latin squares of orders 16 and 18. Later, Maenhaut, Wanless and Webb(2007) showed that if n is odd and divisible by 3, then an N_∞-square of order n exists. (See the bibliography of that paper for some papers on this topic not mentioned here.) Combining this with the earlier paper of Andersen and E.Mendelsohn, we are able to conclude that N_∞-squares exist of all odd orders. However, so far as the present author is aware, the situation regarding even orders is still not completely determined.

CHAPTER 10

Alternative versions of orthogonality

Very many modifications of the concept of orthogonality have been studied over the years such as perpendicularity of symmetric latin squares. Here we define and discuss a number of these. We also introduce the concept of a power set of orthogonal latin squares.

10.1 Variants of orthogonality

(a) *r-orthogonal latin squares*

We remind the reader of the definition. Two latin squares of the same order n and defined on the same set Q are called *r-orthogonal* if, when they are juxtaposed, exactly r different ordered pairs of the set $Q \times Q$ occur among the n^2 ordered pairs of cells. Such squares were discussed in detail in Chapter 6 of [DK2] so here we only draw attention to the fact that, more recently than when [DK2] was published, the spectrum of existence of pairs (or larger sets) of such squares has been determined in a series of papers. The papers concerned are Colbourn and Zhu(1995), Zhu and Zhang(2001,2003) and, for the case of r-orthogonal latin squares which are idempotent, Xu(2015).

(b) *Near-orthogonal latin squares*

A pair of latin squares of order n is said to be *near-orthogonal* if the two squares are orthogonal except that they have a common 2×2 subsquare. In consequence of work by Yamamoto(1954), Horton(1974) and Heinrich(1977), it is known that such a pair exists for all $n \geq 6$. Details are in [DK2], Theorem 4.7 of Chapter 4, where Heinrich denotes such a pair by $IPOLS(n; 2)$. The connection with r-orthogonality is discussed on pages 170 and 190 of [DK2]. In Brouwer(1984), four pairwise near-orthogonal latin squares of order 10 were constructed.

(c) *Nearly orthogonal latin squares*

Two latin squares of the same order n and based on the elements $0, 1, \ldots, n-1$ have been called *nearly orthogonal* if, when they are juxtaposed, every ordered pair (a, b) of *distinct* elements except those for which $b - a \equiv n/2 \pmod{n}$ occurs exactly once and each ordered pair for which $b - a \equiv n/2 \pmod{n}$ occurs twice but pairs (a, a) do not occur. This concept was first introduced in connection with the design of statistical experiments by Raghavarao(1971). Since, in the definition, $n/2$ must be an integer, the concept is only defined for squares of even

Latin Squares and their Applications. http://dx.doi.org/10.1016/B978-0-444-63555-6.50010-6

order. An example for $n = 4$ is given in Figure 10.1.1. Note that the squares in this example are cyclic.

$$
\begin{array}{|cccc}
0_1 & 1_2 & 2_3 & 3_0 \\
1_3 & 2_0 & 3_1 & 0_2 \\
2_0 & 3_1 & 0_2 & 1_3 \\
3_2 & 0_3 & 1_0 & 2_1
\end{array}
$$

FIG. 10.1.1.

In Raghavarao, S.S.Shrikhande and M.S.Shrikhande(2002), these authors showed that, for any such integer $n = 2m$, a pair of nearly orthogonal latin squares exists. They also proved that if t is the maximum number of squares of order $2m$ which could exist in a mutually nearly orthogonal set then $t \leq m+1$ when m is odd and $t \leq m$ when m is even. They constructed a set of three mutually nearly orthogonal latin squares (MNOLS) of order 6. Shortly after this, Burns Pasles made the topic the subject of her PhD thesis. In Burns Pasles and Raghavarao(2004), the two authors showed that no set of four MNOLS of order 6 exists and so the upper bound cannot be attained for this order. Later, in Li and van Rees(2007), the latter authors obtained two recursive constructions for MNOLS and were thence able to prove that a set of at least three MNOLS exists for all orders $n \geq 358$.

(d) *k-plex orthogonality of latin squares*

This is a relatively recent concept due to Liang(2012) and is a natural generalization of orthogonality because it reduces to the latter when $k = 1$ and the k-plex becomes a transversal (see Section 3.5). The present author anticipates that the concept will be developed much further in future years.

DEFINITION. Let L_1 and L_2 be two latin squares of order n, based on sets S_1 and S_2 respectively, and let k be a divisor of n. Then L_1 and L_2 are called *k-plex orthogonal* if the following two conditions are satisfied: (i) there is a partition of the set S_1 into blocks $B_1, B_2, \ldots, B_r$ $(r = n/k)$ of k elements each such that the cells of L_2 which correspond to the cells of L_1 that contain elements of B_i form a k-plex in L_2 for each choice of i; and (ii) there is likewise a partition of the set S_2 into blocks $B_1^\star, B_2^\star, \ldots, B_r^\star$ of k elements each such that the cells of L_1 which correspond to the cells of L_2 that contain elements of $B_i^\star$ form a k-plex in L_1 for each choice of i.

The k-plex orthogonality of L_1, L_2 is unaffected if the symbols of S_1 or S_2 are permuted or if the rows and/or columns of both squares are permuted simultaneously. It is possible to have several latin squares which are pairwise (or mutually) k-plex orthogonal.

In contrast to these facts, it is important to note that pairs of latin squares exist which satisfy condition (i) but not condition (ii) or vice versa. For example, the squares L_1 and l_2 of order 6 exhibited in Figure 10.1.2 are 3-plex orthogonal relative to the partitions $B_1 = \{1,5,6\}$, $B_2 = \{2,3,4\}$ and $B_1^\star = \{1,2,5\}$, $B_2^\star = \{3,4,6\}$ respectively of $S = \{1,2,\ldots,6\}$. In Figure 10.1.2, the cells of L_2 marked with a prime form a 3-plex of L_2 and correspond to the cells of L_1 which contain 1, 5 or 6. The cells of L_1 marked with a prime form a 3-plex of L_1 and correspond to the cells of L_2 which contain 1, 2 or 5. The unprimed entries in L_2 and L_1 also form 3-plexes and correspond in the same way to the blocks B_2 and $B_2^\star$ respectively. On the other hand, the squares L_1 and L_3 satisfy the condition (ii) relative to the blocks $B_1^\star$ and $B_2^\star$ but do not satisfy condition (i) relative to any partition of S into two equally sized blocks. The proof of the latter statement requires checking each of the 20 possible partitions as follows.

Let T_i denote the set of all cells which have i as entry in L_1. Then the left-hand and right-hand tables in Figure 10.1.3 indicate the number of cells of T_i which contain j as entry in L_2 and L_3 respectively. In the left-hand table, it is easy to check that $T_1 \cup T_5 \cup T_6$ contains each element of S exactly thrice and so the cells of L_1 which contain the elements 1, 5 and 6 define a 3-plex of L_2. The remaining cells of L_1 define a second 3-plex of L_2. However, in the right-hand table, none of the unions of three of the T_i contains each element of S thrice and so L_3 is not 3-plex orthogonal to L_1. (In fact, the required union can only contain the symbol 3 if it is $T_2 \cup T_4 \cup T_5$ or includes T_3 but none of the foregoing. The union can only contain the symbol 5 if it is $T_1 \cup T_2 \cup T_5$ or includes T_3 but again none of the foregoing. This is contradictory.)

$$
L_1 = \begin{vmatrix}
1' & 2' & 3 & 4 & 5' & 6 \\
2 & 3' & 6 & 1' & 4 & 5' \\
3' & 6' & 2 & 5 & 1 & 4' \\
4' & 5 & 1' & 2 & 6' & 3 \\
5 & 1 & 4' & 6' & 3 & 2' \\
6 & 4 & 5' & 3' & 2' & 1
\end{vmatrix}
\qquad
L_2 = \begin{vmatrix}
1' & 2 & 3 & 4 & 5' & 6' \\
3 & 1 & 4' & 2' & 6 & 5' \\
2 & 5' & 6 & 3' & 4' & 1 \\
5 & 3' & 1' & 6 & 2' & 4 \\
4' & 6' & 5 & 1' & 3 & 2 \\
6' & 4 & 2' & 5 & 1 & 3'
\end{vmatrix}
$$

$$
L_3 = \begin{vmatrix}
1 & 2 & 3 & 4 & 5 & 6 \\
3 & 5 & 4 & 2 & 6 & 1 \\
5 & 1 & 6 & 3 & 4 & 2 \\
2 & 4 & 5 & 6 & 1 & 3 \\
4 & 6 & 2 & 1 & 3 & 5 \\
6 & 3 & 1 & 5 & 2 & 4
\end{vmatrix}
$$

FIG. 10.1.2.

LEMMA. *If $n = 2m$ is even, any two nearly orthogonal latin squares of order n are 2-plex orthogonal.*

PROOF. If $S_1 = S_2 = \{0,1,\ldots,2m-1\}$, we choose the 2-block partition of S so that two integers in S are in the same block if and only if their difference is m. □

$j =$	1 2 3 4 5 6
T_1	2 1 1 1 0 1
T_2	1 2 1 0 0 2
T_3	1 1 2 1 1 0
T_4	1 0 0 2 2 1
T_5	0 1 2 1 2 0
T_6	1 1 0 1 1 2

$j =$	1 2 3 4 5 6
T_1	1 1 0 2 1 1
T_2	0 2 1 0 1 2
T_3	0 0 3 0 3 0
T_4	0 3 1 1 0 1
T_5	2 0 1 2 1 0
T_6	3 0 0 1 0 2

FIG. 10.1.3.

LEMMA. *Two latin squares of order n which are k-plex orthogonal are also ks-plex orthogonal for any integer s such that ks divides n.*

PROOF. This follows from the fact that s disjoint k-plexes of a latin square L together form a ks-plex of L.   □

Because Raghavarao *et al* (2002) had earlier shown that a pair of nearly orthogonal latin squares exists for any even integer n, it follows from the preceding lemma that there exists a pair of k-plex orthogonal latin squares of order $n = 2m$ for any even integer k such that $k|n$.

Liang(2012) has proved the stronger result that "for any integers k and n such that $k|n$, there exists a pair of k-plex orthogonal latin squares of order n except when $n = 2$ or 6 and $k = 1$". This generalizes the equivalent statement for orthogonal squares proved by Bose, S.S. Shrikhande and Parker, see Section 5.1.

Liang has also proved appropriate generalizations for well-known theorems about existence of orthogonal pairs of latin squares due to Maillet, MacNeish and Mann. For full details, see the original paper.

(e) *Quasi-orthogonal latin squares*

This concept was introduced by Bedford(1998a) and is as follows:

DEFINITION. Two latin squares of order n defined on the same symbol set are *quasi-orthogonal* if, when the two squares are juxtaposed, each unordered pair (u, v) of distinct elements occurs exactly twice and each pair (u, u) occurs exactly once.

We illustrate this definition by means of the triple of mutually quasi-orthogonal latin squares (MQOLS) of order four shown in Figure 10.1.4.

It is well-known that a group based latin square has an orthogonal mate if and only if the group possesses a complete mapping or orthomorphism. Similarly, a group based latin square has a quasi-orthogonal mate if and only if the group possesses a quasi-complete mapping (defined below).

A group which has no complete mappings may nonetheless have quasi-complete mappings. This fact also is demonstrated by the triple of mutually quasi-orthogonal latin squares of order four based on the cyclic group of order four (which has no complete mappings) and shown in Figure 10.1.4. They were first given in Bedford and Whitaker(2000b,2003). Because there exist statistical experiments for which a pair of quasi-orthogonal latin squares may serve as well as an orthogonal pair, the former concept may be practically useful. In that connection, Bedford has shown (see below) that quasi-orthogonal latin squares of order six exist, although these are not group based.

In order to introduce the concept of quasi-complete mapping, we shall need the following definition.

DEFINITION. Let G be a finite group of order n with identity element e. A listing $a_1, a_2, \ldots, a_n$ of some of the elements of G (with repetitions allowed) is called a *quasi-ordering* of G if and only if (i) the list contains the identity element and each element of order two exactly once and, (ii) for every element $g \in G$ such that $g^2 \neq e$, the list contains two occurrences of g and none of g^{-1} or one occurrence of each of g and g^{-1} or two occurrences of g^{-1} and none of g.

Hence we get:

DEFINITION. Let G be a finite group of order n with $G = \{g_1, g_2, \ldots, g_n\}$. A mapping θ of G into G is a *quasi-complete mapping* if the mapping ϕ defined by $\phi(x) = x\theta(x)$ is a permutation of G and θ is such that $\theta(g_1), \theta(g_2), \ldots, \theta(g_n)$ is a quasi-ordering of G. The permutation ϕ is then a *quasi-orthomorphism* of G.

Bedford(1998a) gave the following example of a quasi-complete mapping of the group $(\mathbf{Z}_4, +)$ and the corresponding quasi-orthomorphism.

$$\theta = \begin{pmatrix} 0\ 1\ 2\ 3 \\ 0\ 2\ 3\ 3 \end{pmatrix} \qquad \phi = \begin{pmatrix} 0\ 1\ 2\ 3 \\ 0\ 3\ 1\ 2 \end{pmatrix}$$

Thus, for example, $\phi(3) = 3 + \theta(3) = 3 + 3 \equiv 2 \bmod 4$.

If ψ_1 and ψ_2 are permutations of a group $(G, \cdot)$, where $G = \{g_1, g_2, \ldots, g_n\}$, such that $\psi_1(g_1)^{-1}\psi_2(g_1), \psi_1(g_2)^{-1}\psi_2(g_2), \ldots, \psi_1(g_n)^{-1}\psi_2(g_n)$ is a quasi-ordering of G, then ψ_1 and ψ_2 will be called *quasi-orthogonal permutations* of G. (If, on the other hand, the above sequence is an ordering of G, then ψ_1 and ψ_2 are *orthogonal permutations* of G.)

Using this concept, Bedford proved the following theorem and corollary.

THEOREM 10.1.1 *Let L be the Cayley table of the finite group $(G, \cdot)$ and let ϕ_1 and ϕ_2 be permutations of G. Let L_{ϕ_1} and L_{ϕ_2} be obtained from L by permuting its columns according to ϕ_1 and ϕ_2 respectively. Then, if ϕ_1 and ϕ_2 are quasi-orthogonal, L_{ϕ_1} and L_{ϕ_2} are quasi-orthogonal.*

PROOF. Let $A = (L_{\phi_1}, l_{\phi_2})$ denote the array obtained by juxtaposing L_{ϕ_1} and L_{ϕ_2}. Then the (i, j)th cell of A contains the ordered pair $[g_i\phi_1(g_j), g_i\phi_2(g_j)]$. For any ordered pair (u, v) of elements of G, let $d = u^{-1}v$ denote the *difference* between u and v. Then the jth column of A contains all the ordered pairs whose difference is $\phi_1(g_j)^{-1}\phi_2(g_j)$. For a particular ordered pair (u, v) of elements of G, there are two possibilities:
(i) if $d^2 = e$ then, since ϕ_1 and ϕ_2 are quasi-orthogonal, there exists a unique element g_k in G such that $\phi_1(g_k)^{-1}\phi_2(g_k) = d$. So the kth column of A contains

all ordered pairs whose difference is d and these include both (u, v) and (v, u) since $d = d^{-1}$.

(ii) if $d^2 \neq e$, then there are exactly two distinct elements g_k and g_l of G such that $\phi_1(g_k)^{-1}\phi_2(g_k)$ and $\phi_1(g_l)^{-1}\phi_2(g_l)$ are either both equal to d, both equal to d^{-1} or one equal to d and the other equal to d^{-1}.

In the first case of (ii), because u occurs once in the kth column of L_{ϕ_1} and also once in the lth column of L_{ϕ_1}, the ordered pair (u, v) occurs both in the kth column and in the lth column of A. In the second case, because u occurs once in the kth column of L_{ϕ_2} and also once in the lth column of L_{ϕ_2}, the ordered pair (v, u) occurs both in the kth column and in the lth column of A. From these arguments, it is easy to see that, in the third case, the ordered pairs (u, v) and (v, u) each occur just once, one in the kth column of A and the other in the lth column of A.

It follows that L_{ϕ_1}, L_{ϕ_2} are quasi-orthogonal. Moreover, if ϕ_1 and ϕ_2 are orthogonal permutations, only the third possibility occurs and so L_{ϕ_1} and L_{ϕ_2} are orthogonal in that case.   □

COROLLARY. *If $(G, \cdot)$ is a finite group which possesses a quasi-complete mapping θ with corresponding quasi-orthomorphism ϕ, then L and L_ϕ are quasi-orthogonal.*

PROOF.   We need to show that the identity permutation I on G and ϕ are quasi-orthogonal. However, this is indeed the case since $I(x)^{-1}\phi(x) = x^{-1}\phi(x) = \theta(x)$ and we know that $\theta(g_1), \theta(g_2), \ldots, \theta(g_n)$ is a quasi-ordering of G by definition of θ.   □

Bedford(1998a) observed that the following two quasi-orthomorphisms of $(\mathbf{Z}_4, +)$ are quasi-orthogonal.

$$\phi_1 = \begin{pmatrix} 0\ 1\ 2\ 3 \\ 0\ 3\ 1\ 2 \end{pmatrix} \qquad \phi_2 = \begin{pmatrix} 0\ 1\ 2\ 3 \\ 0\ 2\ 3\ 1 \end{pmatrix}$$

Hence, from the above theorem and corollary, the latin squares $L, L_{\phi_1}, L_{\phi_2}$ shown in Figure 10.1.4 are a set of MQOLS.

$$L = \begin{array}{|cccc} 0 & 1 & 2 & 3 \\ 1 & 2 & 3 & 0 \\ 2 & 3 & 0 & 1 \\ 3 & 0 & 1 & 2 \end{array} \qquad L_{\phi_1} = \begin{array}{|cccc} 0 & 3 & 1 & 2 \\ 1 & 0 & 2 & 3 \\ 2 & 1 & 3 & 0 \\ 3 & 2 & 0 & 1 \end{array} \qquad L_{\phi_2} = \begin{array}{|cccc} 0 & 2 & 3 & 1 \\ 1 & 3 & 0 & 2 \\ 2 & 0 & 1 & 3 \\ 3 & 1 & 2 & 0 \end{array}$$

FIG. 10.1.4.

REMARKS.

(i) The argument of the above theorem remains valid when ϕ_1 and ϕ_2 are neither quasi-orthomorphisms nor orthomorphisms but, in that case, the corollary no longer applies.

(ii) When ϕ_1 and ϕ_2 are both orthomorphisms (rather than quasi-orthomorphisms), the latin squares L_{ϕ_1}, L_{ϕ_2} and L are mutually orthogonal.

We illustrate these remarks by means of Figure 10.1.5. In that figure, ψ_1, ψ_2 are orthogonal permutations but ψ_1 is neither an orthomorphism nor a quasi-orthomorphism of $Z_2 \times Z_2 = \langle a, b : a^2 = b^2 = (ab)^2 = e \rangle$ so L is not orthogonal or quasi-orthogonal to L_{ψ_1}. (In fact, those ordered pairs which appear when L and L_{ψ_1} are juxtaposed all appear four times.) However, ψ_2 is an orthomorphism with corresponding complete mapping $\theta_2 = \begin{pmatrix} e\ a\ b\ ab \\ a\ b\ e\ ab \end{pmatrix}$ and ψ_1, ψ_2 are orthogonal permutations so each of L, L_{ψ_2} and L_{ψ_1}, L_{ψ_2} are orthogonal pairs of latin squares.

$$\psi_1 = \begin{pmatrix} e\ a\ b\ ab \\ a\ e\ ab\ b \end{pmatrix} \qquad \psi_2 = \begin{pmatrix} e\ a\ b\ ab \\ a\ ab\ b\ e \end{pmatrix} \qquad (L, L_{\psi_1}) = \begin{vmatrix} e_a & a_e & b_{ab} & ab_b \\ a_e & e_a & ab_b & b_{ab} \\ b_{ab} & ab_b & e_a & a_e \\ ab_b & b_{ab} & a_e & e_a \end{vmatrix}$$

$$(L, L_{\psi_2}) = \begin{vmatrix} e_a & a_{ab} & b_b & ab_e \\ a_e & e_b & ab_{ab} & b_a \\ b_{ab} & ab_a & e_e & a_b \\ ab_b & b_e & a_a & e_{ab} \end{vmatrix} \qquad (L_{\psi_1}, L_{\psi_2}) = \begin{vmatrix} a_a & e_{ab} & ab_b & b_e \\ e_e & a_b & b_{ab} & ab_a \\ ab_{ab} & b_a & a_e & e_b \\ b_b & ab_e & e_a & a_{ab} \end{vmatrix}$$

FIG. 10.1.5.

0	1	2	3	4	5		5	1	3	0	2	4		4	1	0	2	5	3
1	0	5	4	3	2		0	4	1	3	5	2		0	5	2	1	3	4
2	3	0	5	1	4		4	0	2	5	3	1		1	2	3	5	4	0
3	2	4	0	5	1		1	3	4	2	0	5		5	0	4	3	1	2
4	5	3	1	2	0		3	2	5	4	1	0		3	4	1	0	2	5
5	4	1	2	0	3		2	5	0	1	4	3		2	3	5	4	0	1

FIG. 10.1.6.

In Bedford and Whitaker(2000b), these authors used a computer and the connection with equidistant permutation arrays to help them find all sets of MQOLS of largest size for orders $n \leq 6$. In particular, they obtained sets of three MQOLS for each of the orders 4 and 6. In the case of the latter order, the squares are not group-based. One triple of order 4 is that given in Figure 10.1.4 and one of order 6 is shown in Figure 10.1.6. This raises the question "What is the upper bound $N_q(n)$ on the number of squares in a set of MQOLS of order n?". It is well-known that the upper bound for a set of MOLS is $n - 1$ (see Theorem 5.1.2) but no similar argument can be used for MQOLS because a permutation of symbols in any one member of the set may, and often does, destroy its property of being quasi-orthogonal to the remaining members of the set. In Bedford and Whitaker(2001), the authors reported their own efforts to solve this problem and also mentioned the very weak bound $N_q(n) \leq (n-1)(n-2)$ obtained earlier by Wild(1997). In particular, it is not known whether, for some n, $n > 6$, $N_q(n) > n - 1$.

In a further paper, Bedford and Whitaker(2003), these authors discuss connections between quasi-orthogonal latin squres and various other combinatorial designs. In particular, they consider connections with orthogonal Steiner triple systems (see Section 2.3 and Section 5.4) and with Room squares (see Section 6.4).

First, they point out and prove that there exists a skew Room square of side n if and only if there exists a pair of quasi-orthogonal, symmetric and idempotent latin squares of order n.

DEFINITION. A Room square in normalized form is said to be *skew* if, when cell (a, b) of the square is filled, cell (b, a) is empty.

In the construction of a Room square from a pair of Room quasigroups which we described in Section 6.4, the multiplication tables of the quasigroups have the properties of being symmetric and idempotent and if, when they are juxtaposed, the ordered pair a, b occurs in row u and column v (and in row v and column u by the symmetry), then the unordered pair u, v occurs as the entry in the cell (a, b) of the Room square. When the juxtaposed squares are quasi-orthogonal, the ordered pair b, a cannot occur because the ordered pair a, b has already occurred twice and, consequently, when cell (a, b) of the square is filled, cell (b, a) is empty as required.

Bedford and Whitaker define two Steiner triple systems to be quasi-orthogonal if the Cayley tables of their associated Steiner quasigroups are quasi-orthogonal. O'Shaugnessy's construction of Room squares from orthogonal Steiner triple systems is just a special case of the construction discussed above in which the Room pair of quasigroups are both Steiner quasigroups and so are totally symmetric. Consequently, when the Steiner quasigroups are quasi-orthogonal, the Room square is skew. In that case, the orthogonal Steiner triple systems, say S and S', have the additional property that when (a, b, u) and (c, d, v) both occur in S and (a, b, v) and (c, d, w) occur in S', then $w \neq u$. The above authors point out that Dukes and E.Mendelsohn(1999) have introduced such pairs of Steiner triple systems independently and have called them skew-orthogonal rather than quasi-orthogonal. The latter authors have investigated for which orders such pairs of systems exist (or, equivalently, for which orders pairs of quasi-orthogonal totally symmetric idempotent latin squares exist).

Quite recently, Liang(201?) has studied standardized (defined as first row in natural order) quasi-orthogonal latin squares and has shown that some classical results of Parker, MacNeish and Sade can be modified to apply to such squares.

Since the pairs of latin squares which we have just been considering have all been perpendicular pairs, this seems a convenient point at which to discuss such squares more generally.

If a symmetric latin square is of odd order, the elements of its main diagonal are necessarily all different (see Theorem 1.5.4) and so it can be made idempotent. If a second idempotent square of the same odd order $2m + 1$ is juxtaposed, the number of distinct ordered pairs then arising can be $(1 + 2 + \ldots + 2m) + (2m + 1) = (m + 1)(2m + 1)$ but not more. When the latter number is achieved, the squares are perpendicular. Symmetric squares of even order cannot be made idempotent so the same definition is not valid. In Gross, Mullin and Wallis(1973a,b), the maximum number $\nu(r)$ of idempotent symmetric latin squares which can occur in a pairwise perpendicular set has been investigated. The procedure used by these authors makes use of the connection with Room squares mentioned above and in Section 6.4. See also Section 4.3.

Steiner quasigroups (defined in Section 2.3) also have multiplication tables which are idempotent symmetric latin squares. Consequently, orthogonal Steiner triple systems and their associated perpendicular Steiner quasigroups provide examples of perpendicular idempotent latin squares. Perpendicular commutative quasigroups have been investigated from this point of view in O'Shaughnessy(1968), N.S.Mendelsohn(1970), Lindner and N.S.Mendelsohn(1973), Mullin and Németh (1969), Radó(1974) and Steedley(1974). (See also Section 5.4.)

(f) *Mutually orthogonal partial latin squares*

Two partial latin squares (not necessarily distinct) are *orthogonal* if, when they are juxtaposed, no ordered pair of elements appears more than once. A collection of partial latin squares is called *r-compatible* if each has r occupied cells and these occupied cells are in corresponding positions.

It follows that, in particular, a partial latin square of order n which has only n of its cells occupied is orthogonal to and n-compatible with itself. Thus, most interest is in partial $n \times n$ latin squares which have more than n cells filled. This concept, which is due to Abdel-Ghaffar(1996), arose in connection with coding theory: namely, in minimizing the retrieval time for items belonging to a large file of data stored on several disks.

Let $M_n(r)$ denote the maximum number of pairwise orthogonal r-compatible partial Latin squares. For the above application, Abdel-Ghaffar was interested in finding bounds for $M_n(r)$ when $r > n$. He showed that, for $n + 1 \leq r \leq n^2$,

$$M_n(r) \leq \lfloor \frac{r(r-1)}{\lfloor r/n \rfloor (2r - n - n\lfloor r/n \rfloor)} \rfloor - 2$$

Note that this result implies in particular that $M_n(n^2) \leq n - 1$. Abdel-Ghaffar showed further that, for $n + 1 \leq r \leq 2n$, $M_n(r) = \lfloor r(r-1)/2(r-n) \rfloor - 2$ and he constructed squares which meet the latter bound.

In the coding theory application mentioned earlier, the above bounds on $M_n(r)$ give bounds on the best possible retrieval time.

10.2   Power sets of latin squares

A set of MOLS is called a *power set* if the squares can be expressed in the form $L, L^2, L^3, \ldots, L^r$, where, if $L_1 = (\alpha_1, \alpha_2, \ldots \alpha_n)$ and $L_2 = (\beta_1, \beta_2, \ldots \beta_n)$ are two latin squares of order n with rows represented as permutations from natural order of the symbols by α_i and β_i respectively (for $i = 1, 2, \ldots, n$), we define their product $L_1 L_2$ to be the square $(\alpha_1 \beta_1, \alpha_2 \beta_2, \ldots \alpha_n \beta_n)$.

This notion of multiplication was originally introduced by D.A.Norton(1952a) in connection with row latin squares. In the same paper, Norton observed that the members of a latin power set are mutually orthogonal latin squares.

Later, Dénes became intrigued by this concept. In Dénes(1997) he pointed out that, if a group G is R_h-sequenceable (see[DK2] or Keedwell(1983a) for the definition) then a latin power set of h members exists based on G. In the same note and in an earlier joint paper published in 1994, he had conjectured that a latin power set of at least two members exists for all orders $n \geq 4$ except $n = 2$ or 6. This was proved by Damm(2011) who also verified an earlier result of Wanless(2001) to the effect that no latin power set of squares of order 10 containing more than two members exists. In the latter paper, Wanless also disproved several other conjectures which had been proposed earlier in joint papers involving Dénes.

In Dénes, Mullen and Suchower(1994), these authors proved that (i) if n is a prime, there is a latin power set of $n-1$ squares; (ii) if n is not a prime, then no complete latin power set containing $n - 1$ squares based on a group exists; (iii) if $n \geq 7$ and $n \equiv 0$ or 1 (mod 3), then there exists a latin power set containing at least two squares. They also made the conjecture mentioned above.

In Dénes and Owens(1999), examples of two special types of latin power set were constructed. A latin square $L = (\alpha_1, \alpha_2, \ldots \alpha_n)$ is of *D-type* if each of the permutations α_i representing its rows has cycle type $(1, n - 1)$. It is of *C-type* if the permutation α_1 representing its first row is the identity [that is, of type (1^n)] and if $\alpha_2, \alpha_3, \ldots, \alpha_n$ are all single cycles [that is, of type (n)]. Dénes and Owens conjectured that, for all orders $n \geq 7$, there exists a D-type latin square L not based on a group such that L^2 is also a latin square and they made a similar conjecture for a C-type latin square. In his paper already mentioned, Wanless (with computer aid) showed that the first conjecture is false for $n = 7, 8$ and 10 and the second for $n = 7, 8, 9$ and 10 despite the fact that Dénes and Owens had constructed such a square of order 11. This leaves open the question as to what happens in the case of larger values of n.

Finally, let us mention that Belyavskaya(2002) translated the concept of latin power set into the equivalent concept for quasigroups which she called a *quasigroup power set*. She was able to construct such sets, in particular one of three members of order 21, using pairwise balanced designs and what she called *complete S-systems* of quasigroups.

CHAPTER 11

Miscellaneous topics

The first two sections of this final chapter are devoted to the important subjects of orthogonal arrays and direct products, both of which can be used to construct sets of orthogonal latin squares of larger order from ones of smaller orders. In the third section of the chapter, we explain the Kézdy-Snevily conjecture and the fact that its truth would imply the truth of both Brualdi's and Ryser's conjectures. In the fourth section, we introduce the relatively new concept of the chromatic number of a latin square. The next topics considered are the practical applications of latin squares to coding, experimental designs and games tournaments. Lastly, we introduce latin triangles and also we comment on the use of computers in the solution of latin square problems.

11.1 Orthogonal arrays and latin squares

We mentioned in Section 5.6 that a set of $h - 2$ orthogonal latin squares is equivalent to the existence of a particular type of orthogonal array. Our main task in the present section is to confirm the validity of the last result by formally demonstrating the equivalence between a set of $h - 2$ mutually orthogonal latin squares of order n and an orthogonal array of h rows (or constraints), n^2 columns (n levels), strength 2, and index unity.

We shall begin with a definition.

DEFINITION. Let $A = ||a_{ij}||$ and $B = ||b_{ij}||$ be two $r \times s$ matrices whose entries are the numbers $1, 2, \ldots, n$, where r and s are integers such that $rs = n^2$. The matrices A and B are said to be n^2-orthogonal (cf. Chapter 6 of [DK2]) if, for every ordered pair (a, b) of the integers $1, 2, \ldots, n$, there is (exactly) one pair of indices i, j such that $a_{ij} = a$ and $b_{ij} = b$.

For example, the two $n \times n$ matrices R and C given in Figure 11.1.1 are clearly n^2-orthogonal.

Likewise, the two $1 \times n^2$ matrices

$$R^* = (1\,1\ldots1\,2\,2\ldots2\,3\,3\ldots3\ldots\ldots\ldots n\,n\ldots n)$$

and

$$C^* = (1\,2\ldots n\,1\,2\ldots n\,1\,2\ldots n\ldots\ldots\ldots 1\,2\ldots n)$$

obtained from R and C respectively by writing their rows consecutively are n^2-orthogonal. Evidently, two n^2-orthogonal $n \times n$ matrices which are such that each of $1, 2, \ldots, n$ occurs exactly once in each row and once in each column are orthogonal latin squares.

Latin Squares and their Applications. http://dx.doi.org/10.1016/B978-0-444-63555-6.50011-8

$$R = \begin{bmatrix} 1 & 1 & \ldots & 1 \\ 2 & 2 & \ldots & 2 \\ \cdot & \cdot & \ldots & \cdot \\ \cdot & \cdot & \ldots & \cdot \\ n & n & \ldots & n \end{bmatrix} \qquad C = \begin{bmatrix} 1 & 2 & \ldots & n \\ 1 & 2 & \ldots & n \\ \cdot & \cdot & \ldots & \cdot \\ \cdot & \cdot & \ldots & \cdot \\ 1 & 2 & \ldots & n \end{bmatrix}$$

FIG. 11.1.1.

THEOREM 11.1.1 *The existence of $k - 2$ mutually orthogonal latin squares of order n implies and is implied by the existence of k mutually n^2-orthogonal $n \times n$ matrices whose entries are the numbers $1, 2, \ldots, n$.*

PROOF. It is clear from the definition that the n^2-orthogonallty of two matrices is unaffected by any permutation of the n^2 cells of all the matrices simultaneously or by relabelling all of the symbols $1, 2, \ldots, n$ simultaneously as, say $1', 2', \ldots, n'$, where $1', 2', \ldots, n'$ is some permutation of the symbols $1, 2, \ldots, n$.

Now suppose that k mutually n^2-orthogonal matrices $A_1, A_2, \ldots, A_k$ are given. Since A_1 and A_2 are n^2-orthogonal, every ordered pair (i, j), $i = 1, 2, \ldots, n$, $j = 1, 2, \ldots, n$, occurs just once among the n^2 cells of these matrices. Consequently, by a suitable permutation of these n^2 cells, the two matrices can be transformed simultaneously into the matrices R and C. If this rearrangement of cells is carried out on all the matrices simultaneously, their n^2-orthogonality is not affected thereby, but the transforms of the matrices $A_3, A_4, \ldots, A_k$ are necessarily latin squares: for, since the entries in each row of R are all the same, the entries in each row of any matrix which is n^2-orthogonal to R must be all different. Likewise, since the entries in each column of C are all the same, the entries in each column of any matrix which is n^2-orthogonal to C must be all different. Thus, the transforms of the matrices $A_3, A_4, \ldots, A_k$ are a set of $k - 2$ mutually orthogonal latin squares.

Conversely, suppose that a set $L_1, L_2, \ldots, L_{k-2}$ of $k - 2$ mutually orthogonal latin squares is given. Since the entries in each row of the square L_i ($i = 1, 2, \ldots, k - 2$) are all different, L_i is an $n \times n$ matrix which is n^2-orthogonal to the matrix R. Similarly, since the entries in each column of L_i are all different, L_i is n^2-orthogonal to the matrix C. Thus, $R, C, L_1, L_2, \ldots, L_{k-2}$ are a set of k mutually n^2-orthogonal $n \times n$ matrices. □

COROLLARY. *The existence of $k - 2$ mutually orthogonal latin squares of order n implies and is implied by the existence of a $k \times n^2$ matrix S whose entries are the numbers $1, 2, \ldots, n$ and such that each two rows of S are n^2-orthogonal submatrices.*

PROOF. The corollary follows by observing that, if the rows of each of a set of k mutually n^2-orthogonal $n \times n$ matrices are written consecutively so as to form a $1 \times n^2$ row matrix (as, for example, when the matrix R is transformed to R^*), the set of matrices, say $R, C, L_1, L_2, \ldots, L_{k-2}$ becomes equivalent to a $k \times n^2$ matrix of the kind described, with R^* and C^* as its first two rows.   □

DEFINITION. A $k \times n^2$ matrix S of the type described in the corollary is called an *orthogonal array of k constraints, n levels, strength 2, and index 1*, as already mentioned in Section 5.6. In the present section, we shall denote such an orthogonal array by the symbol $OA(n, k)$.

THEOREM 11.1.2 *If there exists both an $OA(n_1, s)$ and an $OA(n_2, s)$ then an $OA(n_1 n_2, s)$ can be constructed.*

PROOF.   The proof we give here provides an actual construction for an $OA(n_1 n_2, s)$ whose entries are the integers $1, 2, \ldots, n_1 n_2$ from an $OA(n_1, s)$ with entries $1, 2, \ldots, n_1$ and an $OA(n_2, s)$ with entries $1, 2, \ldots, n_2$. Let $OA(n_1, s)$ be the matrix $A = ||a_{ij}||$, $i = l, 2, \ldots, s$, $j = 1, 2, \ldots, n_1^2$, and let $OA(n_2, s)$ be the matrix $B = ||b_{ij}||$, $i = l, 2, \ldots, s$, $j = 1, 2, \ldots, n_2^2$. Form a new matrix $C = ||c_{ir}||$, where, for $r = p + n_1^2(q - 1)$ with $1 \le p \le n_1^2$, $1 \le q \le n_2^2$, $c_{ir} = a_{ip} + n_1(b_{iq} - 1)$. We shall show that C is an $OA(n_1 n_2, s)$.

In the first place, since $1 \le a_{ip} \le n_1$ and $0 \le b_{iq} - 1 \le n_2 - 1$, each c_{ir} is one of the integers $1, 2, \ldots, n_1 n_2$. Secondly, we easily show that, for any choice of k and l, the kth and lth rows of C are $(n_1 n_2)^2$-orthogonal matrices. Let u and v be any two integers between 1 and $n_1 n_2$. Then $u = u_1 + n_1(u_2 - 1)$ and $v = v_1 + n_1(v_2 - 1)$ for suitable integers u_1, v_1 between 1 and n_1 and u_2, v_2 between 1 and n_2. Since the kth and lth rows of A are n_1^2-orthogonal matrices, there exists one column, say the pth, whose entries a_{kp} and a_{lp} are respectively equal to u_1 and v_1. Similarly, for a suitable integer q, we have $b_{kq} = u_2$ and $b_{lq} = v_2$. But then, $c_{kr} = u$ and $c_{ir} = v$, where $r = p + n_1^2(q - 1)$, and so the kth and lth rows of C are $(n_1 n_2)^2$-orthogonal. This completes the proof.   □

COROLLARY. *If there exists a set of t MOLS of order n_1 and a set of t MOLS of order n_2, then there exists a set of t MOLS of order $n_1 n_2$.*

PROOF.   The corollary is an immediate consequence of the fact that the existence of a set of t MOLS of order n implies and is implied by the existence of an $OA(n, t + 2)$.   □

As an example, we shall outline the construction of a pair of orthogonal latin squares of order 15 (or equivalently, an $OA(15, 4)$) from the pairs of orthogonal latin squares of orders 3 and 5 given in Figure 11.1.2, using the method given in our proof above.

We first form sets of four 3^2-orthogonal 3×3 matrices and 5^2-orthogonal 5×5 matrices by appending the matrices R and C of appropriate size to the squares given. Then, writing the rows of each matrix in a single row, we get an $OA(3, 4)$ and an $OA(5, 4)$ as in Figure 11.1.3.

From these, we derive an $OA(15, 4)$ of 255 columns, and, of these columns, we exhibit only the 1st to the 18th and the 37th to 54th in Figure 11.1.4.

$$
\begin{array}{|ccc}
1 & 2 & 3 \\
2 & 3 & 1 \\
3 & 1 & 2
\end{array}
\qquad
\begin{array}{|ccc}
1 & 2 & 3 \\
3 & 1 & 2 \\
2 & 3 & 1
\end{array}
$$

$$
\begin{array}{|ccccc}
1 & 2 & 3 & 4 & 5 \\
4 & 3 & 1 & 5 & 2 \\
5 & 1 & 4 & 2 & 3 \\
2 & 4 & 5 & 3 & 1 \\
3 & 5 & 2 & 1 & 4
\end{array}
\qquad
\begin{array}{|ccccc}
1 & 2 & 3 & 4 & 5 \\
3 & 5 & 2 & 1 & 4 \\
2 & 4 & 5 & 3 & 1 \\
5 & 1 & 4 & 2 & 3 \\
4 & 3 & 1 & 5 & 2
\end{array}
$$

FIG. 11.1.2.

```
1 1 1 2 2 2 3 3 3
1 2 3 1 2 3 1 2 3
1 2 3 2 3 1 3 1 2
1 2 3 3 1 2 2 3 1
```

```
1 1 1 1 1 2 2 2 2 2 3 3 3 3 3 4 4 4 4 4 5 5 5 5 5
1 2 3 4 5 1 2 3 4 5 1 2 3 4 5 1 2 3 4 5 1 2 3 4 5
1 2 3 4 5 4 3 1 5 2 5 1 4 2 3 2 4 5 3 1 3 5 2 1 4
1 2 3 4 5 3 5 2 1 4 2 4 5 3 1 5 1 4 2 3 4 3 1 5 2
```

FIG. 11.1.3.

We reorder these 225 columns in such a way that the first two rows of the matrix represent the rows, written consecutively, of two 15×15 matrices R and C and the remaining two rows then represent, in order, the rows of the two desired orthogonal latin squares of order 15. Hence the latter can be written down.

The result stated in Theorem 11.1.2 has probably been more used than any other in constructions of sets of MOLS. In particular, it leads at once to a property of the function $N(n)$ which was first pointed out by MacNeish(1922): namely, $N(p_1^{r_1}p_2^{r_2}\ldots p_s^{r_s}) \geq \min_i(p_i^{r_i} - 1)$

The theorem asserts that there exist at least $t = \min(p_i^{r_i} - 1)$ MOLS of order

```
1 1 1 2 2 2 3 3 3 1 1 1 2 2 2 3 3 3 ......
1 2 3 1 2 3 1 2 3 4 5 6 4 5 6 4 5 6 ......
1 2 3 2 3 1 3 1 2 4 5 6 5 6 4 6 4 5 ......
1 2 3 3 1 2 2 3 1 4 5 6 6 4 5 5 6 4 ......
```

```
...... 1  1  1  2  2  2  3  3  3  4  4  4  5  5  5  6  6  6 ......
...... 13 14 15 13 14 15 13 14 15 1  2  3  1  2  3  1  2  3 ......
...... 13 14 15 14 15 13 15 13 14 10 11 12 11 12 10 12 10 11 ......
...... 13 14 15 15 13 14 14 15 13 7  8  9  9  7  8  8  9  7 ......
```

FIG. 11.1.4.

$n = p_1^{r_1} p_2^{r_2} \ldots p_s^{r_s}$. Since we know by Theorem 5.2.3 that there are at least $p_i^{r_i} - 1$ MOLS of order $p_i^{r_i}$ for each i, $i = 1, 2, \ldots, r$, the result follows by repeated application of the corollary to Theorem 11.1.2. For an alternative proof, see Section 11.2.

MacNeish's theorem gives a lower bound for $N(n)$. As already mentioned in Section 5.3, MacNeish himself conjectured that this was also an upper bound because, for example, this is certainly the case when n itself is a prime power. Further evidence for this conjecture seemed to have been provided when, some twenty years later, Mann(1942) proved Theorem 7.2.2.

The MacNeish conjecture was not disproved until 1959 when Parker showed (as a counter-example) that there exists a set of at least four mutually orthogonal latin squares of order 21. See Section 5.3 for the further history of this topic.

We end this brief section by mentioning that J.W.Brown(1972) has used the medium of orthogonal arrays to describe an extension of Theorem 5.1.6 to the special case of a triple of MOLS of order ten.

For a survey of recursive and other constructions of orthogonal latin squares used up to the date of publication of that paper, see Colbourn and Dinitz(2001).

11.2   The direct product and singular direct product of quasigroups

A much-used means of obtaining pairs (or larger sets) of orthogonal latin squares from given sets of MOLS of smaller orders is by constructing the direct product or generalized direct product of the quasigroups which are defined by the squares of the given orders. We have already mentioned this approach earlier in this book. For example, MacNeish's theorem (Section 5.3 and Section 11.1) may be translated into the language of quasigroups by means of the concept of direct product.

We have the following theorem:

THEOREM 11.2.1 *Let* $(G, \cdot)$ *and* $(G, \odot)$, $(H, +)$ *and* $(H, \oplus)$ *be pairs of orthogonal quasigroups (Section 5.3) and let us define the operations* $(\times)$ *and* $(\otimes)$ *on the cartesian product* $F = G \times H$ *by the statements* $(x_1, y_1) \times (x_2, y_2) = (x_1 \cdot x_2, y_1 + y_2)$ *and* $(x_1, y_1) \otimes (x_2, y_2) = (x_1 \odot x_2, y_1 \oplus y_2)$, *where* $x_1, x_2 \in G$ *and* $y_1, y_2 \in H$. *Then the groupoids* $(F, \times)$ *and* $(F, \otimes)$ *are quasigroups and are orthogonal.*

PROOF.   Let us show first that the groupoid $(F, \times)$ is a quasigroup. The proof that $(F, \otimes)$ is also a quasigroup is similar.

Note firstly that the equation $(x_1, y_1) \times (x_2, y_2) = (x_1, y_1) \times (x_3, y_3)$ can be written $(x_1 \cdot x_2, y_1 + y_2) = (x_1 \cdot x_3, y_1 + y_3)$ and is equivalent to the equations $x_1 \cdot x_2 = x_1 \cdot x_3$ and $y_1 + y_2 = y_1 + y_3$, which imply $x_2 = x_3$ and $y_2 = y_3$, since $(G, \cdot)$ and $(H, +)$ are quasigroups. Therefore, $(x_2, y_2) = (x_3, y_3)$ from which we conclude that the equation $(x_1, y_1) \times (x_2, y_2) = (X, Y)$ is uniquely soluble for (x_2, y_2). It is equally easy to show that it is uniquely soluble for (x_1, y_1) and so $(F, \times)$ is a quasigroup.

To show the orthogonality of $(F, \times)$ and $(F, \otimes)$ it is necessary and sufficient to show that $(x_1, y_1) \times (x_2, y_2) = (x_3, y_3) \times (x_4, y_4)$ and $(x_1, y_1) \otimes (x_2, y_2) =$

$(x_3, y_3) \otimes (x_4, y_4)$ together imply $(x_1, y_1) = (x_3, y_3)$ and $(x_2, y_2) = (x_4, y_4)$. The equations can be written $(x_1 \cdot x_2, y_1 + y_2) = (x_3 \cdot x_4, y_3 + y_4)$ and $(x_1 \odot x_2, y_1 \oplus y_2) = (x_3 \odot x_4, y_3 \oplus y_4)$. That is, $x_1 \cdot x_2 = x_3 \cdot x_4$, $x_1 \odot x_2 = x_3 \odot x_4$, $y_1 + y_2 = y_3 + y_4$, $y_1 \oplus y_2 = y_3 \oplus y_4$. The first two of the latter equations imply $x_1 = x_3$ and $x_2 = x_4$ by the orthogonality of the quasigroups $(G, \cdot)$ and $(G, \odot)$ and the last two similarly imply $y_1 = y_3$ and $y_2 = y_4$. The result now follows.   □

If G has n_1 elements and H has n_2 elements, then F has $n_1 n_2$ elements. If there exist $N(n)$ mutually orthogonal latin squares of order n, then there exist $N(n)$ mutually orthogonal quasigroups of that order. Hence, by forming the direct products of $\min[N(n_1), N(n_2)]$ pairs of quasigroups of orders n_1 and n_2 (those of the same order n_1 or n_2 being orthogonal), we can obtain an equal number of mutually orthogonal quasigroups of order $n_1 n_2$. Thus, $N(n) \geq \min[N(n_1), N(n_2)]$, as we obtained in Section 11.1 by an argument in terms of orthogonal arrays. In particular, if $n = \prod_{i=1}^{r} p_i^{\alpha^i}$ where the p_i are distinct primes, we have $N(n) \geq \min_{i=1}^{r}(p_i^{\alpha^i} - 1)$, which is MacNeish's theorem.

By use of a generalized form of the above direct product method which he called "produit direct singulier", Sade has shown how a latin square of order $m + lk$ may be constructed from given latin squares L_g, L_r and L_h, where L_g is a latin square of order $m + l$ which contains a latin subsquare of order m and L_r, L_h have orders l, k respectively. Taking advantage of the fact that the direct product of two (or more) pairs of orthogonal quasigroups is again a pair of orthogonal quasigroups, as proved above, he has thence been able to obtain sets of orthogonal latin squares of order $m + lk$ in cases when the values of the integers m, l, k are suitable. In particular, the method provides counter-examples to the Euler conjecture. For the details, see Sade(1960a). We shall give only a summary of Sade's results.

Let $(G, \cdot)$ be a quasigroup of order $n = m + l$ which contains a subquasigroup $(Q, \cdot)$ of order m and let $(R, \otimes)$ be a quasigroup of order l defined on the set $R = G - Q$. Let $(H, \oplus)$ be an idempotent quasigroup of order k. The set $T = Q \cup (R \times H)$ is then of order $m + lk$ and we define an operation $(\circ)$ on it in such a way that the structure $(T, \circ)$ is a quasigroup, called by Sade the *singular direct product* of the four given quasigroups. [Note that T consists of elements $q_u \in Q$ and ordered pairs $(r_v, h_w) \in (R \times H)$.]

The operation $(\circ)$ is defined by the following five statements:

(i) for all $q_1, q_2 \in Q$, $q_1 \circ q_2 = q_1 q_2$;

(ii) for all $q \in Q, r \in R, h \in H$, $q \circ (r, h) = (qr, h)$ and $(r, h) \circ q = (rq, h)$;

(iii) for $r_1, r_2 \in R, h_1, h_2 \in H, h_1 \neq h_2$, $(r_1, h_1) \circ (r_2, h_2) = (r_1 \otimes r_2, h_1 \oplus h_2)$;

(iv) for $r_1, r_2 \in R$ with $r_1 r_2 \in R$, and for $h \in H$ $(r_1, h) \circ (r_2, h) = (r_1 r_2, h)$;

(v) for $r_1, r_2 \in R$ with $r_1 r_2 \in Q$, and for $h \in H$, $(r_1, h) \circ (r_2, h) = r_1 r_2$.

By way of explanation of these definitions, let us remark firstly that, whereas $r_1 \otimes r_2 \in R$ for all choices of $r_1, r_2 \in R$, we cannot make the same statement about the products $r_1 r_2$. We have $r_1 r_2 \in G$ but, in general, some of the products will be in Q. However, as may clearly be seen from Figure 11.2.1, we always have

$(\cdot)$	q_1	q_2	$\cdots$	q_m	r_1	r_2	$\cdots$	r_l
q_1	$\cdot$	$\cdot$	$\cdots$	$\cdot$	$\cdot$	$\cdot$	$\cdots$	$\cdot$
q_2	$\cdot$	$\cdot$	$\cdots$	$\cdot$	$\cdot$	$\cdot$	$\cdots$	$\cdot$
$\cdot$	$\cdot$	$\cdot$		$\cdot$	$\cdot$	$\cdot$		$\cdot$
$\cdot$	$\cdot$	$\cdot$	Q	$\cdot$	$\cdot$	$\cdot$	A	$\cdot$
$\cdot$	$\cdot$	$\cdot$		$\cdot$	$\cdot$	$\cdot$		$\cdot$
q_m	$\cdot$	$\cdot$	$\cdots$	$\cdot$	$\cdot$	$\cdot$	$\cdots$	$\cdot$
r_1	$\cdot$	$\cdot$	$\cdots$	$\cdot$	$\cdot$	$\cdot$	$\cdots$	$\cdot$
r_2	$\cdot$	$\cdot$	$\cdots$	$\cdot$	$\cdot$	$\cdot$	$\cdots$	$\cdot$
$\cdot$	$\cdot$	$\cdot$		$\cdot$	$\cdot$	$\cdot$		$\cdot$
$\cdot$	$\cdot$	$\cdot$	B	$\cdot$	$\cdot$	$\cdot$	C	$\cdot$
$\cdot$	$\cdot$	$\cdot$		$\cdot$	$\cdot$	$\cdot$		$\cdot$
r_l	$\cdot$	$\cdot$	$\cdots$	$\cdot$	$\cdot$	$\cdot$	$\cdots$	$\cdot$

FIG. 11.2.1.

$qr \in R$ and $rq \in R$ whatever the choices of $q \in Q$ and $r \in R$: for, since Q is a quasigroup and since each element of Q is to occur exactly once in each row and each column of the multiplication table of G, no element of Q can occur in either of the subsquares A or B shown in Figure 11.2.1. Again from Figure 11.2.1, we see that the ordered pairs (qr_i, h) and (rr_i, h) for values of r such that $rr_i \in R$ just cover the set $R \times \{h\}$ as q and r vary through the sets Q and R respectively with $r_i \in R$ kept fixed. [We have $qr_i \notin Q$ for all $q \in Q$. But gr_i for $g \in G$ covers G, so the products qr_i (with $q \in Q$) together with the products rr_i for which $rr_i \notin Q$ together cover $G - Q = R$.] The same is true of the ordered pairs (r_iq, h) and (r_ir, h).

We are now able to prove, as in Sade(1960a),

THEOREM 11.2.2 *The groupoid $(T, \circ)$ defined above is a quasigroup.*

THEOREM 11.2.3 *Let $(G, \cdot)$ and $(G, \odot)$ be orthogonal quasigroups defined on the set G of $n = m + l$ elements and suppose that these quasigroups contain subquasigroups $(Q, \cdot)$ and $(Q, \odot)$ respectively of order m which are themselves orthogonal. Let $(R, \times)$ and $(R, \otimes)$ be orthogonal quasigroups of order l defined on the set $R = G - Q$ and let $(H, +)$ and $(H, \oplus)$ be orthogonal idempotent quasigroups of order k. Then, if $T = Q \cup (R \times H)$, the singular direct product quasigroups $(T, \circ)$ and $(T, \bullet)$ (where $\bullet$ stands for the operation $\circ$ circled) are themselves orthogonal.*

The proofs of both the theorems just stated involve consideration of many cases and Sade does not give detailed proofs. From them, we may deduce the following one.

THEOREM 11.2.4 *If r mutually orthogonal latin squares of order $n = m + l$, each containing a latin subsquare of order m such that the r latin subsquares are also mutually orthogonal, are given and if there exist s mutually orthogonal* *latin squares of order l and t mutually orthogonal latin squares of order k, then there exist at least $u = \min(r, s, t-1)$ mutually orthogonal latin squares of order $m + lk$.*

PROOF.   Each of the r latin squares may be used to define the multiplication table of a quasigroup $(G, \cdot)$ of order n which contains a subquasigroup $(Q, \cdot)$ of order m, and each of the s latin squares of order l may be used to specify the multiplication table of a quasigroup $(R, \otimes)$ defined on the set $R = G - Q$. Also, by Theorem 5.4.4, the t mutually orthogonal latin squares of order k give rise to $t - 1$ idempotent quasigroups of order k, all defined on the same subset H. Thence, using Theorem 11.2.2 and Theorem 11.2.3, we can construct u mutually orthogonal quasigroups (and latin squares) of order $m + lk$ on the set $T = Q \cup (R \times H)$. $\square$

By putting $m = 1$ in the above theorem, we may deduce that
$$N(1 + lk) \geq \min[N(1 + l), N(l), N(k) - 1].$$
It follows that
$$N(22) = N(1 + 3.7) \geq \min[N(4), N(3), N(7) - 1] = \min(3, 2, 5) = 2,$$
so there exist at least two orthogonal latin squares of order 22 which is a counter-example to Euler's conjecture. (See Section 5.1.)

Lindner has made use of the singular direct product to obtain a number of interesting results, and he has also generalized the construction in several ways. In his earliest paper on the subject, Lindner(1971d), he has proved that the singular direct product preserves the Stein identity $x(xy) = yx$ [identity (11) of Section 2.1]. As was pointed out on page 184, quasigroups which satisfy the Stein identity (called *Stein quasigroups*) are orthogonal to their own transposes (that is, they are *anti-abelian*) and, using this fact, Lindner has constructed several new classes of quasigroups having this property. In Lindner(1971g,1972a,b), he has investigated the general question of what kinds of quasigroup identity are preserved by the singular direct product. In two further papers, Lindner(1971e,f), he has generalized the singular direct product in two different ways. For more details, see [DK1] or the original papers. In another paper, Lindner(1971c), he has used a third generalization of the singular direct product which combines the two generalizations just mentioned to construct a large number of latin squares each of which is orthogonal to a given latin square L and no two of which are isomorphic.

Yamamoto's construction, mentioned on page 287, may also be regarded as a generalized direct product.

Versions of the singular direct product have been used for (i) constructing pairs of perpendicular Steiner quasigroups [Lindner and N.S.Mendelsohn(1973)]; (ii) constructing so-called self-orthogonal latin squares (see Section 5.5) [Crampin and Hilton(1975a,b)]; (iii) constructing sets of orthogonal diagonal latin squares (see Section 6.1) [Lindner(1973), Crampin and Hilton(1975b), Hilton(1975b,c), Heinrich and Hilton(1983)]; (iv) constructing skew Room squares [Mullin(1980)].

In more recent years, analogues of the singular direct product have been derived for Steiner triple systems [Ling, Colbourn, Grannell and Griggs(2000), Dinitz and Stinson(2002)] and Steiner quadruple systems [Hartman(1982), Ji and Zhu(2003)]. Also, quite recently Liang(201?) has extended the original singular direct product construction of Sade to make it apply to standardized quasi-orthogonal latin squares.

11.3   The Kézdy-Snevily conjecture

A conjecture (which we shall call the *K-S conjecture*, see below) concerning covering of the elements of a metric space by spheres would, if true, prove both the conjecture that every latin square of odd order has a transversal (attributed to Ryser) and the conjecture that every latin square of order n has a partial transversal of length at least $n - 1$ (attributed to Brualdi).

Let S_n denote the symmetric group formed by the $n!$ permutations of n elements, say $1, 2, \ldots, n$. Regard these as the vertices of a metric space where the distance between any two permutations is their Hamming distance: that is, the number of places in which they differ.

If a subset T of permutations of S_n is given, its *covering radius* $\mathrm{cr}(T)$ is the smallest r such that S_n is covered by the spheres of radius r whose centres are the permutations of T.

Suppose that T is a sharply transitive subset of permutations of S_n. Then $|T| = n$ since exactly one member of T maps a selected integer i to the integer j for each $j \in \{1, 2, \ldots, n\}$. The permutations which define the rows of a latin square L as re-arrangements from natural order form such a subset T and, conversely, every such sharply transitive subset defines a latin square L.

Suppose further that $\rho \in S_n$ has distance $n - 1$ from every member of T. Then, for each $i \in \{1, 2, \ldots, n\}$, there is exactly one member $\tau \in T$ such that $i\tau = i\rho = k$ say, because τ and ρ agree in only one place. Also, because T is sharply transitive, i is different for each $\tau \in T$. Each i defines a column of L and each k defines an entry in that column. These entries form a transversal of L because they are all different and are in different rows and columns of L.

Example. In Figure 11.3.1, let the permutations which define the rows of L be denoted by τ_h for $h = 1, 2, 3, 4$. Each of the permutations ρ and ρ' has distance three from the permutations of T and defines a transversal of L of which the elements are indicated by suffices a and b respectively.

In Cameron and Wanless(2005), the following theorem is proved.

THEOREM 11.3.1 *Let T be a sharply transitive subset of S_n. Then T has covering radius at most $n - 1$ with equality if and only if the corresponding latin square has a transversal.*

PROOF.   For any position i, any permutation of S_n must agree at i with some member of T, so the covering radius cannot exceed $n-1$. If equality holds, let ρ be

$$L = \begin{vmatrix} 1_a & 2 & 3 & 4_b \\ 2 & 1_b & 4 & 3_a \\ 3_b & 4_a & 1 & 2 \\ 4 & 3 & 2_{ab} & 1 \end{vmatrix}$$

$$\rho = \begin{pmatrix} 1\,2\,3\,4 \\ 1\,4\,2\,3 \end{pmatrix} \qquad \rho' = \begin{pmatrix} 1\,2\,3\,4 \\ 3\,1\,2\,4 \end{pmatrix}$$

FIG. 11.3.1.

a permutation at distnce $n-1$ from every permutation of T.[1] Such a permutation defines a transversal of L as explained above. Conversely, a transversal of L gives rise to a permutation at distance $n-1$ from the permutations of T. ☐

Corollary. Let T be a sharply transitive subset of S_n such that the corresponding latin square has no transversal. Then every permutation of S_n has distance less than $n-1$ from some member of T and so $\mathrm{cr}(T) < n-1$.

Let $f(n,s)$ denote the smallest cardinal for which there is a subset T of permutations of S_n such that the spheres with centres at the members of T and radius $n-s$ cover S_n. For example, $f(n,n) = n!$ since each sphere of zero radius covers only the permutation which is its centre. Because each two permutations of S_n differ in at least two places, each sphere of radius one likewise covers only the permutation which is its centre, so we also have $f(n, n-1) = n!$. Since any permutation of S_n differs in at most n places from a given permutation ρ, a single sphere of radius n and centre at ρ covers S_n: that is, $f(n,0) = 1$.

Cameron and Wanless(2005) have shown that $f(n,1) = \lfloor n/2 \rfloor + 1$ though they attribute this result to A.E. Kézdy and H.S. Snevily. The value of $f(n,2)$ is not known but the authors just mentioned have conjectured that $f(n,2) > n$ if n is odd and $f(n,2) = n$ if n is even. For more information about this function and the Kézdy-Snevily conjecture, see Cameron and Wanless(2005) and Wanless and Zhang(2013).

Suppose that the conjecture $f(n,2) > n$ when n is odd turns out to be true. Then, for every set of n spheres of radius $n-2$ in the metric space S_n, there is at least one permutation ρ which is uncovered. So ρ has distance $n-1$ or greater from the centres of all the spheres. In the case when the centres of the spheres are the permutations which define the rows of a latin square L, they form a sharply transitive set T and so no permutation has distance more than $n-1$ from any one of them. So there exists a permutation ρ whose distance is exactly $n-1$ from each of the n permutations of L. Hence, ρ defines a transversal of L. Thus, validity of the K-S conjecture when n is odd would prove Ryser's conjecture.

Validity of the same conjecture would also prove Brualdi's conjecture.

---

[1] If none exists, every permutation of S_n has distance less than $n-1$ from some member of T and so the covering radius can be reduced.

PROOF. Suppose on the contrary that that there is an $n \times n$ latin square L which has no partial transversal of $n - 1$ cells and suppose that its rows, when regarded as a sharply transitive set of permutations, form a subset T of S_n as before. Then, any permutation of S_n intersects at most $n - 2$ of the rows of L in exactly one cell. Such a permutation intersects one of the remaining rows thrice and the other not at all or else intersects both of the remaining rows twice.

To see this, consider the following two examples.

In the first example (Figure 11.3.2), the latin square shown has a partial transversal (indicated by the entries with suffix a) of length five which cannot be extended. The missing entry u in the permutation ρ is necessarily 1 and so ρ intersects the second row of L twice ($2 \to 3$, $6 \to 1$) and the last row not at all.

$$L = \begin{vmatrix} 1 & 2 & 3 & 4_a & 5 & 6 \\ 2 & 3_a & 4 & 5 & 6 & 1 \\ 3 & 4 & 5_a & 6 & 1 & 2 \\ 4 & 5 & 6 & 1 & 2_a & 3 \\ 5 & 6 & 1 & 2 & 3 & 4 \\ 6_a & 1 & 2 & 3 & 4 & 5 \end{vmatrix}$$

$$\rho = \begin{pmatrix} 1\,2\,3\,4\,5\,6 \\ 6\,3\,5\,4\,2\,u \end{pmatrix}$$

FIG. 11.3.2.

In the second example (Figure 11.3.3), the latin square shown has a partial transversal (indicated by the entries with suffix b) of length four which cannot be extended. The missing entries u and v in the permutation ρ are necessarily 1 and 2 and so ρ intersects the third and sixth rows of L twice if $u = 1$ and $v = 2$ ($2 \to 1$, $3 \to 2$) or the first and second rows twice if $u = 2$ and $v = 1$ ($2 \to 2$, $3 \to 1$) and the remaining rows not at all. Clearly, there may also be examples in which ρ intersects one row thrice and another not at all.

$$L = \begin{vmatrix} 1 & 2 & 3 & 4_b & 5 & 6 \\ 2 & 3 & 1 & 5 & 6_b & 4 \\ 6 & 4 & 2 & 3 & 1 & 5_b \\ 4 & 5 & 6 & 1 & 2 & 3 \\ 5 & 6 & 4 & 2 & 3 & 1 \\ 3_b & 1 & 5 & 6 & 4 & 2 \end{vmatrix}$$

$$\rho = \begin{pmatrix} 1\,2\,3\,4\,5\,6 \\ 3\,u\,v\,4\,6\,5 \end{pmatrix}$$

FIG. 11.3.3.

We return to our proof.

Let us now append an additional symbol, say the symbol $n + 1$, to each of the rows of L to give an $n \times (n + 1)$ rectangle L' whose rows may be regarded as forming a set T' of cardinal n of permutations of S_{n+1}. We shall show that every permutation $\sigma' \in S_{n+1}$ intersects at least one member of T' twice (or more) and so has distance at most $(n + 1) - 2 = n - 1$ from some member of T'.

Suppose that the conjecture $f(n + 1, 2) \geq n + 1$ (for n either odd or even) is true. Then there is no subset T^* of permutations of S_{n+1} of cardinal n such that the spheres with centres at the permutations of T^* and radius $(n + 1) - 2$ cover S_{n+1}. But this contradicts the fact (about to be proved) that every permutation $\sigma' \in S_{n+1}$ has distance at most $n - 1$ from some member of T'. Consequently, no latin square of order n has a partial transversal which cannot be extended to one of $n - 1$ cells.

If $(n+1)\sigma' = n+1$, then certainly σ' intersects some member of T' twice since every permutation of S_n intersects some row of L at least once (in view of the fact that the rows of L form a sharply transitive subset of S_n). Suppose therefore that $(n + 1)\sigma' = s < n + 1$ and that $t\sigma' = n + 1$. Let σ be the permutation of S_n such that $w\sigma = w\sigma'$ for all $w \neq t$ and $t\sigma = s$.

If σ intersects some row of L in three or more places, then σ' intersects some row of L' in two or more places as we wished to prove. If not, then σ intersects each of two rows of L (say the rows r_1 and r_2 represented by the permutations ρ_1 and ρ_2) in two places. But $t\rho_i = s$ for at most one of $i = 1, 2$, say $i = 2$. Then σ' intersects the row r_1 of L' in two places, which completes the proof.

Remark concerning the first example for the KS-conjecture.

Since any permutation which intersects a particular row just once defines a partial transversal which includes the entry intersected, any permutation of S_6 other than ρ must intersect at most five rows just once. In that case, It would also intersect the remaining row just once and L would have a transversal. We conclude that no permutation can inersect more than four $(= n - 2)$ rows exactly once.

11.4   The chromatic number of a latin square

In Bose(1963), that author introduced the concept of the *latin square graph* $\Gamma(L)$ of a latin square L. (See page 312 of [DK1].) The vertices of this graph are the ordered triples (i, j, l_{ij}) which define L (see Section 1.4). Two vertices are adjacent if their triples have one component in common. Thus, an independent set of vertices represents a set of triples which differ in all three components and so define a partial transversal of L. The *chromatic number* $\chi(L)$ of L is defined to be the chromatic number of $\Gamma(L)$ and so is the minimal number of partial transversals which cover the cells of L.[2]

---

[2]Cavenagh and Kuhl(2015) had earlier represented the triples of L as the edges of a hypergraph and called this number the chromatic *index* of L.

It follows immediately that, since each partial transversal of a latin square L of order n uses at most n cells, $\chi(L) \geq n$ for every such latin square and, if L has an orthogonal mate, then $\chi(L) = n$. It also follows that $\chi(L)$ is an invariant of the main class of L.

Suppose that $\chi(L) \leq n + 1$. Since $n + 1$ partial transversals each of length $n - 1$ cannot cover all the cells of L because $(n + 1)(n - 1) < n^2$, that would imply that L must have at least one partial transversal of length n. Thus, if we could show that $\chi(L) \leq n + 1$ for all latin squares of odd orders n, that would imply the truth of Ryser's conjecture that every latin square of odd order has a transversal (see Section 1.5).

Suppose that $\chi(L) \leq n + 2$. Since $n + 2$ partial transversals each of length $n - 2$ cannot cover all the cells of L because $(n + 2)(n - 2) < n^2$, that would imply that L must have at least one partial transversal of length $n - 1$. So, if we could show that $\chi(L) \leq n + 2$ for all latin squares of order n, that would imply the truth of the Brualdi-Stein conjecture that every latin square has a partial transversal of length $n - 1$ (again see Section 1.5).

Further, it follows that, for a latin square (of order n) with no transversals (of which there are many of all even orders), $\chi(L) \geq n + 2$ because $n + 1$ partial transversals each of length $n - 1$ or less cannot cover all the cells of L.

A bachelor latin square L_b (that is, one without an orthogonal mate, see Section 9.1) has at most $n - 1$ full-length transversals but $n - 1$ full-length transversals and one partial transversal cannot cover all the cells of L_b, so $\chi(L_b) \geq (n - 1) + 2 = n + 1$. Wanless and Webb(2006) have shown that such squares exist of all orders $n \geq 3$.

Consequently, there exist latin squares of all even orders for which $\chi(L) \geq n + 2$ and latin squares of all orders $n \geq 3$ for which $\chi(L) \geq n + 1$ so the conjectures of Brualdi-Stein and Ryser cannot be bettered.

Up to the present time, this alternative way of expressing these conjectures has not provided additional insight which might lead to their resolution.

For a latin square L of order n, the triple (R, C, S) is adjacent to $n - 1$ other triples (r, c, s) for which $r = R$ (that is, triples in row R), adjacent to $n - 1$ others for which $c = C$ (triples from column C) and adjacent to $n - 1$ for which $s = S$ (triples representing cells containing symbol S), so $\Gamma(L)$ is regular of valency (or degree) $3n - 3$. It follows that $\chi(L) \leq 3n - 2$.

The above observations and upper bound for $\chi(L)$ are effectively given both in Cavenagh and Kuhl(2015) and in Besharati, Goddyn, Marmoodian and Morteza-eefar(2016). The first authors have also pointed out that, if we make use of Brook's theorem[3], we can improve the above bound to $\chi(L) \leq 3n - 3$.

If the latin square L (of order n) is row complete (see Section 2.6), the bounds $\chi(L) \leq 3n - 2$ or $3n - 3$ can be reduced. Let h, k be an adjacent pair of entries

---

[3]This theorem states that, for all connected simple graphs G except an odd length cycle or a complete graph, $\chi(G)$ is equal to the maximum of the valencies of the vertices of G.

in row r_1 of L so that (r_1, c_1, h) and $(r_1, c_1 + 1, k)$ are both triples of L. We give the vertex of $\Gamma(L)$ which is labelled by (r_1, c_1, h) colour k.

Now suppose that the vertices (r_1, c_1, h_1) and (r_2, c_2, h_2) are both given colour k so that $(r_1, c_1 + 1, k)$ and $(r_2, c_2 + 1, k)$ are both triples. Since symbol k cannot occur twice in any row or column, $r_2 \neq r_1$ and $c_2 \neq c_1$. Also, since the ordered pair of symbols h_1, k can only occur once among the rows of L, $h_2 \neq h_1$. Thus, the $n - 1$ vertices of $\Gamma(L)$ which are coloured k by this prescription form a partial transversal of L. We require n further colours for the triples (vertices) which represent cells of the last column of L. So, for a row complete latin square of order n, $\chi(L) \leq n + n = 2n$.

The above argument was first given in Besharati et al (2016). The authors of that paper pointed out further that the upper bound $\chi(L) \leq 3n - 2$ could also be reduced if L has a partition into plexes. (See Section 3.5.)

A k-plex is a set of kn cells of L of which k are in each row, k are in each column and k contain each symbol. L is said to have a $(k_1, k_2, \ldots, k_h)$-partition if it is the union of a k_1-plex, a k_2-plex, $\ldots$ and a k_h-plex. The subgraph of $\Gamma(L)$ whose vertices are those of a k_i-plex is a regular subgraph of valency $3k_i - 3$ (by the same argument as we used to show that $\Gamma(L)$ itself is regular of valency at most $3n - 3$ so we can colour the vertices of this subgraph with at most $3k_i - 2$ colours. (Note that Brook's theorem is not applicable if any $k_i = 1$ since then the corresponding subgraph consists of isolated vertices. Nor is it applicable if any $k_i = 2$ and L has odd order.) Since, by hypothesis, the vertex set of $\Gamma(L)$ is the union of the vertex sets of these subgraphs if L has a $(k_1, k_2, \ldots, k_h)$-partition, we get the upper bound $\chi(L) \leq \sum_{i=1}^{h}(3k_i - 2) = 3n - 2h$ for such a latin square.

Th authors of both the papers previously mentioned went on to discuss upper bounds for the chromatic number $\chi(Z_n)$ of the Cayley table of a cyclic group Z_n. Cavenagh and Kuhl used constructive arguments by which they actually showed how to cover the Cayley table with partial transversals. By this means, they proved that $\chi(Z_{2m}) = 2m + 2$ except when m is odd and divisible by 3. For the latter case, they were only able to establish the bound $\chi(Z_{2m}) \leq 2m + 3$. [Since every group of odd order has an orthogonal mate, the result $\chi(Z_{2m-1}) = 2m - 1$ was already known.]

Besharati et al approached the problem in a different constructive way which they described graphically and which was not dependent on whether m is divisible by 2 and/or 3. They were able to improve the above result to show that $\chi(Z_{2m}) = 2m + 2$ in all cases.

The latter authors also calculated (with the aid of a computer) the values of $\chi(L)$ for all main classes of latin squares up to and including order 8. They found that $\chi(L) \leq n + 2$ for all latin squares of these orders and that $\chi(L) \leq n + 1$ for those of odd orders.

Finally, both sets of authors obtained an asymptotic bound for the chromatic number of any latin square L of order n to the effect that, when $n \to \infty, \chi(L) \leq n + o(n)$.

11.5   Practical applications of latin squares

(a) *Latin squares and coding*

Orthogonal latin squares have been used in coding theory for a number of different purposes: among others for designing error-detecting and correcting codes and in authentication. Early work on this topic, in particular that of Hamming(1950) and Golomb and Posner(1964), was described in [DK1]. Also Chapter 9 of [DK2] gave an account of developments in which latin squares play a role up to 1990. The many aspects of both coding theory and cryptography have now become the subject matter of several books so it seemed inappropriate to attempt a further update here. Instead, the reader is invited to look at Chapter 3 of Joyner and Kim(2011) for some unsolved problems.

(b) *Latin squares as experimental designs*

Since the 1930s, when the idea of doing so was pioneered by R.A. Fisher, latin squares and other related designs have been much used in the design of statistical experiments whose subsequent interpretation is to be effected with the aid of the procedure known as the "analysis of variance".[4] We have already referred to such statistical applications in several earlier chapters of the present book: notably in Section 2.6, Section 5.6 and Section 6.4, where the uses of complete latin squares, of latin cubes and hypercubes, and of Room designs for this purpose were mentioned. By means of an example, we shall explain next the use of orthogonal latin squares for the same purpose.

Suppose that it is required to test the effect of five different treatments on yarn. Five looms are available for spinning the yarn and it is believed that both the loom used and the operator employed may affect the final texture of the spun yarn. If time were of no account, it would be desirable ideally to test the effect of every treatment with every operator and on every loom. However, by making use of the statistical technique of analysis of variance, a satisfactory test result can be obtained provided that each type of treated yarn is spun once on each loom and also once by each operator. If five operators are employed, the necessary set of twenty-five experiments can be specified by means of the 5×5 latin square illustrated in the left-hand diagram of Figure 11.5.1. Here, the columns $0_1, 0_2, 0_3, 0_4, 0_5$ specify the operator and the rows L_1, L_2, L_3, L_4, L_5 specify which loom is to be used for the particular experiment. The entries Y_0, Y_1, Y_2, Y_3, Y_4 in the body of the square indicate which type of treated yarn is to be used. (Usually, one sample of yarn, called the control, is left untreated and the suffix 0 is used for this type of treatment.) Thus, in the first experiment, yarn Y_1 is to be spun by operator 0_1, on loom L_1. In another experiment, yarn Y_3 is to be spun by operator 0_2 on loom L_4. If it were also the case that the efficiency of the loom operators varied with the day of the week (five weekdays), this extra source of variation could be allowed for by using a 5×5 latin square which was orthogonal to the first to specify the day of the week on which each experiment

---

[4] A few much earlier examples of this use exist but systematic use dates from the publication of the first edition of Fisher(1966).

was to be performed. Thus, in the right-hand square illustrated in Figure 11.5.1, the integers 0, 1, 2, 3, 4 are used to specify the five days of a week. Then each loom, each operator, and each type of treated yarn are associated with each day of the week just once.

	O_1 O_2 O_3 O_4 O_5		1 2 4 3 0
L_1	Y_1 Y_4 Y_0 Y_2 Y_3		3 4 1 0 2
L_2	Y_3 Y_1 Y_2 Y_4 Y_0		2 3 0 4 1
L_3	Y_2 Y_0 Y_1 Y_3 Y_4		0 1 3 2 4
L_4	Y_0 Y_3 Y_4 Y_1 Y_2		4 0 2 1 3
L_5	Y_4 Y_2 Y_3 Y_0 Y_1		

FIG. 11.5.1.

The use of a latin square design for an experiment of the above kind is somewhat restrictive. It might happen, for example, that only four operators were available. This situation could be accommodated by using the same latin square with its last column deleted. It would still be the case that each type of yarn would be spun by each of the four remaining operators but now only four of the five different yarns would be spun on any one of the looms. A development of this observation gives rise to the concept of a *balanced incomplete block design*.

DEFINITION. A *balanced incomplete block design* comprises a set of v *varieties* (or *treatments*) arranged in b *blocks* (or *rows*) in such a way that (i) each block has the same number k of treatments $(k < v)$, no treatment occurring twice in the same block; (ii) each treatment occurs in exactly r $(= \lambda_1)$ blocks; (iii) each pair of treatments occur together in exactly $\lambda(= \lambda_2)$ blocks.

Balanced incomplete block designs were first introduced by F. Yates in Yates (1936). An example of such a design is given by the latin square used above when the last column is deleted. This has parameters $b = 5, v = 5, k = 4, r = 4, \lambda = 3$.

A design for which property (iii) is satisfied is called *pairwise balanced*. More generally, if each set of t treatments occurs in exactly λ_t blocks, the design is called *t-wise balanced* or, more briefly, a *t-design*. A very well-known result is the following:

THEOREM 11.5.1 *The parameters b, v, r, k, λ of a (pairwise) balanced incomplete block design (usually written briefly as a BIBD) satisfy the relations $bk = vr$, $\lambda(v - 1) = r(k - 1)$, $r > \lambda$, $b \geq v$.*

PROOF. Each side of the first relation is an expression for the total number of treatments to be found in all the blocks. Each side of the second relation is an expression for the total number of treatments (other than v_i) which occur in the various blocks which contain an assigned treatment v_i. Next, if $r > \lambda$ were untrue, then with any given treatment every other treatment would occur in every block, contradicting $k < v$. Finally, to show that $b \geq v$ it is easiest to make use of the *incidence matrix* of the design. Let a *BIBD* with parameters

b, v, r, k, λ be given and let $n_{ij} = 1$ or 0 according as the ith treatment v_i does or does not occur in the jth block. The $v \times b$ matrix so obtained is the incidence matrix of the design. Clearly,

$$\sum_{j=1}^{b} n_{ij}^2 = r \text{ and } \sum_{j=1}^{b} n_{ij}n_{hj} = \lambda \ (h \neq i)$$

by definition of r and λ. If possible, let $b < v$ and consider the $v \times v$ matrix

$$N = \begin{vmatrix} n_{11} & n_{12} & \ldots & n_{1b} & 0 & 0 & \ldots & 0 \\ n_{21} & n_{22} & \ldots & n_{2b} & 0 & 0 & \ldots & 0 \\ \cdot & \cdot & \ldots & \cdot & & \ldots & \cdot \\ \cdot & \cdot & \ldots & \cdot & & \ldots & \cdot \\ n_{v1} & n_{v2} & \ldots & n_{vb} & 0 & 0 & \ldots & 0 \end{vmatrix}$$

where the last $v - b$ columns consist of zeros. It follows from the relations above that

$$NN^T = \begin{vmatrix} r & \lambda & \lambda & \ldots & \lambda \\ \lambda & r & \lambda & \ldots & \lambda \\ \cdot & \cdot & \cdot & \ldots & \cdot \\ \cdot & \cdot & \cdot & \ldots & \cdot \\ \lambda & \lambda & \lambda & \ldots & r \end{vmatrix}$$

and so

$$\det(NN^T) = [r + \lambda(v - 1)] \begin{vmatrix} 1 & 1 & 1 & \ldots & 1 \\ \lambda & r & \lambda & \ldots & \lambda \\ \cdot & \cdot & \cdot & \ldots & \cdot \\ \cdot & \cdot & \cdot & \ldots & \cdot \\ \lambda & \lambda & \lambda & \ldots & r \end{vmatrix}$$

by addition of rows. Then, subtracting the first column from each other one, we get

$$\det(NN^T) = [r + \lambda(v - 1)](r - \lambda)^{v-1} = kr(r - \lambda)^{v-1}$$

since $\lambda(v - 1) = r(k - 1)$. But $\det(NN^T) = (\det N)(\det N^T) = 0$ which implies that $r = \lambda$. This contradiction shows that the supposition $k < v$ is impossible. Therefore, $b \geq v$ as required.   □

A *BIBD* for which $b = v$ and consequently $r = k$ is said to be *symmetric* and such a design may also be referred to as a (v, k, λ)-*design*.

Two obvious problems concerning *BIBD*'s are to provide methods of constructing them and to determine for what values of the parameters such designs exist. The second problem is not completely solved. As regards the first, many methods have been devised. Here, we mention that a symmetric *BIBD* is a finite projective plane or a complete set of *MOLS*. (See Section 5.2.) Also, a *BIBD* with $k = 3$ and $\lambda = 1$ is a *Steiner triple system*. (See Section 2.3.)

Chapter 10 of [DK2] is devoted to a quite detailed explanation of the appropriateness of the use of latin squares and other block designs in the design of experiments of various kinds and to an account of the analysis of the results, so we shall not repeat that account here.

An earlier account of the same topic is that of Hedayat and Shrikhande(1971).

(c) *Designing games tounaments with the aid of latin squares*

In Chapter 6, we discussed Room squares and their use for scheduling Bridge tournaments and we mentioned the related concept of a referee square first used for scheduling Rugby, which we shall now elaborate.

Such squares were first discussed by I. Anderson, Hamilton and Hilton in an unpublished manuscript.

Suppose that $n = 2m + 1 \geq 3$ teams wish to compete in a league, each team playing every other team exactly once. The $m(2m + 1)$ games required are to be scheduled for $2m + 1$ days with m games on each day. Each team will have just one free day if it plays only once a day. It will be convenient if the schedule is such that the ith team has the ith day as its free day. Suppose further that each team selects/nominates a referee from among its own members or elsewhere.

If we number the days, the teams and the referees by $1, 2, \ldots, 2m+1$ (so that referee i is the one selected by team i), is it possible to arrange the fixtures in such a way that (i) each referee referees each other team exactly once; (ii) no referee referees his own team; (iii) no referee has to referee two games on the same day; and (iv) the referee's team plays on days when he is free to watch? Conditions (i) and (ii) imply that each referee referees m games.

In papers by Liaw(1998) and Dinitz and Ling(2001), it is proved that the answer is YES for all $m > 0$ except $m = 2$ and the solution given by these authors is such that referee i is always in action on his team's rest day: that is, on day i. A solution of the latter kind gives rise to the concept of a referee square.

DEFINITION. A *referee square* of order $2m + 1$ is a $(2m + 1) \times (2m + 1)$ array R using symbols from the set $S = \{1, 2, \ldots, 2m + 1\}$ such that
(1) each cell is either empty or contains an unordered pair of distinct symbols from S;
(2) each $i \in S$ occurs precisely once in each row except the ith and precisely once in each column except the ith and does not occur in the ith row or the ith column;
(3) each unordered pair of distinct elements from S occurs in exactly one cell of R; and
(4) the main diagonal cells are non-empty.

Such a square provides a solution to our scheduling problem when teams a and b play with referee i on day j if and only if $\{a, b\}$ occurs in cell (i, j). An example for $m = 3$ taken from Liaw(1998) is given in Figure 11.5.2.

Liaw's main result is that such a square exists whenever $2m + 1$ is composite. To obtain that result, he proved among other things that (i) if a Room square of side n exists, then referee squares of orders $3n$ and $5n$ exist; and (ii) if a Room square of side s, a referee square of side t and two MOLS of order t exist, then

6, 7	4, 5		2, 3			
3, 5	7, 1			4, 6		
		5, 6		2, 7	1, 4	
	3, 6	1, 2	5, 7			
		4, 7		1, 3		2, 6
2, 4					3, 7	1, 5
			1, 6		2, 5	3, 4

FIG. 11.5.2.

a referee square of side st can be constructed. Dinitz and Ling completed the proof of existence for all odd orders except five by using Liaw's results and also several other combinatorial structures not discussed in the present book.

It is surprising how many games tournament requirments can be met with the aid of constructions using latin squares. In a survey paper of the present author, Keedwell(2000), solutions are provided for Round robin tournaments, Tennis on unequal courts, Home and away football, Balance in carry-over effects from one game in a tournament to the next, Mixed doubles tournaments, Spouse-avoiding mixed doubles, and Duplicate bridge and Whist. To save space, we shall not repeat the contents of that paper here but we shall draw attention to several papers not mentioned in it.

The first of these, Bedford, Ollis and Whittaker(2001), concerns balancing carry-over effects in a bipartite tournament between two teams of equal size where each player has to compete against each of the members of the opposing team and the solution makes use of a certain kind of directed terrace. (See Section 2.6 for the definition of the latter.)

The second, due to Preece and Phillips(2002), concerns the problem of scheduling bowls tournaments. A bowls team consists of $4n$ players, $n \geq 4$, n playing lead position, n playing second position, n playing third position and n playing skip position. The problem is to assign the players to n rinks in n successive games in such a way that each rink contains one player for each position and no two players appear together more than once in any given rink. Solving this problem is equivalent to finding three mutually orthogonal Latin squares of order n. Adding the restriction that each player appears exactly once in each rink yields an enhanced problem that is equivalent to finding four mutually orthogonal Latin squares of order n.

The third paper, Berman and Smith(2012), discusses what the authors call a *Mitchell tournament* which is a variant of the famous type of tournament known as a spouse-avoiding mixed doubles round robin tennis tournament. The conditions for a Mitchell tournament differ from those for the type just mentioned only in the fact that a tournament of the latter type is not spouse-avoiding. For the construction, the authors use one square A which is column-latin and a second square B which is row-latin and such that A and B are orthogonal.

Finally, it is worth mentioning that scheduling a Round robin tournament for $2m$ teams or players is equivalent to finding a 1-factorization(see Section 8.3 for the definition) of the complete graph K_{2m} as has been pointed out in Gelling and Odeh(1974). The teams are the vertices of K_{2m}, the games are the edges, the rounds are the 1-factors.

The topic of devising games tournaments is still an active one so the above situations are not the only ones. In particular, I.Anderson(1997) has written a book on the subject and Berman has written several relevant papers on tournaments in addition to that mentioned above. [See, for example, Berman and Smith(2013) and D.R.Berman's computer home page.]

11.6   Latin triangles

There have been several attempts to define a latin triangle. The most natural and appropriate of these in the opinion of the present author is that of Halberstam, Hoffman and Richter(1986) who gave a definition equivalent to the following:

A *triangle* of order n is an array in the shape of an equilateral triangle having n rows in each of the directions parallel to a side of the triangle with row i $(1 \leq i \leq n)$ having $n + 1 - i$ entries. The horizontal rows are called a-rows. The rows in the \-direction are called b-rows and the rows in the /-direction are called c-rows.

For the definition of a *latin triangle* of order n, or $LT(n)$ for brevity, we need to distinguish between the cases n odd and n even. Let $x \in \{a, b, c\}$. For odd positive integers n, the *line* given by $x = i$ for $2 \leq i \leq (n + 1)/2$ is the union of the x-rows $x = i$ and $x = n + 2 - i$. For even positive integers n, the line given by $x = i$ for $1 \leq i \leq n/2$ is the union of the x-rows $x = i$ and $x = n + 1 - i$.

A triangle is *latin* if and only if each of its a-lines, b-lines and c-lines contains each of a set S of distinct symbols exactly once. For an $LT(2m)$ and for an $LT(2m + 1)$, we need $2m + 1$ symbols.

For instance, every $LT(5)$ is equivalent to that shown in the left-hand array of Figure 11.6.1 and, in the right-hand array of Figure 11.6.1, we give an example of an $LT(8)$.

Halberstam *et al* have constructed examples of latin triangles of all orders $n \leq 11$ except $n = 4, 6, 10$ and have shown non-existence for the latter orders. They have also given a general construction valid for all orders $n = 3^m$ and $n = 3^m - 1$.

In Halberstam and Richter(1989), examples of an $LT(12)$, $LT(13)$ and $LT(15)$ have been obtained.

In a later paper by Alves, Gurganus, McLaurin and Smith(1991), these authors have obtained an $LT(14)$ and have provided a construction of latin triangles of all odd orders n such that $n + 2$ is prime. They have thus obtained an $LT(17)$ and an $LT(21)$. They have also discussed how to define orthogonality of latin

```
                                        9
                                      4   5
              w                     5   9   4
          v       x               6   4   5   3
       y     w     u            7   8   9   1   2
     x    u    y     v        2   3   1   8   6   7
   u    v    w    x     y    3   1   2   9   7   8   6
                            1   2   3   4   5   6   7   8
```

FIG. 11.6.1.

triangles. Halberstam *et al* had earlier proposed the definition that two latin triangles of the same odd order are orthogonal if, when they are superimposed, each unordered pair of symbols appears just once. (For even orders, their requirement is that each unordered pair of *distinct* symbols appears just once.) However, this property is not preserved if the symbols in one of the squares are permuted and it distinguishes between odd and even orders n, so Alves *et al* have proposed instead:

DEFINITION. Two latin triangles are *orthogonal* if, when they are superimposed, no ordered pair of symbols occurs more than once.

This definition includes the former one, does not distinguish beween odd and even orders and is unaffected by permutations of symbols. Alves *et al* have constructed a set of three latin triangles of order 7 which are mutually orthogonal with respect to the latter definition.

Many questions about this topic remain to be answered:

Do latin triangles exist of all odd orders and of all even orders except 4, 6 and 10? Alternatively, is there an integer n_0 such that, for all $n > n_0$, an $LT(n)$ exists?

For a given order n, how many different $LT(n)$'s exist? In this context, how should we define "different"?

What is the maximum number of mutually orthogonal $LT(n)$'s that can exist?

11.7   Latin squares and computers

When the first edition of this book was published, digital computers had been available for academic use to a limited extent for about 15 years and the idea of applying them in Algebra and Combinatorics was still quite novel. A first conference to discuss such applications had been held in Oxford in 1967 under the auspices of the Atlas Computing Laboratory, see Leech(1970). So the final chapter of [DK1] was devoted to describing the use of such computers as a new aid to the solution of latin square problems. In particular, techniques such as the back-track method for enumeration of latin squares and for the construction of sequencings of groups were described[5]. Also, the use of computers by Donald Knuth (later, the inventor of T_EX) to enumerate all isomorphically distinct division rings of 32 elements and by one of the present authors to enumerate all D-neofields up to order 17, and a sample of order 18, was reported. To provide present-day readers with the flavour of that Chapter and because the problem raised remains unsolved, we reproduce the following short extract.

"Computers have been used as an aid in essentially two kinds of orthogonal latin square problems. In the first place, they have been used in searching for sets of mutually orthogonal latin squares of an assigned order n which shall contain as many squares as possible. In the second place, they have been used in attempts to construct new non-desargusian finite projective planes and hence complete sets of mutually orthogonal latin squares which would be non-equivalent to any previously known set."

"The earliest search of the first kind known to the authors was that of Paige and Tomkins(1960).[6] These authors attempted to find a counter-example to the Euler conjecture by finding a pair of orthogonal latin squares of order 10, ten being the smallest integer greater than two and six which is an odd multiple of 2."

"The first method they contemplated was to take a randomly generated 10×10 latin square and to use their SWAC computer to search directly for an orthogonal mate where, without loss of generality, the first rows of the two squares could be assumed to be the same. They estimated from trial runs that such a complete search would take of the order of 4.8×10^{11} machine hours for each initially selected square and they pointed out the difficulties of devising a method which would reject equivalent pairs of orthogonal squares and hence eliminate duplication of trials. They called two pairs *equivalent* if by rearranging rows, rearranging columns, or renaming the symbols of both squares of one pair simultaneously it would become either formally identical to the second pair or would differ from the second pair only in having its two squares oppositely ordered. (In the latter case, if L_1 and L_2 were the squares of one orthogonal pair, L_2 and L_1 would be those of the other.)"

"They then discussed a method by which only pairs of squares having a combinatorial pattern of such a kind that orthogonality would be possible would be constructed. This method involved identifying some of the symbols used in each latin square and is said to have been originally proposed by E. Seiden and W. Munro. Let us suppose that the symbols used in a given 10×10 latin square are the digits 0 to 9, and let each of the digits 0 to 4 be replaced by α and each of the digits 5 to 9 be replaced by β. In the resulting square, the symbols α and β each occur five times in each row and five times in each column. If there exists

---

[5]A search for sequencings had been carried out by the original first author of this book in conjunction with É. Török.

[6]However, see S. Cairns(1954) for a comment on an attempted computer investigation carried out in the early 1950s.

a second latin square orthogonal to the first which is similarly treated and if the two squares are then placed in juxtaposition, we shall get a 10×10 square matrix each element of which is one of the ordered pairs of symbols (or *digraphs*) $\alpha\alpha$, $\alpha\beta$, $\beta\alpha$, or $\beta\beta$, and with the following properties. Each of the four digraphs will occur all together 25 times in the matrix and in each row of the matrix there will be five digraphs whose first member is α and five whose first member is β, likewise five digraphs whose second member is α and five whose second member is β. The same statements will be true for the columns of the matrix. Such a square matrix was called *admissible* by Paige and Tomkins. They obtained further conditions which would have to be satisfied by a matrix constructed from two orthogonal latin squares in this way. The essentials of their computational procedure were first to compute an admissible 10×10 matrix and then to reverse the process just described by replacing the first member of each digraph by one of the digits 0 to 9 so as to get a latin square (α being replaced by one of the digits 0 to 4 and β by one of the digits 5 to 9) and subsequently to use the computer again to assign digits to all the cells of the partially specified orthogonal mate. That is, to enter one of the digits 0 to 4 in each α-cell and one of the digits 5 to 9 in each β-cell until one of the orthogonality conditions was violated. When a violation occurred, the programme backtracked in the usual manner to an earlier cell."

"Unfortunately neither this nor the previous method was successful in producing an orthogonal pair of 10×10 latin squares despite the fact that more than 100 hours of computing time was used and, shortly afterwards, indeed even before the publication of Paige and Tomkins paper, the Euler conjecture was disproved by theoretical means. (See Chapter 11.[7])"

"Almost immediately after the first counter-example to the Euler conjecture was obtained, E.T. Parker found a pair of orthogonal latin squares of order 10 by a construction which made use of statistical designs, and this sequence of events was regarded by many as a triumph for mathematical theory over computer search. However, not long afterwards, E.T. Parker himself initiated a new computer search for latin squares of order 10 which have orthogonal mates using a faster computer and a more sophisticated computer programme. The essential refinement was to generate and store all the transversals of a given 10×10 latin square and to search for disjoint sets of ten transversals (equivalent to finding an orthogonal mate) rather than to attempt to fill the cells of a partial orthogonal mate individually. E.T. Parker has given a full description of his method and results in Parker(1962a,1963). He discovered that 10×10 latin squares with orthogonal mates are not, in fact, particularly scarce and he also showed that there exist such squares with a large number of alternative orthogonal mates. His most striking result concerns the square displayed in Fig. 13.2.1[8] which has 5504 transversals and an estimated one million alternative orthogonal mates (that is, sets of 10 disjoint transversals). However, Parker was able to show by a partly

---

[7]of [DK1].

[8]of [DK1]. This square is also exhibited on page 15 of [DK2].

theoretical argument that no two of these alternative orthogonal mates are themselves orthogonal and so, much to his own disappointment, he was not able to obtain a triad of mutually orthogonal 10×10 latin squares. The existence or non-existence of such triads remains an open question."

It is remarkable that, 40 years later, this is still an open question though recent work, especially that in McKay, Meynert and Myrvold(2007), has made existence of such triads very unlikely.

Comment on the Problems

The original edition of this book ended with a section labelled "Problems". In this revised edition, we list these same problems but after each one we state the present situation so far as we know it. If the problem has been solved, we indicate where the solution may be found or, if not, we give a summary of what (to the best of our knowledge) is currently known about it. The statement "unsolved" means that we are not aware of any solution having been published. In certain of the cited references, a letter such as a, b, c, etc. appears after the publication date. For example, in Problem 1.4, A.D.Keedwell(1983c) is cited. This is to provide consistency with the bibliography of the present book and/or with [DK2]. We hope that this convention will be helpful to readers of both [DK2] and the present book.

Chapter 1

PROBLEM 1.1. ⟨*What is the maximum number of associative triples which a quasigroup may have but still not be a group?* (page 20)⟩

The earliest results relevant to this question were those of D.A. Norton(1960), A. Wagner(1962) and A.K. Austin(1966). Norton proved: (1) If $(Q, \cdot)$ is a tri-associative quasigroup [that is, for every triple of distinct elements x, y, z of Q, $x(yz) = (xy)z$] such that both Q and $Q^2 = \{q^2 : q \in Q\}$ contain at least 17 elements, then $(Q, \cdot)$ is diassociative [that is, for every pair of distinct elements x, y of Q, $x(xy) = (xx)y$]; (2) If $(Q, \cdot)$ is a diassociative quasigroup, then $(Q, .)$ is mono-associative (also called "power associative"). Thus, every triassociative quasigroup with sufficient elements is a group. Wagner proved the same result independently without restriction on the number of elements. More precisely, he proved: (1) If $(Q, \cdot)$ is a finite or infinite triassociative quasigroup, then $(Q, \cdot)$ is a group. (He showed also that the result remains true if the axiom concerning unique solubility of equations is weakened to the requirement that they have at least one solution.); (2) If $(Q, \cdot)$ is a finite quasigroup of order n, it is sufficient to test approximately $3n^3/8$ appropriately chosen associative statements to ensure that Q is a group. Austin proved: (1) A quasigroup which contains an associative element is a loop. [An element a is an associative element of the loop $(L, \cdot)$ if, for all elements $x, y, z \in L$ such that $x(yz) = a$, we have $(xy)z = a$.] (2) If a loop contains an associative element, then its left, middle and right nuclei coincide. (Not all loops contain associative elements.)

Most of the papers concerned with problem 1.1 have expressed it in the form: What is the minimal possible number $\sigma(n)$ of non-associative triples of elements in a non-associative quasigroup of order n?

A. Drápal(1983) has proved that, if n is an even integer, then $\sigma(n) = 16n - 64$ for $n \geq 168$, that $3n^2/32 \leq \sigma(n) \leq 16n - 64$ for $6 \leq n \leq 166$ and that $\sigma(4) = 32$. These results provide a partial solution to the problem for the case when the order of the quasigroup is even. To obtain his partial solution, Drápal first proved the following results:

Let $G_i = (G, \otimes_i)$ be a group and let $Q = (G, \circ)$ be a quasigroup defined on the same set G of order n. Let $\text{dist}(G_i, Q)$ denote the number of pairs $(x, y) \in G \times G$

such that $x \otimes_i y \neq x \circ y$ and let $t = \min \text{dist}(G_i, Q)$ as G_i runs through all groups defined on the set G. Let $s = s(Q)$ denote the number of non- associative triples in the quasigroup $(Q, \circ)$. Then $\sigma(n)$ is the minimum value of s as Q runs through all non-associative quasigroups of order n. Drápal showed that $4tn - 2t^2 - 24t \leq s \leq 4tn$. Also, if $1 \leq s < 3n^2/32$, then $3tn < s$ holds as well.

The following results contribute further to the problem:

In T.Kepka(1981b), the author proved that a non-associative commutative Moufang loop of order n has at most $313n^3/729$ associative triples of elements.

In T.Kepka(1981a), the author proved that the number $a(n)$ of associative triples in a non-associative commutative quasigroup isotopic to a group of order n satisfies the inequalities $n^2 \leq a(n)$ for n odd or $n \equiv 0 \pmod 4$, $n^2 + 2n \leq a(n)$ for $n \equiv 2 \pmod 4$ and $a(n) \leq n^3 - 4n^2 + 6n$ provided that $n \geq 3$ is odd, $a(n) \leq n^3 - 4n^2 + 8n$ provided that n is even. Further, in A. Drápal and T.Kepka(1981) the authors showed that the number $a(n)$ of associative triples in a non-associative and possibly non-commutative quasigroup isotopic to a group of order n again satisfies the inequalities $a(n) \leq n^3 - 4n^2 + 6n$ provided that $n \geq 3$ is odd and $a(n) \leq n^3 - 4n^2 + 8n$ provided that n is even. In A.Kotzig and C.Reischer(1983), these latter authors called $a(n)$ *the associativity index* of the quasigroup and studied its bounds in more detail for various particular types of quasigroup.

In a more recent paper, J. Ježek and A. Drápal(1990) denoted the number of associative triples in any finite quasigroup $(Q, \cdot)$ of order n by $a(n)$ and the number of non-associative triples by $b(n)$ so that $a(n) + (b(n) = n^3$. They denoted by $a_{max}(n)$ the maximum and by $a_{min}(n)$ the minimum of the numbers $a(Q)$ for Q running over all the non-associative quasigroups of order $n \geq 3$ and they defined the numbers $b_{max}(n)$ and $b_{min}(n)$ similarly so that $b_{max}(n) = n^3 - a_{min}(n)$ and $b_{min}(n) = n^3 - a_{max}(n)$. They obtained precise values for these numbers when $n = 3, 4, 5$ and 6 and they also proved the following: $b_{min}(n) \leq 16n - 64$ and so $a_{max}(n) \geq n^3 - 16n + 64$ for n even and ≥ 6. Also, $b_{max}(n) \geq n^3 - n^2$ for $n \geq 3$ and not singly even (that is, $n \neq 4k + 2$); $b_{max}(n) \geq n^3 - 2n^2$ for n singly even and ≥ 6. Using results obtained by Drápal, in particular in A.Drápal(1983), and known properties of the function $\text{gdist}(n)$ which we defined in Section 3.2 they derived a number of further results. For example, let n be even and $6 \leq n \leq 166$, then $3n^2/32 \leq b_{min}(n)$.

Further relevant papers are A. Drápal and T. Kepka(1983, 1985, 1989), W.D. Frazer(1973), C.-M. Fu, H.-L. Fu. and S.-H. Guo(1991) and M. Niemenmaa(1988).

[Austin A.K.
    (1966) A note on loops. Pacific J. Math. 18, 209-212.
Drápal A.
    (1983) On quasigroups rich in associative triples. Discrete Math. 44, 251-255.
Drápal A. and Kepka T.
    (1981) A note on the number of associative triples in quasigroups isotopic to groups. Comment. Math. Univ. Carolinae 22, 735-743.

(1983) Group modifications of some partial groupoids. In Proc. Internat. Conf. on Combinatorial Geometries and their Applications, Ann. Discrete Math. 18, pp. 319-332.

(1985) Group distances of latin squares. Comment. Math, Univ. Carolinae 26, 275-283.

(1989) On a distance of groups and latin squares. Comment. Math. Univ. Carolinae 30, 621-626.

Frazer W.D.

(1973) On testing a binary relation for associativity. In "Combinatorial Algorithms" (Courant Comput. Sci. Sympos., No. 9, 1972), pp. 77-90. Algorithmic Press, New York.

Fu C.-M., Fu H.-L. and Guo S.-H.

(1991) The intersections of commutative latin squares. Ars Combinatoria 32, 77-96.

Ježek J. and Drápal A.

(1990) Notes on the number of associative triples. Acta Universitatis Carolinae 31, 15-19.

Kepka T.

(1981a) Notes on associative triples of elements in commutative groupoids. Acta Univ. Carolinae Math. Phys. 22, 39-47.

(1981b) A note on the number of associative triples in finite commutative Moufang loops. Comment. Math. Univ. Carolinae 22, 745-753.

Kotzig A. and Reischer C.

(1983) Associativity index of finite quasigroups. Glasnik Math. 18, 243-253.

Niemenmaa M.

(1988) Upper bounds on the distance between groups and quasigroups. Europ. J. Combin. 9, 391-393.

Norton D.A.

(1960) A note on associativity. Pacific J. Math. 10, 591-595.

Wagner A.

(1962) On the associative law of groups. Rend. Mat. e Appl. 21, 60-76.]

PROBLEM 1.2. ⟨*Find necessary and sufficient conditions on a loop G in order that every loop isotopic to G be isomorphic to G.* (page 25)⟩

UNSOLVED. By Corollary 2 of Theorem 1.3.4, associativity is sufficient. Also, Fenyves has shown that isotopic extra loops are isomorphic. See page 44 of the present book.

PROBLEM 1.3. ⟨*Do there exist quasigroups of odd order which have no complete mappings?* (page 32)⟩

UNSOLVED. H.J. Ryser(1967) conjectured that the answer is No and gave a proof that this is the case when the quasigroup is commutative. He also noted that all quasigroups of order 5 have complete mappings. Since all quasigroups of order 5 are isotopic either to the cyclic group of order 5 or to one of the five loops listed on page 145 of [DK1], this is easy to check. In fact, A. Sade(1960) had pointed out much earlier than 1967 that the identity mapping is a complete mapping for every commutative quasigroup of odd order. It was pointed out by D.A. Robinso(1966) that the same is true for Bol loops of odd order. See also Section 1.5 of the present book.

More recently, it has been shown (with the aid of a computer) that all latin squares of odd order up to and including order 9 have transversals (and hence complete mappings). See B.D. McKay, J.C. McLeod and I.M. Wanless(2006).

[McKay B.D., McLeod J.C. and Wanless I.M.
   (2006) The number of transversals in a latin square. Designs, Codes Crytogr. 40, 269-284.
Robinson D.A.
   (1966) Bol loops. Trans. Amer. Math. Soc. 123, 341-354.
Ryser H.J.
   (1967) Neuere Probleme der Kombinatorik, Vortrage uber Kombinatorik Ober-wolfach, 24-29 Juli 1967. Mathematisches Forschungsinstitut Oberwolfach.
Sade A.
   (1960) Produit direct-singulier de quasigroupes orthogonaux et anti-abeliens. Ann. Soc. Sci. Bruxelles, Ser,I, 74, 91-99.]

PROBLEM 1.4. ⟨Do there exist any finite groups which are not P-groups? (page 35)⟩

The answer is "NO". For the complete solution, see J. Dénes and P. Hermann(1982). Note that their proof makes use of the Feit-Thompson theorem.

Later, P. Yff obtained a more elementary proof which does not use this theorm. Unfortunately, the latter proof was never published. For more details, see page 40 of [DK2].

A related problem asks Which finite groups are super P-groups? For information on this question, see A.D. Keedwell(1983c,1984a).

Some further problems connected with this topic are listed in J. Dénes and P. Yff(1992). This paper also outlines the history of the original problem.

[Dénes J. and Hermann P.
   (1982) On the product of all the elements of a finite group. Annals of Discrete Math. 15, 105-109.
Dénes J. and Yff P.
   (1992) Some research problems on finite groups. PUMA A3, 109-115.
Keedwell A.D.
   (1983c) On the existence of super P-groups. J. Combin. Theory A35, 89-97.
   (1984a) More super P-groups. Discrete Math. 49, 205-207.]

PROBLEM 1.5. ⟨Do there exist non-soluble groups whose Sylow 2-subgroups are non-cyclic and such that, despite this, the groups have no complete mappings? (pages 36, 37)⟩

SOLVED. All finite non-soluble groups have non-cyclic Sylow 2-subgroups and have complete mappings. See Section 2.5 of the present book.

PROBLEM 1.6. ⟨*Is the necessary condition given in Theorem 1.4.5 for the existence of a complete mapping also sufficient?* (page 38)⟩

The answer is "YES". See Section 2.5 of this book for the details.

PROBLEM 1.7. ⟨*Is it true that, for all sufficiently large n, there exist quasigroups of order n which contain no proper sub-quasigroups?* (page 44)⟩

The problem as stated is trivial: there exist quasigroups of all orders $n \geq 3$ which have no proper sub-quasigroups. [Since idempotent quasigroups exist for all $n \geq 3$, latin squares in which the main diagonal is a transversal exist for all $n \geq 3$. If the rows and columns of the square are arranged so that this diagonal is $2, 3, \ldots, n, 1$, then the square can be bordered so as to be the Cayley table of a quasigroup $(Q, \otimes)$ for which $i \otimes i = i + 1$ and $n \otimes n = 1$. It then follows that every element generates the whole subgroup. This was pointed out to the authors of [DK1] in a letter from N.S. Mendelsohn dated 12/10/87.]

The problem should have read "Is it true that, for all sufficiently large n, there exist latin squares of order n which contain no proper latin subsquares?" The problem was originally posed in this form by A.J.W. Hilton and appeared in print in Hilton(1977b).

Such squares are called N_∞-squares and we state what is presently known about such squares at the end of Section 9.4 of this book.

As regards the original problem about quasigroups, T. Kepka(1978) has proved that every countable quasigroup with at least three elements is isotopic to a quasigroup without proper sub-quasigroups.

[Andersen L.D. and Mendelsohn E.

(1982) A direct construction for latin squares without proper subsquares. Annals of Discrete Math, 15, 27-53,

Elliott J.R. and Gibbons P.B.

(1992) The construction of subsquare-free latin squares by simulated annealing. Australs. I Combin. 5, 209-228.

Heinrich K.

(1980) Latin squares with no proper subsquares. J. Combin. Theory A29, 346-353.

Hilton A.J.W.

(1977b) On the Szamkolowicz-Doyen classification of Steiner triple systems. Proc. London Math. Soc. 34, 102-116.

Kepka T.

(1978) A note on simple quasigroups. Acta Univ. Carolinae Math. Phys. 19, 59-60.

Maenhaut B.M., Wanless I.M. and Webb B.S.

(2007) Subsquare-free Latin squares of odd order. Europ. J. Combin. 28, 322-336.]

PROBLEM 1.8. ⟨*Given an arbitrary integer n, does there always exist a latin square of order n which contains a latin subsquare of every order m such that $m \leq n/2$?* (page 55)⟩

The answer is "NO". For the complete solution, see K. Heinrich(1977d) who has proved that a latin square of order n with latin subsquares of every order m such that $m \leq n/2$ exists only if $1 \leq n \leq 7$ or if $n = 9, 11$ or 13. For a full discussion of this problem and the similar problem concerning quasigroups, see Section 3 of Chapter 4 in [DK2].

[*Heinrich K.*

(1977d) Subsquares in latin squares. Proc. Eighth S.E. Conf. on Combinatorics, Graph Theory and Computing, 1977. Congressus Numerantium 19, 329-344.]

PROBLEM 1.9. ⟨*If n is any positive integer and $n = n_1 + n_2 + \ldots + n_k$ any fixed partition of n, is it possible to find a quasigroup Q_n, of order n which contains subquasigroups $Q_1, Q_2, \ldots, Q_k$ of orders $n_1, n_2, \ldots, n_k$ respectively whose set theoretical union is Q_n?* (page 56)⟩

A partial solution has been obtained, the main results being those of K. Heinrich(1977c,1982). For a full discussion of the problem and the results so far obtained, see Section 3 of Chapter 4 in [DK2].

[*Heinrich K.*

(1977c) Latin squares composed of four disjoint subsquares. In "Combinatorial Math. V" (Proc. Fifth Austral. Conf., Melbourne, 1976), pp. 118-127. Lecture Notes in Math. No.622, Springer Verlag, Berlin.

(1982) Disjoint subquasigroups. Proc. London Math. Soc. (3)45, 547-563.]

PROBLEM 1.10. ⟨*The same question as in problem 1.9 but this time Q_n is required to satisfy the condition for every partition of n.* (page 56)⟩

UNSOLVED. A solution to this problem pre-supposes a complete solution to problem 1.9.

Chapter 2

PROBLEM 2.1. ⟨*Do there exist row-complete latin squares which cannot be made column-complete by permutation of their rows?* (page 83)⟩

The answer is "YES". See P.J. Owens(1976) for a general construction for such squares. However, for latin squares which are the Cayley tables of groups, the answer is "NO". See A.D. Keedwell(1975,1976c) or page 45 of [DK2] for the proof. More recently, a construction alternative to that of P.J. Owens has been given by D. Cohen and T. Etzion(1991).

[*Cohen D. and Etzion T.*

(1991) Row complete latin squares which are not column complete. Ars Combin. 32, 193-201.

Keedwell A.D.

(1975) Row complete squares and a problem of A. Kotzig concerning P-qausigroups and Eulerian circuits. J. Combin. Theory A18, 291-304.

(1976c) Latin squares, P-quasigroups and graph decompositions. (Symposium on Quasigroups and Functional Equations, Belgrade, 1974.) Recueil des Travaux de lInstitute Mathematique, Belgrade, N.S. 1(9), 41-48.
Owens P.J.
(1976) Solutions to two problems of Dénes and Keedwell. J. Combin. Theory A21, 299-308.]

PROBLEM 2.2. ⟨*For which integers n do there exist sets of n − 1 latin squares in which each ordered triad of distinct elements occurs just once in a row of some member of the set?* (page 84)⟩

It has been shown in H. Niederreiter(1993) that such sets exist whenever n is a prime power and, moreover, that the latin squares are mutually orthogonal.

[*Niederreiter H.*
(1993) Proof of Williams conjecture on experimental designs balanced for pairs of interacting residual effects. Europ. J. Combin. 14, 55-58.]

PROBLEM 2.3. ⟨*For which odd integers n do complete latin squares exist?* (page 89)⟩

UNSOLVED. For odd values of n for which non-abelian groups exist, it is probably true that complete latin squares exist (see the conjecture mentioned under problem 2.4). For other odd values of n which are not prime, it has been proved by J. Higham(1998) that row complete latin squares exist. As regards prime values of n, it is known that no complete or row-complete latin squares of orders 3, 5 or 7 exist but, for other prime orders, the question is completely open. For more details, see page 48 of [DK2] and the Introduction to J. Higham's paper.

[*Higham J.*
(1998) Row-complete latin squares of every composite order exist. J. Combin. Designs 6, 63-77.]

PROBLEM 2.4. ⟨*What are the necessary and sufficient conditions that a non-abelian group be sequenceable?* (page 89)⟩

The present conjecture is that all non-abelian groups of orders greater than 8 are sequenceable. In support of this conjecture, it has been shown that all non-abelian groups except the dihedral groups D_3, D_4 and the quaternion group Q of orders up to and including 32 are sequenceable [B.A. Anderson(1987a,c)]. Also, all soluble groups with a unique element of order 2 have symmetric sequencings except the quaternion group Q [B.A. Anderson and E.C. Ihrig(1993a)]. As regards non-soluble groups with a unique element of order 2, B.A. Anderson and E.C. Ihrig (1993b) have shown that whether or not they have symmetric sequencings is dependent on whether or not the groups A_7 (alternating), $PSL(2,q)$ and $PGL(2,q)$, where $q > 3$ is an odd power, have 2-sequencings.

See Section 2.6 of the present book for further classes of non-abelian groups which are sequenceable.

[*Anderson B.A.*

(*1987a*) *A fast method for sequencing low order non-abelian groups. Annals of Discrete Math. 34, 27-42.*

(*1987b*) *Sequencings of dicyclic groups. Ars Combinatoria 23, 131-142.*

(*1987c*) S_5, A_5 *and all non-abelian groups of order 32 are sequenceable. Congressus Numerantium 58, 53-68.*

(*1988*) *Sequencings of dicyclic groups II. J. Combin. Math. and Combin. Comput. 3, 5-27.*

(*1990*) *All dicyclic groups of order at least 12 have symmetric sequencings. Contemporary Math. (AMS) 111, 5-21.*

Anderson B.A, and Ihrig E.C.

(*1993a*) *Every finite solvable group with a unique element of order two, except the quaternion group, has a symmetric sequencing. J. Combin. Designs 1, 3-14.*

(*1993b*) *Symmetric sequencings of non-solvable groups. Congressus Numerantium 93, 73-82.*

Isbell J.

(*1990*) *Sequencing certain dihedral groups. Discrete Math. 85, 323-328.*

Keedwell A.D.

(*1981a*) *On the sequenceability of non-abelian groups of order pq. Discrete Math. 37, 203-216.*

Li P.

(*1997*) *Sequencing the dihedral groups D_{4h}. Discrete Math. 175, 271-276.*

Wang C.

(*2002*) *Complete latin squares of order p^n exist for odd primes p and $n > 2$. Discrete Math. 252, 189-201.*]

PROBLEM 2.5. ⟨*Do all complete latin squares satisfy the quadrangle criterion?* (page 89)⟩

The answer is "NO". P.J. Owens(1976) has constructed counter-examples.

[*Owens P.J.*

(*1976*) *Solutions to two problems of Dénes and Keedwell. J. Combin. Theory A21, 299-308.*]

PROBLEM 2.6. ⟨*Is it true that every finite non-abelian group of odd order on two generators is sequenceable?* (page 89)⟩

It is conjectured that the answer is "YES". See the comments on Problem 2.4 and also Section 2.6 of this book.

Chapter 3

PROBLEM 3.1. ⟨*If n is a given odd integer, what is the maximum value of m_n such that a complete latin rectangle of m_n rows and n columns exists?* (page 98)⟩

UNSOLVED. The corresponding problem for a row-complete latin rectangle is partly solved. (See also problem 2.3.) In particular, H. Taylor (1991) has obtained partial solutions to the more general problem of finding m_n in the case when each ordered pair of elements is required to occur t-apart (not just 1-apart) exactly once in the rows. He has noted that, in this case, $m_p \geq p - 1$ and $m_{p-1} = p - 1$ for all primes $p > 2$ (also shown earlier by S.W. Golomb), that $m_8 = 7$ (proved by S. Alquaddoomi using a computer search) and that $m_9 \geq 4$.

[*Taylor H.*

(1991) Florentine rows or left-right shifted permutation matrices with cross-correlation values ≤ 1. Discrete Math. 93, 247-260.]

PROBLEM 3.2. ⟨*For which values of k_n is it true that an arbitrary complete latin rectangle of k_n rows and n columns ($k_n < n$) can be completed to a complete latin square of order n?* (page 98)⟩

UNSOLVED. The corresponding problem for a row-complete latin rectangle is partly solved. See problems 2.1 and 2.3.

PROBLEM 3.3. ⟨*What is the largest number of cells of an arbitrarily chosen latin rectangle of given size which can be contained in a single partial transversal?* (page 99)⟩

If we suppose that the latin rectangle has h rows and k columns and is defined on n symbols, then clearly the answer is $\min(h, k)$ if h and k are small compared with n. When h and k are close to n, the results of Koksma, Drake, Wang, Brouwer *et al* , Woolbright and Shor (see also Problem 3.4) become relevant. The problem remains unsolved.

PROBLEM 3.4. ⟨*Is it true that every latin square of order n has a partial transversal containing at least $n - 1$ cells?* (page 103)⟩

This is known as *Brualdi's conjecture*. See the main text of this book.

It is strongly believed that the answer is "YES" for all latin squares which are isotopic to the multiplication tables of groups and conjectured that it is "YES" for all latin squares. The answer is certainly "YES" for all latin squares which represent quasigroups possessing a complete or near-complete mapping. [A quasigroup Q is said to have a *near-complete* mapping θ if there exists a one-to one mapping θ from $Q - \{a\}$ to $Q - \{b\}$ such that the mapping $\phi(g) = g\theta(g)$ has the same property. See Section 7.6.]

Every abelian group with a unique element of order 2 is sequenceable (Theorem 2.6.3 of this book) and so has a near-complete mapping. All other abelian groups have $\prod g = e$ and so they have complete mappings. [See Theorem 2.5.3 of this book or L.J. Paige(1947).]

As regards non-abelian groups, it is conjectured that all such groups except the dihedral groups D_3, D_4 and the quaternion group Q are sequenceable (see the comments on Problem 2.4) and so they have near-complete mappings. D_4 and Q both have complete mappings. D_3 has the near-complete mapping (see Section 7.6) whose corresponding near-orthomorphism is $\phi = [b\, a^2](e\, a\, ba^2\, ba)$.

For a discussion of quasigroups which have complete mappings, see Section 1.5.

Chapter 6 of B. Smetaniuk's Ph.D. thesis (Univ. of Sydney, 1983) contains additional evidence to support the belief that the answer to the problem is YES.

A number of authors have obtained lower bounds for the length of (that is, number of cells in) a partial transversal of a latin square of order n. Those obtained prior to 1991 are discussed in detail in Section 2 of Chapter 2 of [DK2]. For more recent results, see Section 3.5 and Section 11.3 of the present book.

Some generalizations of the problem are discussed in S. Stein(1975). Also, see P. Erdös, D.R. Hickerson *et al* (1988), P.J. Cameron and I.M. Wanless(2005), I.M. Wanless(2011), I.M. Wanless and X. Zhang(2013).

[*Cameron P.J. and Wanless I.M.*
    (2005) Covering radius for sets of permutations. Discrete Math. 293, 91-109.
Erdös P., Hickerson D.R., Norton D.A., and Stein S.K.
    (1988) Has every latin square of order n a partial latin transversal of size $n - 1$? *Amer. Math. Monthly 95, 428-430.*
Stein S.K.
    (1975) Transversals of latin squares and their generalizations. Pacific J, Math. 59, 567-575.
Wanless I.M.
    (2011) Transversals in Latin squares: a survey. Surveys in combinatorics 2011, 403-437, London Math. Soc. Lecture Note Ser. No. 392, Cambridge Univ. Press.
Wanless I.M. and Zhang X.
    (2013) Transversals of Latin squares and covering radius of sets of permutations. European J. Combin. 34 , no. 7, 1130-1143.]

PROBLEM 3.5. ⟨*If k elements are deleted at random from the Cayley table of a finite abelian group G of order n, what is the greatest value which k may have in order that the remaining part of the table still determines G up to isomorphism?* (page 106)⟩

PARTIALLY SOLVED, A. Drápal(1992) has proved that groups of order n are isomorphic if their Hamming distance apart is $\leq n^2/9$.

Two papers which consider problems closely related to problem 3.5 are A. Drápal and T. Kepka(1989) and P. Vojtěchovský and I.M. Wanless(2012). Also, the problem is related to that of finding so-called critical sets in latin squares which we discuss in Section 3.2 of this book. See, for example, A.D. Keedwell(1996,2004) and J.A. Bate and G.H.J. van Rees(1999). In particular, for the Cayley table of a cyclic group, k is bounded above by $3n^2/4$.

[*Bate J.A. and van Rees G.H.J.*
    (1999) The size of the smallest strong critical set in a latin square. Are Combinatoria 53, 73-83.
Drápal A.
    (1992) How far apart can the group multiplication tables be? European J. Combin. 13, 335-343.

Drápal A. and Kepka T.

    (1989) On a distance of groups and Latin squares. Comment. Math. Univ. Carolin. 30, 621-626.

Keedwell A.D.

    (1996) Critical sets for latin squares, graphs and block designs: a survey. Congressus Numerantium 113, 231-245.

    (2004) Critical sets in latin squares and related matters: an update. Utilitas Math. 65, 97-131.

Vojtěchovský P. and Wanless I.M.

    (2012) Closest multiplication tables of groups. J. Algebra 353, 261-285.]

PROBLEM 3.6. ⟨*What is the maximum number h of squares which a set of latin squares satisfying the quadrangle criterion and all of the same order n can contain if each pair of squares in the set are to differ from each other in at most m places?* (page 111)⟩

UNSOLVED. This problem has some connections with Problem 1.1. See the papers of Drápal and Kepka listed there. Also, the results in the paper of P. Vojtěchovský and I. M. Wanless(2012) show when $h \geq 2$ for given values of m.

PROBLEM 3.7. ⟨*Is it possible to place $n-1$ elements chosen from the set $\{0, 1, 2, \ldots, n-1\}$ in an $n \times n$ matrix in such a way that no two elements of any row or column of the matrix are equal and so that, despite this condition, it is not possible to complete the matrix to an $n \times n$ latin square?* (page 115)⟩

The answer is "NO", (as was originally conjectured by T. Evans(1960)). In R.Häggkvist(1978), that author gave a proof for all $n \geq 1111$. Subsequently, two complete solutions have been given: one by L.B. Andersen and A.J.W. Hilton(1983), which is very lengthy, and the other by B. Smetaniuk(1981). For a detailed discussion of these and an explanation of the delayed publication of the former one, see Section 10 of Chapter 8 in [DK2] and Section 3.4 of this book.

[*Andersen L.D. and Hilton A.J.W.*

    (1983) Thank Evans. Proc. London Math. Soc. (3) 47, 507-522.

Evans T.

    (1960) Embedding incomplete latin squares. Amer. Math. Monthly 67, 958-961.

Häggkvist T.

    (1978) A solution of the Evans conjecture for latin squares of large size. In "Combinatorics" Colloq. Math. Soc. Janos Bolyai, Vol.18; edited by A. Hajnal and V.T. Sós, pp. 495-514. North Holland, Amsterdam.

Smetaniuk B.

    (1981) A new construction on latin squares I. A proof of the Evans conjecture. Ars Combinatoria 11, 155-172.]

PROBLEM 3.8. ⟨*How many elements of a latin square of even order n and which satisfies the quadrangle criterion can be located arbitrarily subject only to the*

condition that no row or column shall contain any element more than once?
(page 115)⟩

This problem has some connections with Problem 1.1; more particularly with the results of A. Drápal and T. Kepka(1981) and A. Drápal(1983) mentioned there. In a private communication to the author, Wanless claimed to have solved the problem and details have subsequently been given in a joint paper with Webb. The number h of elements which can be located arbitrarily in a latin square of order n which satisfies the quadrangle criterion is as follows: for $n = 1$ and 2, $h = 1$; for $n = 3$, $h = 2$; for $n = 4$ or $n > 3$ and odd, $h = 3$; for $n = 6$ or $n \equiv 2$ or 4 mod 6 and > 4, $h = 5$; for $n \equiv 0$ mod 6 and > 6, $h = 6$.

[*Wanless I.M. and Webb B.S.*
(2017) Small partial latin squares that cannot be embedded in a Cayley table.
Australas. J. Combin. 67,352-363.]

PROBLEM 3.9. ⟨*Can a pair of $n \times n$ incomplete latin squares which are orthogonal (insofar as the condition for orthogonality applies to the incomplete squares) be respectively embedded in a pair of $t \times t$ orthogonal latin squares; and, if so, what is the smallest value of t for each value of n?* (page 116)⟩

As long ago as 1976, C.C. Lindner proved that a finite embedding is possible but, not until 2014, was it proved that a polynomial size embedding is obtainable. The latter result was achieved by joint work of D. Donovan and E.Ş. Yazici. The main theorem of these authors is that two orthogonal partial latin squares of order n and of the same size and shape can be embedded in a pair of orthogonal latin squares of order at most $16n^4$. A partial latin square is said to be of *size* h if h of its cells are filled and the *shape* of such a square is determined by the configuration of its filled cells.

Note. The term "size" dates back at least as far as Nelder(1977) and has been used by many authors since then. Much later, Mahmoodian introduced the term "volume" to denote the same concept. As well as causing confusion (perpetuated by Donovan and Yazici, who use the latter term), the term "volume" to describe a subset of a two-dimensional array is, in the opinion of the present author, wholly inappropriate.

[*Donovan D. and Yazici E.Ş.*
(2014) A polynomial embedding of pairs of partial orthogonal partial latin squares. J. Combin. Theory A126, 24-34.
Lindner C.C.
(1976) Embedding orthogonal partial latin squares. Proc. Amer. Math. Soc. 59, 184-186.]

PROBLEM 3.10. ⟨*Can every Brandt groupoid B be embedded in a quasigroup Q which is group-isotopic? Alternatively, can every Brandt groupoid be embedded in a group?* (pages 119, 120)⟩

UNSOLVED.

Chapter 4

PROBLEM 4.1. ⟨*Is it true that no two distinct sets of n disjoint transversals of a latin square of odd order have any transversal in common?* (page 149)⟩

FALSE. See Figure 4.5.4 and Figure 4.5.5 of this book for counter-examples.

PROBLEM 4.2. ⟨*What is the widest class of latin squares of order n with the property that their number of transversals is a multiple of n?* (page 149)⟩

UNSOLVED. It follows from a theorem due to G.B. Belyavskaya and A.F. Russu(1975) that the number of transversals of a latin square L in reduced form is congruent to zero modulo the number of elements in the left (or right) nucleus of the loop of which L is the multiplication table. So, in particular, if a quasigroup is isotopic to an admissible group of order n, its number of transversals is a multiple of n. For more information, see Section 3 of Chapter 2 in [DK2] and Section 4.5 of this book.

[*Belyavskaya G.B. and Russu A.F.*
   (1975) On the admissibility of quasigroups. (Russian) Mat. Issled. 10, No.1 (35), 45- 57.]

PROBLEM 4.3. ⟨*Find formulae for the numbers R(5, n) and V(5, n) of 5 × n reduced and very reduced latin rectangles.* (page 152)⟩

UNSOLVED. Numerical values for $R(k, n)$ for $n \leq 11$ have been obtained and are listed in Figure 4.4.3 and Figure 4.4.4 of this book. Also, D.S. Stones has computed $R(5, n)$ for $n \leq 28$ and $R(6, n)$ for $n \leq 13$. For more details, see Section 4.4.

Chapter 5

PROBLEM 5.1. ⟨*Do there exist complete sets of mutually orthogonal latin squares of non-prime power order?* (page 160)⟩

UNSOLVED. It is conjectured that the answer is "NO". However, the only general class of non-prime powers for which existence is ruled out remains those excluded by the Bruck-Ryser theorem (Theorem 5.2.5). It has been shown with the aid of very extensive computer power that there is no projective plane of order 10 (the smallest positive integer not ruled out by the Bruck-Ryser theorem) which proves non-existence for this order. For the details, see C.W.H. Lam, L. Thiel and S. Swiercz(1989) and C.W.H. Lam(1991) or [DK2], pages 377-378. A summary of what is known is in Section 5.2 of this book.

[*Lam C.W.H.*
   (1991) The search for a finite projective plane of order 10. Amer. Math. Monthly 98, 305-318.
Lam C.W.H., Thiel L. and Swiercz S.
   (1989) The non-existence of finite projective planes of order 10. Canad. J. Math. 41, 1117-1123.]

PROBLEM 5.2. ⟨*Is it true that there do not exist sets of* $n-1$ *mutually orthogonal latin squares based on a cyclic group of order* n *unless* n *is prime?* (page 170)⟩

UNSOLVED except when n is even and for a few sporadic odd values of n. In the former case the answer is "YES" because a cyclic group of even order has no complete mappings. The Bruck-Ryser theorem implies that the answer is "YES" for certain odd values of n (see Section 5.2 in the main text of this book) and some results due to W. De Launey(1984) and reported in T. Evans(1992) imply that the answer is "YES" for some further odd values: namely ones for which a so-called generalized Hadamard matrix does not exist.

[*De Launey W.*
   (1984) On the nonexistence of generalised Hadamard matrices. J. Statist. Plann. Inference 10, 385-396.
Evans A.B.
   (1992) Orthomorphism graphs of groups. Lecture Notes in Mathematics, No.1535. (Springer-Verlag, Berlin.) viii+114 pp. ISBN: 3-540-56351-2.]

PROBLEM 5.3. ⟨*If* $\Lambda(n)$ *denotes the maximum number of mutually orthogonal* $n \times n$ *latin squares none of which satisfies the quadrangle criterion, what are the values of* $\Lambda(n)$? (page 170)⟩

UNSOLVED. It is obvious that $\Lambda(1) = \Lambda(2) = \Lambda(3) = \Lambda(4) = 0$ and that $\Lambda(6) = 1$. Also, $\Lambda(9) = 8$ as is shown by Figure 8.4.3 on page 285 of [DK1].[9] (The squares listed in Fig. 8.4.3 represent the dual translation plane, not the Hughes plane.) $\Lambda(10) \geq 2$, $\Lambda(14) \geq 3$, and $\Lambda(20) \geq 4$. The last three results follow from the work of D.F. Hsu and A.D. Keedwell(1984) and D. Bedford (1993) which shows that the squares constructed by L. Weisner(1963), A.I. Lyamzin(1963), and D.T. Todorov(1985,1989) all implicitly make use of the left neofield construction (as defined by Bedford). In fact, it is true that the additive loop of a cyclic left neofield of even order cannot be group-isotopic unless its order is a power of two. [In a cyclic left neofield $(N, +, \cdot)$ of even order, we have $1+1 = 0$, see Section 7.6 of this book or A.D. Keedwell(1983e), so it follows from the left distributive law that every non-zero element of N has additive order 2. Thus, $(N, +)$ cannot be group isotopic unless its order is a power of 2.] So far as the authors are aware, little else is known except that $\Lambda(5) \geq 1$ and $\Lambda(7) \geq 2$. The latter fact is shown by the following pair of orthogonal latin squares constructed by D. Bedford(1993) using presentation function number $3'$ for the cyclic group C_6 in Table 1 of his paper.

---

[9]Schönheim had earlier conjectured (in an oral communication to J. Dénes) that $\Lambda(n) < n-1$ for $n > 2$.

$$
\begin{array}{ccccccc}
6 & 0 & 1 & 2 & 3 & 4 & 5 \\
1 & 5 & 2 & 4 & 0 & 6 & 3 \\
2 & 4 & 0 & 3 & 5 & 1 & 6 \\
3 & 6 & 5 & 1 & 4 & 0 & 2 \\
4 & 3 & 6 & 0 & 2 & 5 & 1 \\
5 & 2 & 4 & 6 & 1 & 3 & 0 \\
0 & 1 & 3 & 5 & 6 & 2 & 4
\end{array}
\qquad
\begin{array}{ccccccc}
6 & 1 & 2 & 3 & 4 & 5 & 0 \\
0 & 5 & 4 & 6 & 3 & 2 & 1 \\
1 & 2 & 0 & 5 & 6 & 4 & 3 \\
2 & 4 & 3 & 1 & 0 & 6 & 5 \\
3 & 0 & 5 & 4 & 2 & 1 & 6 \\
4 & 6 & 1 & 0 & 5 & 3 & 2 \\
5 & 3 & 6 & 2 & 1 & 0 & 4
\end{array}
$$

[Bedford D.

(1993) Construction of orthogonal latin squares using left neofields. Discrete Math.115, 17-38.

Hsu D.F and Keedwell A.D.

(1984) Generalized complete mappings, neofields, sequenceable groups and block designs I. Pacific J. Math 111, 317-332.

Keedwell A.D.

(1983e) The existence of pathological left neofields. Ars Combinatoria 16B, 161-170.

Lyamzin A.I.

(1963) An example of a pair of orthogonal latin squares of order ten. (Russian) Uspehi Mat. Nauk. 18, 173-174.

Todorov D.T.

(1985) Three mutually orthogonal latin squares of order 14. Ars Combin. 20, 45-48.

(1989) Four mutually orthogonal latin squares of order 20. Ars Combin. 27A, 63-65.

Weisner L.

(1963) Special orthogonal latin squares of order 10. Canad. Math. Bull. 6, 61-63.]

PROBLEM 5.4. ⟨*Do there exist triples of mutually orthogonal latin squares of order 10? What is the value of* $N(10)$? (page 173)⟩

UNSOLVED. Many orthogonal pairs of such squares have been constructed, but no pairwise orthogonal triple. E.T. Parker(1978) found one latin square of order 10 with 5504 transversals and he estimated it to have about one million orthogonal mates. [Much later, it was proved that the actual number of orthogonal mates is 12,265,168. See B.M. Maenhaut and I.M. Wanless(2004).] A. Brouwer(1984) constructed four pairwise almost orthogonal squares. The four squares share a common 2×2 latin subsquare and so each pair is $(n^2 - 2)$-orthogonal. A.D. Keedwell(1980) had earlier constructed an $(n^2 - 4)$-orthogonal triple.

In a forthcoming paper, J. Egan and I.M. Wanless exhibit two MOLS of order 10 which have seven common transversals.

There have been many other attempts to solve this problem. More details are given in Section 5.3 of this book. However, as a consequence of a massive computer search reported in B.D. McKay, A. Meynert and W. Myrvold(2007), it seems very unlikely that triples exist.

[*Brouwer A.E.*

   (1984) Four MOLS of order 10 with a hole of order 2. J. Statist. Planning and Inf. 10, 203-205.
Brown J.W. and Parker E.T.

   (1985) An attempt to construct three MOLS of order 10 with a common transversal. Proc. Conf. on groups and geometry, Part A (Madison, Wis., 1985). Algebras Groups Geom. 2, 258-262.
Egan J. and Wanless I.M.

   (201?) Enumeration of MOLS of small order. Math. Comp. To appear.
Keedwell A.D.

   (1980) Concerning the existence of triples of pairwise almost orthogonal 10 × 10 latin squares. Ars Combin. 9, 3-10.
McKay B.D., Meynert A. and Myrvold W.

   (2007) Small latin squares, quasigroups and loops. J. Combin. Designs 15, 98-119.
Maenhaut B.M. and Wanless I.M.

   (2004) Atomic latin squares of order eleven. J. Combin. Designs. 12, 12-24.
Parker E.T.

   (1978) A collapsed image of a completion of a "turn-square" J. Combin. Theory. A24, 128-129.]

PROBLEM 5.5. ⟨*For each order n which is not a prime power, what is the maximum number $T(n)$ of latin squares in a set with the property that all the squares in the set have a transversal in common?* (page 173)⟩

Clearly, $T(p^r) \geq p^r - 2$. Can it exceed this bound? Another obvious remark to make is that, if there exists a quasigroup of order n which has a complete mapping, then $T(n) \geq 1$. The author is not aware that any work has been done on this problem.

PROBLEM 5.6. ⟨*Can a complete classification of Stein quasigroups be given?* (page 177)⟩

The author is not aware that any work has been done on this problem.

PROBLEM 5.7. ⟨*What is the maximum number of m-dimensional permutation cubes of order n in a variational set?* (page 186)⟩

The problem has been considered in terms of multiquasigroups in paragraphs 2.1 and 2.2 of a paper by G. Čupona, V. Stojaković and J. Ušan(1981).

[*Čupona G., Stojakoviú V. and Ušan J.*

   (1981) On finite multiquasigroups. Publ. de l'Institut Math. (Belgrade) 29(43), 53-59.]

PROBLEM 5.8. ⟨*What is the maximum number k of constraints possible in an orthogonal array of n levels, strength t and index λ?* (page 191)⟩

The author does not know whether this problem has been solved.

PROBLEM 5.9. ⟨*Is it true that orthogonal Steiner triple systems exist for all orders v congruent to 1 modulo 6?* (page 192)⟩

The problem has been solved completely, see C.J. Colbourn *et al* (1994). where it is shown that such systems exist for all orders $v \equiv 1$ or 3 (mod 6) with $v \geq 7$ except $v = 9$.

Previously, it was known that the answer is "YES" for all except 29 values of $v \equiv 1$ mod 6. For the latter values, the problem remained unsolved. It was also known that orthogonal Steiner triple systems exist for orders $v \equiv 3$ mod 6 such that $v > 27363$ and for all except at most 918 values of $v \equiv 3$ mod 6 with $v \leq 27363$. For the latter values, existence had remained in doubt. For more details, see D.R. Stinson and L. Zhu(1991) and the papers cited therein.

[*Colbourn C.J., Gibbons P.B., Mathon B., Mullin R.C. and Rosa A.*

(1994) The spectrum of orthogonal Steiner triple systems. Canad. J. Math.46, 239-252.

Stinson D.R. and Zhu L.

(1991) Orthogonal Steiner triple systems of order 6m + 3. Ars Combin. 31, 33-63.]

Chapter 6

PROBLEM 6.1. ⟨*Do there exist orthogonal pairs of diagonal latin squares of every order n distinct from 2, 3, and 6?* (page 212)⟩

The answer is "YES". For the complete solution, see J.W. Brown, F. Cherry, L. Most, M. Most, E.T. Parker and W.D. Wallis(1992). In B. Du(1991), a stronger result has been obtained: namely, for all integers n except 2, 3, 4, 5, 6 and possibly also 10, 14, 15, 18, 21, 22, 26, 30, 33, 34 and 46, there exist at least three pairwise orthogonal diagonal latin squares of order n. Moreover, in B. Du(1992) it is shown that, with the possible further exceptions of n = 20, 24, 28, 35, 36, 38, 39, 42, 44, 45, 48, 52, 54, 55, 62, 66, 68 and 69, there exist at least four pairwise orthogonal diagonal latin squares of order n.

It is known also that, for all integers n except 2, 3, 4, 6 and possibly also 10, 18, and 26, there exist at least three pairwise orthogonal idempotent (that is, left semi-diagonal) latin squares of order n. See R.J.R. Abel, X. Zhang and H. Zhang(1996), F.E. Bennett, K.T. Phelps, C.A. Rodger, and L. Zhu(1992) and X. Zhang and H. Zhang(1997). It is worth noting that, in this topic particularly, order of publication has not coincided with order of discovery.

In more recent papers, the existence of self-orthogonal and parastrophic orthogonal diagonal latin squares has been investigated. See, for example, K.J. Danhof, N.C.K. Phillips and W.D. Wallis(1990), F.E. Bennett, B. Du and H. Zhang(2001), H. Cao and W. Li(2011) and Y. Zhang, K. Chien, N. Cao and H. Zhang(2014).

[Abel R.J.R., Zhang X. and Zhang H.

(1996) Three mutually orthogonal idempotent Latin squares of orders 22 and 26. Shanghai Conference Issue on Designs, Codes, and Finite Geometries, Part 1 (Shanghai, 1993). J. Statist. Planning and Inference 51, 101-106.

Bennett F.E., Du B. and Zhang H.

(2001) Existence of (3,1,2)-conjugate orthogonal diagonal Latin squares. J. Combin. Des. 9, 297-308.

Bennett F.E., Phelps K.T., Rodger C.A. and Zhu L.

(1992) Constructions of perfect Mendelsohn designs. Discrete Math. 103, 139-151.

Brown J. W, Cherry F, Most L., Most M., Parker E. T and Wallis W. D.

(1992) Completion of the spectrum of orthogonal diagonal latin squares. In Graphs, Matrices, and Designs, pp.43-49, Marcel Dekker, New York, 1992.

Cao H. and Li W.

(2011) Existence of strong symmetric self-orthogonal diagonal Latin squares. Discrete Math. 311, 841-843.

Danhof K. J., Phillips N. C. K and Wallis W. D.

(1990) On self orthogonal diagonal latin squares. J. Combin. Math. and Combin. Comput. 8, 3-8.

Du B.

(1991) Some constructions of pairwise orthogonal diagonal latin squares. J. Combin. Math. and Combin. Comput. 9, 97-106.

(1992) Four pairwise orthogonal diagonal latin squares. Utilitas Math. 42, 247-254.

Zhang Y., Chen K., Cao N. and Zhang H.

(2014) Strongly symmetric self-orthogonal diagonal Latin squares and Yang Hui type magic squares. Discrete Math. 328, 79-87.

Zhang X. and Zhang H.

(1997) Three mutually orthogonal idempotent latin squares of order 18. Ars Combin. 45, 257-261.]

PROBLEM 6.2. ⟨*For what orders n do pandiagonal magic squares whose n^2 entries are consecutive integers exist?* (page 214)⟩

The earliest papers in which the theory of pandiagonal magic squares is discussed are those of A.H. Frost(1866, 1867, 1878, 1895-96). (This fact was drawn to the attention of the author by N. Biggs.) Frost called them "Nasik" squares after the town in Hindustan where he first encountered such squares (of order 4), There are no Nasik squares of order 3 but there are 48 distinct ones of order 4.

B. Lucas(1882) proposed to call these squares "diaboliques", see the preface of that journal, page xvii. The term pandiagonal was introduced by E.M. McClintock(1897), page 99. The latter author discussed such squares in detail and, in particular, proved that no pandiagonal magic squares of singly-even order $(n = 4m - 2)$ exist. The same result was proved again by C. Planck(1919), G.

Tarry(1903) showed that, if $n/3$ is an odd integer not divisible by 3, there exists a pandiagonal magic square of order n.

Further details of the early history of this topic will be found in "Encyclopédie des Sciences mathematiques", Gauthier Villars, Paris,1906.

M. Kraitchik(1930) gave a method, which he attributed to Margossian, for constructing pandiagonal magic squares of any doubly-even order ($n = 4m$) or any order which is an odd multiple of 3 (except 3 itself) using a row-latin square and a column-latin square which are mutually orthogonal. [For the details, see M. Kraitchik(1930), pages 148-452 or W.W.R. Ball(1939), pages 208-210,] A construction valid for all orders relatively prime to 6 (that is, for all odd orders which are not multiples of 3) has been described on pages 204-206 of W.W.R. Ball(1939) under the title "Generalization of De la Lobère's Rule". Together these constructions solve the problem completely.

Two more recent papers on this topic are A.O.L. Atkin, L. Hay and R.G. Larson(1977,1983).

[It is assumed throughout the foregoing that the magic squares discussed have entries which are consecutive integers. A construction of pandiagonal magic squares of singly-even order in which this requirement is dropped has been given by C. Planck(1919).]

The problem has also been discussed in great detail by B. Rosser and R. J. Walker(1939) who used the adjective "diabolic" instead of "pandiagonal". These authors produced an algebraic theory for diabolic magic squares. They again proved non-existence for $n = 3$ and $n \equiv 2$ mod 4 but showed existence for all other orders. In particular, they showed that there are 28,800 different diabolic magic squares of order 5 all of a type which they called "regular" and that the number of regular squares of order p, p an odd prime, is $(p!)^2(p - 3)(p - 4)$. In a supplement to their paper, they also established existence or non-existence of a diabolic cube of any assigned order.

A more recent paper on the subject is that of R. Sun(1989) who claims that a pandiagonal magic square of order n whose entries are consecutive integers exists if and only if $n = 2m + 3$ or $n = 4m$, where m is any integer greater than zero. This contradicts the results given above in that pandiagonal magic squares exist of all odd orders $n \equiv 1$ mod 2 as well as all orders $n \equiv 3$ mod 2 (except $n = 3$).

For the most up-to-date information, see the website [www.multimagie.com] of C. Boyer.

[Atkin A.O.L., Hay L. and Larson R.G.

(1977) Construction of Knut Vik designs. J. Statist. Plann. Inference.1, 289-297. MR 58(1979)#10526.

(1983) Enumeration and construction of pandiagonal latin squares of prime order. Comput. Math. Appl. 9, 267-292. MR 85i:05048.
Ball W.W.R.

(1939) Mathematical recreations and essays. Macmillan, New York, Eleventh Ed. Revised by H. S. M. Coxeter.

Frost A.H.

   (1866) Invention of magic cubes and construction of magic squares possessing additional properties. Quart. J. Pure Appl. Math.7, 92-102.

   (1867) Supplementary note on Nasik cubes. Quart. J. Pure Appl. Math. 8, 74.

   (1878) On the general properties of Nasik squares. Quart. J. Pure Appl. Math. 15, 34-49.

   (1878) On the general properties of Nasik cubes. Quart. J. Pure Appl. Math. 15, 93-123.

   (1895-96) The construction of Nasik squares of any order. Proc. London Math. Soc.27, 487-518.

Kraitchik M.

   (1930) "Le Matématique des Jeux" (Brussels).

Lucas E.

   (1882) "Récréations Math 1" (Paris).

McClintock E.

   (1897) On the most perfect forms of magic squares, with methods for their production. Amer. J. Math. 19, 99-120.

Planck C.

   (1919) "The Monist"(Chicago) 19, 307-316.

Rosser B. and Walker R.J.

   (1939) The algebraic theory of diabolic magic squares. [With an unpublished supplement.] Duke Math. J. 5, 705-728. MR 1(1940), page 133.

Sun R.

   (1989) On existence of pandiagonal magic squares. Ars Combin. 27, 56-60. MR 90j:05026.

Tarry G.

   (1903) Carrés panmagiques de base 3n. C. R. Assoc. Française Avance. Sci. 32, 130-142.]

PROBLEM 6.3. ⟨*For what orders n do addition-multiplication magic squares exist?* (page 215)⟩

Using the result of Problem 6.1 above, P. J. Liang., R. G. Sun, T. Ku and L. Zhu(1992) have proved that an addition-multiplication magic square of order $n = rs$ exists for all choices of positive integers r, s such that $r, s \notin \{1, 2, 3, 6\}$. This left the problem open for the orders 24, 27, 54, p, $2p$, $3p$ and $6p$, where p is a prime integer, except that existence was known for $n = 9$, see Section 6.3 of this book. More recently, existence has been shown for the order 27 by R. Sun(1993) and for the order 18 by J. Zhu, R. Sun and M. Cheng(1993), Also, non-existence for order 3 has been shown both by R. Sun(1993) and, independently, by Shi-De Zhang(1993). For the most up-to-date information, again see the website of C. Boyer [www.multimagie.com].

[In D. Borkovitz and F.K. Hwang(1983), the much simpler problem of constructing multiplicative magic squares is discussed and the problem of minimizing the constant product is addressed.]

[*Borkovitz D. and Hwang F.K.*
   (1983) Multiplicative magic squares. Discrete Math. 47, 1-11. MR 85a:05018.
Liang P, Sun R., Ku T and Zhu L.
   (1992) A construction of addition-multiplication magic squares using orthogonal diagonal latin squares. J. Combin. Math. and Combin. Comput. 11, 173-181. MR1160074.
Sun R.
   (1993) On the existence of addition-multiplication magic squares of order 27. Unpublished.
Zhang S.-D.
   (1993) On the non-existence of addition-multiplication magic squares of order 3. Unpublished.
Zhu J., Sun R. and Cheng M.
   (1996) A construction of addition-multiplication magic squares of order 18. J. Statist. Plann. Inference 51, 331-337. MR1397540.]

PROBLEM 6.4. ⟨*For which orders do patterned Room squares not exist?* (page 225)⟩

UNSOLVED. So far as the author has been able to discover, the most recent relevant paper is that of Gross and Leonard(1975).

[*Gross K.B. and Leonard P. A.*
   (1975) The existence of strong starters in cyclic groups. Utilitas Math. 7, 187-195.]

PROBLEM 6.5. ⟨*How many non-equivalent Room designs of order 2n exist?* (page 228)⟩

UNSOLVED except for very small orders. See page 233 of the main text.

PROBLEM 6.6. ⟨*How many non-isomorphic Room designs of order 2n exist?* (page 228)⟩

UNSOLVED except for very small orders. See page 233 of the main text.

Chapter 7

PROBLEM 7.1. ⟨*What is the maximum number of latin squares of order 12 in a pairwise orthogonal set?* (page 233)⟩

UNSOLVED. No set of pairwise orthogonal 12×12 latin squares of cardinal greater than 5 has been found. Considerable work has been done on trying to show non-existence of a projective plane of order 12 (cf. Problem 13.1). For details of the latter work, see Section 2 of Chapter 11 in [DK2].

PROBLEM 7.2. ⟨*Do any triads of mutually orthogonal latin squares of order 10 constructed by the method described in A. D. Keedwell [4] exist?* (pages 235, 478)⟩

The problem[10] is well within the capacity of present-day computers but until recently no-one had attempted to resolve it. D. Bedford has stated in a private communication to the author that he has completed the search and found that the answer is "NO".

PROBLEM 7.3. ⟨*Do there exist property D neofields of all finite orders r except 6?* (page 248)⟩

UNSOLVED. So far as the author is aware, the only person who has worked on the resolution of this and the following two problems in recent years is S. Lacey(2014).

[*Lacey S.*
(2014) Unpublished work in connection with M.Sc.]

PROBLEM 7.4. ⟨*Can it be proved that both commutative and non-commutative D-neofields exist for all r > 14 and that the number of isomorphically distinct D-neofields of assigned order r increases with r?* (page 248)⟩

UNSOLVED.

PROBLEM 7.5. ⟨*Do there exist planar D-neofields which are not fields?* (page 248)⟩

UNSOLVED.

Chapter 8

PROBLEM 8.1a. ⟨*Are all projective planes of prime order isomorphic and consequently Galois planes?* (page 276)⟩

UNSOLVED. There have been several attempts to tackle this problem but these have been unsuccessful. For details, see Section 3 of Chapter 11 in [DK2].

PROBLEM 8.1b. ⟨*If the answer to problem 8.1a is "NO", how many inequivalent sets of p − 1 mutually orthogonal latin squares exist of a given prime order p?* (page 276)⟩

UNSOLVED.

PROBLEM 8.2. ⟨*How many non-equivalent complete sets of mutually orthogonal latin squares of order 9 exist?* (page 280)⟩

SOLVED COMPLETELY. P.J. Owens and D.A. Preece have shown that there are 19 non-equivalent complete sets (which we will assume are based on the first n natural numbers) where two complete sets S and S' are called equivalent if there are permutations θ and ϕ of the first n natural numbers such that the following transformation converts S into S': Permute the rows of every square so that row i becomes row $i\theta$, $1 \leq i \leq n$. Permute the columns of every square so that column j becomes column $j\phi$, $1 \leq j \leq n$. Then permute the symbols, in each square separately, so that the new first rows are finally all in natural order.

---

[10] A. D. Keedwell[4] is Keedwell(1966) in the bibliography of this book.

[*Owens P.J. and Preece D.A.*

(1995) Complete sets of pairwise orthogonal latin squares of order 9. J. Combin. Math. and Combin. Comput. 18, 83-96.]

PROBLEM 8.3. ⟨*For what orders n do digraph complete sets of latin squares which are not mutually orthogonal exist? That is, for what orders n do finite projective planes which contain K-configurations exist?* (page 295)⟩

UNSOLVED. For the definition of "digraph complete", see L.J. Paige and C. Wexler(1953).

[*Paige L.J. and Wexler C.*

(1953) A canonical form for incidence matrices of finite projective planes and their associated latin squares. Portugaliae Math. 12, 105-112.]

PROBLEM 8.4. ⟨*For what orders n and for which values of k do k-nets which contain K-configurations exist?* (page 295)⟩

UNSOLVED.

PROBLEM 8.5. ⟨*Which types of finite and infinite projective planes contain K-configurations?* (page 295)⟩

As stated in [DK1], R.B. Killgrove (private communication to the author) has shown that K-configurations exist in the Hughes plane of order 9 and also in the Moulton plane but the present author is not aware of any more recent work on this topic.

PROBLEM 8.6. ⟨*Do all projective planes with characteristic p have p-power order?* (page 296)⟩

PARTIALLY SOLVED. It seems that no work more recent than that of the present author in 1970 has been done on this problem. He showed that planes of characteristic p, p prime, which satisfy an additional condition do have prime power order.

[*Keedwell A.D.*

(1971) A note on planes with characteristic, in "Atti del Convegno di Geometria Combinatoria e Sue Applicazioni", Perugia, pages 307-318.]

PROBLEM 8.7. ⟨*Do there exist non-desarguesian projective planes with characteristic p?* (page 296)⟩

The conjecture that all projective planes which have characteristic p are desarguesian dates from the 1950s and was probably originally proposed by L. Lombardo-Radice. It remains UNSOLVED except in the case when $p = 2$. See the end of section 8.2 of the main text for details.

Chapter 9

PROBLEM 9.0.[11] ⟨*Can a complete directed graph with an odd number $n = 2m - 1$ of vertices be separated into n disjoint Hamiltonian paths?* (page 306)⟩

[11]This problem is mentioned in the text of [DK1] but was not listed among the problems.

This question, posed by E.G. Straus in the late 1960s [see page 1 of N.S. Mendelsohn(1968)], is equivalent to asking whether the complete directed graph on $n + 1 = 2m$ vertices can be separated into n disjoint Hamiltonian circuits. Much later, the same question was posed again by J.C. Bermond and V. Faber(1976). T.W. Tillson(1980) has answered both questions in the affirmative for all $m \geq 4$.

The question is also connected with that of the existence of row-complete latin squares of odd order $2m - 1$ (cf. the comments on Problem 2.3), since the existence of such a square of a particular order implies an affirmative answer for that order. See Section 3.1 of this book.

[*Bermond J.C. and Faber V.*
   (1976) Decomposition of the complete directed graph into k-circuits. J. Combi. Theory B21, 146-155.
Mendelsohn N.S.
   (1968) Hamiltonian decomposition of the complete directed n-graph. In "Theory of Graphs", Proc. Colloq. Tihany, 1966, pp. 237-241. Academic Press, New York.
Tillson T.W.
   (1980) A Hamiltonian decomposition of K_{2m}^*, $2m \geq 8$. J. Combin. Theory B29, 68-74.]

PROBLEM 9.1. ⟨*For which odd integers n do there exist decompositions of the complete undirected graph K_n on n vertices into n nearly linear factors with the property that the union of every two of them is a Hamiltonian path of K_n?* (page 306)⟩

This problem is closely connected with that of deciding for which even integers $2m$ the edges of the complete undirected graph K_{2m} have a decomposition into a so-called "perfect 1-factorization". A 1-factorization of K_{2m} is said to be *perfect* if the union of each pair of 1-factors is a Hamiltonian circuit of the graph. Suppose that such a 1-factorization exists. Let us delete one vertex of the graph and all its incident edges. We are left with the complete graph K_{2m-1}. Also, each 1-factor of K_{2m} is replaced by a nearly linear factor of K_{2m-1}. The union of every pair of these is now a Hamiltonian path of K_{2m-1}.

As regards values of m for which the edges of K_{2m} have a decomposition into a perfect 1-factorization, see E. Seah(1991) and D.G. Wagner(1992) for information regarding what was known at that date and I.M. Wanless(2005) for more up-to-date information. Also, such 1-factorizations have been counted for K_{14} and they exist for $m = 26$ [see A.J. Wolfe(2009)].

[*Dinitz J.H. and Garnick D.K.*
   (1996) There are 23 nonisomorphic perfect one-factorizations of K_{14}. J. Combin. Des. 4, 1-4.
Seah E.
   (1991) Perfect one-factorizations of the complete graph: a survey. Bull. Inst. Combin. Appl. 1, 59-70.

Wagner D.G.

   (1992) On the perfect one-factorization conjecture. Discrete Math. 104, 211-215.

Wanless I.M.

   (2005) Atomic latin squares based on cyclotomic orthomorphisms. Electron. J. Combin. 12, Research Paper 22, 23 pp.

Wolfe A.J.

   (2009) A perfect one-factorization of K_{52}. J. Combin. Designs 17, 190-196.]

PROBLEM 9.2. ⟨Determine the number of isomorphically distinct symmetric latin squares of order $2m - 1$ for arbitrary $m > 4$. (page 306)⟩

This problem is equivalent to that of asking in how many ways can the edges of the complete undirected graph K_{2m-1} be partitioned into nearly linear factors. See Theorem 8.3.4 in the main text of this book. Alternatively, the problem is equivalent to asking how many isomorphically distinct 1-factorizations $\mathcal{F}$ does the complete graph K_{2m} possess. To see this, delete one vertex v from K_{2m}. Then the 1-factors of $\mathcal{F}$ become nearly linear factors of $K_{2m} - \{v\}$. Let these nearly linear factors be denoted by $F_1, F_2, \ldots, F_{2m-1}$, where the labelling is chosen so that F_h is the 1-factor which has $[h, v]$ as an edge in K_{2m}.

Now construct a $(2m - 1) \times (2m - 1)$ matrix by placing h in the cell (i, j) if the edge $[i, j]$ occurs in the nearly linear factor F_h. Also, place h in the cell (h, h). The resulting matrix will be a symmetric latin square. In the first place, h will appear both in the cell (i, j) and in the cell (j, i) so the square will be symmetric. Also, h will appear just once in the ith row since only one edge of F_h is incident with the vertex i. Likewise, h will appear just once in the jth column since only one edge of F_h is incident with the vertex j. This proves the claim.

As regards the equivalent problem of finding the number M_m of isomorphically distinct 1-factorizations of K_{2m}, this number has been counted/computed for $m \leq 7$. L.E. Dickson and F.H. Safford(1906) proved that $M_4 = 6$;[12] E.N. Gelling and R.E. Odeh(1974) obtained $M_5 = 396$; J.H. Dinitz, D.K. Garnick and B.D. McKay(1994) obtained $M_6 = 526,915,620$ and P. Kaski and P.R.J. Östergård(2008) obtained $M_7 = 1,132,835,421,602,062,347$.

[Dickson L.E. and Safford F.H.

   (1906) Solution to problem 8 (group theory). Amer. Math. Monthly 13, 150-151.

Dinitz J.H., Garnick D.K. and McKay B.D.

   (1994) There are 526,915,620 nonisomorphic one-factorizations of K_{12}. J. Combin. Designs. 2, 273-285.

Gelling E.N. and Odeh R.E.

   (1974) On 1-factorizations of the complete graph and the relationship to round robin schedules. Proc. Third Manitoba Conference on Numerical Mathematics (Winnipeg, Man., 1973), pp.213-221. Utilitas Math., Winnipeg, Man., 1974.

---

[12]On page 306 of [DK1], we mentioned that A. Kotzig obtained $M_4 = 7$ but this value was not correct.

Kaski P. and Östergård P.R.J.

(2009) There are 1,132,835,421,602,062,347 nonisomorphic one-factorizations of K_{14}. J. Combin. Designs 17, 147-159.]

PROBLEM 9.3. ⟨*For what values of n does a P-quasigroup exist which defines a partition of the complete undirected graph G_n into a single closed path?* (page 307)⟩

SOLVED COMPLETELY. N.P. Korovina(1984) has proved that a P-quasigroup of order n defining an Eulerian cycle exists for every odd n except $n = 5$. Earlier, it had been proved by Hilton and Keedwell that such decompositions exist for all values of $n \equiv 3 \pmod 4$, that is $n = 4r + 3$, except possibly when $r \equiv 127 \pmod{595}$ and for a considerable number of values of $n \equiv 1 \pmod 4$ which are less easy to specify. See also Section 8.3 of the present book.

[Hilton A.J.W. and Keedwell A.D.

(1976) Further results concerning P-quasigroups and complete graph decompositions. Discrete Math. 14, 311-318.

Keedwell A.D.

(1975) Row complete squares and a problem of A. Kotzig concerning P-quasigroups and Eulerian circuits. J. Combinatorial Theory 18, 291-304.

Korovina N.P.

(1984) Complete solution of the Kotzig problem on the existence of an Euler cycle in P-quasigroups. (Russian) Uspekhi Mat. Nauk 39, 163-164.]

PROBLEM 9.4. ⟨*What are the distinguishing features of a P-quasigroup of the type mentioned in problem 9.3?* (page 307)⟩

UNSOLVED so far as the author is aware, but see the paper of Korovina mentioned above.

PROBLEM 9.5. ⟨*What criteria must a colour graph satisfy if it is to be the Cayley colour graph of a quasigroup?* (page 315)⟩

SOLVED COMPLETELY. W. Dörfler(1974) has proved that every regular graph is a quasigroup graph.

[Dörfler W.

(1974) Every regular graph is a quasigroup graph. Discrete Math.10,181-183.]

Chapter 10

PROBLEM 10.1. ⟨*For which values of q, n and d can the Joshibound be attained?* (page 356)⟩

The author is not in touch with recent developments in this topic.

PROBLEM 10.2. ⟨*What is the least number $\rho(k,n)$ of elements which a covering set of an abelian group G which is the direct sum of k cyclic groups each of order n can contain?* (page 362)⟩

The following paper by S. Szabó(1988) is relevant to this problem.

[*Szabó S.*

   (1988) Certain coverings of finite homocyclic abelian groups. Periodica Poly-technica 32, 69-72.]

PROBLEM 10.3. ⟨*Find a set, as large as possible, of k-digit binary code words such that, for a given small positive integer m, every check sum of up to m different code words is distinct from every other sum of m or fewer code words, and so that each such check sum can be decomposed uniquely, apart from the order of the words, into the m code words which were used to form that sum?* (page 366)⟩

   The author understands that some information with regard to this problem may be found in the book of W.W. Wu(1985).

[*Wu W.W.*

   (1985) Elements of Digital Satellite Communication. Vol. 2. Computer Science Press, New York.]

PROBLEM 10.4. ⟨*Can E. Knuth's upper bounds be attained for non-prime values of n?* (pages 367, 368)⟩

   In E. Knuth(1967), that author discussed three problems concerning a set of pairs of orthogonal latin squares which had arisen in connection with the design of telephone exchanges. Details are on pages 366-368 of [DK1]. The present author is not aware of any more recent work having been done on these problems.

[*Knuth E.*

   (1967) Egy ortogonális latin négyzetekkel kapcsolatos problémáról. Közlemények No. 2 (1967), 26-40. In Hungarian. [On a problem connected with orthogonal latin squares.]]

PROBLEM 10.5. ⟨*For what values of the parameters b, v, r, k, λ do balanced incomplete block designs exist?* (page 375)⟩

   The problem remains unsolved in general. For a fairly recent listing of all known BIBDs with $r < 42$, see R. Mathon and A. Rosa(1990). This listing has been updated in "The CRC Handbook of Combinatorial Designs" [Colbourn and Dinitz(1966,2006)]

[*Mathon R. and Rosa A.*

   (1990) Tables of parameters of BIBDs with $r \leq 41$ including existence, enumeration and resolvability results: an update. Ars Combin. 30, 65-96.]

Chapter 11

PROBLEM 11.1. ⟨*What are the exact values of v_n, for $n = 3, 4, 5, \ldots$?* (page 425)⟩

   Hanani defined the integer v_r as being the smallest positive integer with the property that $N(v) \geq r$ for every $v > v_r$. In particular, $v_2 = 6$ and $v_3 \leq 23$. The value $v_3 \leq 46$ was obtained by Wilson(1974) but this has subsequently been improved to that just stated. For more information on this topic, see Section 5.3

of this book and "The CRC Handbook of Combinatorial Designs" [Colbourn and Dinitz(1966,2006)].

[*Wilson R.M.*
 (1974) Concerning the number of mutually orthogonal Latin squares. Discrete Math. 9, 181-198.]

PROBLEM 11.2. ⟨*What are the exact values of $N(n)$ for $n \geq 15$, n not a prime power?* (page 426)⟩

UNSOLVED, but see Section 5.3 of this book for present knowledge on this subject.

Chapter 12

PROBLEM 12.1. ⟨*For what values of m and h are E. Barbette's problems soluble? Is it always possible to find solutions which consist of superimposing a row latin square on its own transpose as in Fig. 12.5.2?* (page 463)⟩

In E. Barbette(1898), that author proposed a generalization of Euler's 36 officer problem (see Section 5.1) and he provided solutions for some particular cases. Details are in [DK1], pages 463-464. The specific questions stated in the problem have not since been addressed.

[*Barbette E.*
 (1898) Sur le probléme d'Euler dit des 36 officiers. Généralisation. Interméd. Math. 5, 83-85.]

PROBLEM 12.2. ⟨*Which groups contain a latin square group as a subgroup?* (page 466)⟩

See problem 12.3 for information about this problem.

PROBLEM 12.3. ⟨*How can the latin square groups be characterized?* (page 466)⟩

In J.J. Carroll *et al* (1973), these authors posed the question: "Given a permutation group G_p acting transitively on the symbols $N = \{1, 2, \ldots, n\}$, when is it possible to find a subset of n permutations $g_1 = e, g_2, \ldots, g_n$ such that, for each $i \in N$, the symbols $g_1(i), g_2(i), \ldots, g_n(i)$ are all distinct?" They called such a sequence *a driving sequence*.

Suppose that a driving sequence exists. Then the $n \times n$ array whose (k, l)th entry is $g_k(l)$ is a latin square whose rows are the permutations $g_1 = e, g_2, \ldots, g_n$ of the driving sequence.

Conversely, let L be a latin square of order n with first row in natural order. Then the permutations which define the rows of L are a driving sequence of the permutation group which they generate.

In particular, if L is the Cayley representation of a group G, then the rows of L define a driving sequence and they generate G. If, on the other hand, L is not group isotopic then its row permutations define a driving sequence for the group of order greater than n which they generate. Evidently this group is a subgroup of the symmetric group S_n. Carroll *et al* called such a group *a latin square group* and they raised the question "Which groups are latin square groups?" They stated further that, among others, there is a permutation group K of order 12 on 6 symbols which does not contain a latin square subgroup. We remark here that K cannot contain a subgroup of order 6 since the permutations of such a subgroup would form a latin square so K must be isomorphic to the alternating group A_4. More generally, a sufficient (but not necessary) condition for a permutation group on n symbols to be a latin square group is that it has a subgroup of order n.

Problems 12.2 and 12.3 remain unsolved in general except for small and very large orders. Carroll *et al* themselves investigated small orders. Also, A. Drápal and T. Kepka(1989) have proved that, for $n \neq 2, 4, 5$, the alternating group A_n is a latin square group. P.J. Cameron(1992) has proved that, "for almost all loops", the left multiplication group is symmetric or alternating. This is closely related to the problem because a latin square becomes the multiplication table of a loop when it is suitably bordered and the row permutations of a loop generate its left multiplication group

[*Cameron P.J.*

(1992) Almost all quasigroups have rank 2. (A collection of contributions in honour of Jack van Lint.) Discrete Math. 106/107, 111-115.
Carroll J.J., Fisher G.A., Odlyzko A.M. and Sloane N.J.A.

(1973) Research Problems: What are the Latin Square Groups? Amer. Math. Monthly 80, 1045-1046.
Drápal A. and Kepka T.

(1989) Alternating groups and Latin squares. European J. Combin 10, 175-180.]

Chapter 13

PROBLEM 13.1. ⟨*Do projective planes of orders 12 or 15 exist?* (page 484)⟩

UNSOLVED. There have been several attempts to tackle this problem by considering, for example, what kind of collineation group such a plane would have to possess but these attempts have so far been unsuccessful. For details, see Section 2 of Chapter 11 in [DK2].

PROBLEM 13.2. ⟨*How many right quasifields of order 16 with GF[2] as nucleus exist? How many isomorphically distinct projective planes of order 16 do these give rise to?* (page 484)⟩

The problem, so far as the author has been able to discover, is unsolved. However, as a contribution towards its solution, it is known that there are just eight isomorphically distinct translation planes of order 16 altogether and the structure of these planes is described in U. Dempwolff and A. Reinfart(1983). See also the earlier paper of these authors listed below.

[*Dempwolff U. and Reinfart A.*

(1982) Translation planes of order 16 admitting a Baer 4-group. J. Combin. Theory Ser. A32, 119-124.

(1983) The classification of the translation planes of order 16. I. Geom. Dedicata 15, 137-153.]

New Problems

Chapter 1.

1.1 How can the latin subsquares which occur in the Cayley tables of Bruck, Bol and other classes of loops be characterized? (page 28).

1.2 What is the largest number of $s \times s$ subsquares for fixed s $(1 < s < n)$ which can be avoided by a given $n \times n$ latin square L? Does this number depend on the structure of L: for example, whether L is group-based? Also, the same question when s is allowed to take all values $(1 < s < n)$. (page 36)

Chapter 2.

2.1. Which classes of groups other than those discussed on pages 75 to 78 of [DK2] are super P-groups? (page 67)

2.2. Is it true that if L is the multiplication table of a non-soluble group then, not only does L have at least one, and possibly many, square roots $\sqrt{L}$, but at least one of these square roots is a latin square. (page 70).

Chapter 3.

3.1. Does there exist an odd integer n such that a weakly completable critical set of size less than $\lfloor n^2/4 \rfloor$ exists for the cyclic latin square of order n? (page 95)

3.2. What is the largest size (in terms of n) for a critical set in a cyclic latin square of order n? (page 95)

3.3. What are the sizes of minimal critical sets for the various main classes of non-cyclic latin squares? (page 95)

3.4. What is the size of a minimal critical set for a Sudoku latin square L whether group-based or not? (page 96)

3.5. What is the largest size which a critical set in a latin square may have? (page 98)

3.6. How can latin bitrades of genus 1 (or any other genus greater than zero) be characterized? (page 103)

3.7. 'What is the minimal order of a latin square into which a given separated (or non-separated) latin trade T of size h may be embedded? (page 104)

3.8. Are there combinatorial structures other than latin bitrades which can be represented topologically and hence assigned a genus? (page 105)

3.9. Is it true that every $\frac{1}{4}\epsilon$-dense partial latin square is completable? (page 106)

3.10. Is Rodney's conjecture that every latin square contains a duplex true? (page 121)

Chapter 4.

4.1. Is it true that, for odd $n \geq 7$, every transversal of L_n can be extended in at least two distinct ways to a decomposition of L_n into transversals? (page 157)

4.2 Do there exist latin squares of orders $n \neq 3^h$ which contain as many as $\frac{1}{18}n^2(n-1)$ 3×3 subsquares? (page 158)

Chapter 5.

5.1. What are the conditions for two latin directed triple systems to be orthogonal? (page 188)

5.2. Can the concept of orthogonality be modified to cover the case of non-latin directed triple systems? (page 188)

5.3. Do pairs of maximal orthogonal $r \times n$ latin rectangles exist for all integers r such that $n/3 < r < n$ when n is sufficiently large? (page 194)

Chapter 6.

6.1. What is the largest number of totally diagonal latin squares that can exist in a pairwise orthogonal set? (page 219)

6.2. Can a 25×25 bimagic square be constructed by a modification of the method described in Keedwell(2011c)? (page 222)

6.3. What is the size of a minimal critical set for a 4×4 magic square? Is this the same for all such squares in the same class? Does it vary according to which of the twelve classes a particular square belongs? (page 224)

6.4. Is it possible to construct an $n^2 \times n^2$ Sudoku latin square $(n \geq 4)$ all of whose $n \times n$ subsquares are magic squares of which no two are equivalent? (page 224)

6.5. How many non-isomorphic and non-equivalent Room designs of order $2n$ exist? (page 233)

Chapter 8.

8.1. Can a complete directed graph with a prime number p of vertices be separated into p disjoint Hamiltonian paths? (page 277)

Chapter 9.

9.1. For which orders n not divisible by 8, if any, do 4-homogeneous latin squares exist? (page 285)

9.2. Find the complete spectrum of integers $n-k$ for which $(n-k)$-homogeneous latin squares exist. (page 285)

9.3. For which integers h do orthogonal h-homogeneous latin squares exist or, alternatively, can it be proved that they do not exist for some values of h? (page 285)

9.4. Do atomic latin squares of composite but non-prime-power orders exist? (page 294)

9.5. For which even orders do N_∞-squares exist? (page 294)

Chapter 10.

10.1. Can the bounds $t \leq m+1$ when m is odd and $t \leq m$ when m is even for the number t of latin squares of order $2m$ in a mutually nearly orthogonal set be attained when $m \neq 3$? (page 296)

10.2. What is the upper bound $N_q(n)$ on the number of squares in a set of MQOLS of order n? (page 301)

10.3. Is it true that, for some n, $n > 6$, $N_q(n) > n - 1$. (page 302)

10.4. For which orders $n > 10$ do D-type and/or C-type latin power sets of at least two members exist? (page 304)

Chapter 11.

11.1. Do latin triangles exist of all odd orders and of all even orders except 4, 6 and 10? (page 325)

11.2. Alternatively, is there an integer n_0 such that, for all $n > n_0$, an $LT(n)$ exists? (page 325)

11.3. For a given order n, how many different $LT(n)$'s exist? (page 325)

11.4. In this context, how should we define "different"? (page 325)

11.5. What is the maximum number of mutually orthogonal $LT(n)$'s that can exist? (page 325)

Bibliography

Notes on the Bibliography

(1) The italicized numbers which follow each bibliographic item indicate the pages on which that item is referred to in the text.

(2) Only papers actually cited in the book are included as the number of published papers on, or related to, latin squares is now so large that to provide a comprehensive bibliography is no longer possible.

Abdel-Ghaffar K.A.S.

(1996) On the number of mutually orthogonal partial Latin squares. Ars Combin. 42, 259-286. MR 97f:05025. [303]

Acketa D.M. and Matić-Kekić S.

(1995) An attempt for construction of a triple of pairwise mutually orthogonal Latin squares on 10 elements. Zb. Rad. Prirod.-Mat. Fak. Ser. Mat. 25, 141-153. MR 97i:05015. [178]

Aczél J.

(1964) Kvázicsoportok - hálózatok - nomogramok. Mat. Lapok 15, 114-162. In Hungarian. [Quasigroups - nets - nomograms.] MR 30(1965)#3171. [48,57,253,265, 267]

(1965) Quasigroups, nets and nomograms. Advances in Math. 1, 383-450. MR 33(1967)#1395. [48,57,253,265,267]

(1969) Conditions for a loop to be a group and for a groupoid to be a semigroup. Amer. Math. Monthly 76, 547-549. Not reviewed in MR. [5]

Aczél J., Pickert G. and Radó F.

(1960) Nomogramme, Gewebe und Quasigruppen. Mathematica (Cluj) (2) 25, 5-24. MR 23(1962)#A1679. [253]

Adams P. and Khodkar A.

(2001) Smallest weak and smallest totally weak critical sets in the latin squares of order at most seven. Ars Combin. 61, 287-300. MR 2002h:05027. [94,96]

Afsarinejad K.

(1986) Self-orthogonal Knut Vik designs. Statist. Probab. Lett. 4, 289. MR 87j:62136. [221]

(1987) On mutually orthogonal Knut Vik designs. Statist. Probab. Lett. 5, 323-324. MR 88i:62138. [221]

Ahrens W.

(1901) Mathematische Unterhaltungen und Spiele (Chapter 12) Leipzig. [205]

Akar A.

(1895) Deuxième réponse à la question numéro 261. Interméd. Math. 2, 79-80. [165]

Albert A.A.

(1943) Quasigroups I. Trans. Amer. Math. Soc. 54, 507-519. MR 5(1944), page 229. [9,12,14,57]

(1944) Quasigroups II. Trans. Amer. Math. Soc. 55, 401-419. MR 6(1945), page 42. [9,57,141]

Alekseev V.E.

(1974) The number of Steiner triple systems. (In Russian) Mat. Zametki 15, 769-774. MR 50(1975)#12754. [58]

Alimena B.S.

(1962) A method of determining unbiassed distribution in the Latin square. Psychometrika 27, 315-317. MR 26(1963)#865. [74]

Alter R.

(1975)Research Problems: How Many Latin Squares are There? Amer. Math. Monthly 82, 632-634. MR 1537769. [148]

Alves C., Gurganus K., McLaurin S. and Smith D.

(1991) On Latin triangles. Ars Combin. 32, 129-141. MR 92i:05048. [324]

Andersen L.D.

(2013) Latin squares. In "Combinatorics: Ancient and Modern", pp.251-283. Eds. R.J. Wilson and J. Watkins. (Oxford Univ. Press) [1,205]

Andersen L.D. and Mendelsohn E.

(1982) A direct construction for Latin squares without proper subsquares. Algebraic and geometric combinatorics, 27-53, North-Holland Math. Stud., 65, North-Holland, Amsterdam. MR 85m:05021. [294]

Anderson B.A.

(1987a) A fast method for sequencing low order non-abelian groups. Annals of Discete Math. 34, 27-42. MR 89f:20033. [80]

(1987b) Sequencings of dicyclic groups. Ars Combin. 23, 131-142. MR 88j:20022. [81]

(1991) A product theorem for 2-sequencings. Discrete Math. 87, 221-236. MR 92a:05025. [80]

Anderson B.A. and Ihrig E.C.

(1992) All groups of odd order have starter-translate 2-sequencings. Australas. J. Combin. 6 , 135-146. MR 93j:20049. [81]

(1993a) Every finite solvable group with a unique element of order two, except the quaternion group, has a symmetric sequencing. J. Combin. Des. 1, 3-14. MR 95h:20031. [80]

(1993b) Symmetric sequencings of nonsolvable groups. In Proc. 24th SE Conf. on Combinatorics, GraphTheory and Computing (Boca Raton, 1993). Congr. Numer. 93, 73-82. MR 94m:20054. [80]

Anderson I.

(1997) Combinatorial designs and tournaments. Oxford Lecture Series in Mathematics and its Applications, 6. (The Clarendon Press, Oxford University

Press, New York.) MR 99h:05008. [324]

Anderson I. and Preece D.A.
(2006) Logarithmic terraces. Bull. Inst. Combin. Appl. 46, 49-60.
MR 2006i:11023. [81]
(2008a) A general approach to constructing power-sequence terraces for Z_n.
Discrete Math. 308, 631-644. MR 2009b:11047. [81]
(2008b) Some Z_{n+2} terraces from Z_n power-sequences, n being an odd prime.
Discrete Math. 308, 4086-4107. MR 2009c:11002. [81]
(2010) Some narcissistic power-sequence Z_{n+1} terraces with n an odd prime
power. Ars Combin. 97A, 33-57. MR 2012c:05062. [81]

Andrews W.S.
(1917) Magic squares and cubes. (Open Court, Chicago) (Dover Reprint,
1960). [205]

Appa, G.
(2003) Private communication to the author regarding unpublished work involving linear programming and computing. [178]

Appa, G.; Magos, D.; Mourtos, I.
(2004) An LP-based proof for the non-existence of a pair of orthogonal Latin
squares of order 6. Oper. Res. Lett. 32, 336-344. MR2057788. [165]

Archbold J.W.
(1960) A combinatorial problem of T. G. Room. Mathematika 7, 50-55. MR
23(1962)#A2334. [227]

Archbold J.W. and Johnson N.L.
(1958) A construction for Room's squares and an application in experimental
design. Ann. Math. Statist. 29, 219-225. MR 21(1960)#950. [227]

[Argunov B.I.] Аргунов Б.И.
(1950) Конфиурационные постулаты и их алгебраические зквиваленты.
Мат.Сборник (Н.С.) 26(68), 425-456. [Configurational postulates and their
algebraic equivalents. Mat. Sb. (N.S.) 26(68), 425-456.] MR 12(1951), page 525.
[261,264,265]

Arkin J.
(1973a) The first solution of the classical Eulerian magic cube problem of
order ten. Fibonacci Quart. 11, 174-178. MR 47(1974)#3199. [198]
(1973b) A solution to the classical problem of finding systems of three mutually orthogonal numbers in a cube formed by three superposed $10 \times 10 \times 10$
cubes. Fibonacci Quart. 11, 485-489, 494. MR 50(1975)#1922. [198]
(1974) A solution of orthogonal triples in four superimposed $10 \times 10 \times 10$
Latin Cubes. Fibonacci Quart. 12, 133-140. MR 51(1976)#7904. [198]

Arkin J. and Straus E.G.
(1974) Latin k-cubes. Fibonacci Quart. 12, 288-292. MR 50(1975)#4337.
[198]

[Arlazarov V.L., Baraev A.M., Gol'fand Ja.Ju. and Faradžev I.A.] Арлазаров В.Л., Бараев А.М., Гольфанд Я.Ю., Фараджев И. А.

(1978) Построение с помощью ЗВМ всех латинских квадратов порядка 8, В сб. "Алгоритм исслед в комбинаторике", Москва, 129-141. [The construction with the aid of a computer of all Latin squares of order eight. In "Algorithmic studies in combinatorics" (Russian), pp.129-141.] MR 80a:05031. [140]

Armstrong E.J.

(1955) A note on latin squares. Math. Gaz. 39, 215-217. Not reviewed in MR. [205]

Asplund J. and Keranen M.S.

(2011) Mutually orthogonal equitable Latin rectangles. Discrete Math. 311, 1015-1033. MR2787313. [194]

Athreya K.B., Pranesachar C.R. and Singhi N.M.

(1980) On the number of Latin rectangles and chromatic polynomials of $L(K_{r,s})$. Europ. J. Combin. 1, 9-17. MR 82d:05033. [149]

Atkin A.O.L., Hay L. and Larson R.G.

(1977) Construction of Knut Vik designs.J. Statist. Plann. Inference.1, 289-297. MR 58(1979)#10526. [221]

(1983) Enumeration and construction of pandiagonal latin squares of prime order. Comput. Math. Appl. 9, 267-292. MR 85i:05048. [221]

Bachet C.G.

(1612) Problèms plaisens et délectable. (First published in 1612. 2nd Ed., Lyons, 1624.) [205]

Baer R.

(1939) Nets and groups I. Trans. Amer. Math. Soc. 46, 110-141. MR 1(1940), page 6. [9,253]

(1940) Nets and groups II. Trans. Amer. Math. Soc. 47, 435-439. MR 2(1941), page 4. [9,253]

Bailey R.A.

(1984) Quasi-complete latin squares: construction and randomization. J. Royal Statist. Soc. B46, 323-334. MR 86i:62161. [81]

Bailey R.A., Ollis M.A. and Preece D.A.

(2003) Round-dance neighbour designs from terraces. In Proc. 18th Brit. Combin. Conf. (Brighton, 2001). Discrete Math. 266, 69-86. MR 2004k:05043. [87]

Balasubramanian K.

(1990) On transversals in Latin squares. Linear Algebra Appl. 131, 125-129. MR 91d:05022. [155]

Ball W.W.R.

(1939) Mathematical recreations and essays. Macmillan, New York, Eleventh Ed.), revised by H. S. M. Coxeter. MR 8(1947), page 440. See also MR 22(1961)

#3649. [*58,205,215,216,219*]

Bammel S.E. and Rothstein J.

(1975) The number of 9 × 9 Latin squares. Discrete Math. 11, 93-95. MR 51(1976)#7882. [*140,147*]

Barbette E.

(1896) Question numéro 767. Interméd. Math. 3, 54; also 12(1905), 101. [*165,215*]

Barra J.R.

(1963) A propos d'un théorème de R. C. Bose. C. R. Acad. Sci. Paris 256, 5502-5504. MR 27(1964)#1383. [*186*]

Barra J.R. and Guérin R.

(1963a) Extension des carrés gréco-latins cycliques. Publ. Inst. Statist. Univ. Paris 12, 67-82. MR 28(1964)#1138. [*2,287*]

(1963b) Utilisation pratique de la méthode de Yamamoto pour la construction systématique de carrés gréco-latins. Publ. Inst. Statist. Univ. Paris 12(1963), 131-136. MR 28(1964)#1139. [*287*]

Bartlett P.

(2013) Completions of ε-dense partial Latin squares. J. Combin. Des. 21, 447-463. MR3090723. [*106*]

Bate J.A. and van Rees G.H.J.

(1999) The size of the smallest strong critical set in a latin square. Ars Combin. 53, 73-83. MR 2000g:05034. [*95*]

Bateman P.T.

(1950) A remark on infinite groups. Amer. Math. Monthly 57, 623-624. MR 12(1951), page 670. [*70*]

Baumgartner L.

(1921) Gruppentheorie. (De Gruyter, Berlin). [*29*]

Beals R., Gallian J.A., Headley P. and Jungreis D.

(1991) Harmonious groups. J. Combin. Theory A56, 223-238. MR 93a:20033. [*81*]

Bean R., Bidkhori H., Khosravi M. and Mahmoodian E.S.

(2005) k-homogeneous Latin trades. Bayreuth. Math. Schr. No.74, 7-18. MR 2006k:05034. [*105*]

Bedford D.

(1993) Construction of orthogonal Latin squares using left neofields. Discrete Math. 115, 17-38. MR 94c:05016. [*156,179,243,289*]

(1995) Orthomorphisms and near orthomorphisms of groups and orthogonal Latin squares: a survey. Bull. Inst. Combin. Appl. 15, 13-33. MR 96k:05026. [*286*]

(1998a) Quasi-orthogonal Latin squares and related designs. J. Combin. Math. Combin. Comput. 26 , 213-224. MR 98i:05024. [*298,299,300*]

(1998b) A note on non-cyclic latin squares with cyclic properties. Private communication to the author, unpublished. [*287,290*]

Bedford D. and Johnson M.

(2000) Weak uniquely completable sets for finite groups. Bull. Lond. Math. Soc. 32, 155-162. MR 2001c:05032. [*95,96*]

(2001) Weak critical sets in cyclic latin squares, Australas. J. Combin. 23, 301-316. MR 2001i:05041. [*96*]

Bedford D., Ollis M.A. and Whitaker R.M.

(2001) On bipartite tournaments balanced with respect to carry-over effects for both teams. In Proc. 17th Brit. Combin. Conf. (Canterbury, 1999). Discrete Math. 231, 81-87. MR 2002a:05120. [*323*]

Bedford D. and Whitaker R.M.

(1999) Enumeration of transversals in the Cayley tables of the non-cyclic groups of order 8. In Proc. 16th Brit. Combin. Conf. (London, 1997). Discrete Math. 197/198, 77-81. MR 99i:05200. [*156*]

(2000a) New and old values for maximal MOLS(n). Ars Combin. 54, 255-258. MR 2000j:05023. [*179*]

(2000b) Sets of mutually quasi-orthogonal Latin squares. (Papers in honour of Ernest J. Cockayne.) J. Combin. Math. Combin. Comput. 33, 299-310. MR 2001d:05024. [*299,301*]

(2001) Bounds on the maximum number of Latin squares in a mutually quasi-orthogonal set. In Proc.17th Brit. Combin. Conf. (Canterbury, 1999). Discrete Math. 231, 89-96. MR 2002a:05047. [*302*]

(2003) Quasi-orthogonal Latin squares and their applications. Bull. Inst. Combin. Appl. 39, 41-52. MR 2004f:05030. [*299,302*]

Behrooz Bagheri, Gh. and Mahmoodian E.S.

(2011) On the existence of k-homogeneous Latin bitrades. Utilitas Math. 85, 333-345. MR 2012h:05052. [*95*]

Beineke L.W.

(1964) Decomposition of complete graphs into forests. Publ. Math. Inst. Hung. Acad. Sci. 9, 589-593. MR 32(1966)#4031. [*71*]

[Belousov V.D.] Белоусов В.Д.

(1958) Ассоциативные системы квазигрупп.Успехи Мат. Наук 13, 243. [Associative systems of quasigroups. Uspehi Mat. Nauk. 13(1958), 243.] Not reviewed in MR. [*55,56*]

(1961) Ассоциативные в целом системы квазигрупп. Мат. Сборник (Н. С.) 55(97), 221-236. [Globally associative systems of quasigroups. Mat. Sb. (N. S.) 55 (97)(1961), 221-236.] MR 27(1964)#2578. [*56*]

(1962) Замкнутые системы взаимно ортогональных квазигрупп. IV. Всесоюзное совещание по общей алгебре (Киев, 16-22 Мая 1962 г.) Успехи Мат. Наук. 17, 202-203. [Closed systems of mutually orthogonal quasigroups. Usp. Nat. Nauk. 17(1962), No. 6(108), 202-203.] Not reviewed in MR. [*185*]

(1965) Системы квазигрупп с обобщенными тождествами. Успехи Мат. Наук 20, No.1(121),75-146. [Systems of quasigroups with generalized identities

Usp. Mat. Nauk. 20, No.1(121), 75-146.] Translated as Russian Math. Surveys 20(1965), 73-143. MR 30(1965)#3934. [*15,37,46,55,57,183,184*]

(1966) У равновшенные тождества в квазигруппах. Мат. Сборник (Н. С.) 70(112), 55-97. [Balanced identities in quasigroups. Mat. Sb. (N. S.) 70(112), 55-97.] MR 34 (1967)#2757. [*48*]

(1967a) Неассоциативные бинарные системы В Сб. "Алгебра. Топология. Геометрия. 1965." (Итоги науки) Москва(1967), 63-81. [Non-associative binary systems. Algebra, Topology, Geometry 1965, 63-81. Akad. Nauk SSSR Inst. Nauěn. Tehn. Informacii, Moscow, 1967.] MR 35(1968)#5537. [*16,57*]

(1967b) Основы теории квазигрупп и луп. Издат. Наука, Москва. [Foundations of the theory of quasigroups and loops. Izdat. Nauka, Moscow 1967.] MR 36(1968)#1569. [*16,22,23,45,57,125,253,265*]

(1967c) Продолжения квазигрупп. Изв. А. Н. Молд. ССР No.8, 3-24. [Extensions of quasigroups. Bull. Akad. Stiince RSS Moldoven (1967), No.8., 3-24.] MR 38(1969)#4592. [*23*]

(1967d) Преобразования в сетях. Мат.Зап.Уральский УН-Т6, Тетрадь 1, 3-20. [Transformations in nets. Ural. Gos. Univ. Mat. Zap. 6, Tetrad' 1,(1967), 3-20.] MR 36(1968)#3906. [*268*]

(1968) Системы ортогональных оператзий. Мат. Сборник (Н. С.) 77 (119), 38-58. [Systems of orthogonal operations. Mat. Sb. (N. S.) 77(119)(1968), 38-58.] Translated as Mathematics of the USSR. Sbornik 6(1968), 33-52. MR 38(1969)#1200. [*183*]

(1971) Алгебраические сети и квазигруппы. Издательство Штиинца, Кишинев. [Algebraic nets and quasigroups. Izdat. Stiince Kishinev, 1971.] MR 49(1975)#5214. [*253,265*]

(1972) n-арные квазигруппы. Издательство Штиинтза, Кишинев. [n-ary quasigroups. Izdat. Stiince, Kishinev 1972.] MR 50(1975)#7396. [*253*]

(1983a) Одна теорема об уравновешенных тождествах, Мат. Исслед. 71, 22-24. [A theorem on balanced identities. Mat. Issled. 71(1983), 22-24. MR 84m:20075. [*49*]

(1983b) Парастрофно-ортогоналные квазигруппы. [Parastrophic-orthogonal quasigroups] Akad. Nauk Moldav. SSR, Inst. Mat. s Vychisl. Tsentrom, Kishinev, 51 pp. MR 86f:20086. [*190,192*]

(2005) Parastrophic-orthogonal quasigroups. [Translated from the 1983 Russian original by A.D. Keedwell and P. Syrbu based on the original Russian and on an earlier English translation supplied to Keedwell by Belousov himself, see MR 86f:20086.] Quasigroups Related Systems 13, 25-72. MR 2006i:20077. [*190,191,192*]

[Belousov V.D. and Belyavskaya G.V.] Белоусов В.Д. и Белявская Г.Б.

(1968) О продолжении квазигрупп. Тезисы 1. Всесоюзного симпоз по теории квазигрупп и её приближениям, Сухуми. [On prolongations of quasigroups. Thesis I. All-Union Symposium on the theory of quasigroups and their applications, Suhumi, 1968.] Not reviewed in MR. [*23*]

[Belousov V.D. and Gvaramiya A.A.] Белоусов В.Д. и Гварамия А.А.

(1966) О квазигруппах Стейна. Сообщ. А. Н. Груз ССР, 44, 537-544.
[Stein quasigroups. Sakharth SSR. Mecn. Akad. Moombe 44(1966), 537-544.] MR
34(1967)#2758. [185]

[Belousov V.D. and Ryzkov V.V.] Белоусов В.Д. и Рыжкоб В.В.

(1966) Об одном способе получения фигур замыкания. Мат.Исслед
1,140-150. [On a method of obtaining closure figures. Mat. Issled 1(1966), No.
2., 140-150.] MR 36(1968)#6526. [268]

[Belyavskaya G. В.] Белявская Г.Б.

(1969) О сужении квазигрупп. Десятый всесоюзный алгебраический
коллоквиум. Новосибирск. Резюме докладов. ТОМ 2, 106-107. [On contraction of quasigroups. Tenth All-Union Algebra Colloquium. Novosibirsk (1969).
Summaries of Reports. Vol. 2, 106-107.] Not reviewed in MR. [23]

(1970a) Туепно-изотопные квазигруппы. Мат. Исслед 5, No.2, 13-27.
[Chain-isotopic quasigroups. Mat. Issled. 5 (1970), No. 2, 13-27.] MR 44(1972)
#1757. [23]

(1970b) Обобщенное продолжение квазигрупп. Мат. Исслед. 5, No.2,
28-48. [On generalized prolongation of quasigroups. Mat. Issled. 5(1970), No. 2.,
28-48.] MR 44(1972)#1758. [23]

(1970c) Сжатие квазигрупп I. Изв. А. Н. МССР. No.1, 6-12. [Compres-
sions of quasigroups I. Izv. Akad. Nauk Moldav. SSR (1970), No.1, 6-12.] MR
43(1972)#4942. [23]

(1970d) Сжатие квазигрупп II. Изв. А. Н. МССР. No.3, 3-17. [Com-
pressions of quasigroups II. Izv. Akad. Nauk Moldav. SSR (1970), No.3, 3-17.]
MR 44(1972)#1756. [23]

(1971) Алгоритмы решения некоторых задач теории квазигрупп. "Во-
просы теории квазигрупп и луп." Ред В. Д. Белусов. Акад.Наук Мол-
давской ССР Кишинев, 20-30. [Algorithms for solving some problems in the
theory of quasigroups. From Questions in the theory of quasigroups and loops.
Edited by V. B. Belousov. Acad. Nauk Moldav. SSR. Kishinev, 1971, 20-30.] Not
reviewed in MR. [155]

(1979) Построение (n^2-2)-ортогональных квазигрупп четного порядка
n, где $n-1 \not\equiv 0 \pmod 3$. Квазигруппы и лупы. Мат. Исслед. [Construction
of $(n^2 - 2)$-orthogonal quasigroups of even order n, where $n - 1 \not\equiv 0 \pmod 3$.
Quasigroups and loops. Mat. Issled. No. 51 (1979), 23-26, 162.] MR 80m:20053.
[156]

(2002) Quasigroup power sets and cyclic S-systems. [in English] Quasigroups
Related Systems 9, 1-17. MR 2003j:20103. [304]

[Belyavskaya G.B. and Russu A.F.] Белявская Г.Б. и Рущу А.Ф.

(1975) О допустимости квазигрупп. Мат. Исслед 10, No.1(35), 45-57.
[On the admissibility of quasigroups. Mat. Issled 10, No.1(35), 45-57.] MR
52(1976)#3407. [155]

Bennett F.E., Du B. and Zhang H.

(1997) Existence of conjugate orthogonal diagonal Latin squares. J. Combin. Des. 5, 449-461. MR 98i:05026. [*219*]

(1998) Existence of self-conjugate self-orthogonal diagonal Latin squares. J. Combin. Des. 6, 51-62. MR 98m:05022. [*219,242*]

(2001) Existence of (3, 1, 2)-conjugate orthogonal diagonal Latin squares. J. Combin. Des. 9, 297-308. MR 2002d:05031. [*219*]

Bennett F.E., Zhang H. and Zhu L.

(1996) Self-orthogonal Mendelsohn triple systems. J. Combin. Theory A73, 207-218. MR 97b:05020. [*188*]

Bennett F.E. and Zhu L.

(1992) Conjugate-orthogonal Latin squares and related structures. In "Contemporary design theory" (Wiley-Intersci. Ser. Discrete Math. Optim., Wiley, New York), pp.41-96. MR1178500. [*188,190,192*]

Berge C.

(1962) The Theory of Graphs and its Applications. Translated by A. Doig. (Methuen, London and J. Wiley, New York.) MR 24(1962)#A2381. [*84*]

Berlekamp E.R., Conway J.H. and Guy R.K.

(2004) Winning ways for your mathematical plays. Vol. 4. Second edition. A K Peters, Ltd., Wellesley, MA. pp. i-xvi and 801-1004. ISBN: MR 2004m:91050. [*224*]

Berman D.R. and Smith D.D.

(2012) Mitchell tournaments, Bull. ICA 65, 33-42. MR3052006. [*324*]

(2013) Balanced equitable mixed doubles round-robin tournaments. Bull. ICA 68, 90-101. MR3136866. [*321*]

Besharati N.,Goddyn L.,Mahmoodian E.S. and Mortezaeefar M.

(2016) On the chromatic number of latin square graphs. Discrete Math.339, 2613-2619. MR 3518411. [*317,318*]

Beth T.

(1983) Eine Bemerkung zur Abschätzung der Anzahl orthogonaler lateinischer Quadrate mittels Siebverfahren. (German) [A remark on estimating the number of orthogonal Latin squares using sieve methods] Abh. Math. Sem. Univ. Hamburg 53, 284-288. MR 86f:05032. [*179*]

Betten D.

(1983) Zum Satz von Euler-Tarry. (German) Math. Nat. Unt. 36, 449-453. Not reviewed in MR. [*165*]

(1984) Die 12 lateinischen Quadrate der Ordnung 6. (German) [The 12 Latin squares of order 6] Mitt. Math. Sem. Giessen No. 163, 181-188. MR 86j:05038. [*165*]

Beynon G.W.

(1943) Bridge Directors Manual. (Coffin Publishing Co., Massachusetts.) Not reviewed in MR. [*226*]

(1944) Duplicate Bridge Direction. (Stuyvesant Press, New York.) Not reviewed in MR. [226,227]

Blackburn S.R. and McCourt T.A.
(2013) Triangulations of the sphere, bitrades and abelian groups. Combinatorica, to appear. [103,105]

Blaschke W.
(1928) Topologisehe Fragen der Differentialgeometrie I. Thomsens Sechseckgewebe Zueinander diagonale Netze. Math. Z. 28, 150-157. [253]

Blaschke W. and Bol G.
(1938) Geometrie der Gewebe. Grundlehrung der Math. Wiss. 49, Berlin. [253]

Blokhuis A., Korchmáros G. and Mazzocca F.
(2011) On the structure of 3-nets embedded in a projective plane. J. Combin. Theory A118, 1228-1238. MR 2012e:51011. [268]

Bol G.
(1937) Topologische Fragen der Differentialgeometrie LXV. Gewebe und Gruppen. Math. Ann. 114, 414-431. [253,264,265]

Bogart K.P. and Longyear J.Q.
(1976) Counting 3 by n Latin rectangles. Proc. Amer. Math. Soc. 54, 463-467. MR 52(1976)#10449. [149]

Bondesen A.
(1969) Er det en gruppetavle? Nordisk Mat. Tidskr. 17, 132-136. In Danish. [Is it a group table?] MR 41(1971)#7016. [5]

Bose R.C.
(1938) On the application of the properties of Galois fields to the construction of hyper-Graeoo-Latin squares. Sankhyā 3, 323-338. [166,172]
(1963) Strongly regular graphs, partial geometries and partially balanced designs. Pacific J. Math. 13, 389-419 MR 28(1964)#1137. [281,316]

Bose R.C. and Bush K.A.
(1952) Orthogonal arrays of strength 2 and 3. Ann. Math. Statist. 23, 508-524. MR 14(1953), page 442. [202]

Bose R.C., Chakravarti I.M. and Knuth D.E.
(1960) On methods of constructing sets of mutually orthogonal latin squares using a computer I. Technometrics 2, 507-516. MR 23(1962)#A3099. [237]
(1961) On methods of constructing sets of mutually orthogonal latin squares using a computer II. Technometrics 3, 111-117. MR 23(1962)#A3100. [237]
(1978) Errata: "On methods of constructing sets of mutually orthogonal Latin squares using a computer. II" [In Technometrics 3(1961), 111-117]. Technometrics 20, No. 2, 219. MR 80a:05033. [237]

Bose R.C. and Nair K.R.
(1941) On complete sets of latin squares. Sankhyā 5, 361-382. MR 4(1943), page 33. [160,161,174,272]

Bose R. C. and Shrikhande S. S.

(1959) On the falsity of Euler's conjecture about the non-existence of two orthogonal latin squares of order $4t + 2$. Proc. Nat. Acad. Sci. USA. 45, 734-737. MR 21(1960)#3343. [165]

(1960) On the construction of sets of mutually orthogonal latin squares and the falsity of a conjecture of Euler. Trans. Amer. Math. Soc. 95, 191-209. MR 22(1961)#2557. [166]

Bose R.C., Shrikhande S.S. and Parker E.T.

(1960) Further results on the construction of mutually orthogonal latin squares and the falsity of Euler's conjecture. Canad. J. Math. 12, 189-203. MR 23(1962)#A69. [166,178,209,235]

Boyer C.

(2006) Les ancêtres français du Sudoku. Pour la Science. No.344. June 2006, 8-11. [222]

(2012) Addition-Multiplication magic squares. http://www.multimagie.com/English/AddMult1. [221]

Bradley J.V.

(1958) Complete counterbalancing of immediate sequential effects in a latin square design. J. Amer. Statist. Assoc. 53, 525-528. Not reviewed in MR. [72]

Brandt H.

(1927) Verallgemeinierung des Gruppenbegriffs. Math. Ann. 96, 360-366. [62,117]

Brant L.J. and Mullen G.L.

(1985) A note on isomorphism classes of reduced Latin squares of order 7. Utilitas Math. 27, 261-263. MR 87a:05037. [140]

Brayton R.K., Coppersmith D. and Hoffmann A.J.

(1974) Self-orthogonal Latin squares of all orders $n \neq 2, 3, 6$. Bull. Amer. Math. Soc. 80, 116-118. MR 48#5886. [190]

(1976) Self-orthogonal Latin squares. (Italian summary) Colloq. Internaz. sulle Teorie Combinatorie (Rome, 1973), Tomo II, pp. 509-517. (Atti dei Convegni Lincei, No. 17, Accad. Naz. Lincei, Rome.) MR 57(1979)#16101. [190]

Brocard H.

(1896) Problème des 36 officiers (Deuxième réponse à la question numéro 453.) Tnterméd. Math. 3, 90. [165]

Brouwer A.E.

(1984) Four MOLS of order 10 with a hole of order 2. J. Statist. Plann. and Inf. 10, 203-205. MR 86c:62091. [295]

Brown J.W.

(1968) Enumeration of latin squares with application to order 8. J. Combin. Theory 5, 177-184. MR 37(1969)#5111. [17,140]

(1972) An extension of Mann's theorem to a triple of mutually orthogonal latin squares of order 10. J. Combin. Theory A12, 316-318. MR 45(1973)#8544. [309]

Brown J.W., Cherry F., Most L., Most M., Parker E.T., Wallis W.D.

(1993) Completion of the spectrum of orthogonal diagonal latin squares. In "Graphs, Matrices and Designs", pp. 43-49. Lecture Notes in Pure and Appl. Math., 139. (Dekker, New York). MR 93i:05002. [217,218]

Brown J.W. and Parker E.T.

(1985) An attempt to construct three mols 10 with a common transversal. Proc. Conf. on groups and geometry, Part A (Madison, Wis., 1985). Algebras Groups Geom. 2, 258-262. MR 87g:05045. [178]

Browning J.M., Cameron P.J. and Wanless I.M.

(2014) Bounds on the number of small Latin subsquares. J. Combin. Theory A124, 41-56. MR3176190. [158]

Browning J., Stones D.S. and Wanless I.M.

(2013) Bounds on the number of autotopisms and subsquares of a Latin square. Combinatorica 33, 11-22. MR3070084. [125,158]

Brownlee K.A. and Loraine P.K.

(1948) The relationship between finite groups and completely orthogonal squares, cubes and hypercubes. Biometrika 35, 277-282. MR 10(1949), page 313. [200]

Bruck R. H.

(1944) Some results in the theory of quasigroups. Trans. Amer. Math. Soc. 55, 19-52. MR 5(1944), page 229. [19,23,48]

(1946) Contributions to the theory of loops. Trans. Amer. Math Soc 60, 245-354 MR 8(1947) page 134. [12,30]

(1951) Finite nets I. Numerical invariants. Canad. J. Math. 3, 94-107. MR 12(1951), page 580. [68,253,269]

(1955) Difference sets in a finite group. Trans. Amer. Math. Soc. 78, 464-481. MR 16(1955), page 1081. [175]

(1958) A survey of binary systems. (Springer Verlag, Berlin). MR2O(1959)#76. [4,12,14,20,57,59,117,118,126,253]

(1963a) Finite nets. II. Uniqueness and embedding. Pacific J. Math. 13, 421-457. MR 27(1964)#4768. [269,281]

(1963b) Studies in modern algebra. Chapter 4, What is a loop? Edited by A. A. Albert. Math. Assoc. of America and Prentice-Hall. MR 26(1963)#3750. [20,59,227,228,266]

Bruck R.H. and Ryser H.J.

(1949) The non-existence of certain finite projeetive planes. Canad. J. Math. 1, 88-93. MR 10(1949), page 319. [166,175]

Bruckheimer M., Bryan A.C. and Muir A.

(1970) Groups which are the union of three subgroups. Amer. Math. Monthly 77, 52-57. MR 41(1971)#322. [31]

Bryant D., Maenhaut B.M. and Wanless I.M.

(2006) New families of atomic Latin squares and perfect 1-factorisations. J. Combin. Theory A113, 608-624. MR 2006m:05195. [289,292]

Bryant B.F. and Schneider H.

(1966) Principal loop isotopes of quasigroups. Canad. J. Math. 18, 120-125. MR 32(1966)#5772. [*12,141*]

Bryant D., Egan J., Maenhaut B. and Wanless I.M.

(2009) Indivisible plexes in Latin squares. Des. Codes Cryptogr. 52, 93-105. MR 2010g:05055. [*122*]

Buekenhout F.

(2001) Generalized elliptic cubic curves. I. Finite Geometries, pp.35-48. Dev. Math. 3. (Kluwer Acad. Publ., Dordrecht) MR 2005a:51015. [*59*]

Bugelski B.R.

(1949) A note on Grant's discussion of the latin square principle in the design of experiments. Psychological Bulletin 46, 49-50. Not reviewed in MR. [*71*]

Burger A.P., Kidd M.P. and van Vuuren J.H.

(2010) Enumeration of isomorphism classes of self-orthogonal Latin squares. Ars Combin. 97, 143-152. MR 2012b:05051. [*145*]

Burgess D.R.B.

(2000) Weakly completable critical sets in latin squares. J. Combin. Math. Combin. Comput. 34, 65-69. MR 2001b:05043. [*96*]

Burgess D.R.B. and Keedwell A.D.

(2001) Weakly completable critical sets for proper vertex and edge colourings of graphs. Australas. J. Combin. 24, 35-45. MR 2002e:05052. [*274*]

Burns Pasles E. and Raghavarao D.

(2004) Mutually nearly orthogonal Latin squares of order 6. Utilitas Math. 65, 65-72. MR2048412. [*296*]

Bush K.A.

(1952a) A generalization of a theorem due to MacNeish. Ann. Math. Statist. 23, 293-295. MR 14(1953), page 125. [*202*]

(1952b) Orthogonal arrays of index unity. Ann. Math. Statist. 23, 426-434. MR 14(1953), page 125. [*202*]

Byleen K.

(1970) On Stanton and Mullin's construction of Room squares. Ann. Math. Statist. 41, 1122-1125. Not reviewed in MR. [*231*]

Byleen K. and Crowe D.W.

(1971) An infinite family of cyclic Room designs. Submitted to Rocky Mountain J. Math. (unpublished). [*230*]

Cairns S.

(1954) Computational attacks on discrete problems. Proc. Symp. on Special Topics in Appl. Math. Supplement to Amer. Math. Monthly 61, 29-31. MR 16(1955), page 77. [*326*]

Cameron P.J.

(1994) How few entries determine any Latin square? Bull. Inst. Combin. Appl. 10, 63-65. MR 94k:05037. [*105*]

Cameron P.J., Hilton A.J W. and Vaughan E.R.

(2012) An analogue of Ryser's theorem for partial Sudoku squares. J. Combin. Math. Combin. Comput. 80, 47-69. MR2918777. [85]

Cameron P.J. and Wanless I.M.

(2005) Covering radius for sets of permutations. Discrete Math. 293, 91-109. MR 2005k:05187. [121,313,314]

Carlitz L.

(1953a) Congruences connected with three line latin rectangles. Proc. Amer. Math. Soc. 4, 9-11. MR 14(1953), page 726. [150]

(1953b) A note on abelian groups. Proc. Amer. Math. Soc. 4, 937-939. MR 15(1954), page 503. [67]

Cavenagh N.J.

(2003) The size of the smallest latin trade in a back-circulant latin square. Bull. ICA 38, 11-18. MR 2004c:05037. [98]

(2006) A uniqueness result for 3-homogeneous Latin trades. Comment. Math. Univ. Carolin. 47, 337-358. MR 2007m:05040. [105]

(2008) The theory and application of Latin bitrades: a survey. Math. Slovaca 58, 691-718. MR 2009i:05045. [105]

(2010) Avoidable partial Latin squares of order 4m+1. Ars Combin. 95, 257-275. MR 2011f:05051. [36]

Cavenagh N.J., Donovan D. and Drápal A.

(2004) Constructing and deconstructing Latin trades. Discrete Math. 284, 97-105. MR 2005b:05037. [104]

(2005a) 3-homogeneous Latin trades. Discrete Math. 300 , No.1-3, 57-70. MR 2006j:05027. [105]

(2005b) 4-homogeneous Latin trades. Australas. J. Combin. 32 , 285-303. MR 2006e:05027. [105]

Cavenagh N.J., Hämäläinen C. and Nelson A.M.

(2009) On completing three cyclically generated transversals to a Latin square. Finite Fields Appl. 15, 294-303. MR 2010f:05030. [121]

Cavenagh N.J. and Kuhl J.

(2015) On the chromatic index of latin squares. Contrib. Discrete Math. 10, 22-30. MR 3499074. [316,317]

Cavenagh N.J. and Wanless I.M.

(2009) Latin trades in groups defined on planar triangulations. J. Algebraic Combin. 30, 323-347. MR 2010m:05047. [103]

(2010) On the number of transversals in Cayley tables of cyclic groups. Discrete Appl. Math. 158, 136-146. MR 2011a:05052. [155,289]

Cayley A.

(1877/8) On the theory of groups. Proc. Lond. Math. Soc. 9, 126-133. [1]

(1878a) Desiderata and suggestions No.1. The theory of groups. Amer. J. Math. 1, 50-52. [1]

(1878b) Desiderata and suggestions No.2. The theory of groups: graphical representation. Amer. J. Math. 1(1878), 174-176. [*274*]

(1889) On the theory of groups. Amer. J. Math. 11, 139-157. [*135*]

(1890) On latin squares. Messenger of Math. 19, 135-137. [*136*]

Cazalas E.

(1934) Carrés magiques au degré n. (Herman, Paris). [*222*]

Chetwynd A.G. and Rhodes S.J.

(1997) Avoiding partial Latin squares and intricacy. Discrete Math. 177, 17-32. MR 98j:05033. [*36*]

Chien O., Pflugfelder H.O. and Smith J.D.H.

(1990) Quasigroups and loops: theory and applications. Edited by O. Chein, H. O. Pflugfelder and J. D. H. Smith. Sigma Series in Pure Mathematics, 8. Heldermann Verlag, Berlin. MR 93g:20133. [*254*]

Choudhury A.C.

(1948) Quasigroups and non-associative systems I. Bull. Calcutta Math Soc 40, 183-194 MR 10(1949), page 591. [*282*]

(1949) Quasigroups and non-associative systems II. Bull. Calcutta Math. Soc. 41, 211-219. MR 11(1950), page 417. [*282*]

(1957) Quasigroups and non-associative systems III. Bull. Calcutta Math. Soc. 49, 9-24. MR 20(1959)#6478.[*282*]

Chowla S., Erdös P. and Strauss E.G.

(1960) On the maximal number of pairwise orthogonal latin squares of a given order. Canad. J. Math. 12, 204-208. MR 23(1962)#A70. [*218*]

Clark D. and Lewis J.T.

(1997) Transversals of cyclic Latin squares. Proc. 28th S.E. Conf. on Combinatorics, Graph Theory and Computing (Boca Raton, FL, 1997). Congr. Numer. 128, 113-120. MR 98i:05029. [*156*]

Cohen D. and Etzion D.

(1991) Row-complete latin squares which are not column-complete. Ars Combin. 32, 193-201. MR 92j:05029. [*72*]

Colbourn C.J. and Dinitz J.H.

(2001) Mutually orthogonal latin squares: a brief survey. J. Statist. Plann. Inference 95,9-48. MR1829097. [*309*]

Colbourn C.J. and Dinitz J.H. Eds.

(1996,2006) The CRC handbook of combinatorial designs. CRC Press Series on Discrete Mathematics and its Applications. (CRC Press, Boca Raton, FL.) ISBN: 0-8493-8948-8. 2nd Edition, 2006. MR 97a:05001. [*156,178,233,254*]

Colbourn C.J., Gibbons P.B., Mathon R., Mullin R.C. and Rosa A.

(1994) The spectrum of orthogonal Steiner triple systems. Canad. J. Math. 46, 239-252. MR 95e:05019. [*188*]

Colbourn C.J. and Rosa A.

(1992) Directed and Mendelsohn triple systems. In "Contemporary Design Theory", Eds. J.H.Dinitz and D.R. Stinson. pp. 97-136. (Wiley, New York.) MR 94c:05001. [60]

(1999) Triple systems. (Clarendon Press, Oxford Univ. Press, New York.) MR 2002h:05024. [58,59]

Colbourn C.J. and Zhu L.

(1995) The spectrum of r-orthogonal latin squares. In: Combinatorics Advances (Colbourn C.J. and Mahmoodian E.S.), Kluwer Academic Press, Dordrecht, 49-75. MR1366841. [295]

Cole F.N., White A.S. and Cummings L.D.

(1925) Complete classification of triad systems on fifteen elements. Mem. Nat. Acad. Sci. 14, Second memoir, 89. [58]

Cooper J., Donovan D. and Seberry J.

(1981) Latin squares and critical sets of minimal size. Australas. J. Combin. 4, 113-120. MR 92i:05049. [95]

Coxeter H.S.M. and Moser W.O.J.

(1965) Generators and relations for discrete groups. (2nd Edition, Springer-Verlag, Berlin.) MR 30(1965)#4818. [143]

Crampin D.J. and Hilton A.J.W.

(1975a) Remarks on Sade's disproof of the Euler conjecture with an application to Latin squares orthogonal to their transpose. J. Combin. Theory A18, 47-59. MR51(1976)#197. [312]

(1975b) On the spectra of certain types of Latin square. J. Combin. Theory A19, 84-94. MR51#12563.[312]

Csima J.

(1972) Restricted patterns. J. Combin. Theory, A12, 346-356. MR 46(1973)#8867. [115]

Curran D. and van Rees G.H.J.

(1979) Critical sets in latin squares. In Proc. Eighth Manitoba Conf. on Numerical Math. and Comput. (Manitoba, Winnipeg, 1978), pp.165-168. Congressus Numerantium XXII, Utilitas Math. MR 80j:05022. [95]

Damm H.M.

(2011) Half quasigroups and generalized quasigroup orthogonality. Discrete Math. 311, 145-153. MR 2011m:05063. [304]

Danhof K.J., Phillips N.C.K. and Wallis W.D.

(1990) On self-orthogonal diagonal latin squares. J. Combin. Math. Combin. Comput. 8, 3-8. MR 91j:05026. [218]

Davenport H.

(2008) The Higher Arithmetic. An introduction to the theory of numbers. Eighth edition. With editing and additional material by James H. Davenport.

Cambridge University Press, Cambridge. x+239 pp. ISBN: 978-0-521-72236-011-01. MR 2009j:11001. [*175*]

Davies I.J.

(1962) Enumeration of certain subgroups of abelian p-groups. Proc. Edinburgh Math. Soc. 13, 1-4. MR 26(1963)#3780. [*144*]

Dellanoy H. and Barbette E.

(1898) Probléme des 36 officiers. Interméd. Math. 5, 252. [*165*]

Dénes J.

(1961) Candidature of Mathematical Sciences Thesis. (Budapest, Hungary.) [*205,223*]

(1962) On a problem of L. Fuchs. Acta Sci. Math. (Szeged) 23, 237-241. MR 27(1964)#1493. [*2,106,108*]

(1967a) О таблицах умножения конечных квазигрупп и полугрупп. Мат. Исслед 2, 172-175. [Multiplication tables for finite quasigroups and semigroups. Mat. Issled. 2(1967), No. 2, 172-175.] Translated as a University of Surrey preprint, 1967. MR 37(1969) #6391. [*113*]

(1967b) Algebraic and combinatorial characterization of latin squares I. Mat. Časopis Sloven. Akad. Vied. 17, 249-265. MR 38(1969)#3164. [*27*]

(1997) Unsolved Problems: When Is There a Latin Power Set? Amer. Math. Monthly 104, no. 6, 563-565. MR1543037. [*304*]

Dénes J. and Hermann P.

(1982) On the product of all elements in a finite group. In "Algebraic and Geometric Combinatorics", pp. 105-109. (North Holland Math. Studies 65, Amsterdam.) MR 86c:20024. [*67*]

Dénes J. and Keedwell A.D.

(1988) Latin squares and one-factorizations of complete graphs. II. Enumerating one-factorizations of the complete directed graph K_n using MacMahon's double partition idea. Utilitas Math. 34 , 73-83. MR 90d:05046. [*289*]

(1989) A new conjecture concerning admissibility of groups. Europ. J. Combin. 10, 171-174. MR 90c:20033. [*69*]

(1991)) On two conjectures related to admissible groups and quasigroups. Proc. Second Internat. Math. Mini-Conf. Part II (Budapest, 1988). Period. Polytech. Transportation Energy 19, 33-35. MR 93f:20030. [*69*]

Dénes J., Mullen G.L. and Suchower S.J.

(1994) A note on power sets of Latin squares. J. Combin. Math. Combin. Comput. 16, 27-31. MR 95j:05045. [*304*]

Dénes J. and Owens, P.J.

(1999) Some new Latin power sets not based on groups. J. Combin. Theory A85, 69-82. MR 2000b:05029. [*305*]

Dénes J. and Pásztor K.

(1963) A kvázicsoportok néhány problémájáról. Magyar Tud. Akad. Mat. Fiz. Oszt. Közl. 13(1963), 109-118. In Hungarian. [Some problems on quasigroups.] MR 29(1965)#180. [*2,18,23,27,29,30,31,85,113*]

Dénes J. and Török É.

(1970) Groups and graphs. Combinatorial Theory and its Applications, pp.257-289. (North Holland, Amsterdam.) MR 46(1973)#91. [*67,72,77,78,275*]

Deriyenko I.I.

(2011) Private communication to the author. [*23*]

Deriyenko I.I. and Deriyenko, A.I.

(2009) Prolongations of quasigroups by middle translations. Quasigroups Related Systems 17, 177-190. MR 2011d:20128. [*23*]

Deriyenko L.L. and Dudek W.A.

(2008) On prolongations of quasigroups. Quasigroups Related Systems 16, 187-198. MR 2009m:05019. [*23*]

(2013) Contractions of quasigroups and latin squares. Quasigroups Related Systems 21, 167-176. Not yet reviewed in MR. [*23*]

Derksen N., Eggermont C. and van den Essen A.

(2007) Multimagic squares. Amer. Math. Monthly 114, 703-713. MR 2009a:05027. [*222*]

Devidé V.

(1955) Über eine Klasse von Gruppoiden. Hrvatsko Prirod. Društvo. Glasnik Mat.-Fiz. Astr. Ser. II. 10, 265-286. MR 18(1957), page 872. [*56*]

Dinitz J.H. and Ling A.C.H.

(2001) The existence of referee squares. Discrete Math. 232, 109-112. MR 2001k:05042. [*322*]

Dinitz, J.H. and Stinson, D.R.

(1992) Room squares and related designs. In "Contemporary design theory", pp. 137-204, (Wiley-Interscience. Ser. Discrete Math. Optim., Wiley, New York). MR 94c:05001. [*233*]

(2002) A singular direct product for bicolorable Steiner triple systems. In "Codes and designs" (Columbus, OH, 2000), pp.87-97. [Ohio State Univ. Math. Res. Inst. Publ., 10, de Gruyter, Berlin, 2002.] MR 2003m:05034. [*313*]

Donovan D.

(1999) Critical sets in latin squares of orders less than 11. J. Combin. Math. Combin. Comput. 29, 223-240. MR 99k:05038. [*98*]

Donovan D. and Bean R.

(2000) Closing a gap in the spectrum of critical sets. Australas. J. Combin. 22, 191-200. MR 2000f:05030. [*95*]

Donovan D. and Cooper J.

(1996) Critical sets in back circulant latin squares. Aequationes Math. 52, 157-179. MR 97g:05032. [*95*]

Donovan D. and Howse A.

(2000) Towards the spectrum of critical sets. Australas. J. Combin. 21, 107-130. MR 2000m:05041. [*98*]

Donovan D., Howse A. and Adams P.
(1997) A discussion of latin interchanges. J. Combin. Math. Combin. Comput. 23, 161-182. MR 98b:05019. [98]

Donovan D. and Mahmoodian E.S.
(2002) An algorithm for writing any latin interchange as a sum of intercalates. Bull. Inst. Combin. Appl. 34(2002), 90-98. MR 2002j:05028. [104]
(2003) Correction to a paper on critical sets: "An algorithm for writing any Latin interchange as a sum of intercalates" [Bull. Inst. Combin. Appl. 34 (2002), 90-98; MR1880972 (2002j:05028)]. Bull. Inst. Combin. Appl. 37, 44. MR 2003m:05039. [104]

Donovan D., Oates-Williams S. and Praeger Ch.E.
(1997) On the distance between distinct group latin squares. J. Combin. Designs 5, 235-248. MR 98e:05017. [111]

Dougherty, S.T.
(1994) A coding-theoretic solution to the 36 officer problem. Designs Codes Cryptogr. 4, 123-128. MR 95b:05037. [165]

Doyen J.
(1970a) On the number of non-isomorphic Steiner systems $S(2, m, n)$. Combinatorial Structures and their Applications. (Proc. Calgary Internat. Conf., Calgary, Alta., 1969), 63-64. (Gordon and Breach, New York). MR 43(1972)#4695. [58]
(1970b) Sur la croissance du nombre de systèmes triples de Steiner non isomorphes. J. Combin. Theory 8, 424-441. MR 41(1971)#1555. [58]

Doyen J. and Rosa A.
(1973) A bibliography and survey of Steiner systems. Bol. Un. Mat. Ital. (4) 7, 382-419. MR 48(1974)#118. [58]
(1978) An extended bibliography and survey of Steiner triple systems. In Proc. Seventh Manitoba Conf. on Numerical Math. and Computing (Univ. Manitoba, Winnipeg, 1977), pp.297-361. Congressus Numerantium 20, Utilitas Math. MR 80g:51009. [58]
(1980) Bibliography and survey of Steiner systems. (An updated bibliography and survey of Steiner systems.) In "Topics on Steiner triple systems" Eds. C.C. Lindner and A. Rosa. pp. 317-349. MR 84d:05039. [58]

Doyen J. and Valette G.
(1971) On the number of non-isomorphic Steiner triple systems. Math. Z. 120, 178-192. MR 43(1972)#4695. [58]

Drake D.A.
(1977) Maximal sets of latin squares and partial transversals. J. Statist. Planning and Inf. 1, 143-149. MR58(1979)#5272. [155,164,283]

Drake D.A. and Myrvold W.
(2004) The non-existence of maximal sets of four mutually orthogonal Latin squares of order 8. Des. Codes Cryptogr., 63-69. MR 2005d:05031. [179]

Drake D.A., van Rees G.H.J. and Wallis W.D.

(1999) Maximal sets of mutually orthogonal Latin squares. Discrete Math. 194, 87-94. MR 99k:05039. [179]

Drápal A.

(1991) On a planar construction of quasigroups. Czech. Math. J. 41, 538-548. MR 93c:20114. [99]

(1992) How far apart can the group multiplication tables be? Discrete Math. 13, 335-343. MR 93e:20035. [111]

(2001) Hamming distances of groups and quasigroups. In "Combinatorics, Prague, 1998". Discrete Math. 235, 189-197. MR 2002c:20037. [99,111]

(2003) On geometrical structure and construction of latin trades. Unpublished manuscript circulated at the conference "Loops '03" held in Prague. [101,104]

(2004) On mimimum distances of latin squares and the quadrangle criterion. Acta Sci. Math. (Szeged) 70, 3-11. MR 2005b:05041. [111]

Drápal A. and Griggs T.S.

(2010) Homogeneous toroidal Latin bitrades. Ars Combin. 96, 343-351. MR 2011f:05056. [105]

Drápal A., Griggs T.S. and Kozlik A.R.

(2014) Triple systems and binary operations. Discrete Math. 325, 1-11. MR3181226. [60]

(2015) Basics of DTS quasigroups: Algebra, geometry and enumeration. J. Algebra Appl. 14, No.6, 1550089 (24 pages). MR3338085. [60]

Drápal A., Hämäläinen C. and Vitězslav K.

(2010) Latin bitrades, dissections of equilateral triangles, and abelian groups. J. Combin. Designs. 18, 1-24. MR 2011g:05046. [103,104]

Drápal A. and Kepka T.

(1983) Exchangeable partial groupoids I. Acta Univ. Carolin. Math. Phys. 24, 57-72. MR 85f:08003. [94,98]

(1985) Exchangeable partial groupoids II. Acta Univ. Carolin. Math. Phys. 26, 3-9. MR 87e:08004. [98]

(1989) On a distance of groups and latin squares. Comment. Math. Univ. Carolin. 30, 621-626. MR 91c:05040. [98]

Drápal A., Kozlik A. and Griggs T.S.

(2012) Latin directed triple systems. Discrete Math. 312, 597-607. MR2854805. [60]

van Driel M.J.

(1936) Magic Squares of $(2n + 1)^2$ cells. Rider & Co., London. MR 1(1940), page 290. [205]

(1939) A supplement to Magic Squares of $(2n+1)^2$ cells. Rider & Co., London. MR 1(1940), page 290. [205]

Du B.L.

(1991) Some constructions of pairwise orthogonal diagonal Latin squares. J. Combin. Math. Combin. Comput. 9, 97106. MR 92c:05031. [218]

(1993) New bounds for pairwise orthogonal diagonal latin squares. Australas. J. Combin. 7, 87-99. MR 94a:05025. [*219*]

(1996a) Some constructions of pairwise orthogonal diagonal Latin squares. (Chinese) Gaoxiao Yingyong Shuxue Xuebao A11, 113-120. MR 97h:05032.[*219*]

(1996b) On a conjecture concerning self-conjugate self-orthogonal diagonal Latin squares. J. Combin. Math. Combin. Comput. 22, 65-66. MR 97f:05030. [*219*]

(1998) Constructing self-conjugate self-orthogonal diagonal Latin squares. Acta Math. Appl. Sinica (English Ser.) 14, 324-327. MR 99h:05015. [*219*]

Du B. and Wallis W.D.

(1999) The existence of self-conjugate self-orthogonal idempotent diagonal latin squares. Ars Combin. 53, 97-109. MR 2000h:05041. [*218*]

Dudeney H.E.

(1917) Amusements in Mathematics. (Nelson, London.) [*205*]

Dukes P. and Mendelsohn E.

(1999) Skew-orthogonal Steiner triple systems. J. Combin. Des. 7, 431-440. MR 2001d:05021. [*302*]

Dulmage A.L. and McMaster G.E.

(1975) A formula for counting three-line Latin rectangles. Proc. Sixth S.E. Conf. on Combinatorics, Graph Theory and Computing (Florida Atlantic Univ., Boca Raton, Fla., 1975), pp. 279-289. Congressus Numerantium, No. XIV, Utilitas Math., Winnipeg, Man. MR 52(1976)#13428. [*149*]

Egan J.

(2011) Bachelor Latin squares with large indivisible plexes. J. Combin. Des. 19, 304-312. MR 2012f:05048. [*120*]

Egan J. and Wanless I.M.

(2011) Indivisible partitions of Latin squares. J. Statist. Plann. Inference 141, 402-417. MR 2012i:05036. [*119,120,122*]

(2012) Latin squares with restricted transversals. J. Combin. Des. 20, 124-141. MR2868854. [*157*]

Elliott J.R. and Gibbons P.B.

(1992) The construction of subsquare free Latin squares by simulated annealing. Australas. J. Combin. 5, 209-228. MR 93a:05033. [*294*]

Elspas B., Minnick R. C. and Short R. A.

(1963) Symmetric latin squares. IEEE. Trans. on Elec. Comput. EC-12, 130-131. Not reviewed in MR. [*23,40*]

Erdös P. and Ginzburg A.

(1963) On a combinatorial problem in latin squares. Magyar Tud. Akad. Mat. Kutató Int. Közl. 8, 407-411. MR 29(1965)#2197. [*9*]

Erdös P. and Kaplansky I.

(1946) The asymptotic number of latin rectangles. Amer. J. Math. 68, 230-236. MR 7(1946), page 407. [*140,151*]

Erdös P., Rényi A. and Sös V.

(1970) Combinatorial Theory and its Applications. Colloquia Mathematica Societatis János Bolyai 4. (North Holland, Amsterdam). MR 45(1973)#4981. [114]

Erdös P. and Szekeres G.

(1934/5) Über die Anzahl der Abelschen Gruppen gegebener Ordnung und über ein verwandtes zahlentheoretisches Problem. Acta Sci. Univ. Szeged 7, 95-102. [143]

Etherington I.M.H.

(1962/63) Note on quasigroups and trees. Proc. Edinburgh Math. Soc. (2) 13, 219-222. MR 28(1964)#157. [41]

(1964/65) Quasigroups and cubic curves. Proc. Edinburgh Math. Soc. (2) 14, 273-291. MR 33(1967)#4170. [48,53,59]

Euler L.

(1776) De quadratis magicis. [Memoir presented to the Academy of Sciences of St. Petersburg on 17th October, 1776.] Published as (a) Mémoire de la Société de Flessingue, Commentationes arithmeticae collectae (elogé St. Petersburg 1783), 2(1849), 593-602; (b) Opera postuma 1(1862), 140-151; (c) Leonardi Euleri Opera Omnia, Série 1, 7(1923), 441-457. [165,205]

(1779) Recherches sur une nouvelle espèce de quarrés magiques. [Memoir presented to the Academy of Sciences of St. Petersburg on 8th March, 1779.] Published as (a) Verh. Zeeuwsch. Genootsch. Wetensch. Vlissengen 9(1782), 85-239; (b) Mémoire de la Société de Flessingue, Commentationes arithmetica collectae (elogé St. Petersburg 1783), 2(1849), 302-361; (c) Leonardi Euleri Opera Omnia, Série 1, 7(1923), 291-392. [18,135,153,165,205,215,219]

Evans A.B.

(1990) On strong complete mappings. Congr. Numer. 70, 323-328. MR 91c:05042. [18]

(1991) Maximal sets of mutually orthogonal Latin squares. I. European J. Combin. 12, 477-482. MR 92k:05028. [179]

(1992) Maximal sets of mutually orthogonal Latin squares. II. European J. Combin. 13 , no. 5, 345-350. MR 93i:05030. [179]

(2006) Latin squares without orthogonal mates. Des. Codes Cryptogr. 40, 121-130. MR 2007b:05031. [120,283]

(2009) The admissibility of sporadic simple groups. J. Algebra 321, 105-116. MR 2009i:20028. [69]

Evans T.

(1950) A note on the associative law. J. Lond. Math. Soc. 25, 196-201. MR 12(1951), page 75. [48,56]

(1960) Embedding incomplete latin squares. Amer. Math. Monthly. 67, 958-961. MR 23(1962)#A68. [27,85,114,116,117]

(1973) Latin cubes orthogonal to their transposes - A ternary analogue of Stein quasigroups. Aequationes Math. 9, 296-297. MR 48(1974)#3763. [190]

(1975) Algebraic structures associated with Latin squares and orthogonal arrays. Proc. Conf. on Algebraic Aspects of Combinatorics (Univ. Toronto, Toronto, Ont., 1975), pp. 31-52. (Congressus Numerantium, No. XIII, Utilitas Math., Winnipeg.) MR 52(1976)#13429. [*190,191*]

Falconer E.
(1971) Isotopes of some special quasigroup varieties. Acta Math. Acad. Sci. Hungar. 22, 73-79. MR 44(1972)#5400. [*48*]

Faragó T.
(1953) Contribution on the definition of a group. Publ. Math. Debrecen 3, 133-137. MR 15(1954), page 851. [*7*]

Feit W. and Thompson J.G.
(1963) Solvability of groups of odd order. Pacific J. Math. 13, 775-1029. MR 29(1965)#3538. [*67*]

Fenyves F.
(1968) Extra loops I. Publ. Math. Debrecen 15, 235-238. MR 38(1969)#5976. [*43*]
(1969) Extra loops II. Publ. Math. Debrecen 16, 187-192. MR 41(1971)#7017. [*44*]

Fiala N.C.
(2007) Short identities implying a quasigroup is a loop or group. Quasigroups Related Systems 15, 263-271. MR2383952. [*7*]

Finney D.J.
(1945) Some orthogonal properties of the 4 × 4 and 6 × 6 latin squares. Ann. Eugenics 12, 213-219. MR 7(1946), page 107. [*121*]

Fisher R.A.
(1942a) Some combinatorial theorems and enumerations connected with the numbers of diagonal types of a latin square. Ann. Eugenics 11, 395-401. MR 4(1943), page 183. [*138*]
(1945) A system of confounding for factors with more than two alternatives giving completely orthogonal cubes and higher powers. Ann. Eugenics 12, 283-290. MR 7(1946), page 107. [*199*]
(1966) The design of experiments. (Oliver & Boyd, Edinburgh, 8th edition). Not reviewed in MR. [*199,319*]

Fisher R.A. and Yates F.
(1934) The 6 × 6 latin squares. Proc. Camb. Phil. Soc. 30, 492-507. [*165*]

Fleisher E.
(1934) On Euler Squares. Bull. Amer. Math. Soc. 40, 218-219. [*165*]

Fog D.
(1934) Gruppentafeln und abstrakte Gruppentheorie. Skand. Mat. Kongr. Stockholm, 376-384. [*2*]

Ford G.G.

(1970) Remarks on quasigroups obeying a generalized associative law. Canad. Math. Bull. 13, 17-21. MR 41(1971)#5533. [56]

Franklin M.F.

(1984a) Cyclic generation of orthogonal Latin squares. Ars Combin. 17, 129-139. MR 85m:05022. [285]

(1984b) Cyclic generation of self-orthogonal Latin squares. Utilitas Math. 25, 135-146. MR 85h:05025. [285]

Freeman G.H.

(1979a) Some two-dimensional designs balanced for nearest neighbours. J. Roy. Statist. Soc. Ser.B 41, 88-95. MR 81j:62153. [80]

(1979b) Complete Latin squares and related experimental designs. J. Roy. Statist. Soc. Ser.B 41, 253-262. MR 81m:05032. [80]

(1985) Duplexes of 4 × 4 , 5 × 5 and 6 × 6 latin squares. Utilitas Math. 27, 5-24. MR 87d:0504. [121]

Frink O.

(1955) Symmetric and self-distributive systems. Amer. Math. Monthly 62, 697-707. MR 17(1956), page 458. [48]

Frisch S.A.

(1997) On the minimal distance between group tables. Acta Sci. Math. (Szeged) 63, 341-351. MR 98m:05023. [111,113]

Frolov M.

(1884) Le Problème d'Euler et les carrés magiques. (Gauthier-Villars, Paris.) [205]

(1890a) Recherches sur les permutations carrés. J. Math. Spéc. (3) 4, 8-11. [5,6,136]

(1890b) Recherches sur les permutations carrés. J. Math. Spéc. (3) 4, 25-30. [6,136]

Fu H-L., Lin S-C. and Fu C-M.

(2002) The length of a partial transversal in a Latin square. J. Combin. Math. Combin. Comput. 43, 57-64. MR 2003j:05027. [119]

Fuchs L.

(1958) Abelian Groups. (Akadémiai Kiadó, Budapest.) MR 21(1960)#5672. [106,143]

Gagola III S.M.

(2011) How and why Moufang loops behave like groups. Quasigroups Related Systems 19, 1-22. MR2850316. [21,28]

Gagola III S.M. and Hall J.I.

(2005) Lagrange's theorem for Moufang loops. Acta Sci. Math. (Szeged) 71, 45-64. MR 2006f:20079. [20]

Gallagher P.X.

(1967) Counting finite groups of given order. Math. Z. 102, 236-237. MR 36(1968)#5210. [*144*]

Gelling E.N. and Odeh R.E.

(1974) On 1-factorizations of the complete graph and the relationship to round robin schedules. Proc. Third Manitoba Conf. on Numerical Math. (Winnipeg, Man., 1973), pp. 213-221. (Utilitas Math., Winnipeg, Manitoba.) MR 50(1975)#171. [*324*]

Gergely E.

(1974a) A simple method for constructing doubly diagonalized latin squares. J. Combin. Theory, A16, 266-272. MR 48(1974)#10851. [*208*]

(1974b) A remark on doubly diagonalized orthogonal latin squares. Discrete Math. 10, 185-188. MR 50(1975)#1923. [*218*]

Gessel I.M.

(1985) Counting three-line Latin rectangles. In "Combinatoire énumérative" (Montreal, Quebec, 1985), pp.106-111, Lecture Notes in Math., No.1234, Springer, Berlin, 1986. MR 89b:05012. [*149*]

(1987) Counting Latin rectangles. Bull. Amer. Math. Soc. (N.S.) 16, 79-82. MR 88m:05019. [*151*]

Ghurye S.G.

(1948) A characteristic of species of 7 × 7 latin squares. Ann. Eugenics 14, 133. MR9(1948), page 559. MR0024873. [*139*]

Gilbert E.N.

(1965) Latin squares which contain no repeated digrams. SIAM Rev. 7, 189-198. MR 31(1966)#3346. [*74,144*]

Ginzburg A.

(1960) Systèmes multiplicatifs de relations. Boucles quasi-associatives. C. R. Acad. Sci. Paris 250, 1413-1416. MR 22(1961)#4793. [*9*]

(1964) A note on Cayley loops. Canad. J. Math. 16, 77-81. MR 29(1965)#2287. [*9*]

(1967) Representation of groups by generalized normal multiplication tables. Canad. J. Math. 19, 774-791. MR 35(1968)#5503. [*9*]

Ginzburg A. and Tamari D.

(1969a) Representation of binary systems by families of binary relations. Israel J. Math. 7, 21-32. MR 40(1970)#1315. [*9*]

(1969b) Representation of generalized groups by families of binary relations. Israel J. Math. 7, 33-45. MR 40(1970)#1316. [*9*]

Glaubermann C.

(1964) On loops of odd order I. J. Algebra 1, 374-396. MR 31(1966)#267. [*20*]

(1968) On loops of odd order II. J. Algebra 8, 393-414. MR 36(1968)#5250. [*20*]

Gleason A.M.

(1956) Finite Fano planes. Amer. J. Math. 78, 797-807. MR 18(1957), page 593. [273]

Glynn D.G.

(2011) A short proof of the transversal theorem for Latin squares of even order. Bull. Inst. Combin. Appl. 63, 73-76. MR2951565. [155]

Godsil C.D. and McKay B.D.

(1990) Asymptotic enumeration of Latin rectangles. J. Combin. Theory B48, 19-44. MR 91j:05009. [148,152]

Golomb S.W. and Posner E.C.

(1964) Rook domains, latin squares, affine planes, and error- distributing codes. IEEE Trans. Information theory. IT-10, 196-208. MR 29(1965)#5657. [319]

Goodaire E.G. and Robinson D.A.

(1990) Some special conjugacy closed loops. Canad. Math. Bull. 33, 73-78. MR 91a:20077. [21]

Gordon B.

(1961) Sequences in groups with distinct partial products. Pacific J. Math. 11, 1309-1313. MR 24(1962)#A3193. [74,77]

Govaerts P., Jungnickel D., Storme L. and Thas J.A.

(2003) Some new maximal sets of mutually orthogonal Latin squares. Proc. Conf. on Finite Geometries (Oberwolfach, 2001). Des. Codes Cryptogr. 29, 141-147. MR 2004f:05032. [179]

Graham G.P. and Roberts C.E.

(1991) Maximal orthogonal sets of self-orthogonal Latin squares. Proc. Twenty-second S. E. Conf. on Combinatorics, Graph Theory, and Computing (Baton Rouge, LA, 1991). Congr. Numer. 83, 125-128. MR 92k:05029. [193]

(2002) Complete sets of orthogonal self-orthogonal Latin squares. Ars Combin. 64, 193-198. MR 2003e:05027. [193]

(2006) Enumeration and isomorphic classification of self-orthogonal Latin squares. J. Combin. Math. Combin. Comput. 59 (2006), 101-118. MR2277343. [145]

(2007) Projective planes and complete sets of orthogonal, self-orthogonal Latin squares. Proc. Thirty-Eighth S. E. Conf. on Combinatorics, Graph Theory, and Computing. Congr. Numer. 184, 161-172. MR 2009d:05034. [193]

Grannell M.J., Griggs T.S. and Quinn K.A.S.

(1999) Mendelsohn directed triple systems. Discrete Math. 205, 85-96. MR 2000g:05027. [60]

(2009) Smallest defining sets of directed triple systems. Discrete Math. 309, 4810-4818. MR 2011a:05047. [60]

Greco D.

(1951) I gruppi finiti che sono somma di quattro sottogruppi. Rend. Accad. Sci. Fis. Mat. Napoli (4)18, 74-85. MR 14(1953), page 445. [*31*]

(1953) Su alcuni gruppi finiti che sono somma di cinque sottogruppi. Rend. Sem. Mat. Univ. Padova 22, 313-333. MR 15(1954), page 503. [*31*]

(1956) Sui gruppi che sono somma di quatro o cinque sottogruppi. Rend. Accad. Sci. Fis. Mat. Napoli (4)23, 49-59. MR 20(1959)#75. [*31*]

Greenberg L. and Newman M.

(1970) Some results on solvable groups. Arch. Math. (Basel) 21, 349-352. MR 42(1971)#6109. [*144*]

Gridgeman N.T.

(1972) Magic squares embedded in a latin square. J. Recreational Math. 5, 250. Not reviewed in MR. [*224*]

Griggs T.S.

(2011) Steiner triple systems and their close relatives. Quasigroups Related Systems 19, 23-68. MR 2012j:05076. [*62*]

Grishkov A.N. and Zavarnitsine A.V.

(2005) Lagrange's theorem for Moufang loops. Math. Proc. Camb. Philos. Soc. 139, 41-57. MR 2006d:20122. [*20*]

Gross K.B., Mullin R.C. and Wallis W.D.

(1973a) The number of pairwise orthogonal symmetric latin squares. Utilitas Math. 4,239-251. MR 48(1974)#10852. [*145,303*]

(1973b) Corrigenda to: "The number of pairwise orthogonal symmetric Latin squares" (Utilitas Math. 4 (1973), 239-251). Utilitas Math. 6, 349. MR 50(1975)#9620. [*303*]

Gruenther M.

(1933) Duplicate Contract complete. (Bridge World Inc., New York.) [*226*]

Guérin R.

(1963a) Aspects algbraiques du probléme de Yamamoto. C. R. Acad. Sci. Paris 256, 583-586. MR 26(1963)#3620. [*178,287*]

(1963b) Sur une généralisation de la méthode de Yamamoto pour la construction des carrés latins orthogonaux. C. R. Acad. Sci. Paris 256, 2097-2100. MR 26(1963)#4934. [*178,287*]

(1964) Sur une généralisation de Bose pour la construction de c.l.m.o. I. M. S. Berne . Not reviewed in MR. [*287*]

(1966a) Existence et propriétés des carrés latins orthogonaux I. Publ. Inst. Statist. Univ. Paris 15, 113-213. MR 35(l968)#73. [*287*]

(1966b) Existence et propriétés des carrés latins orthogonaux II. Publ. Inst. Statist. Univ. Paris 15, 215-293. MR 35(1968)#4118. [*287*]

Guha H.C. and Hoo T.K.

(1965) On a class of quasigroups. Indian J. Math. 7, 1-7. MR 32(1966)#5773. [*43*]

Gunther S.

(1876a) Mathematisch-historische Miscellen. II. Die magischen Quadrate bei Gauss. Z. Math. Phys. 21, 61-64. [137]

(1876b) Geschechte der Mathematischen Wissenschaften. (Chapter 4.) Liepzig. [205]

Haber S. and Rosenfeld A.

(1959) Groups as unions of proper subgroups. Amer. Math. Monthly 66, 491-494. MR 21(1960)#2692. [30,31]

Hajiabolhassan H., Mehrabadi M.L., Tusserkani R. and Zaker, M.

(1999) A characterization of uniquely vertex colorable graphs using minimal defining sets. Discrete Math. 199, 233-236. MR 99k:05137. [274]

Halberstam F.Y., Hoffman D.G. and Richter R.B.

(1986) Latin triangles. Ars Combin. 21, 51-58. MR 87h:05047. [324]

Halberstam F.Y. and Richter R.B.

(1989) A note on Latin triangles. In "Combinatorial Mathematics", Proc. Third Internat. Conf. (New York, 1985), pp.213-215, Ann. New York Acad. Sci., No.555, New York. MR1018626. [324]

Hall M.

(1943) Projective planes. Trans. Amer. Math. Soc. 54, 229-277. MR 5(1944), page 72. [269,272]

(1945) An existence theorem for latin squares. Bull. Amer. Math. Soc. 51, 387-388. MR 7(1946), page 106. [83]

(1948) Distinct representatives of subsets. Bull. Amer. Math. Soc. 54, 922-926. MR 10(1949), page 238. [143]

(1949) Correction to "Projective planes". Trans. Amer. Math. Soc. 65(1949), 473-474. MR 10(1949), page 618. [269,272]

(1952) A combinatorial problem on Abelian groups. Proc. Amer. Math. Soc. 3, 584-587. MR 14(1953), page 350. [67]

(1953) Uniqueness of the projective plane with 57 points. Proc. Amer. Math. Soc. 4, 912-916. MR 15(1954), page 460. [174]

(1954) Correction to "Uniqueness of the projective plane with 57 points". Proc. Amer. Math. Soc. 5, 994-997. MR 16(1955), page 395. [174]

(1959) The Theory of Groups. (Macmillan, New York). MR 21(1960)#1996. [68]

(1967) Combinatorial Theory. (Blaisdell, Massachusetts.) MR 37(1969)#80. [58,164,174]

Hall M. and Paige L.J.

(1955) Complete mappings of finite groups. Pacific J. Math. 5, 541-549. MR 18(1957), page 109. [67]

Hall M. and Senior J.K.

(1964) The groups of order 2^n ($n \leq 6$). (Macmillan, New York.) MR 29(1965) #5889. [144]

Hall M. and Swift J.D.

(1955) Determination of Steiner triple systems of order 15. Math. Tables Aids Comput. 9, 146-152. MR 18(1957), page 192. [58]

Hall M., Swift J.D. and Walker R.J.

(1956) Uniqueness of the projective plane of order eight. Math. Tables Aids Comput. 10, 186-194. MR 18(1957), page 816. [174]

Hall P.

(1928) A note on soluble groups. J. London Math. Soc. 3, 98-105. [69]

(1935) On representation of subsets. J. London Math. Soc. 10(1935), 26-30. [83]

Hammel A.

(1958) Verifying the associative property for finite groups. Math. Teacher 61, No.2, 136-139. Not reviewed in MR. [5]

Hamming R.W.

(1950) Error detecting and error correcting codes. Bell System Tech. J. 29, 147-160. MR 12(1951), page 35. [319]

Hanani H.

(1970) On the number of orthogonal latin squares. J. Combin. Theory 8, 247-271. MR 40(1970)#5466. [178]

Harary F.

(1960) Unsolved problems in the enumeration of graphs. Magyar Tud. Akad. Mat. Kutat Tnt. Kzl. 5, 63-95. MR 26(1963)#4340. [274]

Hartman A.

(1982) A general recursive construction for quadruple systems. J. Combin. Theory A33, 121-134. MR 83k:05018. [313]

Hatami P. and Shor P.W.

(2008) A lower bound for the length of a partial transversal in a Latin square. J. Combin. Theory A115, 1103-1113. MR 2009h:05039. [119]

Hedayat A.

(1972) An algebraic property of the totally symmetric loops associated with Kirkman-Steiner triple systems. Pacific J. Math. 40, 305-309. MR 46(1973)#3340. [157]

(1977) A complete solution to the existence and non-existence of Knut Vik designs and orthogonal Knut Vik designs. J. Combin. Theory A22, 331-337. MR 55(1978)#12548. [220]

Hedayat A. and Federer W.T.

(1969) An application of group theory to the existence and non-existence of orthogonal Latin squares. Biometrika 56, 547-551. MR 41(1971)#6373. [153,155]

(1975) On the non-existence of Knut Vik designs for all even orders. Ann. Statist. 3, 445-447. MR 51(1976)#4577. [219]

Hedayat A., Parker E.T. and Federer W.T.

(1970) The existence and construction of two families of designs for two successive experiments. Biometrika 57, 351-355. MR 42(1971)#3936. [*178*]

Hedayat A. and Seiden E.

(1971) On a method of sum composition of orthogonal Latin squares. In "Atti del Convegno di Geometria Combinatoria e sue Applicazioni" (Univ. Perugia, Perugia, 1970), pp. 239-256. Ist. Mat., Univ. Perugia, Perugia. MR 49(1975)#4804. [*287*]

(1974) On the theory and application of sum composition of Latin squares and orthogonal Latin squares. Pacific J. Math. 54 , no. 2, 85-113. MR 51(1976)#10125. [*287*]

Hedayat A. and Shrikhande S.S.

(1971) Experimental designs and combinatorial systems associated with latin squares and sets of mutually orthogonal latin squares. Sankhyā Ser.A 33, 423-443. MR 47(1974)#1214. [*322*]

Hedayat A.S., Sloane N.J.A. and Stufken J.

(1999) Orthogonal arrays: theory and applications. (Springer-Verlag, New York.) MR 2000h:05042. [*202*]

Heffter B.

(1896) Problème des 36 officiers. Interméd. Math. 5, 176. [*165*]

Heinrich K.

(1977) Subsquares in latin squares. Proc. Eighth S.E. Conf. on Combinatorics, Graph Theory and Computing (Louisiana State Univ.). Congressus Numerantium 19, 329-344. MR 58(1979)#10503. [*28,295*]

(1979) Pairwise orthogonal row complete Latin squares. Proc. Tenth S.E. Conf. on Combinatorics, Graph Theory and Computing (Florida Atlantic Univ., Boca Raton). Congressus Numerantium 23/24, 501-510. MR 81m:05033. [*160*]

Heinrich K. and Hilton A.J.W.

(1983) Doubly diagonal orthogonal latin squares. Discrete Math. 46, 173-182. MR 84i:05030. [*218,312*]

Heinrich K., Kim K. and Prasanna Kumar V.K.

(1992) Perfect Latin squares. Discrete Appl. Math. 37/38, 281-286. MR 93h: 05025. [*97*]

Heinrich K. and Wallis W.D.

(1981) The maximum number of intercalates in a latin square. Combin. Math. VIII (Geelong, 1980), pp.221-233. Lecture Notes in Math. 884, Springer, Berlin-New York. MR 84g:05034. [*34,35,158,284,285*]

Heppes A. and Révész P.

(1956) A latin négyzet és az ortogonális latin négyzet-pár fogalmának egy új általánosítása és ennek felhasználása kísérletek tervezésére. Magyar Tud. Akad. Mat. Int. Közl. 1, 379-390. In Hungarian. [A new generalization of the method of latin squares and orthogonal latin squares and its application to the design of experiments.] Not reviewed in MR. [195,196]

Hessenburg G.

(1905) Beweis des Desarguessehen Satzes aus dem Pascalschen. Math. Ann. 61, 161-172. [264]

Higham J.

(1997) A product theorem for row-complete latin squares. J. Combin. Des. 5, 311-318. MR 98i:05032. [80]

(1998) Row-complete latin squares of every composite order exist. J. Combin. Des. 6, 63-77. MR 98j:05034. [80,276]

Higman G.

(1960) Enumerating p-groups I. Inequalities. Proc. London Math. Soc. (3)10, 24-30. MR 22(1961)#4779. [144]

Hilton A.J.W.

(1972) A simplification of Moore's proof of the existence of Steiner triple systems. J. Combin. Theory 13, 422-425. MR 46(1973)#1616. [58]

(1973) On double diagonal and cross latin squares. J. London Math. Soc. 6, 679-689. MR47(1974)#8328. [208]

(1974) On the number of mutually orthogonal double diagonal latin squares of order m. Sankhyā B36, 129-134. MR 52(1976)#13432. [218]

(1975a) On the number of mutually orthogonal double diagonal latin squares of order m. In "Infinite and Finite Sets" (Colloq. Keszthely, 1973; dedicated to P. Erdös on his 60th birthday) Vol.II, pp. 867-874. Colloq. Math. Soc. Janos Bolyai, Vol.10, North Holland, Amsterdam. MR 51(1976)#7905. [218]

(1975b) Some simple constructions for doubly diagonal orthogonal latin squares. In "Infinite and Finite Sets" (Colloq. Keszthely, 1973; dedicated to P. Erdös on his 60th birthday) Vol.II, pp. 887-904. Colloq. Math. Soc. Janos Bolyai, Vol.10, North Holland, Amsterdam. MR 51(1976)#12564. [218,312]

(1975c) Embedding incomplete double diagonal Latin squares. Discrete Math. 12, 257-268. MR 51(1976)#12565. [312]

(1977) Embedding incomplete latin rectangles and extending the edge colourings of graphs. Nanta Math. 10, 201-206. MR 80k:05058. [27,84]

Hilton A.J.W. and Johnson P.D.Jnr.

(1988) A variation of Ryser's theorem and a necessary condition for the list-colouring problem. In "Graph Colourings" (Milton Keynes,1998) Pitman Research Notes Math, Ser. 218, (Longman Sci. Tech., Harlow, 1990), pp.135-143. MR 93k:05072. [85]

Hilton A.J.W. and Keedwell A.D.

(1976) Further results concerning P-quasigroups and complete graph decompositions. Discrete Math. 14, 311-318. MR 55(1978)#7834. [281]

Hilton A.J.W. and Scott S.H.

(1974) A further construction of double diagonal orthogonal latin squares. Discrete Math. 7, 111-127. MR 48(1974)#8266. [218]

Hiner F.P. and Killgrove R.B.
   (1970) Subsquare complete latin squares. AMS Notices 17, 677-05, 758. [32]
   (19**) Subsquares of latin squares, Unpublished preprint.[32,34]

Hirschfeld, J.W.P.
   (1998) Projective geometries over finite fields. (Second edition. Oxford Mathematical Monographs. The Clarendon Press, Oxford University Press, New York.) MR 99b:51006. [174]

Hobby C., Rumsey H. and Weichsel P. M.
   (1960) Finite groups having elements of every possible order. J. Washington Acad. Sci. 50, No.4, 11-12. MR 26(1963)#1356. [28]

Hobbs A. and Kotzig A.
   (1983) Groups and homogeneous Latin squares. Proc. 14th S.E. Conf. on combinatorics, graph theory and computing (Florida Atlantic Univ.,Boca Raton). Congressus Numerantium 40, 35-44. MR 85i:05051. [285]

Hobbs A., Kotzig A. and Zaks J.
   (1982) Latin squares with high homogeneity. Proc. 13th S.E. Conf. on combinatorics, graph theory and computing (Florida Atlantic Univ.,Boca Raton). Congressus Numerantium 36, 333-345. MR 85e:05035. [284,285]

Höhler P.
   (1970) Eine Verallgemeinerung von orthogonalen lateinischen Quadraten auf höhere Dimensionen. Abhandlung zur Erlangung der Würde eines Doktors der Mathematik der Eidgenössischen Technischen Hochschule Zürich. Diss. Dokt. No.4522. Eidgenössische Technische Hochschule, Zurich, 1970. 58 pp. MR 45(1973) #8545. [195]

Horák P., Rosa A. and Širáň J.
   (1997) Maximal orthogonal Latin rectangles. Ars Combin. 47, 129-145. MR 98j:05035. [194]

Horner W.W.
   (1952) Addition-multiplication magic squares. Seripta Math. 18, 300-303. Not reviewed in MR. [221]
   (1955) Addition-multiplication magic squares of order 8. Scripta Math. 21, 2-27. MR 17(1956), page 227. [221]

Horton J.D.
   (1974) Sub-latin squares and incomplete orthogonal arrays. J. Combin. Theory A16, 23-33.MR 50(1975)#143. [295]

Hosszú M.
   (1959) Belouszov egy tételéröl és annak néhány alkalmazásáról. Magyar Tud. Akad. Mat. Fiz. Oszt. Közl. 9, 51-56. In Hungarian. [Concerning a theorem of Belousov and some of its applications.] MR 21(1960)#4198. [56]

Houston T.R.
   (1966) Sequential counterbalancing in latin squares. Ann. Math. Statist. 37, 741-743. MR 34(1967)#905. [72]

Hsu D.F.

(1980) Cyclic neofields and combinatorial designs. Lecture Notes in Mathematics, 824. Springer-Verlag, Berlin-New York, 1980. vi+230 pp. ISBN: 3-540-10243-4 MR 84g:05001. [*252*]

Hsu D.F. and Keedwell A.D.

(1984) Generalized complete mappings, neofields, sequenceable groups and block designs. I. Pacific J. Math. 111, 317-332. MR 85m:20031. [*156,252*]

Hughes D.R.

(1955) Planar division neo-rings. Trans. Amer. Math. Soc. 80, 502-527. MR 17(1956), page 451. [*272*]

(1957) A class of non-desarguesian projective planes. Canad. J. Math. 9, 378-388. MR 19(1958), page 444. [*272*]

Hughes D.R. and Piper F.C.

(1973) Projective planes. (Graduate Texts in Mathematics, Vol. 6. Springer-Verlag, New York-Berlin.) MR 48(1974)#12278. [*174*]

Hughes N.J.S.

(1957) A theorem on isotopic groupoids. J. London Math. Soc. 32, 510-511. MR 19(1958), page 634. [*12*]

Hulpke A., Kaski P. and Östergård P.R.J.

(2011) The number of Latin squares of order 11. Math. Comp. 80, 1197-1219. MR 2011m:05064. [*141*]

Humblot L.

(1971) Sur une extension de la notion de carrés latins. C. R. Acad. Sci. Paris 273, 795-798. MR 44(1972)#3892.[*198*]

Hung S.H.Y. and Mendelsohn N.S.

(1973) Directed triple systems. J. Combin. Theory Ser.A 14, 310-318. MR 47(1974)#3190. [*60*]

[Ibragimov S. G.] Ибрагимов С.Г.

(1967) Из предыстории теории квазигрупп. (О забытых работах Эрнста Шредера в XIV в.) [On the history of quasigroup theory. (From the neglected work of Ernst Schröder, 19th Century.) Abstracts of lectures: First All-Union Conference on quasigroups. Suhumi (1967), 15-16.] Not reviewed in MR. [*2*]

Indlekofer K-H.

(1973) A remark on solvable groups. Arch. Math. (Basel) 24, 57-58. MR 49(1975)#2923. [*144*]

Isbell J.

(1990) Sequencing certain dihedral groups. Discrete Math. 85,323-328. MR 91j:20067. [*80*]

Ishihara T.

(2006) Construction of Knut Vik designs and orthogonal Knut Vik designs. J. Math. Univ. Tokushima 40, 1-7. MR 2007i:05033. [*221*]

Jacob S.M.

(1930) The enumeration of the latin rectangle of depth three by means of a formula of reduction, with other theorems relating to non-clashing substitutions and latin squares. Proc. London Math. Soc. (2) 31, 329-354. [137,147]

Ji L. and Zhu L.

(2003) Constructions for Steiner quadruple systems with a spanning block design. Papers on the occasion of the 65th birthday of Alex Rosa. Discrete Math. 261, 347-360. MR 2003m:05022. [313]

Johnson D.M., Dulmage A.L. and Mendelsohn N.S.

(1961) Orthomorphisms of groups and orthogonal latin squares I. Canad.J. Math. 13, 356-372. MR 23(1962)#A1544. [17,18,237]

Joyner D. and Kim Jon-L.

(2011) Selected unsolved problems in coding theory. Applied and Numerical Harmonic Analysis. Birkhuser/Springer, New York. xii+200 pp. ISBN: 978-0-8176-8255-2. MR 2012i:94003. [319]

Jungnickel D.

(1996) Maximal sets of mutually orthogonal Latin squares. Finite fields and applications (Glasgow, 1995), pp.129-153, London Math. Soc. Lecture Note Ser., No.233. (Cambridge Univ. Press, Cambridge.) MR 98a:05032. [179,281]

Jungnickel D. and Storme L.

(2003) Maximal partial spreads in PG(3,4) and maximal sets of mutually orthogonal Latin squares of order 16. Papers on the occasion of the 65th birthday of Alex Rosa. Discrete Math. 261, 361-371. MR 2004d:05039. [179]

Kaplansky I.

(1943) Solution of the problème des ménages. Bull. Amer. Math. Soc. 49, 784-785. MR 5(1944), page 86. [148,151]

Kaplansky I. and Riordan J.

(1946) Le problème des ménages. Scripta Math. 12, 113-124. MR 8(1947), page 365. [148]

Karski P., Östergård P.R., Pottonen O. and Kiviluoto L.

(2009) A catalogue of the Steiner triple systems of order 19. Bull. Inst. Combin. Appl. 57, 35-41. MR 2010j:05058. [58]

Kárteszi F.

(1963) Incidenciageometria. Mat. Lapok 14, 246-263. In Hungarian. [Incidence Geometry.] MR 31(1966)#3905. [174]

(1972) Bevezetés a véges geometriákba. (Disquisitiones mathematicae Hungaricae. Akadémiai Kiadó, Budapest.) [174] English translation follows:

(1976) Introduction to finite geometries. (Translated from the Hungarian by L. Vekerdi. North-Holland Texts in Advanced Mathematics, Vol. 2. North-Holland Publishing Co., Amsterdam-Oxford; American Elsevier Publishing Co., Inc., New York.) MR 54(1977)#11156. [174]

Keedwell A.D.

(1963) On the order of projective planes with characteristic. Rend. Mat. e Appl. (5) 22, 498-530. MR 30(1965)#5213. [274]

(1964) A geometrical proof of an analogue of Hessenberg's theorem. J. London Math. Soc. 39, 424-426. MR 29(1965)#6359. [264]

(1965) A search for projective planes of a special type with the aid of a digital computer. Math. Comp. 19, 317-322. MR 31(1966)#3921. [32,174]

(1966) On orthogonal latin squares and a class of neofields. Rend. Mat. e Appi. (5) 25, 519-561. MR 36(1968)#3664; erratum MR 37(1969), page 1469. [166,178,235,240,243,246,247,249]

(1967) On property D neofields. Rend. Mat. e Appl. (5)26, 383-402. MR 37(1969)#5112.[235,249]

(1970) On property D neofields and some problems concerning orthogonal latin squares. Computational Problems in Abstract Algebra (Proc. Conf., Oxford 1967, pp.315-319. Pergamon Press, Oxford. MR 41(1971)#88. [249]

(1971) A note on planes with characteristic. In "Atti del Convegno di Geometria Combinatoria e sue Applicazioni", (Univ. Perugia, Perugia, 1970) pp.307-318. Ist. Mat., Univ. Perugia, Perugia. MR 49(1975)#7911.[274]

(1974) Some problems concerning complete latin squares. In "Combinatorics" (Proc. Brit. Combinatorial Conf., Aberystwyth, 1973), pp.89-96. London Math.Soc. Lecture Notes No.13, Camb. Univ. Press, London. MR 51(1976)#2943. [87]

(1975) Row-complete squares and a problem of A. Kotzig concerning P-quasigroups and Eulerian circuits. J. Combin. Theory A18, 291-304. MR 51(1976) #2982. [72,281]

(1976a) Some connections between latin squares and graphs. Colloq. Internaz. Teorie Combin., (Roma,1973), Tomo 1, pp.321-329. Accad. Naz. Lincei, Rome. MR 55(1978)#7806. [281]

(1976b) Latin squares ,P-quasigroups and graph decompositions. Sympos. en Quasigroupes et Équations Fonctionnelles (Belgrade-Novi Sad. 1974). Zbornik Rad. Mat. Inst. Beograd (N.S.) 1(9),41-48. MR 55(1978)#2610. [72,281]

(1978) Uniform P-circuit designs, quasigroups, and Room squares. Utilitas Math. 14(1978), 141-159. MR 80c:05053.[233,280,281,293]

(1980) Concerning the existence of triples of pairwise almost orthogonal 10 × 10 Latin squares. Ars Combin. 9, 3-10. MR 81k:05021. [178]

(1981a) On the sequenceability of non-abelian groups of order pq. Discrete Math. 37, 203-216. MR84h:20016. [80]

(1982) Decompositions of complete graphs defined by quasigroups. In "Theory and practice of combinatorics", pp.185-192. (North-Holland Math. Stud., 60, North-Holland, Amsterdam). MR 86m:05069. [281]

(1983a) On R-sequenceability and R_h-sequenceability of groups. In "Combinatorics '81" (Rome, 1981), pp.535-548, Ann. Discrete Math., 18, (North-Holland, Amsterdam-New York). MR 84f:20025. [304]

(1983b) On the existence of super P-groups. J. Combin. Theory A35, 89-97. MR 85b:20034. [67]

(1983c) Sequenceable groups, generalized complete mappings, neofields and block designs. Combin. Math. X (Adelaide,1982), pp.49-71. MR 85g:05035. [*80*]

(1984a) More super P-groups. Discrete Math. 49, 205-207. MR 85m:20032. [*67*]

(1984b) Circuit designs and Latin squares. Ars Combin. 17, 79-90. MR 85i:05036. [*178*]

(1991) Proper loops of order n in which each non-identity element has left order n. In "Universal algebra, quasigroups and related systems" (Jadwisin, 1989). Demonstratio Math. 24, 27-33. MR 93a:20113. [*290*]

(1994) Critical sets and critical partial latin squares. In "Graph Theory, Combinatorics, Algorithms and Applications." (Proc. Third China-USA Internat. Conf., Beijing, June 1993),pp.111-124. MR 96a:05027. [*98,274*]

(1996) Critical sets for latin squares, graphs and block designs: a survey. Congressus Numerantium 113, 231-245. MR 97i:05017. [*96,274*]

(1999) What is the size of the smallest latin square for which a weakly completable critical set of cells exists? Ars Combin. 51, 97-104. MR 99j:05036. [*95*]

(2000) Designing tournaments with the aid of Latin squares: a presentation of old and new results. Utilitas Math. 58, 65-85. MR 2001k:05044. [*190,234,323*]

(2001) A characterization of the Jacobi logarithms of a finite field. Discrete Math. 231, 295-302. MR 2002d:12007. [*252*]

(2004) Critical sets in latin squares and related matters: an update. Utiltas Math. 65, 97-131. MR 2005k:05047. [*95,96,98*]

(2005) Tests for loop nuclei and a new criterion for a loop to be group-based, Europ. J. Combin. 26, 111-116. MR 2005g:05030. [*64*]

(2006a) Defining sets for magic squares. Math. Gazette 90, 417-424. Not reviewed in MR. [*224*]

(2006b) Two remarks about Sudoku squares. Math. Gazette 90, 425-430. Not reviewed in MR. [*224*]

(2007) On Sudoku squares. Bull. Inst. Combin. Appl. 50, 52-60. MR 2007m:05044. [*97*]

(2009a) Realizations of loops and groups defined by short identities. Comment Math. Univ. Carolin. 50, 373-383. MR 2011a:20166. [*7*]

(2009b) Corrigendum to "Realizations of loops and groups defined by short identities". Comment Math. Univ. Carolin. 50, 639-640.. MR 2011f:20150. [*7*]

(2010) Constructions of complete sets of orthogonal diagonal Sudoku squares, Australas J. Combin. 47, 227-238. MR 2011f:05060. [*182*]

(2011a) A short note regarding existence of complete sets of orthogonal diagonal Sudoku squares, Australas J. Combin. 51, 271-273. MR 2012i:05039. [*182*]

(2011b) Gaston Tarry and multimagic squares. Math. Gazette 95, 454-468. Not reviewed in MR. [*222*]

(2011c) Confirmation of a conjecture concerning orthogonal Sudoku and bimagic squares. Bull. Inst. Combin. Applic. 63, 39-47. MR2951561. [*222*]

Kendall D.G. and Rankin R.A.

(1947) On the number of Abelian groups of a given order. Quart. J. Math. Oxford Ser.18, 197-208. MR 9(1948), page 226. [*143*]

Kerewala S.M.

(1941) The enumeration of the latin rectangle of depth three by means of a difference equation. Bull. Calcutta Math. Soc. 33, 119-127. MR 4(1943), page 69. [*147*]

(1947a) The asymptotic number of three-deep latin rectangles. Bull. Calcutta Math. Soc. 39, 71-72. MR 9(1948), page 404. [*151*]

(1947b) Asymptotic solution of the "problème des ménàges". Bull. Calcutta Math. Soc. 39, 82-84. MR 9(1948), page 405. [*151*]

Kerr J.R., Pearce S.C. and Preece D.A.

(1973) Orthogonal designs for three-dimensional experiments. Biometrika 60, 349-358. MR 48(1974)#1399. [*199*]

Kertész A.

(1964) Kvazicsoportok. Mat. Lapok 15, 87-113. In Hungarian. [Quasigroups.] MR 30(1965)#3935. [*57*]

Killgrove R.B.

(1964) Completions of quadrangles in finite projective planes. Canad. J. Math. 16, 63-76. MR 28(1964)#513. [*32,35*]

(1974) Subsquare complete latin squares of order 12. Proc. Fifth SE Conf. on Combinatorics, Graph Theory and Computing, pp.535-547. Congressus Numerantium 10, Utilitas. Math. MR 55(1978)#136. [*32,35*]

(2001) Private communication to the author (December 2001). [*284*]

Kim K. and Prasanna Kumar V.K.

(1993) Latin squares for parallel array access. IEEE Trans. on Parallel and Distributed Systems 4, 361-370. Not reviewed in MR. [*97*]

Kinyon M.J. and Kunen K.

(2004) The structure of extra loops. Quasigroups Related Systems 12, 39-60. MR 2006a:20121. [*21*]

Kinyon M.J., Kunen K. and Phillips J.D.

(2004) Diassociativity in conjugacy closed loops. Comm. Algebra 32, 767-786. MR 2005h:20159. [*21*]

Kinyon M.J. and Wanless I.M.

(2015) Loops with exponent three in all isotopes. Internat. J. Algebra Comput. 25, 11591177. MR3432224. [*158*]

Kirkman Rev. T.P.

(1847) On a problem in combinations. Camb. & Dublin Math. J. 2, 191-204. [*58*]

Kishen K.

(1942) On latin and hyper-graeco cubes and hypercubes. Current Science 11, 98-99. Not reviewed in MR. [*199*]

(1950) On the construction of latin and hyper-graeco-latin cubes and hyper-cubes. J. Indian Soc. Agric. Statistics 2, 20-48. MR 11(1950), page 637. [*200*]

Klingenberg W.

(1952) Beziehungen zwischen einigen affinen Schliessungssätzen. Abh. Math. Sem. Univ. Hamburg 18, 120-143. MR 14(1953), page 786. [*264*]

(1955) Beweis des Desarguesschen Satzes aus der Reidemeisterfigur und verwandte Sätze. Abb. Math. Sem. Univ. Hamburg 19, 158-175. MR 16(1955), page 950. [*264*]

Kolesova G., Lam C.W.H. and Thiel L.

(1990) On the number of 8×8 latin squares. J. Combin. Theory A54, 143-148. MR 91d:05024. [*15,140*]

Koksma K.K.

(1969) A lower bound for the order of a partial transversal in a latin square. J. Combin. Theory 7, 94-95. MR 39(1970)#1342. [*119*]

Korchmaros G., Nagy G.P. and Pace N.

(2014) 3-nets realizing a group in a projective plane. J. Algebraic Combin. 39, 939-966. MR3199033. [*268*]

Korovina N.P.

(1984) Complete solution of the Kotzig problem on the existence of an Euler cycle in P-quasigroups. (Russian) Uspekhi Mat. Nauk 39, No. 2(236), 163-164. MR 85f:20072. [*281*]

Kotzig A.

(1970) Groupoids and partitions of complete graphs. Combinatorial Structures and their Applications. (Proc. Calgary Internat. Conf., Calgary, Alta., 1969), pp.215-221. Gordon and Breach, New York. MR 42(1971)#4446. [*277,279,281*]

Kraitchik M.

(1930) La Mathématique des Jeux. (Brussels.) [*205*]

(1942) Mathematical recreations. (Norton, New York.) (Dover Reprint, New York, 1953) MR 14(1953), page 620. [*205,215*]

Krapež A. and Taylor M.A.

(1991) Quasigroups satisfying balanced but not Belousov equations are group isotopes. Aequationes Math. 42, 37-46. MR 93c:20115. [*50*

Krätzel E.

(1970) Die maximale Ordnung der Anzahl der wesentlich verschiedenen abelschen Gruppen n-ter Ordnung. Quart. J. Math. Oxford Ser.21, 273-275. MR 42(1941)#3171. [*143*]

Kuhl J.S. and Denley T.

(2012) A few remarks on avoiding partial Latin squares. Ars Combin. 106, 313-319. MR2977206. [*36*]

Kunen K.

(2000) The structure of conjugacy closed loops. Trans. Amer. Math. Soc. 352, 2889-2911. MR 2000j:20132. [*21*]

(2006a) Moufang quasigroups. J Algebra 183, 231-234. MR 97f:20096. [44]

(2006b) Quasigroups, loops and associative laws. J. Algebra 185, 194-204. MR 97g:20083. [44]

Lam C.W.H.

(1991) The search for a finite projective plane of order 10. Amer. Math. Monthly 98, 305-318. MR 92b:51013. [174]

Lam C.W.H., Kolesova G. and Thiel L.

(1991) A computer search for finite projective planes of order 9. Discrete Math. 92, 187-195. MR 92j:51012. [174]

Lam C.W.H., Thiel L. and Swiercz S.

(1989) The non-existence of finite projective planes of order 10. Canad. J. Math. 41, 1117-1123. MR 90j:51008. [174]

Laufer P.J.

(1980) On strongly Hamiltonian complete bipartite graphs. Ars Combin. 9, 43-46. MR 81j:05081. [293]

Laugel L.

(1896) Sur le problème d'Euler, dit des 36 officiers. (Réponse à la question numéro 453.) Interméd. Math. 3, 17. [165]

Laywine C.F., Mullen G.L. and Whittle G.

(1995) d-dimensional hypercubes and the Euler and MacNeish conjectures. Monatsh. Math. 119, 223-238. MR 96b:05039. [201]

Lee C.Y.

(1958) Some properties of non-binary error-correcting codes. IEEE Trans. Information Theory. IT-4 (1958), 77-82. MR 22(1961)#13353. [107]

Leech J.

(1970) Computational problems in abstract algebra. Proceedings of a Conference held at Oxford under the auspices of the Science Research Council, Atlas Computer Laboratory, 29th August to 2nd September 1967. Edited by John Leech. With a foreword by J. Howlett. (Pergamon Press, Oxford-New York-Toronto.) MR 40(1970)#5374. [325]

Lefevre J., Donovan D., Cavenagh N. and Drápal A.

(2007) Minimal and minimum size Latin bitrades of each genus. Comment. Math. Univ. Carolin. 48, 189-203. MR 2008f:05026. [104]

Lemoine E.

(1889) Sur le problème des 36 officiers. Interméd. Math. 6, 273. [165]

(1900) Sur le problème des 36 officiers. Interméd. Math. 7, 311-312. [165]

Li P.

(197) Sequencing the dihedral groups D_{4k}. Discrete Math. 175, 271-276. MR 98h:20034. [80]

Li P.C. and van Rees G.H.J.

(2007) Nearly orthogonal Latin squares. J. Combin. Math. Combin. Comput. 62, 13-24. MR 2008i:05022. [296]

Liang, M.
(2012) A natural generalization of orthogonality of Latin squares. Discrete Math. 312, 3068-3075. MR2956099. [296,298]
(201*) On standardized quasi-orthogonality of latin squares. Submitted. [161, 302,313]

Liaw Y.S.
(1998) Construction of referee squares. Discrete Math. 178, 123-135. MR 98i:05033. [322]

Light F.W., Jr.
(1973) A procedure for the enumeration of $4 \times n$ Latin rectangles. Fibonacci Quart. 11, 241-246. MR 48#3764. [150]

Lindner C.C.
(1970a) On completing latin rectangles. Canad. Math. Bull. 13, 65-68. MR 41(1971)#6702. [114,115]
(1970b) Comment on a note of J. Marica and J. Schönheim. Canad. Math. Bull. 13, 539. MR 43(1972)#74. [115]
(1971a) Embedding partial idempotent latin squares. J. Combin. Theory. A10, 240-245. MR 43(1972)#1862. [116]
(1971b) Extending mutually orthogonal partial Latin squares. Acta Sci. Math. (Szeged) 32 , 283-285. MR 48(1974)#126. [116]
(1971c) Quasigroups orthogonal to a given abelian group. Canad. Math. Bull. 14, 117-118. MR 45(1973)#2070. [59,312]
(1971d) Construction of quasigroups satisfying the identity $X(XY) = YX$. Canad. Math. Bull. 14, 57-59. MR 45(1973#434. [312]
(1971e) The generalized singular direct product for quasigroups. Canad. Math. Bull. 14, 61-64. MR 45(1973)#435. [312]
(1971f) Construction of quasigroups using the singular direct product. Proc. Amer. Math. Soc. 29, 263-266. MR 43(1972)#6354. [312]
(1971g) Identities preserved by the singular direct product. Algebra Universalis 1, 86-89. MR 44(1972)#5401. [312]
(1972a) Finite embedding theorems for partial latin squares, quasigroups, and loops. J. Combin. Theory. A13, 339-345. MR 47(1974)#3200. [117,312]
(1972b) Identities preserved by the singular direct product II. Algebra Universalis 2, 113-117 MR 46(1973)#5519. [312]
(1972c) An algebraic construction for Room squares. Siam J. Appl. Math. 22(1972), 574-579. MR 46(1973)#3671. [233]
(1973) Construction of doubly diagonalized orthogonal latin squares. Discrete Math. 5(1973), 79-86. MR 49(1975)#4805. [217,312]
(1974) On constructing doubly diagonalized latin squares. Period. Math. Hungar. 5, 249-253. MR 50(1975)#9621. [208]
(1976) Embedding orthogonal partial Latin squares. Proc. Amer. Math. Soc. 59, 184-186. MR 53(1977)#12987. [116,117]

Lindner C.C. and Mendelsohn N.S.

(1973) Construction of perpendicular Steiner quasigroups. Aequationes Math. 9, 150-156. MR 48(1974)#6305. [190,303,312]

Lindner C.C., Mendelsohn N.S. and Sun S.R.

(1980) On the construction of Schroeder quasigroups. Discrete Math. 32, 271-280. MR 82b:20104. [42]

Ling A.C.H., Colbourn C.J., Grannell M.J. and Griggs T.S.

(2000) Construction techniques for anti-Pasch Steiner triple systems. J. London Math. Soc. (2) 61, 641-657. MR 2001c:05025. [313]

Lorch J.

(2009) Mutually orthogonal families of linear Sudoku solutions. J. Aust. Math. Soc. 87, 409-420. MR 2011k:05046. [182]

(2011) Orthogonal diagonal Sudoku solutions: an approach via linearity. Australas. J. Combin. 51, 139-145. MR 2012m:05077. [182]

(2012) Magic squares and Sudoku. Amer. Math. Monthly 119, 759-770. MR2990934. [224]

Loriga J.D.

(1894) Question numéro 261. Interméd. Math. 1, 146-147. [165]

Lu M.G.

(1985) The maximum number of mutually orthogonal Latin squares. Kexue Tongbao (English Ed.) 30, 154-159. MR 87a:05039. [179]

Lucas É.

(1882) Récréationes Mathématiques. (Gauthier-Villars, Paris). [205]

(1883) Les jeux des demoiselles: les rondes enfantines. In "Récréationes Mathématiques", Vol.II, (Gauthier-Villars, Paris). [72,87]

Lunn A.C. and Senior J.K.

(1934) A method of determining all the solvable groups of given order and its application to the orders 16p and 32p. Amer. J. Math. 56, 319-327. [144]

[Lyamzin A.I.] Лямзин А.И.

(1963) Пример пары ортогоналъных латинских квадратов десятого порядка Мат. Наук. 18, No.5(113), 173-174. [An example of a pair of orthogonal latin squares of order ten. Uspehi Mat. Nauk 18(1963), No.5(113), 173-174.] MR 28(1964)#2979. [240]

MacInnes C.R.

(1907) Finite planes with less than eight points on a line. Amer. Math. Monthly 14, 171-174. [166,174]

MacMahon P.A.

(1898) A new method in combinatory analysis, with applications to latin squares and associated questions. Trans. Camb. Phil. Soc. 16, 262-290. [137]

(1900) Combinatorial Analysis. The foundations of a new theory. Phil. Trans. Royal Soc. London, Ser. A 194, 361-386. [137]

(1915) Combinatory Analysis.Vols. I, II. (Cambridge Univ. Press).[137,143]

Macneish H.F.

(1921) Das Problem der 36 Offiziere. Jber. Deutsch. Math. Verein 30, 151-153. [165]

(1922) Euler squares. Ann. Math. 23, 221-227. [165,178,308]

Maenhaut B.M. and Wanless I.M.

(2004) Atomic Latin squares of order eleven. J. Combin. Des. 12, 12-34. MR 2004i:05023. [293]

Maenhaut B.M., Wanless I.M. and Webb B.S.

(2007) Subsquare-free Latin squares of odd order. European J. Combin. 28, 322-336. MR 2007f:05029. [294]

Mahmoodian E.S. and Mahdian M.

(1997) On the uniquely list colorable graphs. Proceedings of the 28th Annual Iranian Mathematics Conference, Part 1 (Tabriz, 1997), pp.319-326, Tabriz Univ. Ser., 377, Tabriz Univ., Tabriz, 1997. MR 99b:05060. [274]

Mahmoodian E.S., Naserasr R. and Zaker M.

(1997) Defining sets in vertex colorings of graphs and Latin rectangles. In Proc. 15th Brit. Combin. Conf. (Stirling, 1995). Discrete Math. 167/168, 451-460. MR 98b:05044. [274]

Maillet E.E.

(1894a) Réponse à la question numéro 261. Interméd. Math. 1, 262. [165,215]

(1894b) Sur les carrés latins d'Euler. C. R. Assoc. France Av. Sci. 23, 244-252. [153]

(1894c) Sur une application de la théorie des groupes de substitutions à celle des carrés magiques. Mémoires de l'Académie des sciences de Toulouse (9) 6, 258-281. [215]

(1895) Question numéro 453. Interméd. Math. 2, 17. [165]

(1896) Application de la théorie des substitutions à celle des carrés magiques. Quart. J. Pure and Appl. Math. 27, 132-144. [215]

(1906) Figures magiques. L'Encyclopédie des sciences mathématiques pures at appliquées, Édition française t. I, vol. 3, fasc. 1, 62-75 (Paris & Leipzig.) [205]

Mann H.B.

(1942) The construction of orthogonal Latin squares. Ann. Math. Statist. 13, 418-423. MR 4(1943), pages 184, 340. [18,161,236,239,309]

(1944) On orthogonal latin squares. Bull. Amer. Math. Soc. 50, 249-257. MR 6(1945), page 14. [161,166,174]

(1952) On products of sets of group elements. Canad. J. Math. 4, 64-66. MR 13(1952), page 720. [31]

Mano K.

(1960) On the reduced number of the latin squares of the nth order. Sci. Rep. Fac. Lit. Sci. Hirosaki Univ. 7, 1-2. Not reviewed in MR. [143]

Margossian A.

(1931) Carrés Latins et carrés d'Euler modules impairs. Enseignement Math. 30, 41-49. [206]

Marica J. and Schönheim J.

(1969) Incomplete diagonals of latin squares. Canad. Math. Bull. 12, 235. MR 40(1970)#55. [*115*]

Martin G.E.

(1968) Planar ternary rings and latin squares. Matematiche (Catania) 23, 305-318. MR 40(l970)#3424. [*166*]

McCarthy D.

(1976) Transversals in latin squares of order 6 and (7,1)-designs. Ars Combin. 1, 261-265. MR 54(1977)#5007. [*165*]

McClintock E.

(1897) On the most perfect forms of magic squares, with methods for their production. Amer. J. Math. 19, 99-120. [*219*]

McGuire G.

(2012) There is no 16-clue Sudoku: Solving the Sudoku minimum number of clues problem. [For details, see various aricles on the internet via "There is no 16-clue Sudoku".] [*96*]

McKay B.D., McLeod J.C. and Wanless I.M.

(2006) The number of transversals in a Latin square. Des. Codes Cryptogr. 40, 269-284. MR 2007f:05030. [*155,157*]

McKay B.D., Meynert A. and Myrvold, W.

(2007) Small Latin squares, quasigroups, and loops. J. Combin. Designs 15, 98-119. MR 2007j:05030. [*140,178,273,328*]

McKay B.D. and Rogoyski E.

(1995) Latin squares of order 10. Electron. J. Combin. 2, Note 3, 4 pp. approx. (electronic). MR 96f:05039. [*140,146,147,148*]

McKay B.D. and Wanless I.M.

(1999) Most Latin squares have many subsquares. J. Combin. Theory A86 (1999), 322-347. MR 2000c:05031. [*158*]

(2005) On the number of Latin squares. Ann. Combin. 9, 335-344. MR 2006f:05027. [*140,147,148*]

McWorter W.A.

(1964) On a theorem of Mann. Amer. Math. Monthly 71, 285-286. MR 28(l964)#5135. [*31*]

Mendelsohn N.S.

(1968) Hamiltonian decomposition of the complete directed n-graph. In "Theory of Graphs" (Proc. Colloq., Tihany, 1966), pp. 237-241. (Academic Press, New York). MR 38(1969)#4361. [*78,275*]

(1969) Combinatorial designs as models of universal algebras. In "Recent Progress in Combinatorics". (Proc. Third Waterloo Conf. on Combinatorics, 1968), pp.123-132. (Academic Press, New York), MR 41(1971)#85. [*189*]

(1970) Orthogonal Steiner Systems. Aequationes Math. 5, 268-272. MR 44(1972) #1587. [*188,303*]

(1971a) A natural generalization of Steiner triple systems. In "Computers in Number Theory" (Proc. Sci. Res. Council Atlas Sympos. No. 2, Oxford 1969) pp. 323-338. (Academic Press, New York). MR 48(1974)#122. [*60*]

(1971b) Latin squares orthogonal to their transposes. J. Combin. Theory, A11, 187-189. MR 45(1973)#88. [*189*]

(1971c) On maximal sets of mutually orthogonal idempotent latin squares. Canad. Math. Bull. 14, 449. MR 46#8865. [*187*]

(1979) Self-orthogonal Weisner designs. Second International Conference on Combinatorial Mathematics (New York, 1978), pp. 391-396, (Ann. New York Acad. Sci., No. 319, New York Acad. Sci., New York.) MR 82a:05021. [*242*]

Metsch K.

(1991) Improvement of Bruck's completion theorem. Des. Codes Cryptogr. 1, 99-116. MR 92m:51014. [*269,281*]

Milič S.

(1971) A new proof of Belousov's theorem for a special law of quasigroup operations. Publ. Inst. Math. (Beograd) N. S. 11(25), 89-91. MR 46(1973)#285. [*56*]

Miller G.A.

(1903) A new proof of the generalized Wilson's theorem. Annals of Math. 4, 188-190. Also in Collected Works of G.A.Miller Vol.II, pp. 247-250. [*67*]

(1930) Determination of all the groups of order 64. Amer. J. Math. 52, 617-634. [*143,144*]

Moore E.H.

(1893) Concerning triple systems. Math. Ann. 43, 271-285. [*58*]

(1896) Tactical Memoranda I-III. Amer. J. Math. 18, 264-303. [*172*]

Moser W.O.J.

(1967) The number of very reduced $4 \times n$ Latin rectangles. Canad. J. Math. 19, 1011-1017. MR 36(1968)#61. [*151*]

(1982) A generalization of Riordan's formula for $3 \times n$ Latin rectangles. Discrete Math. 40, 311-313. MR 83m:05031. [*148*]

Moufang R.

(1935) Zur Struktur von Alternativkörpern. Math. Ann. 110, 416-430. [*2*]

Mullen G.L.

(1978) Research Problems: How Many i-j Reduced Latin Squares are there? Amer. Math. Monthly 85, 751-752. MR 1538848. [*148*]

Mullen G.L. and Purdy D.

(1993) Some data concerning the number of Latin rectangles. J. Combin. Math. Combin. Comput. 13, 161-165. MR 94e:05061. [*140,148*]

Mullen G.L. and Shiue J.-S.

(1991) A simple construction for orthogonal Latin rectangles. J. Combin. Math. Combin. Comput. 9, 161-166. MR 92g:05045. [*194*]

Mullin R.C.

(1980) A generalization of the singular direct product with applications to skew Room squares. J. Combin. Theory A29, 306-318. MR 82e:05035. [*312*]

Mullin R.C. and Németh E.

(1969a) A counter-example to a direct product construction of Room squares. 3. Combin. Theory 7, 264-265. MR 40(1970)#4391. [*229*]

(1969b) On furnishing Room squares. J. Combin. Theory 7, 266-272. MR41(1971) #5228. [*188,230,303*]

(1969c) An existence theorem for Room squares. Caned. Math. Bull. 12, 493-497. MR 40(1970)#2560. [*230*]

(1970a) On the non-existence of orthogonal Steiner systems of order 9. Canad. Math. Bull. 13, 131-134. MR 41(1971)#3297. [*188*]

(1970b) A construction for self-orthogonal latin squares from certain Room squares. Proc. Louisiana Conf. on Combinatorics, Graph Theory and Computing (Louisiana State Univ., Baton Rouge, La.), pp. 213-226, MR 42(1971)#2957. [*231*]

Mullin R.C. and Stanton R.G.

(1971) Construction of Room squares from orthogonal Latin squares. Proc. Second Louisiana Conf. on Combinatorics, Graph Theory and Computing (Louisiana State Univ., Baton Rouge, La., 1971), pp. 375386. (Louisiana State Univ., Baton Rouge, La). MR 47(1974)#8329. [*231*]

Mullin R. C. and Wallis W. D.

(1975) The existence of Room squares. Aequationes Math. 13, 1-7. MR 53(1977) #2715. [*231,233*]

Murdoch D.C.

(1939) Quasigroups which satisfy certain generalized associative laws. Amer. J. Math. 61, 509-522. [*48*]

(1941) Structure of abelian quasigroups. Trans. Amer. Math. Soc. 49, 392-409. MR 2(1941), page 218. [*48*]

Nechvatal J.

(1981a) The asymptotics of ménages numbers. Ars Combin. 12, 295-301. MR 83m:05006. [*148*]

(1981b) Asymptotic enumeration of generalized Latin rectangles. Utilitas Math. 20, 273-292. MR 84e:05036. [*151*]

Nelder J.

(!977) Critical sets in latin squares. In "Problem Corner", CSIRO Division of Math. and Stats., Newsletter 38. [*91*]

(1979) Private communication to J. Seberry (January 1979). [*94*]

Netto E.

(1893) Zur Theorie der Triplesysteme. Math. Ann. 42, 143-152. [*58*]

(1901) Lehrbuch der Kombinatorik. (2nd Ed., Teubner, Leipzig, 1927). (Chelsea reprint, New York, 1955.) [*58*]

Neumann M.

(1960) Asupra unor teoreme de închidere Lucrăr Şti. Inst. Ped. Timişoara Mat.-Fiz. 1959(1960), 85-93. In Rumanian. [Some incidence theorems.] MR 24(1962) #A150. [*268*]

(1962) Unele consecinte ale unei introduceri geometrice a quasigrupului. Lucrăr Şti. Inst. Ped. Timişoara Mat.-Fiz. 1961(1962), 99-102. In Rumanian. [Some consequences of a geometric approach to quasigroups.] MR 32(1966)#4206. [*268*]

Niederreiter H.

(1993) Proof of Williams' conjecture on experimental designs balanced for pairs of interacting residual effects. Europ. J. Combin. 14, 55-58. MR 94d: 05012. [*73*]

Norton D.A.

(1952a) Groups of orthogonal row-latin squares. Pacific J. Math. 2, 335-341. MR 14(1953), page 235. [*89,90,304*]

(1952b) Hamiltonian loops. Proc. Amer. Math. Soc. 3, 56-65. MR 13(1952), page 720. [*36*]

(1960) A note on associativity. Pacific J. Math. 10, 591-595. MR 22(1961)#6859. [*7*]

Norton D.A. and Stein S.S.

(1956) An integer associated with latin squares. Proc. Amer. Math. Soc. 7, 331-334. MR 17(1956), page 1043. [*187*]

Norton H.W.

(1939) The 7×7 squares. Ann. Eugenics 9, 269-307. MR 1(1940), page 199. [*18,26,137,138,165,174*]

Ollis M.A.

(2002) Sequenceable groups and related topics. Electronic J. Combin. 10, 34 (Dynamic surveys). Updated in 2013. Not reviewed in MR. [*80*]

(2005) On terraces for abelian groups. Discrete Math. 305, 250-263. MR 2006g:05038. [*81*]

(2012) A note on terraces for abelian groups. Australas. J. Combin. 52, 229-234. MR2917931. [*81*]

(2014) New complete Latin squares of odd order. Europ. J. Combin. 41, 35-46. MR3219250. [*80*]

Ollis M.A. and Spiga P.

(2005) Every abelian group of odd order has a narcissistic terrace. Ars Combin. 76 , 161-168. MR 2006d:20098. [*81*]

Ollis M.A. and Whitaker R.M.

(2007) On invertible terraces for non-abelian groups. J. Combin. Designs 15, 437-447. MR 2008e:05023. [*81*]

Ollis M.A. and Willmott D.T.

(2011) On twizzler, zigzag and graceful terraces. Australas. J. Combin. 51, 243-257. MR 2012k:05347. [*81*]

Osborn J.M.

(1961) New loops from old geometries. Amer. Math. Monthly 68, 103-107. MR 23(1962)#A1686. [23]

O'Shaughnessy C.D.

(1968) A Room design of order 14. Canad. Math. Bull. 11, 191-194. MR 37(1969)#3940. [188,232,303]

Ostrom T.G.

(1968) Vector spaces and constructions of finite projective planes. Arch. Math. (Basel) 19, 1-25. MR 37(1969)#2081. [253]

Ostrowski R.T. and Van Duren K.D.

(1961) On a theorem of Mann on latin squares. Math. Comp. 15, 293-295. MR 23(1962)#A1543. [164]

Owens P.J.

(1976) Solutions to two problems of Dénes and Keedwell. J. Combin. Theory A21, 299-308. MR 54(1977)#7285. [72,80]

(1992) Complete sets of pairwise orthogonal Latin squares and the corresponding projective planes. J. Combin. Theory A59, 240-252. MR 93a:05037. [160,174,273]

Owens P.J. and Preece D.A.

(1995) Complete sets of pairwise orthogonal Latin squares of order 9. J. Combin. Math. Combin. Comput. 18, 83-96. MR 96g:05029. [173,272]

(1996) Some new non-cyclic Latin squares that have cyclic and Youden properties. Ars Combin. 44, 137-148. MR 97g:05035. [289,290,291,298]

(1997) Aspects of complete sets of 9 × 9 pairwise orthogonal Latin squares. [In Proc.15th British Combin. Conf. (Stirling, 1995).] Discrete Math. 167/168, 519-525. MR 97m:05045. [173,272]

Paige L.J.

(1947) A note on finite abelian groups. Bull. Amer. Math. Soc. 53, 590-593. MR 9(1948), page 6. [65,67]

(1949) Neofields. Duke Math. J. 16, 39-60. MR 10(1949), page 430. [247]

(1951) Complete mappings of finite groups. Pacific J. Math. 1, 111-116. MR 13(1952), page 203. [65]

Paige L.J. and Tompkins C.B.

(1960) The size of the 10 × 10 orthogonal latin square problem. Proc. Sympos. Appl. Math. 10, 71-83. American Mathematical Society, Providence, R.I. MR 22(1961)#6724. [326]

Paige L.J. and Wexler C.

(1953) A canonical form for incidence matrices of finite projective planes and their associated latin squares. Portugaliae Math. 12, 105-112. MR 15(1954), page 671. [273]

Parker E.T.

(1959a) Construction of some sets of mutually orthogonal latin squares. Proc. Amer. Math. Soc. 10, 946-949. MR 22(1961)#674. [7,166]

(1959b) Orthogonal latin squares. Proc. Nat. Acad. Sci. USA 45, 859-862. MR 21(1960)#3344. [166,240,287]

(1959c) A computer search for latin squares orthogonal to latin squares of order 10. Abstract 564-571. Notices Amer. Math. Soc. 6, 798. [178]

(1961) Computer searching for orthogonal latin squares of order 10. Abstract 61T-292. Notices Amer. Math. Soc. 8, 617. [178]

(1962a) Computer study of orthogonal latin squares of order 10. Computers and Automation 11, 33-35. Not reviewed in MR. [178,327]

(1962b) On orthogonal latin squares. Proc. Sympos. Pure Math. 6, 43-46. American Mathematical Society, Providence, R.I. MR 24(1962)#A2541. [164]

(1963) Computer investigation of orthogonal latin squares of order ten. Proc. Sympos. Appl. Math. 15, 73-81. Amer. Math. Soc., Providence, R. I. MR 31(1966) #5140. [18,160,178,327]

(1971) Pathological latin squares. In "Combinatorics" (Proc. Sympos. Pure Math. 19, Univ. California, Los Angeles, Calif., 1968), pp. 177-181. Amer. Math. Soc., Providence, R.I. MR 48(1974)#5887. [155]

Pedersen R.M. and Vis T.L.

(2009) Sets of mutually orthogonal Sudoku latin squares, College Math. J. 40, 174-180. MR 2010d:05020. [182]

Petersen J.

(1902) Les 36 officiers. Annuaire des mathématiciens. Laisant et Buhl, Paris, pp.413-427. [165]

Pflugfelder H.O.

(1990a) Quasigroups and Loops: Introduction. Sigma Series in Pure Mathematics, 7. (Heldermann, Berlin) MR 93g:20132. [57,64,253]

Phelps, K.T.

(1978) Conjugate orthogonal quasigroups. J. Combin. Theory A25, 117-127. MR 80f:05016. [190]

(1980) On the number of commutative Latin squares. Ars Combin. 10, 311-322. MR 82c:05026. [145]

Phillips J.P.N.

(1964) The use of magic squares for balancing and assessing order effects in some analysis of variance designs. Appl. Statist. 13, 67-73. MR 31(1966)#5299. [225]

Pickert G.

(1954) Sechseckgewebe und potenzassociative Loops. Proc. Internat. Congress Math. Amsterdam, 1954, Vol. 2, 245- 246. Not reviewed in MR. [253]

(1955) Projective Ebenen. Springer Verlag, Berlin, Gottingen, Heidelberg. MR 17(1956), page 399. [253,265,272]

Pierce W.A.

(1953) The impossibility of Fano's configuration in a projective plane with eight points per line. Proc. Amer. Math. Soc. 4, 908-912. MR 15(1954), page 460. [174]

Plackett R.L. and Burman J.P.

(1943-46) The design of optimum multifactorial experiments. Biometrika 33, 305-325. MR 8(1947), page 44. [199,202]

[Postnikov M.M.] Поцтницов М.М.

(1964) Магические Квадраты, Хаыка, Моцкба. [Magic Squares. Nauka, Moscow.] Not reviewed in MR. [205]

Preece D.A.

(1966)) Classifying Youden rectangles. J. Roy. Statist. Soc. Ser. B 28, 118-130. MR 34(1967)#5234. [140]

(1994) Balanced Ouchterlony neighbour designs and quasi-Rees neighbour designs. J. Combin. Math. Combin. Comput. 15, 197-219. MR 95i:05024. [87]

(2008) Some mutually orthogonal power-sequence terraces. Bull. Inst. Combin. Appl. 54, 11-32. MR 2009f:11006. [81]

Preece D.A., Pearce S.C. and Kerr J.R.

(1973) Orthogonal designs for three-dimensional experiments. Biometrika 60, 349-358. MR 48(1974)#1399. [199]

Preece D.A. and Phillips N.C.K.

(2002) Euler at the bowling green. Utilitas Math. 61, 129-165. MR 2003f:05023. [323]

Quattrocchi P.

(1968) S-spazi e sistemi di rettangoli latini. Atti Sem. Mat. Fis. Univ. Modena 17, 61-71. MR 38(1969)#3759. [193]

Quinn, K.A.S.

(1999) Difference matrices and orthomorphisms over non-abelian groups. Ars Combin. 52, 289-295. MR 2000c:05027. [179]

Radó F.

(1974) On semi-symmetric quasigroups. Aequationes Math. 11, 250-255. MR 53(1977)#13459. [303]

Raghavarao D.

(1971) Constructions and combinatorial problems in design of experiments. Wiley Series in Probability and Mathematical Statistics. John Wiley & Sons, Inc., New York-London-Sydney. MR 51(1976)#2187. [295]

Raghavarao D., Shrikhande S.S. and Shrikhande M.S.

(2002) Incidence matrices and inequalities for combinatorial designs. J. Combin. Des. 10, 17-26. MR 2002j:05021. [296,298]

Ramanathan K.G.

(1947) On the product of the elements of a finite abelian group. J. Indian Math. Soc. 11, 44-48. MR 9(1948), page 408. [67]

Rao C.R.

(1946) Hypercubes of strength "d" leading to confounded designs in factorial experiments. Bull. Calcutta Math. Soc. 38, 67-78. MR 8(1947), page 396. [199,202]

(1947) Factorial arrangements derivable from combinatorial arrangements of arrays. Suppl. J. Roy. Statist. Soc. 9, 128-139. MR 9(1948), page 264. [*202*]

(1949) On a class of arrangements. Proc. Edinburgh Math. Soc. (2)8, 119-125. MR 11(1950), page 710. [*202*]

(1961) A combinatorial assignment problem. Nature 191, 100. Not reviewed in MR. [*165*]

Ray-Chaudhuri D.K. and Wilson R.M.

(1970) On the existence of resolvable balanced incomplete block designs. Proc. Colloq. Calgary (1969), In "Combinatorial Structures and their Applications",pp. 331-341. (Gordon and Breach, New York.) MR42(1971)#1678. [*58*]

(1971) Solution of Kirkman's schoolgirl problem. In "Combinatorics". (Proc. Sympos. Pure Math., Vol.XIX, Univ. California, Los Angeles, 1968), pp.187-203. (Amer. Math. Soc., Providence, R.I.) MR 47(1974)#3195. [*58*]

(1973) The existence of resolvable balanced incomplete block designs. In "Survey of combinatorial theory". (Proc. Internat. Sympos., Colorado State Univ., 1971),pp. 361-375. (North Holland, Amsterdam.) MR 50(1975)#12761. [*58*]

Rédei L.

(1947) Das schiefe Produkt in der Gruppentheorie mit Anwendung auf die endlichen nichtkommutativen Gruppen mit lauter kommutativen echten Untergruppen und die Ordnungszahlen, zu denen nur kommutative Gruppen gehören. Comment. Math. Helv. 20, 225-264. MR 9(1948), page 131. [*143*]

Reidemeister K.

(1929) Topologische Fragen der Differentialgeometrie V. Gewebe und Gruppen. Math. Z. 29, 427-435. [*253,261*]

Reiss M.

(1858/9) Über eine Steinersche combinatorische Aufgabe, welche in 45sten Bande dieses Journals, Seite 181, gestellte worden ist. J. reine angew. Math. 56, 326-344. [*58*]

Rényi A.

(1966) A véges geometriák kombinatorikai alkalmazásai. Mat. Lapok 17, 33-76. In Hungarian. [Combinatorial applications of finite geometries I.] MR 36(1968)#2515. [*77*]

Rhemtulla A.R.

(1969) On a problem of L. Fuchs. Studia Sci. Math. Hungar. 4, 195-200. MR 40(1970)#1468. [*67*]

Riordan J.

(1944) Three-line latin rectangles. Amer. Math. Monthly 51, 450-452. MR 6(1945), page 113. [*148,151*]

(1946) Three-line latin rectangles II. Amer. Math. Monthly 53, 18-20. MR 7(1946), page 233. [*146,147,151*]

(1952) A recurrence relation for three-line latin rectangles. Amer. Math. Monthly 59, 159-162. MR 13(1952), page 813. [*146,149,150*]

(1954) Discordant permutations. Seripta Math. 20, 14-23. MR 16(1955), page 104. [*148*]

(1958) An introduction to combinatorial analysis. (Wiley, New York). MR 20(1959)#3077. [*151*]

Robinson D.A.

(1966) Bol loops. Trans. Amer. Math.Soc. 123, 341-354. MR 33(1967)#2755. [*20,43*]

Rokovska B.

(1971) Some remarks on the number of triple systems of Steiner. Colloquium Mathematicum 22, 317-323. Not reviewed in MR. [*58*]

(1972) On the number of different triple systems of Steiner. Prace Nauk. Inst. Mat. Fiz. Teoretycznej Politech. Wroclawskiej 6, 41-57. Not reviewed in MR. [*58*]

Room T.G.

(1955) A new type of magic square. Math. Gaz. 39, 307. Not reviewed in MR. [*225,227*]

Rosa A.

(1974) On the falsity of a conjecture on orthogonal Steiner triple systems. J. Combin. Theory A16, 126-128. MR 48(1974)#10845. [*58,188*]

[Rybnikov A.K. and Rybnikova N.M.] Рыбников А.К. анд Рыбникова Н.М.

(1966) Новое доказательство несуществования проеквтивной плоскости порядка 6. Вестник Москов. Унив. сер. 1. мат., мех. 21, No. 6, 20-24. [A new proof of the non-existence of a projective plane of order 6. Vestnik. Moscov. Univ. Ser. I Mat. Meh. 21, No. 6, 20-24.] MR 34(1967)#4982. [*165,166*]

Ryser H.J.

(1951) A combinatorial theorem with an application to latin rectangles. Proc. Amer. Math. Soc. 2, 550-552. MR 13(1952), page 98. [*83,84*]

(1967) Neuere Probleme der Kombinatorik. Vorträge über Kombinatorik Oberwolfach 24-29 .Juli 1967. Matematischen Forschungsinstitute Oberwolfach. Not reviewed in MR. [*21*]

Sade A.

(1948a) Énumération. des carrés latins de côté 6. Marseiile, 1948. MR 10(1949), page 278. [*137*]

(1948b) Énumération des carrés latins, Application au 7e ordre. Conjecture pour les Ordres Supérieurs. Published by the author, Marseille, 1948. MR 10(1949), page 278. [*139,140*]

(1950) Quasigroupes. Published by the author, Marseille,1950. MR 13(1952), page 203. [*59,186*]

(1951a) An omission in Norton's list of 7 × 7 squares. Ann. Math. Statist. 22, 306-307. MR 12(1951), page 665. [*139,174*]

(1951b) Omission dans les listes de Norton pour les carrés 7 × 7. J. Reine Angew. Math. 189, 190-191. MR 13(1952), page 813. [*137,139,174*]

(1953a) Contribution à la théorie des quasigroupes: diviseurs singuliers. C. R. Acad. Sci. Paris 237, 372-374. MR 15(1954), page 98. [*186*]

(1953b) Contributions à la théorie des quasigroupes: quasigroupes obéissant à la "loi des keys" ou automorphes par certains groupes de permutations de leur support. C. R. Acad. Sci. Paris 237, 420-422. MR 15(1954), page 98. [40]

(1957) Quasigroupes obéissant à certaines lois. Rev. Fac. Sci. Univ. Istanbul. Sér. A. 22, 151-184. MR 21(1960)#4987. [37,39,54,57,59,186]

(1958a) Quelques remarques sur l'isomorphisme et l'automorphisme des quasigroupes. Abh. Math. Sem. Univ. Hamburg 22(1958), 84-91. MR 20(1959)#77. [135]

(1958b) Quasigroupes automorphes par le groupe linéaire et géométrie finie. J. Reine Angew. Math. 199, 100-120. MR 20(1959)#78. [268]

(1958c) Groupoïdes orthogonaux. Publ. Math. Debrecen 5, 229-240. MR 20(1959)#5751. [183,185]

(1959a) Quasigroupes parastrophiques. Expressions et identités. Math. Nachr. 20, 73-106. MR 22(1961)#5688. [15,16,44,46]

(1959b) Système demosien associatif de multigroupoïdes avec un scalaire non-singulier. Ann. Soc. Sci. Bruxelles. Sér. I, 73, 231-234. MR 21(1960)#3502. [56]

(1959c) Entropie demosienne do multigroupoïdes et de quasigroupes. Ann. Soc. Sci. Bruxelles. Sér. I, 73, 302-309. MR 23(1962)#A1569. [48,55]

(1960a) Produit direct-singulier de quasigroupes orthogonaux et anti-abéliens. Ann. Soc. Sci. Bruxelles. Sér. I, 74, 91-99. MR 25(1963)#4017. [20,184,185,189, 310,311]

(1960b) Théorie des systèmes demosiens de groupoïdes. Pacific J. Math. 10, 625-660. MR 25(1963)#2019. [54,55]

(1961) Demosian systems of quasigroups. Amer. Math. Monthly 68, 329-337. MR25(1963)#2020. [56]

(1962) Paratopie et autoparatopie des quasigroupes. Ann. Soc. Sci.Bruxelles. Sér. 1, 76, 88-96. MR 27(1964)#2576. [7,135]

(1963) Isotopies d'un groupoïde avec son conjoint. Rend. Circ. Mat. Palermo (2)12, 357-381. MR 29(1965)#4831. [20]

(1964/65) Critères d'isotopie d'un quasigroupe avec an quasigroupe demi-symétrique. Univ. Lisboa Revista Fac. Ci A(2)11, 121-136. MR 34(1967)#1437. [42]

(1965a) Quasigroupes demi-symétriques. Ann. Soc. Sci. Bruxelles Sér. I, 79, 133-143. MR 34(1967)#2760. [41,42]

(1965b) Quasigroupes demi-symétriques. II. Autotopies gauches. Ann. Soc. Sci. Bruxelles Sér. I. 79, 225-232. MR 34(1967)#2761. [42]

(1967a) Autotopies d'un quasigroupe isotope à un quasigroupe demi-symétrique. Univ. Beograd. Publ. Elektrotehn. Fak. Ser. Mat. Fiz. No. 175-179, 1-8. MR 35(1968) #4326. [42]

(1967b) Quasigroupes demi-symétriques. III. Constructions linéaires, A-maps. Ann. Soc. Sci. Bruxelles Sér. I, 81, 5-17. MR 35(1968)#5539. [42]

(1967c) Quasigroupes demi-symétriques. Isotopies préservant la demi-symétrie. Math. Nachr. 33, 177-188. MR 35 (1968) #5540. [42]

(1967d) Quasigroupes isotopes. Autotopies d'un groupe. Ann. Soc. Sci. Bruxelles Sr. I, 81, 231-239. MR 38 (1969)#5978. [125]

(1968a) Quasigroupes parastrophiques. Groupe des automorphismes gauches. Ann. Soc. Sci. Bruxelles Sér. I, 82, 73-78. MR 38(1969)#260. [42]

(1968b) Autotopies des quasigroupes et des systèmes associatifs. Arch. Math. (Brno) 4, 1 -23. MR 42(1971)#1930. [125,135]

(1970/71) Morphismes de quasigroupes. Tables. Univ. Lisboa Revista Fac. Ci A(2), 13,Fasc. 2, 149-172. MR 47(1974)#3586. [134,141]

Safford F.H.

(1907) Solution of a problem proposed by 0. Veblen. Amer. Math. Monthly 14, 84-86. [166]

[Samoilenko S.I.] Цамойоленко Ц.И.

(1965) Применение магических квадратов для коррекции ошибок. Здектросвязь 20(1965), 11, 24-32. [The application of magic squares in error correcting. Electrical Communication 20, 11, 24-32.] Not reviewed in MR. [224]

Saxena P.N.

(1950) A simplified method of enumerating latin squares by MacMahon's differential operators. I. The 6 × 6 latin squares. J. Indian Soc. Agric. Statist. 2, 161-188. MR 12(1951), page 312. [137]

(1951) A simplified method of enumerating latin squares by MaeMahon's differential operators II. The 7 × 7 latin squares. J. Indian Soc. Agric. Statist. 3, 24-79. MR 13(1952), page 200. [137]

(1960) On the latin cubes of the second order and the fourth replication of the three-dimensional or cubic lattice design. J. Indian Soc. Agric. Statist. 12, 100-140. MR 23(1962)#A3013. [201]

Schauffler R.

(1956) Über die Bildung von Codewörter. Arch. Elektra. Übertragung 10, 303-314. MR 18(1957), page 368. [55]

(1957) Die Assoziativität im Ganzen, besonders bei Quasigruppen. Math. Z. 67, 428-435. MR 20(1959)#1648. [55,56,57]

Schellenberg P.J.; Van Rees G.H.J. and Vanstone S.A.

(1978) Four pairwise orthogonal Latin squares of order 15. Ars Combin. 6 , 141-150. MR 80c:05039. [178]

Schönhardt E.

(1930) Über lateinische Quadrate und Unionen. J. Reine Angew. Math. 163, 183-229. [2,9,123,128,138,236]

Schubert H.

(1898) Mathematische Mussestunden. (Göschen, Leipzig, First edition.) New edition: De Gruyter, Berlin, 1941. [205]

Scorza G.

(1926) I gruppi che possono pensarsi come somme di tre loro sottogruppi. Boll. un Mat. Ital. (1) 5, 216-218. [31]

Seiden E. and Zemach R.

(1966) On orthogonal arrays. Ann. Math. Statist. 37, 1355-1370. MR 33(1967) #5061. [202]

Senior J.K. and Lunn A.C.

(1934) Determination of the groups of orders 101 to 161 omitting order 128. Amer. J. Math. 56, 328-338. MR 1507025 [143]

(1935) Determination of the groups of orders 162 to 215 omitting order 192. Amer. J. Math. 57, 254-260. MR 1507932. [143]

Shah K.R.

(1970) Analysis of Room's square design. Ann. Math. Statist. 41. 743-745. MR 42(1971)#2605. [227]

Shaw J.B.

(1915) On parastrophic algebras. Trans. Amer. Math. Soc. 16, 361-370. [16]

Shee S.C.

(1970) On quasigroup graphs. Nanyang Univ. J. Part I. 4, 44-66. MR 45(1973) #6695. [274]

Shieh Y-P., Hsiang J. and Hsu D.F.

(2000) On the enumeration of abelian K-complete mappings. Proc. 31st S.E. Conf. on Combinatorics, Graph Theory and Computing (Boca Raton, FL, 2000). Congressus Numerantium 144, 67-88. MR 2001m:20032. [155]

Sholander M.

(1949) On the existence of the inverse operation in alternation groupoids. Bull. Amer. Math. Soc. 5, 746-757. MR 11(1950), page 159. [48]

Shrikhande M.

(2010) Euler's conjecture and the Shrikhande graph - The 50th anniversary. In: Pre-Conf. Proc. Inter. Conf. on Recent Trends in Graph Theory and Combin. Cochin, India, Aug. 12-15, 2010; pp.19-26. Not reviewed in MR.[166]

Shrikhande S.S.

(1961) A note on mutually orthogonal latin squares. Sankhyā Ser. A. 23, 115-116. MR 25(1963)#703. [179]

Šik F.

(1951) Sur les décompositions créatrices sur les quasigroupes. Publ. Fac. Sci. Univ. Masaryk 1951, 169-186. MR 15(1954), page 7. [29]

Sims C.C.

(1965) Enumerating p-groups. Proc. London Math. Soc. (3) 15, 151-166. MR 30(1965)#164. [144]

Singer J.

(1938) A theorem in finite projective geometry and some applications to number theory. Trans. Amer. Math. Soc. 43, 377-385. [227]

(1960) A class of groups associated with latin squares. Amer. Math. Monthly 67, 235-240. MR 23(1962)#A1542. [18,155]

Siu M.K.

(1991) Which latin squares are Cayley tables? Amer. Math. Monthly 98, 625-627. MR 92f:05023. [5]

Sloane N.J.A.

(1973) A handbook of integer sequences. Academic Press [A subsidiary of Harcourt Brace Jovanovich, Publishers], New York-London. MR 50(197?)#9760. [155]

(1994) An on-line version of the encyclopedia of integer sequences. Electron. J. Combin. 1 (1994), Feature 1, approx. 5 pp. (electronic). MR 95b:05001. [143,152,155]

Smetaniuk B.

(1981) A new construction on Latin squares. I. A proof of the Evans conjecture. Ars Combin. 11, 155-172. MR 83b:05028. [115]

Specnicciati R.

(1966) Tavole di moltiplicazione ridotte di un gruppo. Bolletino della Unione Mathematica Italiana, pp.86-89. Not reviewed in MR. [9]

Speiser A.

(1927) Theorie der Gruppen von endlicher Ordnung. Springer-Verlag, Berlin, 1927. (3rd ed., 1937). Dover reprint, New York, 1945. [5]

Stanton R.G. and Horton J.D.

(1970) Composition of Room squares. Combinatorial theory and its applications, III (Proc. Colloq., Balatonfüred, 1969), pp. 1013-1021. North-Holland, Amsterdam. MR 46(1973)#73. [229]

(1972) A multiplication theorem for Room squares. J. Combin. Theory. A12, 322-325. MR 45(1973)#6650. [229]

Stanton R.G. and Mullin R.C.

(1968) Construction of Room squares. Ann. Math. Statist. 39, 1540-1548. MR 38(1969)#2904. [230]

Steedley D.

(1974) Separable quasigroups. Aequationes Math. 11, 189-195. MR 53(1977) #8310. [304]

Stein C.M.

(1978) Asymptotic evaluation of the number of Latin rectangles. J. Combin. Theory A25, 38-49. MR 80a:05036. [152]

Stein S.K.

(1956) Foundations of quasigroups. Proc. Nat. Acad. Sci. U.S.A. 42, 545-546. MR 18(1957), page 111. [15]

(1957) On the foundations of quasigroups. Trans. Amer. Math. Soc. 85, 228-256. MR 20(1959)#922. [15,46,183,189]

(1964) Homogeneous quasigroups. Pacific J. Math. 14, 1091-1102. MR 30(1965) #1206. [185]

(1975) Transversals of Latin squares and their generalizations. Pacific J. Math. 59, 567-575. MR 52#7930. [*24,119*]

Steiner J.

(1852/3) Combinatorische Aufgabe. J. Reine Angew. Math. 45, 181-182. [*58*]

Stevens W.L.

(1939) The completely orthogonalized latin square. Ann. Eugenics. 9, 82-93. Not reviewed in MR. [*172*]

Stinson D.R.

(1984) A short proof of the non-existence of a pair of orthogonal Latin squares of order six. J. Combin. Theory A36, 373-376. MR 85g:05039. [*165*]

Stinson D.R. and van Rees G.H.J.

(1982) Some large critical sets. In Proc. Eleventh Manitoba Conf. on Numerical Math. and Comput. (Winnipeg, Manitoba, 1981.), Congr. Numer. 34, 441-456. MR 84g:05036. [*95*]

Stoffel A.

(1976) Totally diagonal latin squares. Stud. Cerc. Mat. 28, 113-119. MR 53(1977)#7814. [*221*]

Stones D.S.

(2010) The many formulae for the number of Latin rectangles. Electron. J. Combin. 17, no. 1, Article 1, 46 pp. MR 2011f:05062. [*148,152*]

Stones D.S. and Wanless I.M.

(2010) Divisors of the number of Latin rectangles. J. Combin. Theory A 117, 204-215. MR 2011c:05065. [*148,151*]

Suschkewitsch A.

(1929) On a generalization of the associative law. Trans. Amer. Math. Soc. 31, 204-214. [*5,56,89*]

Szele T.

(1947) Über die endlichen Ordnungszahlen, zu denen nur eine Gruppe gehört. Comment. Math. Helv. 20, 265-267. MR 9(1948), page 131. [*143*]

Tamari D.

(1949) Les images homomorphes des groupoïdes de Brandt et l'immersion des semi-groupes. C. R. Acad. Sci. Paris 229, 1291-1293. MR 11(1950), page 327. [*8*]

(1951) Représentations isornorphes par des systèmes de relations. Systèmes associatifs. C. R. Acad. Sci. Paris 232, 1332-1334. MR 12(1951), page 583. [*8*]

(1960) "Near-Groups" as generalized normal multiplication tables. Abstract 564-279. Notices Amer. Math. Soc. 7, 77. Not reviewed in MR. [*9*]

Tarry G.

(1899) Sur le problème dEuler des n^2 officiers. Interméd. Math. 6, 251-252. [*137,165*]

(1900a) Le problème des 36 officiers. C. R. Assoc. France v. Sci. 29, 170-203. [*137,165,178*]

(1900b) Sur le problème d'Euler des 36 officiers. Interméd. Math. 7, 14-16. [*137,165*]

(1900c) Les permutations carrées de base 6. Mém. Soc. Sci. Liège Série 3, 2, mémoire No. 7. [*137,165*]

(1900d) Les permutations carrées de base 6. [Extrait des Mémoires de Ia Société royale des Sciences de Liège, Série 3, 2.] Mathésis, Série 2, 10(1900), Supplément 23-30. [*165*]

(1904) Carrés cabalistiques Eufériens de base $8N$. C. R. Assoc. France Av. Sci. 33, part 2, 95-111. [*206,215*]

(1905) Carrés magiques. (Réponse à la question numéro 767). Interméd. Math. 12, 174-176. [*215*]

(1906) Sur un carré magique (aux n premiers dégres, par séries numérales). Comptes-Rendues Hebdomadaires des Séances del'Académie des Sciences, Paris 142, 757-760. [*222*]

Taylor M.A.

(1978) A generalization of a theorem of Belousov. Bull. Lond. Math. Soc. 10, 285-286. MR 80j:20079. [*49*]

Taylor W.

(1972) On the coloration of cubes. Discrete Math. 2, 187-190. MR 45(1973)#6684. [*211*]

Thomas A.D. and Wood G.V.

(1980) Group tables. Shiva Mathematics Series, No.2. Shiva Publishing Ltd., Nantwich; distributed by Birkhuser Boston, Inc., Cambridge, Mass.) MR 81d:20002. [*143,144*]

Thomsen G.

(1930) Topologische Fragen der Differentialgeometrie. XII Schnittpunktsätze in ebenen Geometrie. Abh. Math. Sem. Univ. Hamburg 7, 99-106. [*253*]

Todorov D.T.

(1985) Three mutually orthogonal Latin squares of order 14. Ars Combin. 20, 45-47. MR 87d:05045. [*178*]

Touchard J.

(1934) Sur un problème de permutations. C. R. Acad. Sci. Paris 198, 631-633. [*148*]

(1953) Permutations discordant with two given permutations. Scripta Math. 19, 109-119. MR 15(1954), page 387. [*148*]

Treash C.

(1971) The completion of finite incomplete Steiner triple systems with applications to loop theory. J. Combin. Theory, A10, 259-265. MR 43(1972)#397. [*85*]

Urzúa G.

(2010) On line arrangements with applications to 3-nets. (English summary) Adv. Geom. 10, 287-310. MR 2011d:14097. [*268*] van Rees G.H.J.

(1990) Subsquares and transversals in Latin squares. In Proc.Twelfth British Combin. Conf. (Norwich, 1989). Ars Combin. 29B, 193-204. MR 97m:05047. [119,158,283]

Vaughan-Lee M. and Wanless I.M.

(2003) Latin squares and the Hall-Paige conjecture. Bull. London Math. Soc.35, 191-195. MR 2004h:05023. [69,121]

Veblen 0. and Wedderburn J.H.M.

(1907) Non-desarguesian and non-pascalian geometries. Trans. Amer. Math. Soc. 8, 379-388. [174,272]

Wagner A.

(1962) On the associative law of groups. Rend. Mat. e Appl. (5) 21, 60-76. MR 25(1963)#5120; erratum MR 26(1963), page 1543. [7]

Wall D.W.

(1957) Sub-quasigroups of finite quasigroups. Pacific J. Math. 7, 1711-1714. MR 19(1958), page 1159. [30]

Wallis W.D.

(1984) Three new orthogonal diagonal latin squares. Enumeration and design. (Waterlloo, Ont., 1982), 313-317, Academic Press, Toronto, 1984. MR 86j:05040. [218]

(1986) Three orthogonal Latin squares. Adv. in Math. (Beijing) 15, 269-281. MR 88e:05022. [179]

Wallis W.D, Street A.F. and Wallis J.S.

(1972) Combinatorics: Room squares, sum-free sets, Hadamard matrices. Lecture Notes in Mathematics, Vol. 292. (Springer-Verlag, Berlin-New York). MR 52#13397. [233]

Wallis W.D. and Zhu L.

(1981) Existence of orthogonal diagonal latin squares. Ars Combin. 12, 51-68. MR 83c:05026. [218]

(1982) Four pairwise orthogonal diagonal latin squares of side 12. Utilitas Math. 21, 205-207. MR 84b:05033. [218]

(1983) Some new orthogonal diagonal latin squares. J. Austral. Math. Soc. A34, 49-54. MR 84d:05049. [218]

(1984a) Some bounds for pairwise orthogonal diagonal latin squares. Ars Combin. 17, 353-366. MR 85m:05025. [213]

(1984b) The existence of orthogonal Latin squares with small subsquares. J. Combin. Inform. System Sci. 9, 1-13. MR 86i:05035. [223]

Wang C.D.

(1993) On the harmoniousness of dicyclic groups. Discrete Math. 120, 221-225. MR 94d:20022. [81]

(2002) Complete latin squares of order p^n exist for odd primes p and $n > 2$. Discrete Math. 252, 189-201. MR 2003c:05043. [80]

Wang C.D. and Leonard P.A.

(1994) On R^*-sequenceability and symmetric harmoniousness of groups. J. Combin. Designs 2, 71-78. MR 94j:20023. [81]

(1995) On R^*-sequenceability and symmetric harmoniousness of groups. II. J. Combin. Designs 3, 313-320. MR 96c:20037. [81]

Wanless I.M.

(1999) Perfect factorizations of bipartite graphs and latin squares without proper subrectangles. Electron. J. Combin. 6, Research Paper 9, 16 pp. MR 2000i:05153. [15,290,291,292,294]

(2001) Answers to questions by Dénes on Latin power sets. Europ. J. Combin. 22, 1009-1020. MR 2003b:05035. [70,194,304]

(2002) A generalisation of transversals for Latin squares. Electron. J. Combin. 9 , no. 1, Research Paper 12, 15 pp. (electronic). MR 2003i:05030. [121]

(2003) Private communication to the author. [42]

(2004a) Cycle switches in Latin squares. Graphs Combin. 20, 545-570. MR 2005i:05025. [138,139]

(2004b) Diagonally cyclic Latin squares. European J. Combin. 25, 393-413. MR 2004k:05045. [287,288,289]

(2005) Atomic Latin squares based on cyclotomic orthomorphisms. Electron. J. Combin. 12, Research Paper 22, 23 pp. (electronic). MR 2005m:05039. [294]

(2007) Transversals in latin squares. Quasigroups Related Systems 15, 169-190. MR 2008j:05073. [209]

(2011) Transversals in Latin squares: a survey. Surveys in combinatorics 2011, 403-437, London Math. Soc. Lecture Note Ser. No. 392, Cambridge Univ. Press. MR 2012k:05009. [119,121,122,156]

Wanless I.M. and Ihrig E.C.

(2005) Symmetries that Latin squares inherit from 1-factorizations. J. Combin. Des. 13, 157-172. MR 2005m:05040. [293]

Wanless I.M. and Webb B.S.

(2006) The existence of Latin squares without orthogonal mates. Des. Codes Cryptogr. 40, 131-135. MR 2008e:05024. [119,120,283,317]

Wanless I.M. and Zhang X.

(2013) Transversals of Latin squares and covering radius of sets of permutations. European J. Combin. 34 , no. 7, 1130-1143. MR3055228. [314]

Warrington P.D.

(1973) Graeco-Latin cubes. J. Recreat. Math. 6, 47-53. MR 55(1978)#12544. [195]

Weisner L

(1963) Special orthogonal latin squares of order 10. Canad. Math. Bull. 6, 61-63. MR 26(l963)#3621. [240,242]

(1964) A Room design of order 10. Canad. Math. Bull. 7, 377-378. MR 29(1965)#4707. [230]

Wells M.B.

(1967) The number of latin squares of order eight. J. Combin. Theory 3, 98-99. MR 35(1968)#5343. [140,147]

Wernicke P.

(1910) Das Problem der 36 Offiziere. Jber. Deutsch Math. verein. 19, 264-267. [165]

Wielandt H.

(1962) Arithmetical and normal structure of finite groups. Proc. Sympos. Pure Math. 6, 17-38. American Math. Soc., Providence R.I., 1962. MR 26(1963)#5045. [2]

Wilcox S.

(2009) Reduction of the Hall-Paige conjecture to sporadic simple groups.J. Algebra 321, 1407-1428. MR2010a:20051. [69]

Wild P.

(1997) Private communication to D. Bedford. April 1997. [302]

Williams E.J.

(1949) Experimental designs balanced for the estimation of residual effects of treatments. Australian J. Sci. Research. Ser. A, 2, 149-168. MR 11(1950), page 449. [71,73]

(1950) Experimental designs balanced for pairs of residual effects. Australian J. Sci. Research Ser. A, 3, 351-363. MR 12(1951), page 726. [73]

Wilson R.J. and Watkins J.J.

(2013) Combinatorics: Ancient and Modern. Edited by Robin Wilson and John J. Watkins. Oxford University Press, Oxford, 2013. x+381 pp. ISBN: 978-0-19-965659-2. MR 3204727. [1,58]

Wilson R.M.

(1973/74) Non-isomorphic Steiner triple systems. Math. Z. 135, 303-313. MR 49#4803. [58]

(1974) Concerning the number of mutually orthogonal latin squares. Discrete Math. 9, 181-198. MR 49(1975)#10575. [178]

Witt E.

(1939) Zum Problem der 36 Offiziere. Jber. Deutsch. Math. verein 48, 66-67. [165]

Xu Y-g

(2015) On the spectrum of mutually r-orthogonal idempotent latin squares. Acta Math, Appl. Sin. Engl. 31, 813-822. MR3377795. [295]

Xu C.-X. and Lu Z.-W.

(1995) Pandiagonal magic squares. Computing and combinatorics (Xi'an,1995), pp.388-391. Lecture Notes in Comput. Sci. No.959 (Springer, Berlin). MR 98c:05034. [219]

Yamamoto K.

(1949) An asymptotic series for the number of three line latin rectangles. Sûgaku 2, 159-162; J. Math, Soc. Japan 1(1950), 226-241. MR 12(1951), page 494. [146,151]

(1951) On the asymptotic number of latin rectangles. Jap. J. Math. 21, 113-119. MR 14(1953), page 442. [152]

(1952) Note on the enumeration of 7 × 7 latin squares. Bull. Math. Statist. 5, 1-8. MR 14(1953), page 610. [140]

(1953) Symbolic methods in the problem of three-line latin rectangles. J. Math. Soc. Japan 5, 13-23. MR 15 (1954), page 3. [151]

(1954) Euler squares and incomplete Euler squares of even degrees. Mem. Fac. Sci. Kūysyū Univ. Ser. A, 8, 161-180. MR 16(1955), page 325. [165,295]

(1956) Structure polynomials of latin rectangles and its application to a combinatorial problem. Mem. Fac. Sci. Kūysyū Univ. Ser. A, 10, 1-13. MR 17(1956), page 1174. [151]

(1960/61) On latin squares. Sûgaku 12, 67-79. In Japanese. MR 25(1963)#20. [34]

(1961) Generation principles of latin squares. Bull. Inst. Internat. Statist. 38, 73-76. MR 26(1963)#4933. [23,34,287]

(1969) Asymptotic evaluation of the number of latin rectangles. Res. Rep. Sci. Div. Tokio Women's Univ. 19, 86-97. Not reviewed in MR. [152]

Yates F.

(1936) Incomplete randomized blocks. Ann. Eugenics 7(1936), 121-140.[320]

Zassenhaus H.J.

(1958) The Theory of Groups. (Chelsea Publishing Company, New York, 2nd edition). MR 19(1958), page 939. [8,30,68]

Zhang Li-Qian.

(1963) On the maximum number of orthogonal latin squares I. Shuxue Jinzhan 6, 201-204. In Chinese. MR 32(1966)#4027. [179]

Zhang Li-Qian and Dai Shu-Sen.

(1964) On the orthogonal relations among orthomorphisms of noncommutative groups of small orders. Acta Math. Sinica 14, 471-480 (Chinese); translated as Chinese Math.-Acta 5(1964), 506-515. MR 30(1965)#4690. [156]

Zhang Li-Qian, Xiang Ke-Feng and Dai Shu-Sen.

(1964) Congruent mappings and congruence classes of orthomorphisms of groups. Acta Math. Sinica 14, 747-756 (Chinese); translated as Chinese Math.Acta 6(1964), 141-152. MR 31(1966)#1220. [156]

Zhang X. and Zhang H.

(1997) Three mutually orthogonal idempotent latin squares of order 18. Ars Combin. 45, 257-261. MR 97m:05049. [178]

Zhu L.

(1984a) Orthogonal diagonal latin squares of order 14. J Austral. Math. Soc. A36, 1-3. MR 84k:05021. [218]

Zhu L. and Zhang H.

(2001) A few more r-orthogonal latin squares. Discrete Math. 238, 183-191. MR1843907. [*295*]

(2003) Completing the specrum of r-orthogonal latin squares. Discrete Math. 268, 343-349. MR1983294. [*295*]

INDEX

Printed in the United States
By Bookmasters